## DATE DUE

| DEC - 7 2004 | | |
|---|---|---|
| | | |
| | | |
| | | |
| | | |
| | | |
| | | |
| | | |
| | | |
| | | |
| | | |
| | | |
| | | |
| | | |
| | | |
| | | |
| | | |
| | | |
| GAYLORD | | PRINTED IN U.S.A. |

# WATER POLLUTION

# WATER POLLUTION

Articles from Volumes 4-7 of
ENVIRONMENTAL SCIENCE & TECHNOLOGY

Collected by **Stanton S. Miller,**
Managing Editor

## An ACS Reprint Collection

American Chemical Society
Washington, D.C.      1974

Library of Congress Catalog Card 73-93250

ISBN 8412-0192-7

PRINTED IN THE UNITED STATES OF AMERICA

# Contents

## Business

## Industry

## Technology

# Preface

"Water Pollution" is the fourth in a continuing series of reprinted articles from *ES&T*. It compliments the earlier *ES&T* reprint books on solid waste and air pollution. Reaction to these books was good, indicating that there is a real need for understandable teaching and reference books on environmental topics.

This book is a collection of 106 articles that appeared in *ES&T* from 1970 to mid-1973 and cover most aspects of the subject of water pollution. Federal and state policy are discussed in detail to give a basic understanding of the long and short term goals we are trying to achieve in our water quality. How and to what extent the military, business, and industrial communities are acting to implement the new (and often not so new) laws and standards is covered. Several controversial issues that are still under study are presented in the last section of the book.

With environmental concern growing daily, it is hoped that this collection of articles will bring the reader up to date on the important aspects of the large subject of water pollution. New developments in the field will be covered in current issues of *ES&T*.

STANTON S. MILLER
Managing Editor, *ES&T*

# The business of water pollution

ES&T editors Stan Miller and Carol Lewicke discuss
the roles of leading firms, consulting engineers,
and equipment manufacturers in this growing field

*Reprinted from* ENVIRON. SCI. TECHNOL., **6**, 974 (November 1972)

Times have never been better for the water pollution control industry, one of the fastest growing industries in the U.S. today. Most newer companies in the business have doubled their sales volume in the past five years and will more than likely redouble that volume in the next five years; established firms likewise are doing quite well.

This ES&T special report covers the companies in the business; how their spokesmen view the company's role; equipment suppliers and their trade association, Waste Water Equipment Manufacturers Assoc.; and consulting engineering firms and technical experts, who basically are members of the Water Pollution Control Federation.

Together, this team stands ready to make progress on cleanup of the nation's most vital resource—water. Several indicators point to the fact that business in the field will definitely increase in the short-term future:

• At the beginning of 1972, there was, collectively in the U.S., a construction backlog of about $7 billion worth of waste water treatment facilities for which federal assistance has been committed and for which construction is incomplete

• In 10 years (1971–1980), an expenditure of more than $86 billion will be needed in the water pollution control category, according to the Council on Environmental Quality's third annual report

• The renewed challenge for waste water treatment was incorporated in this year's Federal Water Pollution Control Act amendments.

• In 1971, actual business investment in water pollution control was $1.4 billion, and the planned business investment for 1972 was $2.04 billion, a 42% increase, according to a survey by McGraw-Hill Publishing Co.

But what can be misleading to the casual water pollution control watcher is that business opportunities in this industry lie somewhere between the optimistic, pie-in-the-sky CEQ figures, the congressional promise but undelivered support of the federal construction program, and the hard-fact, in-plant construction figures of the Department of Commerce. Obviously, somewhere within these limits industry leaders see their future.

## In the business

By now, a number of major U.S. companies listed on the major stock exchanges have devoted a portion of their operation to the waste water treatment business along with, in many cases, the companion business of water treatment. Many have design engineering, equipment manufacturing, and construction capabilities. They can and do handle an array of waste water treatment services and products for both industrial and municipal markets. In addition, other companies handle only one or two segments of the three capabilities. Some companies concentrate solely on municipal markets while others concentrate only on industrial markets. Over 300 companies supply equipment for the water industry.

Companies such as Peabody-Galion and Zurn Industries, of course, cut across all segments of the environment, including water pollution control. On the other hand, a group of companies dealing only with water, such as Eco-dyne, Envirotech, Dorr-Oliver, and others, handle all three aspects of engineering, equipment supplies, and construction.

Then again, companies in the companion water treatment industry take available water and make it usable for industrial use. For example, the Graver division of Ecodyne Corp., Cochrane division of Crane, Infilco division of Westinghouse, and the Permutit division of Sybron Corp., to mention a few, are in this end of the business.

Growth is forecast in both municipal and industrial water pollution control markets. Expansion in waste water treatment will most likely occur in the industrial segment before it does in the municipal area. Industries are not hampered by financing arrangements such as the sale of municipal bonds or the vagaries of federal support which confront the municipal users; however, funds for industrial water pollution abatement programs must come out of profits. In other words, it's an unproductive use of money.

Here are a few companies representative of the water pollution control industry: **Calgon Corp.,** a subsidiary of Merck & Co., Inc., markets a number of projects in water pollution control and water reclamation. Calgon's major sales items are water treatment polymers and granular activated carbon for water purification and water pollution control. An adsorption service for waste water treatment was recently announced by Calgon officials. The Calgon Adsorption Service for on-site industrial pollution control, say Calgon officials, promises to reduce dissolved organic content by passing wastes

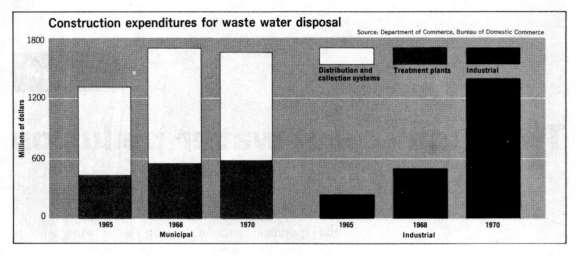

**Construction expenditures for waste water disposal**

Source: Department of Commerce, Bureau of Domestic Commerce

Millions of dollars

1800 — 1200 — 600 — 0

Distribution and collection systems — Treatment plants — Industrial

1965  1968  1970  Municipal

1965  1968  1970  Industrial

through columns of activated carbon. The process avoids large initial capital investment costs associated with other approaches. Calgon currently has a multiple hearth regeneration facility, a $12 million installation for carbon reuse, in the Pittsburgh area which will be ready the first quarter of next year.

With sales reaching $100 million in 1971, 94-year-old **Clow Corp.** (Oakbrook, Ill.) is "devoted to serving the country's water treatment and pollution control needs." They offer products and services for both municipal and industrial water and waste water systems. By product groups, clay pipe and waste treatment equipment accounted for 27% of their 1971 sales, pressure pipes accounted for 45% of sales, and valve products 14%.

One of seven operating divisions of Clow, the waste treatment division is headquartered at Richwood, Ky., and has a main plant at Melrose Park, Ill. In this division, sales have increased at a 6–7% annual rate over the past 10 years. Clow recently increased the plant's capacity by 50% for fabrication and assembly of packaged treatment plants and lift stations. In 1971, Clow spent more than $1 million on R&D; the company maintains one of the country's largest waste treatment test facilities at Rockford, Ill., at which actual sewage can be used to test waste treatment products in a 1-million gal. tank.

Although a precise breakdown of water pollution control activities is not possible for **Crane Co.** (New York City), consolidated sales for 1971 hit $792 million, an increase of 16.4% over the previous year, according to the 1971 annual report. Spokesmen for the Cochrane division (King of Prussia,

Pa.) believe that increasing emphasis on pollution control by municipalities and industries points to future sales.

Crane's Engineered Products Division designs and manufactures fluid control and treatment equipment including a wide range of pumps, meters, valves, and other equipment used in industrial processing, power generation, and municipal water and waste water systems. Formed in 1970, the company's Environmental Systems Division showed increased profitability in 1971, largely due to increased activity in the electrical generating industry, pulp and paper industry, and pollution control markets.

**Dravo Corp.** (Pittsburgh, Pa.) reported that 1971 earnings hit a record $428 million, up 37% from the previous year. Its Process, Construction & Engineering Group (involved in water pollution control) accounted for 23.4% of the total, or $100 million. This group was recently raised to divisional status to "properly recognize the level of activity this operation has achieved."

In addition to deep-bed filter systems for the steel industry, new Dravo bookings in pollution control include facilities to treat wastes from paper and pulp mills and oil refineries as well as a catalytical oxidation process to dispose of liquid waste sludges from industrial and municipal plants. Dravo's Super Aeropack is used widely in removing phosphate and nitrate from waste waters. Other nutrient removal systems are in the planning stage, according to Dravo. For example, the company will supply the filtration sections for the advanced physical-chemical treatment plant at Garland, Tex.

**Dorr-Oliver Inc.** (Stamford, Conn.)

in business for over 50 years, provides systems and equipment for waste water treatment to cope with both industrial and municipal problems. Dorr-Oliver's 1971 sales hit $90 million, according to the annual report, and increasing emphasis on environmental issues by both industry and government resulted in a sharp increase in demands for equipment and systems.

In the private sector, Dorr-Oliver activities include industrial waste treatment, industrial water purification, and in-plant water and waste water recovery. In the public sector, the company handles municipal sewage treatment and municipal water purification. Basically, the company offers systems, equipment, and processes for both municipal and industrial water management.

Two recent Dorr-Oliver customers are Brunswick Pulp and Paper Co.'s mill in Georgia; there, a giant 320-ft diameter Dorr-Oliver clarifier removes 90% of suspended solids from 50 million gal. of plant effluent each day. At Holland, Mich., a Dorr-Oliver fluosolids handling process went on-stream this year for incinerating sludges from a phosphate extraction process.

With sales of $145 million in 1972 (fiscal year ending March 31), **Envirotech Corp.** (Menlo Park, Calif.) markets a broad line of equipment, processes, and expertise for both industrial and municipal water and waste water treatment. The company, founded only three years ago, was listed on the New York Stock Exchange in August. Envirotech has had an annual growth of 15% per year. Municipal equipment sales hit $10 million in 1970 with a $25 million backlog.

Frank Sebastian, vice-president of Envirotech, maintains that the South Lake Tahoe water reclamation plant, which began operations in early 1968, is still perhaps the company's best showcase item. In 1970, the company completed a smaller waste water treatment plant, similar to the Lake Tahoe operation, at Colorado Springs, Colo. However, at Colorado Springs, industry will use some of the water produced by the plant. (Tahoe effluent is used for recreation.)

Envirotech also offers the Z-M process, a physical-chemical waste water treatment system which is now ready for the commercial market. "Using the Z-M process, a 10-mgd plant would cost about a third less to construct than a primary-secondary-tertiary treatment plant and 40% less in amortization and operation costs," Sebastian claims. "If land must be acquired, the capital cost savings may approach 50%."

Envirotech also tackles industrial waste water problems. For example, Envirotech provided the equipment for treating metal finishing wastes at the Guide Lamp Division of General Motors Corp. in Anderson, Ind., which produces plated die castings and molded plastic parts for automobiles. The 2.5-mgd waste treatment plant softens all incoming process water and batch treats metals and cyanide rinse waters.

In June 1970, Envirotech organized a Municipal Equipment Division (MED) which is solely a marketing division for about 40 Envirotech engineers. MED does not design systems for specific municipal installations. Instead, it advises and assists consulting engineers who are typically commissioned to design and develop specifications for competitive bidding on publicly funded sewage treatment systems. Generally, the company sells its equipment to the municipality or the general contractor. For the year ending March 31, 1972, MED sales amounted to $10.3 million and unfilled orders to $24.4 million.

**Ecodyne Corp.** (Chicago, Ill.) claims to be the largest diversified company devoted exclusively to water treatment water cooling, and waste water treatment. Ecodyne's 1971 sales hit $75 million, out of $255 million total for Trans Union.

Ecodyne's president Thomas O'Boyle told ES&T readers in last month's ES&T Interview that the company has a history in the water treatment field running back to the 1920's. "There have been prior pilot plants using the physical-chemical process," O'Boyle further

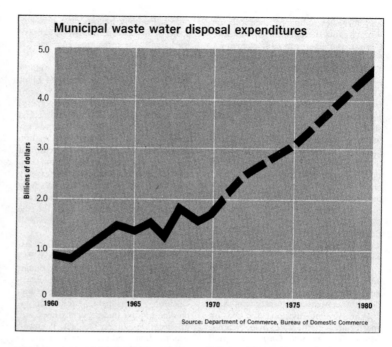

**Municipal waste water disposal expenditures**

Billions of dollars

Source: Department of Commerce, Bureau of Domestic Commerce

explains, "but the Ecodyne plant at Rosemount, Minn. will be the first commercial physical-chemical plant which starts with raw sewage and winds up with water of potable quality except for chlorination."

In the water treatment field, Ecodyne's Powdex (ion exchange) process was considered a breakthrough for ultra high purification of boiler feed water and could become increasingly important as nuclear power plants develop. In addition, Ecodyne's introduction of high-rate solids contact clarifiers, packaged treatment plants, factory-built pump stations and tertiary filters are well recognized in the U.S.

**Environment/One Corp.** (Schenectady, N.Y.) is a small but growing company which devotes part of its business operation to water pollution control. Environment/One makes package sewage treatment plants with capacities up to 100,000 gpd and now plans to market a pressurized sewage transport system utilizing grinder and pump combinations.

**Hercules, Inc.** (Wilmington, Del.) has been actively involved in waste treatment for five years. Basically, Hercules has two environmental operations in its Environmental Systems group. The Environmental Services Division markets water-treating services and chemicals, including flocculants, corrosion inhibitors, and boiler water treatment chemicals. The second divi-

sion is involved in turn-key operations. The 1972 top dollar sales item is the "Hercofloc" flocculant polymer. This item was not sold five years ago, but Hercules expects to triple 1972 sales in the next five years.

Advanced Waste Treatment Systems Inc. (including the recently acquired consulting engineering firm of Black, Crow, and Eidsness) is a partially owned subsidiary of Hercules and is involved with turn-key operations.

Hercules' most outstanding treatment plant is still under construction at Freehold, N.J. The plant, located in the shell of a home similar to one in the community that the plant will serve, will treat 50,000 gpd of waste water. The system will eliminate 99% suspended solids, 98% phosphates, and 95% of BOD.

**Authur G. McKee & Co.** (Cleveland, Ohio) engages in activities involving the process industries. Ninety-nine percent of their $1 million-plus earnings in 1971 was due to industrial contracts, with a mere 1% being municipal jobs. Senior environmental control consultant, W. A. Parsons, explains that McKee specializes in engineering and construction of plants for process industries, and that about 5% of the current cost is allocated to water pollution control facilities.

**Met-Pro Water Treatment Corp.** (Lansdale, Pa.) has five manufacturing centers: Stiles-Kem Corp. blends, mixes,

**3**

and furnishes a line of chemicals for water treatment; Sethco Manufacturing Corp. manufactures in-tank circulators, filter systems, and adsorption equipment for removing suspended and dissolved organic contaminants. Met-Pro Systems Division manufactures domestic and industrial waste water treatment systems and water treatment systems designed for flows up to 250,000 gpd, while Keystone Filter Corp. is a major supplier of pleated fiber for filtration. Sethco products include vertical and horizontal process pumps for use in chemical, petrochemical, paper, steel, automobile, refining, and waste treatment industries. Met-Pro's president Walter Everett believes that the "anti-pollution programs related to water and waste treatment now appear to be gaining momentum."

Until recently, approximately 75% of Met-Pro's sales were to the federal government; however, 80% of its business now consists of commercial sales.

With several divisions focused on specific areas in the water field, **Neptune Meter Co.** (Atlanta, Ga.) is involved with treatment plants for raw water purification, meters for potable water conservation, and tertiary treatment processes for waste water reclamation. As a result, Neptune purification products include Micro-FLOC water treatment systems, tube settlers, and mixed media filtration beds.

Formed July 1, 1970, **Peabody-Galion Corp.** (New York City), an environmental improvement company, derives as much as 70% of its revenues from balanced positions in all major sections—air, water, and land—and reported sales of $90 million in 1971. Peabody-Galion's water group accounted for 17% of the 1971 sales—40% of which was to municipal customers and 60% to industrial. Peabody acquired four companies, each recognized as a division, giving the firm a well-rounded position in the environmental field. Petersen Co. has been in the construction business (sewage and waste water treatment systems) for 11 years. Barnes Co. has 77 years of experience in sludge, slurry, and water pumps. Hart has dealt with packaged industrial waste treatment units for more than 10 years; and Welles has manufactured aeration systems for 13 years.

In 1967, sales from the acquired companies were $10 million, in 1972 they will hit the $22 million mark; and by 1977, they will, Peabody-Galion hopes, again redouble. In August, the company had a booming $12 million backlog of orders. Peabody has no pure turn-key projects in its operations; it always works with the consulting engineer for municipal contracts.

Peabody completed a $1.5 million industrial waste water treatment plant (which the firm considers a show item) for Armour Products (Montgomery, Ala.) earlier this year. In the municipal category, perhaps the best Peabody example is the combined $1 million biological/physical-chemical treatment plant for the city of Bradenton, Fla. The plant treats municipal wastes only and went on-line this spring. Peabody also has a $5 million municipal construction contract under way for the city of Port St. Joe, Fla., on the Gulf Coast.

With total sales of $405 million last year, **Pennwalt Corp.** (Philadelphia, Pa.) estimates that 9% of its 1971 sales relates to environmental control products—chemicals including ferric chloride and calcium hypochlorite; equipment including pumps, meters, centrifuges, and air flotation systems. In fact, Pennwalt was the first (and is the only) domestic producer of the anhydrous type of ferric chloride and the first producer of the liquid type. The material is used as a sludge conditioner, coagulant, and precipitant. The industrial market accounts for 90% of the firm's sales, municipalities for 10%.

Pennwalt's Wallace & Tiernan division markets measuring and feeding devices for the addition of solids, liquids, and gases used in water and waste water treatment. The Sharples Stokes division markets centrifuges for separating waste solids from liquid suspension and dewatering the resulting sludges before disposal.

**Sybron Corp.** (Rochester, N.Y.) with 1971 sales of $355 million, has three major operating divisions—Permutit Co., Leopold Co., and Barnstead Co.—which accounted for 23% of the sales in the water purification and waste treatment markets. In addition, the Pfaudler division handles plating waste treatment systems, and the Ionac Chemical division produces some specialty items.

Permutit Co. (Paramus, N.J.), manufactures "a complete range of water treatment equipment of virtually every treatment process type," according to A. W. Pieper, manager of marketing services. Products include aeration and degasification equipment, controls, instruments, clarification, ion exchange, and reverse osmosis equipment. In 1971, 6% of Sybron's net corporate sales (near $20 million) was for such equipment. A major portion of the sales went to industry, but a respectable volume was attributed to municipal plants in construction of clarification, filtration, and water-softening systems.

On a direct purchase basis, the Permutit and Leopold divisions are most intimately involved in water and waste water treatment. Industrial waste treatment systems are purchased in proportion to new plant construction and also to provide pollution control for existing plants. Permutit has turn-key capability but does very little business of this nature. Leopold, in the municipal water treatment market, produces ceramic tile for filter bottoms and related gravity filter equipment for purifying water for potable use.

**Pollution Control Industries, Inc.** (Stamford, Conn.) is basically an equipment manufacturer of systems for ozone generation, water disinfection, and waste water treatment. Although sales volume in 1972 was only $600,000, company officials expect that to jump to $3 million by 1977. At this point, sales are 70% municipal and 30% industrial contracts.

**Rex Chainbelt Inc.** (Milwaukee Wis.), one of the leading water pollution control equipment companies, reported sales of $339 million in 1971, $22 million of which was attributed to environmental control equipment sales. This equipment-oriented company claims to have a record of 5000 major pollution control system installations in the U.S., Canada, and throughout the world.

During 1971, Rex formed the Ecology division, a research arm of the organization, which serves as a single source for water pollution problem-solving for industrial and municipal planners. One item the team of 20 researchers is looking over is reverse osmosis applications, efficiencies, and economics in the municipal and certain industrial fields.

**Rollins International, Inc.** (Wilmington, Del.) had $135 million sales in 1971, but the company's environmental control system handles only a very small, unidentifiable part of the business. In fact, Rollins-Purle, a subsidiary, handles waste disposal for other companies. Rollins-Purle currently operates three central disposal plants on a regional basis—Logan, N.J., Baton Rouge, La., and Houston, Tex.; future plans call for a network of such disposal centers to be located in industrial centers across the U.S. Rollins International

# Equipment shipments over 5-year period

## Industrial waste water shipments show fourfold increase . . .

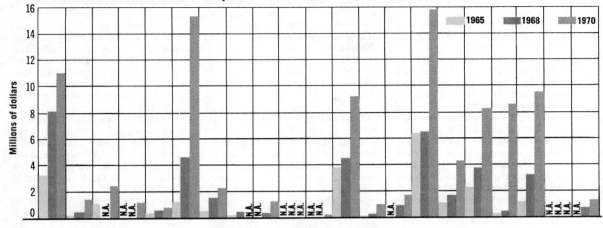

## . . . as municipal waste water shipments increase less than half . . .

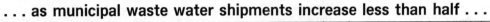

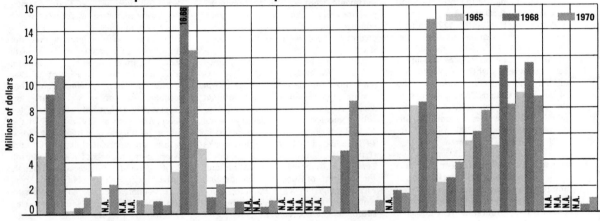

## . . . but industrial water supply treatment also quadruples

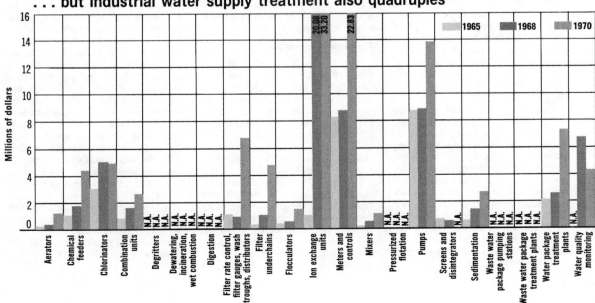

**Source:** Department of Commerce, Bureau of Domestic Commerce

sees four areas of growth in its business activity, one of which is industrial pollution control.

**United States Filter Corp.** (Newport Beach, Calif.) sales in 1971 hit $30 million. Through the first quarter of this year sales were running about $1 million ahead of last year. Pollution control and filtration systems sales have been accelerating this past quarter. U.S. Filter systems are touted as advanced engineering solutions to pollution and filtration problems. The firm's showcase facility is the "only" operating, full-scale advanced waste treatment facility in the U.S.—Lake Tahoe, Calif.

**Zurn Industries, Inc.** (Erie, Pa.) is a total environmental company associated with the waste treatment field since its inception in 1900, Frank Zurn told readers in ES&T's June 1972 Interview. Zurn offers both software (environmental and engineering services) and hardware (products and services) items.

Zurn's total sales hit $178 million when its fiscal year ended in March. Of this total, 76% ($118 million) accounted for sale of services and products; however, slightly less than $78 million was associated with the sales of services, systems, and products in the water pollution control area and related fields. Five years from now this figure could reach $100 million, according to a Zurn spokesman.

In the environmental engineering area alone, 240 projects are on the active job list, constituting a current backlog of about $10 million. Construction backlog is approximately $28 million, with new sales for the current fiscal year expected to reach $33 million.

Generally speaking, 75% of Zurn's business is industrial and 25% municipal, but the percentage has been changing about 1–2% each year in favor of the municipal market. Zurn's municipal plant upgrading project for the city of Cleveland has already been reported in ES&T (September 1972, p 782).

Examples of work recently completed or currently under way by various Zurn environmental engineering affiliates involve a water reclamation plant expansion for Burbank, Calif.; a waste water reclamation plant for Los Angeles and Glendale, Calif. (a system which will provide hydraulic relief to the city's sewerage systems as well as recycling); and consulting services in industrial waste disposal to the Sun Oil Co. for several refinery installations including those of Yabucoa, P.R., Tulsa, Okla., and Corpus Christi, Tex.

## Consulting engineers

Traditionally, in the municipal waste water treatment area, consulting engineers prepare a plant design for a municipality, and then contractors bid competitively for construction based on the design. On the other hand, in the industrial area, construction firms, in addition to private consulting engineers, handle all phases of the contract—engineering design, equipment supply and purchase, and construction—depending usually on the size of the job (the larger the project the greater the tendency toward construction firms).

Fundamentally, the consulting engineer usually specifies the equipment, supplies, processes for new construction projects, and extensions and replacements. In addition, he normally prepares the design and specification. However, certain municipalities operate departments whose municipal engineers prepare design plans and also supervise construction. Often these municipal engineers work with consulting engineers.

The Consulting Engineers Council (CEC), a Washington, D.C.–based organization serving U.S. engineers in private practice, finds from a national survey released in June, that 914 consulting engineering firms designed some $4.5 billion worth of treatment plants in 1970. Prepared by E. Joe Middlebrooks (Utah State University), "A National Survey of Manpower Utilization and Future Needs of Consulting Engineering Firms Engaged in Water Pollution Control" is the first real analysis of the consulting engineer's role in design and construction of U.S. waste water treatment plants. The CEC survey reports that private consultants should readily be able to handle increased work.

## Technical associations

With a membership of 23,000, the Water Pollution Control Federation (Washington, D.C.) is the technical association devoted exclusively to the problems of water pollution control. Its members have the responsibility for technical detail and equipment specification for the construction, operation, and management of waste water collection and treatment systems.

There are 39 so-called WPCF federation member associations in the U.S. and Puerto Rico; these suborganizations are concerned with promotion, management, design, construction, financing, and operation of facilities for the control of water pollution through proper collection and treatment, disposal, or reclamation of domestic and industrial waste waters.

The Water and Waste Water Equipment Manufacturers Assoc. (WWEMA) (Newark, N.J.) is the trade association for some 300 equipment manufacturers and suppliers in the U.S. WWEMA has been operating since 1908. Traditionally, this association has sponsored equipment exhibitions at the annual meetings of the two technical associations in the water category—WPCF and the American Water Works Association. WWEMA membership includes companies that supply a broad line of equipment as well as those that produce only one or a few specialty items including chemicals, instruments, or pipes.

Robert Hughes, WWEMA's public relations manager, says that the association is looking ahead and will sponsor an industrial waste water equipment show and meeting next March, aimed specifically at industrial users of waste water equipment.

By this time one realizes that the largest part of expenditures for a waste water treatment plant is concrete, brick, and mortar, exclusive of the cost of the collection systems. Equipment accounts for only a small part—approximately 18% of the total cost for plants and even as low as 12% in the case of primary sewage treatment plants.

In a paper presented at the National Symposium on Costs of Water Pollution Control, held at Research Triangle Universities (Raleigh, N.C.) in April, K. L. Kollar, director of the Construction, Water Resources, and Engineering Division of the Bureau of Domestic Commerce in the Department of Commerce noted that only 5.4% of the total U.S. waste water expenditures comprise pollution control equipment while about 15.3% of the total treatment plant expenditures (not including interceptor and outfall) comprise water treatment equipment. Kollar grouped the equipment into items (box, p 978) and compiled in-plant shipments of equipment in both the waste water treatment field and the companion water treatment industry. In 1970, the last year for which figures are available, the total for both industrial and municipal equipment in the waste water treatment category hit $184 million, from $77 million in 1965.

For a list of individual companies that supply any one or more of the specific 21 equipment items, the reader is referred to the 1972–73 ES&T Pollution Control Directory, published last month and available from ACS Special Issue Sales.

# Water pollution in the states

Success for the nation's water quality program hinges on state involvement.
Associate editor Stanton Miller canvassed state officials, focused on grass-roots
issues, and found signs for renewed optimism

*Reprinted from* ENVIRON. SCI. TECHNOL., **5**, 120 (February 1971)

The U.S. has more than 3 million miles of streams, 88,633 miles of tidal shoreline, and millions of lakes, ponds, and bayous. Certainly, these waters cannot all be monitored, protected, and restored from Washington. So the key to success in the fight against pollution is state participation.

The action in water pollution control is at the state level. In fact, under existing federal legislation, states have the responsibility. Their programs are the crux of the nation's water pollution control effort. It is at this level that the day-to-day exchange of blows between state officials and their counterparts in industries and municipalities takes place.

Progress on water pollution abatement is being made, but certainly not at the rate envisioned by legislators when they wrote the 1965 and 1966 amendments to the Federal Water Pollution Control Act. Nor, for that matter, is it proceeding at a rate that pleases the public.

The lack of progress mainly is due to understaffing, underfunding, and undervisibility of antipollution programs. Differences of opinion between state and federal officials have developed over the last five years. Deterioration of federal–state relationship has also hampered progress.

There are numerous issues in the water pollution control game. Some are old, some are new; not one has been completely resolved. Basin planning, new financing schemes, permits, and certification requirements are a few of the new issues, while the old

ones of standards, secondary treatment, enforcement, grants, and the need for new legislation are still around. Certainly, the whole list of issues affords ample opportunities for state and federal administrators to argue.

Before delving into the issues, it is important to consider the federal–state relationship. The ray of hope for improved relations between states and the federal government entered the picture only recently. But it is perhaps safe to say that, for the first time in the last five years, things are looking up.

Five years ago, the federal water pollution activity had no visibility. It then existed as the Division of Water Supply and Pollution Control in the Department of Health, Education and Welfare. Prior to the enactment of the 1965 Federal Water Pollution Control Act, the feds were accused of playing footsie with the states people. They were accused of not putting the heat on the states people to act. However, when the program was elevated in stature and moved to the Department of Interior in 1966 as the Federal Water Pollution Control Administration (FWPCA)—based on President Johnson's reorganization plan—problems began.

First, there was severe attrition of key personnel. Many of the old-time Public Health Service employees preferred to maintain their status rather than transfer and accept Civil Service status—despite the fact that practically all of the personnel were seriously

committed and devoted to cleaning up water pollution.

Then, Interior had to put together its team of water pollution experts which, appropriately, was referred to as the "new, green federal team." They had little, if any, practical experience in trading blows with polluters. In those days, the state people dreaded seeing the federal officials coming.

Since then, federal efforts have been a series of faints and thrusts, most coming without warning. These unilateral actions kept the states defensive, sapped the states' manpower and efforts, and perhaps misdirected considerable attention and energies away from more mundane problems.

The feds have made the pitch for improved federal–state relations again and again. The most recent pitch was made at the meeting of the Association of State and Interstate Water Pollution Control Administrators (ASIWPCA), held in Portland, Ore., late last November.

## Basin planning

Today, the biggest problem facing each state is basin planning. No administrator, federal or state, disagrees in principle that the job should be done. State administrators realize that it is a time-consuming and expensive requirement, but what plagues them now is the fact that such planning is a requirement for construction grant funds. FWQA (the Federal Water Quality Administration, formerly FWPCA, and now the Water Quality Office in the Environmental Protection

Agency) has publicly announced in the Federal Register that no funds for waste water treatment facilities would be approved unless they were tied into a comprehensive basin plan. It's a federal requirement much like an earlier requirement for secondary treatment or "no federal funds."

Certainly, state officials do not agree that the requirement for basin planning should be included as an amendment to federal law, as has been suggested by the administration. They argue that such a requirement would merely consume much of the money and manpower needed for more immediate problems, such as day-to-day work on their programs. In their opinion, the commitment of large sums of money for planning is hardly justified at this time.

To do the basin planning job in Virginia alone would cost $15 million, says A. H. Paessler, Virginia's water control official. To do justice to the comprehensive planning that should be done is well-nigh impossible at this time, according to Paessler. Similarly, in Texas, $25 million and several years' work would be involved in the plan, estimates Joe Teller, deputy director of the Texas Water Quality Board. Nevertheless, Texas started its basin planning several years ago. The state contracted for the study and has already spent $2.5 million, but the entire basin planning job would take another 10 to 15 years, Joe Teller says.

State administrators from Arizona, Connecticut, Iowa, and New York come out unequivocally in opposition to the idea of basin planning. Others from Colorado, Missouri, New Hampshire, South Dakota, and Texas admit to the desirability of basin planning, but feel that it is a bit unreasonable at this time.

Arizona's spokesman Joseph E. Obr notes that his state does not have any problems with water pollution control that would appear to lend itself readily to the basin approach, with the exception of the salinity problem in the Colorado River Basin.

California's Kerry W. Mulligan regards river basin planning as essential. This is true, he says, whether it is on the interstate rivers, such as the Colorado and Truckee, or on California's intrastate rivers, which include the Sacramento and San Joaquin.

Although Colorado concurs that basin planning is necessary, Colorado's Frank Rozich comments that, with a total population of 25,000 still in need

**Outfall.** *Polluted waters continuously empty into the nation's waterways*

of adequate sewage treatment, "it seems rather asinine to ask these people now to look into basin planning." Any money could be better spent in construction of treatment facilities, says Rozich. A similar situation exists in Wyoming and New Mexico.

On the positive side, Maryland's W. McLean Bingley, chief of the division of water and sewage, says river basin planning is an important part of the state's entire pollution abatement activity. In his opinion, this approach will help to solve problems that Maryland and other states may have in common (or, for that matter, that any group of political jurisdictions may have in common).

Missouri's Jack K. Smith admits that planning could help overcome some pollution problems, but he cautions that no amount of planning is going to make cooperation between cities a reality.

New Hampshire's C. W. Metcalf, director of municipal services, says that basin planning at this time will result in a slowdown of the state's program.

But a resounding voice of approval for this issue is heard from the New York official, Dwight F. Metzler. He notes that relatively small investments in comprehensive planning have yielded big dividends in improved systems at less cost, and have greatly strengthened the regional concept of solving environmental problems. New York already bases its comprehensive planning on the basin approach, according to Metzler.

**Financing**

Money is often the root of many arguments, and it is no exception in water pollution control. As a nation, the U.S. entered 1970 with a backlog of $4.4 billion in waste treatment needs, according to one FWQA estimate.

The federal promise of $3.4 billion to the states for construction of municipal waste treatment facilities, authorized under the 1966 amendments to the Federal Water Pollution Control Act, simply has not materialized. Slightly more than half ($1.8 billion) has been appropriated by the Congress. Considering that statutory authority for these funds expires this fiscal year, state administrators are wondering what new direction the program will take.

In addition to construction funds, federal funds have been netted out in the past—to the tune of approximately $10 million annually—for states to develop their water pollution control programs. These are provided on a matching basis. By Dec. 1, 1970, for example, some 31 states and three jurisdictions had had their programs approved by the feds; the federal contribution was more than $5.5 million.

One well-founded criticism of the federal role is that construction grant awards were made not on the basis of states' needs, but on a first-come-first-served basis. Another criticism is the lack of guarantee for prefinancing.

Surprisingly, not all states need federal funds. However, a large majority

## State and interstate administrators

**ALABAMA**

Arthur N. Beck, Tech. Sec.
Water Improvement Commission
State Office Bldg., Rm. 328
Montgomery, Ala. 36104

**ALASKA**

James A. Anderegg, Comm.
Department of Health & Welfare
Pouch H
Juneau, Alas. 99801

**ARIZONA**

Edmund C. Garthe, Ass't Comm.
Environmental Health Services
Hayden Plaza West
4019 North 33rd Ave.
Phoenix, Ariz. 85017

**ARKANSAS**

S. L. Davies, Dir.
Arkansas Pollution Control Comm.
1100 Harrington Ave.
Little Rock, Ark. 72202

**CALIFORNIA**

Kerry W. Mulligan, Chairman
State Water Resources Control Board
1416 Ninth St.
Saoramento, Calif. 95814

**COLORADO**

Frank Rozich, Dir.
Water Pollution Control Div.
Colorado Department of Health
4210 East 11th Ave.
Denver, Colo. 80220

**CONNECTICUT**

John J. Curry, Dir.
State Water Resources Comm.
225 State Office Bldg.
165 Capitol Ave.
Hartford, Conn. 06115

**DELAWARE**

John C. Bryson, Acting Dir.
Division of Environmental Control
P.O. Box 916
Dover, Del. 19901

**DISTRICT OF COLUMBIA**

Malcolm C. Hope
Assoc. Dir. of Environmental Health
Government of the District of
Columbia
1875 Connecticut Ave., N.W.
Universal Bldg., North
Washington, D.C. 20009

**FLORIDA**

Vincent D. Patton, Exec. Dir.
Department of Air & Water Pollution
Control
Suite 300, Tallahassee Bank Bldg.
315 South Calhoun St.
Tallahassee, Fla. 32301

**GEORGIA**

R. S. Howard, Jr., Exec. Sec.
Georgia Water Quality Board
47 Trinity Ave., S.W.
Atlanta, Ga. 30334

**GUAM**

O. V. Natarajan, Administrator
Water Pollution Control Program
Public Health & Social Services
Government of Guam
P.O. Box 2816
Agana, Guam 96910

**HAWAII**

Shinji Soneda, Exec. Off.
Environmental Health Div.
Department of Health
P. O. Box 3378
Honolulu, Hawaii 96801

**IDAHO**

Vaughn Anderson, Dir.
Engineering & Sanitation Div.
State Department of Health
Statehouse
Boise, Idaho 837.07

**ILLINOIS**

C. W. Klassen, Tech. Sec.
State Sanitary Water Board
State Office Bldg.
400 South Spring St.
Springfield, Ill. 62706

**INDIANA**

Blucher A. Poole, Tech. Sec.
Stream Pollution Control Board
1330 West Michigan St.
Indianapolis, Ind. 46206

---

do. In fact, many have actually prefinanced the needed construction with the hope that part of the outlay would be refunded by the federal government.

Quite obviously, more funds will be needed in the future, and several estimates have been made of just how much money will be required. The Nixon administration says $10 billion is needed in the next five years. The National Association of Counties says $34 billion for the same time span, while the U.S. Conference of Mayors–National League of Cities says $33 to $37 billion over the next six years. The estimates are not directly comparable since different assumptions are made in each, but the need for considerable sums is apparent.

Most states need federal funds, but a few have more than they can use. One state, Colorado, plans to return money to the federal treasury. Another, Iowa, cannot use all the federal funds available to it. A few are satisfied with their share of federal funding. Generally, these are the small states that are neither heavily urbanized nor industrialized, such as Wyoming, South Dakota, and Utah. But most other states are not satisfied.

For heavily industrialized and urbanized areas, particularly in the coastal states—where 75% of the nation's population lives—the federal share has been too little and too late.

Many states still need federal funds to keep their water pollution control programs on schedule. Certainly, the unofficial federal goal—clean water by 1972—will not be reached.

New Hampshire's Metcalf notes that, even with his state's modest prefinancing program, federal funding continues to be inadequate. New York's Metzler says the state is not getting the authorized 30% federal participation, to say nothing of the maximum of 55% Congress had authorized. Of the 481 municipally owned waste water treatment plants in the state, 236 have been upgraded. The rub is that 358 additional plants will be needed to get the whole job done in New York.

**Permits**

Most states have some type of permit system. But permits differ. Are they for discharges? Construction? Operation? Or all three? For whom? Municipalities or industries, or both?

The states without permit systems include Arizona, California, Colorado, and Wyoming, for example. To say that a state has a permit system is not to imply that the system is exactly the same as that of another state. (Nor, for that matter, does it guarantee that the permit system is enforced, reviewed, or operated at the same level of excellence in any two states.)

But the one big thing that is new to state administrators is a federal requirement for them to pass judgment on construction and operations, so that their state water quality standards are not violated. The new requirement for certification by state officials was contained in the 1970 Water Quality Improvement Act. North Carolina adopted its procedure in this regard on Oct. 13, 1970; other states are in

the throes of writing this language into their state laws. Although Arizona does not issue permits for industrial discharges at the present time, rules and regulations requiring permits are being considered.

Rather than a permit system, California has established requirements for waste discharges which include a monitoring program and quarterly monitoring reports. These requirements are periodically reviewed for consistency. Colorado has no type of permit system; Connecticut reviews its permits at no more than five-year intervals; Missouri reviews its permits annually.

States with actual industrial discharge permits include Iowa, Maryland, and Ohio. Others with permits include South Dakota and Texas. But many states having industrial discharge permits simply do not have the necessary manpower to make the system work.

Other states require permits for construction and operation. Florida's new permit rule, which became effective March 3, 1970, requires both construction and operation permits for new facilities and operating permits for pre-existing sources of air and water pollution. Operating permits for these sources may be issued temporarily when it is found that the facility does not meet the state pollution control codes. Temporary permits are then issued, provided that the facility will take steps to meet the codes within a reasonable period, which is spelled out in the temporary permit.

In New York, construction permits are issued to an industry for a specific

**IOWA**

R. J. Schliekelman, Tech. Sec.
Water Pollution Div.
Iowa State Department of Health
Des Moines, Iowa 50319

**KANSAS**

Melville W. Gray, Chief Eng. and Dir.
Environmental Health Services
Kansas State Department of Health
Topeka, Kan. 66612

**KENTUCKY**

Ralph C. Pickard, Exec. Dir.
Kentucky Water Pollution Control
Comm.
275 East Main St.
Frankfort, Ky. 40601

**LOUISIANA**

R. A. Lafleur, Exec. Sec.
Louisiana Stream Control Comm.
P.O. Drawer FC, University Station
Baton Rouge, La. 70803

**MAINE**

William R. Adams, Dir.
Environmental Improvement Comm.
State House
Augusta, Me. 04330

**MARYLAND**

James B. Coulter, Deputy Sec.
Department of Natural Resources
State Office Bldg.
Annapolis, Md. 21401

Paul W. McKee, Dir.
Maryland Department of Water
Resources
State Office Bldg.
Annapolis, Md. 21401

**MASSACHUSETTS**

Thomas C. McMahon, Dir.
Division of Water Pollution Control
State Office Bldg.
100 Cambridge St.
Boston, Mass. 02202

**MICHIGAN**

Ralph W. Purdy, Exec. Sec.
Water Resources Comm.
Department of Natural Resources
Stevens T. Mason Bldg.
Lansing, Mich. 48926

**MINNESOTA**

John P. Badalich, Exec. Dir.
Minnesota Pollution Control Agency
717 Delaware Ave., S.E.
Minneapolis, Minn. 55440

**MISSISSIPPI**

Glen Wood, Jr., Acting Exec. Sec.
Air & Water Pollution Control Comm.
P.O. Box 827
Jackson, Miss. 39205

**MISSOURI**

Jack K. Smith, Exec. Sec.
Missouri Water Pollution Board
P.O. Box 154
Jefferson City, Mo. 65101

**MONTANA**

Claiborne W. Brinck, Dir.
Division of Environmental Sanitation
State Department of Health
Helena, Mont. 59601

**NEBRASKA**

T. A. Filipi, Sec.
Nebraska Water Pollution Control
Council
State House Station, Box 94757
Lincoln, Neb. 68509

**NEVADA**

Ernest G. Gregory, Chief
Bureau of Environmental Health
Nevada State Health Div.
Nye Bldg., 201 S. Fall St.
Carson City, Nev. 89701

**NEW HAMPSHIRE**

William A. Healy, Exec. Dir.
Water Supply & Pollution Control
Comm.
61 South Spring St.
Concord, N.H. 03301

**NEW JERSEY**

Richard J. Sullivan, Dir.
Department of Environmental
Protection
John Fitch Plaza
P.O. Box 1390
Trenton, N.J. 08625

**NEW MEXICO**

John R. Wright
New Mexico Health & Social Services
Dept.
P.O. Box 2348
Sante Fe, N.M. 87501

**NEW YORK**

Dwight F. Metzler
New York State Dept. of
Environmental Conservation
Albany, N.Y. 12201

project to treat a specific waste water stream. After a treatment facility is constructed, an application is made for an operation permit. The permit is issued if the facility has been constructed in accordance with the approved report and plans. It's good for five years or less, says New York's Metzler. In many cases, permits are limited to one year or less, because of the types of treatment proposed and their lack of proven success.

## Standards

Water quality standards have been a perennial issue, at least since the deadline of June 30, 1967, when they were originally due. All states have standards approved, but they are laced with exceptions. Many basic issues regarding the importance of temperature and dissolved oxygen have not been settled. Consequently, exceptions in standards have not been resolved, despite the fact that resolution was a priority item slated by FWQA for completion by the beginning of 1969.

The first 10 standards were approved with reasonable dispatch. But shortly after the 1967 deadline for state standards, former Secretary of Interior Udall's team of consultants announced its guidelines for criteria (usually referred to as the green book). Then, Interior's interdepartmental task force —comprising other in-house agencies, such as its Bureau of Commercial Fisheries, Bureau of Sport Fisheries and Wildlife, and others—began to look also at state standards. With the green book in hand, Interior's "new, green team" began, in effect, to interpret the

guidelines as standards. In any event, the point has been made repeatedly that the task force became the tail wagging the FWQA dog. The result was that standards were then being approved with many exceptions. The exceptions in many standards for dissolved oxygen, temperature, and salinity exist to this day.

One valid complaint of state administrators is that considerable manpower, time, and money were consumed in the discussion of these issues. Of course, during this time, state energies were directed away from the day-to-day water pollution activities.

The question that remains unanswered—if not down right impossible to answer—is how can state administrators pass judgment on certification issues until the basic issues of temperature, dissolved oxygen, and salinity are resolved? Perhaps not until basin plans are completed can we have a standard for a particular body of water. The consensus maintains that states should be permitted to set stricter standards than those indicated in the federal guidelines, perhaps with certain exceptions. Specifically, the exceptions might be in the situation where an industry is one which operates nationally, and where individual state controls would pose a nonuniform standard for the industry in certain states.

## Secondary treatment

In the unending round of water pollution control gamesmanship, secondary treatment is an old issue. It stems from the now infamous (to state officials) guideline No. 8, which was only

one in a series of unilateral moves that the Johnson administration took. The guideline required treatment of waste waters to a level which would not degrade the quality of the receiving stream. In practice, this has been interpreted by FWQA to mean secondary treatment. Secondary treatment subsequently became a requirement which had to be fulfilled before a state could receive federal construction funds.

Although it was a requirement for funds, secondary treatment is by no means practiced throughout the U.S. Certain coastal states, including California, Massachusetts, and Washington, are to this day exempt from the requirement.

It was the issue of secondary treatment over which the first federal enforcement action was taken against a municipal polluter. As early as 1957, a court decision ordered the city of St. Joseph, Mo., to install secondary treatment facilities for its waste waters. However, according to one official, only 50% of its municipal wastes receive such treatment today.

In an earlier review of water pollution control progress in the U.S. ("Water pollution—coast to coast," ES&T, September 1969, page 804), federal officials noted that several large cities did not provide secondary treatment for their wastes. These included Pittsburgh, Cincinnati, and Louisville. To do so would cost considerable sums of money. There are small towns in Colorado and Utah, for example, that do not presently provide secondary treatment. But on the whole, wastes from these towns are adequately treated.

Colorado's Rozich notes that more than 99% of the state's population connected to some sort of municipal sewer system (1.9 of its 2.2 million people) received secondary treatment with disinfection for its waste before discharge to the streams of Colorado. At the end of 1969, 23 communities with approximately 26,000 people were not providing adequate treatment of their wastes. Other states are not so far along.

Last year, the requirement date for secondary treatment in Missouri was moved up to 1975 from an earlier commitment of 1982. Three of its four largest cities—St. Louis, Kansas City, and St. Joseph—are on the Missouri and Mississippi Rivers and must have secondary treatment by 1975.

In certain instances, secondary treatment is not meeting the Missouri water quality standards. Missouri's official Smith has found instances where effluents from secondary treatment enter underground water supplies. Advanced waste treatment, irrigation, and groundwater recharge are some possible solutions to the problem of discharging secondary treated effluents to low flow and disappearing streams.

## Enforcement

If a state fails to enforce its law, then federal enforcement procedures can be applied. But these procedures are cumbersome and time consuming ("Water polluters: beware the feds!" ES&T, November 1970, page 887). There is a long delay before the federal government can come in and bring a recalcitrant polluter to terms. On the other hand, state officials can generally bring polluters to terms with reasonable dispatch, if they are so inclined.

Enforcement is only one of many priorities of the new EPA. To be sure, enforcement procedures must be firm, fair, and applied uniformly. Perhaps more important is the federal presence, the federal backup, which must become more apparent to state administrators as we move into the seventies—the environmental decade.

Enforcement may be uppermost in the minds of federal officials, but this hardly is true for state administrators. They contend that the yardstick for success of the nation's water pollution abatement effort cannot be measured by the number of enforcement actions that the federal officials file in court. Most state administrators would prefer administrative procedures to the filing of court suits. They argue that the state is the primary enforcement and coordinating agency in dealing with local governments and industry.

Most states are already involved with federal enforcement conference proceedings; nine are not. In the past year, Arizona's enforcement activities included several written orders of abatement. Three involved surface discharge of wastes from septic tank systems serving tourist facilities; others were against a copper rod and wire company, a paper mill, and a trailer park.

One state, California, has enforcement uppermost in its mind. For example, one of the key features of its Porter Cologne Act, which went into operation Jan. 1, 1970, is its strict enforcement provisions. Los Angeles' Hertler notes that the majority of his enforcement actions are secured administratively—that is, by the staff working with the discharger to perform the needed correction. When other actions are necessary, the board issues cease and desist orders, examples of which are recent orders against the Union Pacific Railroad Co. and the Saticoy Meat Packing Co.

Colorado is in the throes of setting abatement dates for violators of state laws and standards. Rozich notes that since the Colorado Water Pollution Control Commission was established in 1966, it has issued 80 to 100 cease and desist orders. It hasn't forced any industry to go out of business, but it has forced them to install adequate waste treatment facilities.

In Florida, there have been two federal–state enforcement conferences; one was called to discuss the interstate problems of Escambia and Perdido Bays in the far western portion of the state. A number of industrial and municipal sources of pollution were singled out, but many were already under notices and orders from the Florida Department of Air & Water Pollution Control. As of last September, state and federal officials were satisfied with progress being made. Satisfaction, however, is not being expressed in the second conference. The conference was called to consider the potential for thermal pollution in Biscayne Bay from a nuclear power generation facility under construction by Florida Power & Light Co. Results of this conference were inconclusive, at best.

**WISCONSIN**

**Thomas G. Frangos,** Administrator
Division of Environmental Protection
Department of Natural Resources
P.O. Box 450
Madison, Wis. 53701

**WYOMING**

**Arthur E. Williamson,** Dir.
Sanitary Engineering Services
Division of Health & Medical
Services
Wyoming Department of Health &
Social Services
State Office Bldg.
Cheyenne, Wyo. 82001

**Interstate administrators**

**DELAWARE RIVER BASIN
COMMISSION**

**James F. Wright,** Exec. Dir.
Delaware River Basin Commission
P.O. Box 360
Trenton, N.J. 08603

**GREAT LAKES COMMISSION**

**L. J. Goodsell,** Exec. Dir.
5104 First Bldg.
2200 N. Campus Blvd.
Ann Arbor, Mich. 48105

**INTERSTATE COMMISSION ON
THE POTOMAC RIVER BASIN**

**Carol J. Johnson,** Exec. Dir.
407 Global Bldg.
1025 Vermont Ave., N.W.
Washington, D.C. 20005

**INTERSTATE SANITATION
COMMISSION**

**Thomas R. Glenn, Jr.,** Dir. and
Chief Eng.
Interstate Sanitation Commission
10 Columbus Circle
New York, N.Y. 10019

**NEW ENGLAND INTERSTATE
WATER POLLUTION CONTROL
COMMISSION**

**Alfred E. Peloquin,** Exec. Sec.
73 Tremont St., Rm. 950
Boston, Mass. 02108

**OHIO RIVER VALLEY WATER
SANITATION COMMISSION**

**Robert K. Horton,** Exec. Dir.
414 Walnut St., Rm. 302
Cincinnati, Ohio 45202

Georgia's most recent enforcement action involved prosecution of a textile firm in 1969. The firm was ordered to clean up or close up. The plant elected to build an adequate treatment plant. Approximately 10 other cases have been referred to the state's attorney general by the Georgia State Water Quality Board.

Normally, the Iowa Water Pollution Control Commission takes action on four to five municipalities monthly, but its spokesman notes that the majority of improvements are made by industry and municipalities "on a voluntary basis."

Missouri's Smith comments that there is a lack of continuity in enforcement. He says that the attorney for the Missouri water pollution control board is a part-time employee of the state attorney general's office. The election of a new attorney general every four years results in the appointment of a new attorney general to work with the board.

New Hampshire's recent statewide effort to abate water pollution includes provisions in its law for review and approval of all sewage disposal facilities to be installed within 1000 ft of surface waters. In New York, assorted cases are routinely referred to the counsel's office within the New York Department of Environmental Conservation. Paper mill and cannery operations are the types of polluters most frequently singled out. These unilateral actions do not involve federal officials. Like Florida, Ohio is involved with two federal–state enforcement conferences, one on Lake Erie and the other on Mahoning River. South Dakota has not taken actions for years.

**Industries**

From state to state across the country, administrators' biggest headaches with specific industries vary. Pulp and paper are particularly troublesome for administrators of Arizona, California, Georgia, New York, Florida, and Oregon. In the midwest and southwest, agriculture is bothersome. Meat packing and milk processing industries were specified by the Iowa official; milk processing, along with mine processing, also was the major concern of the South Dakota official. Agricultural problems have been handled satisfactorily in Texas. Sugar refineries and canning operations are Utah's problems; steel is a big concern in Ohio, with chemicals running a close second, and the pulp and paper industry third.

The majority of Arizona's industry is located in metropolitan Phoenix and Tucson. These cities impose rigid sewer service agreements on their industries before permitting them to discharge into the municipal system. Mining, another major industry in the state, is a good neighbor. In general, the mines hoard their water very carefully with extensive recycling; none has a continuous discharge to any watercourse.

Old refineries, along with industries that handle toxic materials, present the biggest headache for California. Its personnel is working to ensure that industry pays its fair share when a municipality handles its waste.

In Florida, the entire paper industry—with the exception of one mill—the citrus industry, the phosphate mining and processing industry, and most of the chemical industries are on notices and orders of the Florida Department of Air & Water Pollution Control to correct their pollution. Progress is satisfactory, according to state officials.

Georgia's biggest headache comes from Kraft paper and pulp mills, the largest industry in the state. Most plants provide their own treatment, Howard says. Food processors and textile mills are others which present pollution difficulties.

Major industries which present problems in Iowa are the meat packing and milk products industries. Half of the meat packing industries' load (equivalent to a population of 3.8 million persons) is treated in municipal plants. But Iowa's Schliekelman says that there may be some instances in which the industry is not paying an equitable share of treatment cost.

In New York, the paper industry, as a group, is further behind on correcting its pollution than any other industry. There have been few treatment plants built to handle the waste water discharges from the paper industry. The food industry also has problems which involve the dairy and canning groups particularly. The problems arise from unsatisfactory treatment systems as well as from unsatisfactory operation of adequate systems.

Steel, chemicals, and pulp and paper are the major industrial polluters in Ohio. Most chemical companies in the state provide treatment for wastes, but none has closed its water loop.

In Oregon, the pulp and paper industry is again the number-one pollution problem. Last September, only five of the nine mills in the state had secondary treatment and chemical recovery systems. Another had chemical recovery alone. The remainder three have until July 1972 to complete installation of recovery systems and secondary treatment. Then, the overall reduction in BOD will be 90 to 94%.

Agriculture is the root of Texas' problems. But progress is noteworthy. Agricultural wastes are impounded in Texas. No discharge is made to any stream within the state.

In Utah, troublesome wastes are produced by sugar refineries, canning factories, milk processing operations, and slaughterhouse operations. Installations which discharge inorganic materials, which are generally lumped into the term salinity, also create problems for Utah officials.

# Water pollution law—1972 style

*Congress commits the nation to continuing water cleanup; the public must adopt a wait-and-see attitude*

Reprinted from ENVIRON. SCI. TECHNOL., **6**, 1068 (December 1972)

After being vetoed by the President, the Federal Water Pollution Control Act Amendments of 1972 were passed into law by an overwhelming majority in both houses of Congress on the very last day of the 92nd session. It's a far-reaching law, and the most significant piece of environmental water legislation ever to come out of the U.S. Congress.

Without question, the new law is tough, complicated, and costly. The big question now is whether the financial backing for cleanup will be forthcoming. In his letter to Congress, President Nixon made the point that "despite its laudable intent," he was vetoing the bill because of "its unconscionable cost."

But without the new law the U.S. machinery for clean water would have come to an abrupt and screeching halt. Now, it's the only blueprint that the U.S. has for water cleanup, one for which the public has been waiting since the 92nd session of Congress convened. Bottled up in conference since May, P.L. 92-500 extends the Federal Water Pollution Control Act of 1965, as amended in 1966 and 1970. (The 1970 amendments are known as the Water Quality Improvement Act.)

The U.S. now is wed to a new cleanup schedule. Whether the marriage will be successful, only time, money, and commitment can assure. There is something old, something new, something borrowed, and something blue in the new law—old in the sense of continuation of the 1965 law; new with regard to the state permit program and the National Study Commission; borrowed (from previous federal environmental legislation) with regard to the numerous deadlines; and blue since it was passed over the President's veto.

It's not too surprising that the new law contains deadlines, nor that it is modeled closely after the clean air bill, since the water law was given to us basically by the same legislators who gave us the 1970 amendments to the Clean Air Act (ES&T, February 1971, p 106).

### Biggest expenditure

To be sure, the largest authorization goes for publicly owned waste treatment works, a whopping $18 billion over the next three years—three times more than the President had requested for the same time period. What is more, there are $2.75 billion authorized for reimbursements for treatment works already constructed.

In the new law, the definition of "treatment works" has been expanded to include interceptor sewers, sewage collection systems, pumping, power, and other equipment, as well as land site acquisition for either the plant or the site for ultimate disposal of sludges.

The good news for the recipient of "treatment works" grants is the fact that the federal share toward construction funds is now 75%. "Construction" means everything from preliminary planning to actual construction. Another piece of good news is that this share can be paid for treatment works from funds authorized for any federal fiscal year beginning after June 30, 1971.

On the other hand, the cautious—hopefully not bad—news is that certain limitations are spelled out. For example, after June 30, 1974, the EPA administrator cannot make such "treatment works" grants unless the applicant has studied alternative waste management techniques and taken into account the reclaiming and recycling, or otherwise discharging of pollutants. These so-

## Minicalendar for standards watchers, P.L. 92-500

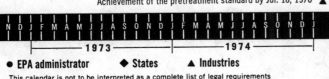

**Water quality standards & implementation**
◆ Must submit plans for any intrastate water quality standard
● Must determine that wqs are consistent with new act, if not notify states
◆ Must adopt, submit intrastate wqs
● Must approve or reject state intrastate wqs plan
● Must promulgate or substitute plan
Must have held at least one public hearing on wqs by Oct. 18, 1975 ◆

**National standards of performance**
● Must publish degree of effluent reduction attainable by secondary treatment
● Must publish list of industries for which new source standards apply
Must publish informa– ● tion on alternative waste treatment management
● Must propose and publish standards
● Must publish non-point source guidelines
Must promulgate standards ● for new industrial sources

**Toxic effluent standards**
● Must publish list for which effluent standards are to be set
● Must propose effluent standards
Must promulgate ● standard with annual updates

**Pretreatment standards**
Must publish ● proposed regulations for introduction of pollutants to works
● Must promulgate standard
Achievement of the pretreatment standard by Jul. 18, 1976 ▲

N D J F M A M J J A S O N D J F M A M J J A S O N D J
├——— 1973 ———┤├——— 1974 ———┤

● EPA administrator    ◆ States    ▲ Industries

This calendar is not to be interpreted as a complete list of legal requirements for the EPA administrator

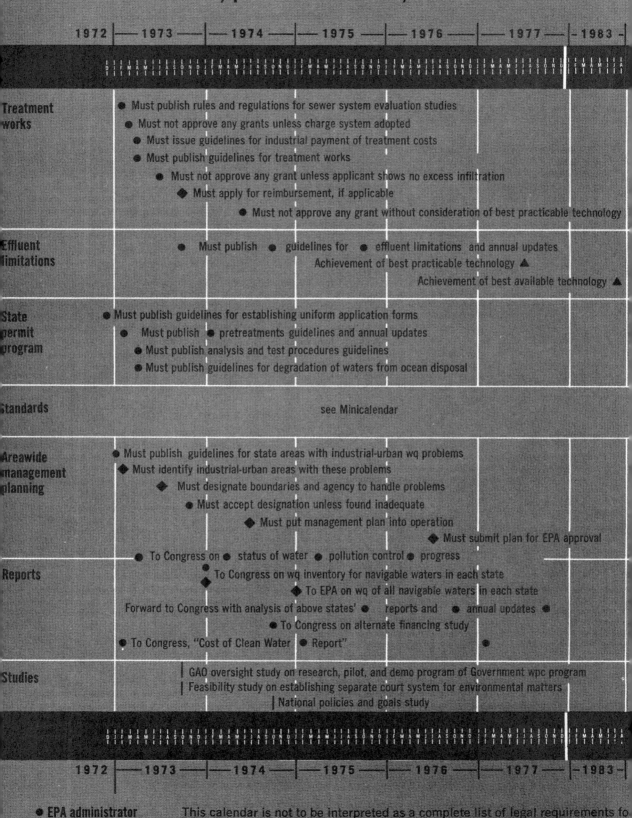

# Calendar for water pollution watchers
## P.L. 92-500, passed on Oct. 18, 1972

1972 —— 1973 —— 1974 —— 1975 —— 1976 —— 1977 —— 1983

**Treatment works**
- Must publish rules and regulations for sewer system evaluation studies
- Must not approve any grants unless charge system adopted
- Must issue guidelines for industrial payment of treatment costs
- Must publish guidelines for treatment works
- Must not approve any grant unless applicant shows no excess infiltration
- Must apply for reimbursement, if applicable
- Must not approve any grant without consideration of best practicable technology

**Effluent limitations**
- Must publish ● guidelines for ● effluent limitations and annual updates
- Achievement of best practicable technology ▲
- Achievement of best available technology ▲

**State permit program**
- Must publish guidelines for establishing uniform application forms
- Must publish ● pretreatments guidelines and annual updates
- Must publish analysis and test procedures guidelines
- Must publish guidelines for degradation of waters from ocean disposal

**Standards**
see Minicalendar

**Areawide management planning**
- Must publish guidelines for state areas with industrial-urban wq problems
- Must identify industrial-urban areas with these problems
- Must designate boundaries and agency to handle problems
- Must accept designation unless found inadequate
- Must put management plan into operation
- Must submit plan for EPA approval

**Reports**
- To Congress on ● status of water ● pollution control ● progress
- To Congress on wq inventory for navigable waters in each state
- To EPA on wq of all navigable waters in each state
- Forward to Congress with analysis of above states' ● reports and ● annual updates ●
- To Congress on alternate financing study
- To Congress, "Cost of Clean Water ● Report" ●

**Studies**
- GAO oversight study on research, pilot, and demo program of Government wpc program
- Feasibility study on establishing separate court system for environmental matters
- National policies and goals study

1972 —— 1973 —— 1974 —— 1975 —— 1976 —— 1977 —— 1983

● EPA administrator
◆ States
▲ Industries

This calendar is not to be interpreted as a complete list of legal requirements for EPA administrator

Copyright 1972, The American Chemical Society

called areawide management plans would spell out the needs of the area for a 20-year period, with annual updates.

**Keyword**

Although terms like effluent limitations do not carry the label "standards" when published within a year after enactment, they will put existing industries on notice, so to speak, that certain technology is being used by some leaders in their industries. The notice, in effect, says that the industries must have "best practicable technology" by July 1, 1977, and "best available technology" by July 1, 1983, with the goal being no discharge of pollutants by 1985.

Standards are the keyword in the new water law, just as they were in the now 2-year-old air law (ES&T, February 1971, p 106); national standards of performance; toxic effluents standards; pretreatment standards; as well as water quality standards and implementation plans for their achievement in intrastate waters.

Of course, the public surely knows by now that standards alone have no effect whatsoever on the achievement of water quality. Only when standards and control technologies are applied will water quality be enhanced.

Under the earlier provisions of the water amendments, states have had most responsibility for water pollution control, but now they have an additional authority to establish a state permit program (under federal guidelines, of course, to ensure some uniformity). The state permit is for the construction and operation of any activity that may result in any discharge into the navigable waters. (Basically, it is a plan similar in concept to that the Corps of Engineers announced two years ago, just prior to the establishment of the EPA.) The permit also applies to effluent limitations, national standards of performance, as well as toxic and pretreatment effluent standards. (See Minicalendar.)

**How reasonable?**

While it is true that the deadline for "best available technology" for point sources is 1983, the National Study Commission, another new creation of this law, must report (within three years) to the Congress specifically on all the social, economic, and environmental effects of achieving or not achieving the effluent limitations specified for 1983. The 15-member commission—comprising senators, congressmen, and members of the public—can contract for studies with the National Academy of Sciences and the National Institute of Ecology, among others.

For the scientist, there will be created an advisory board selected from the scientific community for its expertise on scientific and technical information. The board will advise the EPA administrator on effluent limitations, water quality information and other matters. Eight members are to be appointed by the administrator for a 4-year term and can be reappointed.

Without question, none of the plans is any good without enforcement. Again, modeled after the air legislation, penalties for violation of the new water law are tougher under P.L. 92-500 than under earlier water laws. Conviction for a knowing violation is subject to a penalty of $25,000 per day or imprisonment for one year, or both. For a second knowing violation, the penalty increases to $50,000 per day of violation, imprisonment for two years, or both. What is more, any person who violates a provision of the state permit in regard to standards is subject to a civil penalty of $10,000.

The public's role in water pollution control, which stems from a then unique feature in the air law, has been carried over into the water field: Citizen action suits can be brought against polluters and government officials, but such suits cannot be brought against the EPA administrator in cases where he is given discretion to act. SSM

### Checklist for authorizations followers
### (all figures in millions of dollars)

| | Fiscal 1973 | 1974 | 1975 |
|---|---|---|---|
| CONSTRUCTION GRANTS FOR TREATMENT WORKS | | Subtotal | $21,350 |
| Treatment works construction | $5000 | $6000 | $7000 |
| Reimbursements | ←———— | 2750 | ————→ |
| Basin planning | ←———— | 200 | ————→ |
| Areawide waste treatment management | 50 | 100 | 150 |
| Technical service to agencies | 50 | 50 | |
| RESEARCH AND RELATED PROGRAMS | | Subtotal | $608 |
| Research, investigations, training, and information | 100 | 100 | .. |
| R&D grants | 75 | 75 | .. |
| State pollution control grants | 60 | 75 | .. |
| Mine water demonstration program | ←———— | 30 | ————→ |
| Training grants and contracts | 25 | 25 | .. |
| Great Lakes cleanup demo project | ←———— | 25 | ————→ |
| In-place toxic pollutants removal contracts | ←———— | 15 | ————→ |
| Alaska village demo project | ←———— | 2 | ————→ |
| Lake Tahoe study | ←———— | 0.5 | ————→ |
| STANDARDS AND ENFORCEMENT | | Subtotal | $1,451 |
| Standards' development | 100 | 100 | .. |
| Clean lakes grants to states | 50 | 100 | 150 |
| Revolving oil fund | ←———— | 35 | ————→ |
| National Study Commission | ←———— | 15 | ————→ |
| Alternate financing methods study | ←———— | 1 | ————→ |
| Other than the above sections | 250 | 300 | 350 |
| ADMINISTRATION AND STUDIES | | Subtotal | $105 |
| Environmental financing funding | ←———— | 100 | ————→ |
| National policy and goals study | ←———— | 5 | ————→ |

Not listed $1,086
Total $24,600

# Effluent guidelines are on the way

*With most of its statutory enforcement authority*
*jeopardized by recent court decisions,* EPA *is looking to*
*a better method for controlling industrial water pollution*

Reprinted from ENVIRON. SCI. TECHNOL., **6,** 786 (September 1972)

Although it's not too well known yet, there's been a definite shift in EPA's strategy in the war against industrial water pollution. The agency's Office of Water Programs is readying its guns to fight water pollution at the industrial outfall, using a tactic adopted by the Office of Air Programs to halt air pollution from stationary sources, by promulgating effluent guidelines and performance standards.

The shift in position is tantamount to an admission that the agency's previous control strategy—based on water quality standards—simply has not worked. Enforcement conferences and 180-day notices—the major enforcement tools provided by the Federal Water Pollution Control Act, have proved to be costly and cumbersome remedies, virtually devoid of clout. One agency official, who declines to be quoted, labels them "ludicrous."

Similarly, the utility of the discharge permit provisions of the Rivers and Harbors Act of 1899—the so-called Refuse Act, pressed into service by the Nixon Administration in December 1970 to patch up the holes in EPA's armamentarium—is in serious trouble because of a pair of court decisions. In December 1971, Federal Judge Aubrey E. Robinson of the District Court for the District of Columbia, issued an injunction forbidding the issuance of any discharge permits until environmental impact statements for each major permit could be prepared. Some 20,000 applications for the permits are on file, according to EPA assistant administrator for enforcement and general counsel John R. Quarles, Jr., and preparation of environmental impact statements for each major one would be a "bureaucratic nightmare." EPA has appealed the decision, but until the appeal can be heard and decided, the ban on permits stands.

To make matters worse, in May of this year, the U.S. Court of Appeals in Philadelphia, Pa., ruled in a case involving the Pennsylvania Industrial Chemical Corp., that industrial polluters may not be prosecuted under the Refuse Act unless permits are available. "That

case," Quarles complains, "could stop us dead in our tracks. Something must be done or the Refuse Act—our principal legal weapon against industrial water polluters—may become a dead letter." Quarles has asked the Justice Department to intervene and speed up the appeals in the District of Columbia case.

## Full speed ahead

EPA isn't waiting for a congressional mandate to set effluent guidelines before gearing up to do battle with the industrial polluters, although such a

**Guideline chief Cywin**
*Not waiting for Congress*

mandate is expected as a result of new legislation (see below). For example, there has been a quiet restructuring of the Office of Water Programs—specifically, it appears, to provide a framework for developing effluent guidelines. (See chart, page 779.) The office of Deputy Assistant Administrator for Water Programs, formerly headed by Eugene Jensen, has been split, creating two new positions—deputy assistant administrator for water programs operations and deputy assistant administrator for water programs planning and standards. Jensen has been named to the operations post; the planning and standards position is still vacant. The new effluent guidelines program is headed by division director Allen Cywin, former acting chief of Water Quality Research

for EPA, and is administered out of the planning and standards branch.

There is little doubt that legal authority to operate the effluent guidelines program will soon be forthcoming. Both H.R. 11896 and S. 2770—the House and Senate versions of the omnibus water pollution control amendments being thrashed over in joint committee—have virtually identical language requiring that the EPA administrator adopt and enforce effluent guidelines and standards for industrial point sources. "The Environmental Protection Agency is not waiting for this legislation to be passed before beginning to carry out and develop the necessary guidelines and standards that are required," says Michael B. La Graff, Cywin's second in command. "The time schedules of the legislation are very stringent and we need every minute of time in order to satisfy completely the requirements of the legislation," he adds.

The proposed water legislation (section numbers refer to H.R. 11896) has four separate sections dealing with effluent guidelines and standards:

- Sec. 304 (b): Effluent Limitation Guidelines
- Sec. 306: Standards of Performance for New Sources
- Sec. 307(a): Toxic Pollutant Standards and
- Sec. 307(b): Pretreatment Standards.

## Technology key

Under the effluent limitation guidelines called for in Sec. 304(b), EPA would identify both the "best practicable technology" currently available and the "best available demonstrated technology." Guidelines would be published within one year of the adoption of the new water amendments. By 1976 all industrial waste sources—no matter how old—would be required to adopt, as minimum acceptable treatment, the best practicable technology currently available.

The proposed legislation specifies several parameters to be considered in the development and application of the regulations, including the age of equipment and facilities involved, the processes employed, process changes, engineering aspects of the application of various types of control technology, and the cost of achieving effluent reductions.

Regulations promulgated under Sec. 304(b) would establish the minimum requirements each industrial source would have to meet by 1976 and maximum effluent requirements that could be achieved, where necessary, within the alloted time. Such "practicable" control technology would be based on end-of-the-line treatment techniques rather than on cleaning up the process itself. The technology to be used would be determined purely on technical considerations and not on the quality of the receiving waters. It would be uniform regardless of the local pollution situation.

EPA interprets "currently available," to mean a control technology which in "demonstration projects, pilot plants, and general use has demonstrated a 'reasonable' level of engineering and economic confidence and viability in the process at the time of commencement of actual construction of the control facilities," La Graff says.

On the other hand, "best available demonstrated technology" embraces in-process control technology as well as outfall cleanup. Such levels will be set by technology that "through pilot plant operation or full-scale use has demonstrated both technological performance and economic viability at a level sufficient to reasonably justify the making of investments in such processes," according to La Graff. An important aspect of the legislation provided in Sec. 304(b) is that it would apply to all plants, and not just newer ones.

**Performance standards**

Sec. 306 of the proposed water law would require the administrator to promulgate performance standards within 15 months after passage of the law for new sources or all old sources which undergo "major modification." Such sources would have to incorporate the "best available demonstrated technology" at the time of construction or modification. While "best available" legally means the same for both new and old sources, EPA will probably require lower discharge levels from new plants because of greater opportunity to incorporate in-plant controls. This section of the bill specifically delineates about 30 industrial categories for which standards will be prescribed. EPA will probably end up with more categories than that.

Section 307(a) would require that within 90 days of enactment of the legislation, the administrator must publish a list of pollutants or combination of pollutants that are determined to be toxic. Toxic is distinguished from hazardous, La Graff says, in that a hazardous substance presents an "imminent or substantial danger," while for a substance to be declared toxic "requires evidence that it will cause death, disease, behavioral abnormalities, cancer, . . . or physical deformations in affected organisms or their offspring." Molasses is being considered as a hazardous substance "because when discharged into a body of water it has caused a fish kill," he says. "However, in a normal toxicity study it would require enormous amounts of molasses in order to be toxic, and therefore it would not normally be considered a toxic pollutant per se."

Within six months after publishing the toxic substances list, EPA would publish proposed effluent standards for each toxic pollutant. Such a standard could require a total ban, but standards could also be varied from industry to industry depending on the degree of treatment technology available. Such variances among industries would not be granted simply because it is more expensive for one industry to treat a toxic substance than it is for a different industry to treat the same pollutant.

Section 207(b) would establish two provisions for the publication of effluent pretreatment guidelines. The first is for publication within six months of pretreatment guidelines aimed at aiding municipal systems, with some specific considerations that should be taken into account when accepting industrial wastes for treatment. The second is for publication within six months of pretreatment

standards which will limit discharge of specific pollutants to municipal systems.

Just how EPA will implement the provisions of the water legislation remains open at this time. Clearly the time constraints are formidable, even considering that the bills have not yet become law. The first step for the agency is to develop categories and subcategories which take into account the parameters specified in the bills. A start, at least, has been made in a pilot study with the American Petroleum Institute. Following the subcategorization of refineries by class—A, B, C, D, and E class refineries as defined by the industry—EPA has begun studies aimed at characterizing the effluents—no mean task in itself. The next step will be to assess the treatment technology available for application by each of the various classes. EPA must repeat the process for all industrial sources. The end result will be a set of output standards, in terms of hard numbers such as pounds of BOD/unit of raw material input or production.

The job is complicated by the fact that, depending upon the amount of horse trading that goes on in the conference committee, the law may provide for a program to begin issuing discharge permits upon passage of the statute. "We must be prepared to support this effort immediately upon enactment of the legislation," La Graff says, " and not a year or 15 months following the enactment."

Amid all the uncertainty about specifics, one thing remains clear: With its current legal authority to check industrial water pollution seriously eroded, EPA is pinning its hopes on the effluent guidelines. Enforcement chief Quarles has called on Congress for speedy action.        HMM

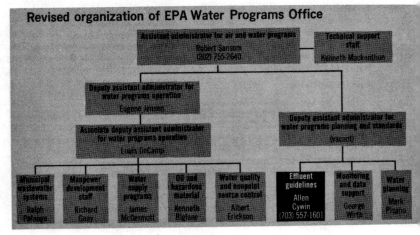

Revised organization of EPA Water Programs Office

# Industrial emissions: at last, some numbers

*Newly proposed EPA "guides" and standards are by no means the final word on regulation, but industry now has something to shoot at*

Reprinted from ENVIRON. SCI. TECHNOL., **5**, 748 (September 1971)

Pollution control requirements for industrial processes are coming into rather sharper focus this month. There are two important developments: First, the Environmental Protection Agency (EPA) has published a list of emission performance standards which will apply to all air emissions from certain types of *new* industrial plants. Second, EPA is providing to its regional offices, a set of effluent reference guides covering the feasibility of controlling liquid industrial wastes.

## Air emission standards

Of the two developments, the promulgation of air emission standards has the weightier push of law behind it. The standards are required by the recent amendments to the federal Clean Air Act (ES&T, February 1970, page 106). These standards—numerically as shown on page 749, if they are not modified as a result of industry comments—will become effective in mid-November.

What the performance standards mean to industry is that any new plants in the first five industries specified (see box) must at least meet the level of emissions control specified in the standards. It is important to recognize, of course, that these standards in no way apply to emissions of plants that are presently operating (though such plants will sooner or later be forced to operate so that their air emissions do not violate an applicable air *quality* standard).

The performance standards were proposed by EPA on Aug. 17, with a 45-day period provided for comments from interested parties. One comment being leveled at the proposed standards, according to Robert Walsh of the Division of Compliance in EPA's Air Programs Office, centers on the sampling method for stack emissions.

An American Society of Mechanical Engineers (ASME) method has been used for many years; it is based on all filterable solids from the stack gas. The proposed EPA method specifies taking a 250° F sample, cooling it to 70° F, and then determining particulate matter. Values obtained with the earlier ASME method are three to four times lower than those found with the newer method. The net result will inevitably be stricter regulation, since controls that could satisfy the old test method would give much higher—and possibly unacceptable—values with the new one.

## Reference guides—liquid effluents

Beginning this month, the Division of Technical Support in EPA's Water Programs Office is sending to all EPA regional offices, a series of effluent reference guides for water-using industries. Although the guides were drawn up under programs mandated by current federal water pollution law, EPA officials take pains to explain that the guides are *not* the same as standards.

The immediate purpose of the guides is to assist regional EPA offices which, under an agreement between EPA and the Corps of Engineers, will be asked to make a determination regarding the acceptability of applications for discharge permits. The permits are required for compliance with the newly resurrected Refuse Act program.

According to Michael La Graff of the Water Programs Office, the guides establish "realistically achievable" levels of control for waste streams in different segments of industry.

What the reference guidelines mean to the 20 water-using industries specified (see box) is that their present levels of discharge will probably be judged in the light of what they realistically could achieve using technology that is relatively well-established for each particular industry. The guides apply to most presently operating plants in the U.S.

The guides for each industry segment consist of two separate groups of waste water constituents. The first group contains the key constituents for which discharge levels are given and which should be regulated. For instance, in a food processing plant, these constituents would be COD, BOD, pH, and suspended solids. In petroleum refining, the constituents would be ammonia, BOD, chromium, COD, oil, pH, phenol, sulfide, suspended solids, temperature, and total dissolved solids.

The numbers and possibly the actual constituents might, however, vary for the various types of plants within a single industry. (The petroleum industry, for example, is divided

cal manufacturing process most representative of actual manufacturing conditions. It also allows differentiation between old and new technology used to make the same product.

### Interpretations

There is a possible danger that the effluent reference guides (which EPA will make available to industry representatives) and published performance standards for air emissions could be interpreted as the absolute last word in control regulations. To make this misinterpretation would be to disregard the overriding importance of water and air *quality* standards which still are at the very heart of current federal legislation. In a sense, the guides for water discharges and standards for air emissions discussed in this story merely reflect minimum requirements; in many practical instances, they may be insufficiently stringent to be acceptable.

At plant sites in remote locations, these guides and performance standards very probably would be acceptable and represent the level of treatment or control required by federal regulations. However, in most other locations, and particularly in densely populated or industrialized areas, a higher level of control will almost certainly be required to meet the applicable water quality or air quality standard.

For example, if an industry installed effluent controls corresponding to EPA's effluent reference guide for that industry, it would still be faced with a requirement to meet a particular water quality standard, cautions Richard Nalesnik, director of EPA's Water Quality Standards Office. Of course, this requirement is built right into the discharge permit program so that any company holding a valid permit would presumably be in compliance with the standard or under a timetable to become so.

Likewise, an industry that builds a new stationary source of air emissions might find that the new performance standards suffice when a plant is built out in the country, but that they are insufficient to meet air quality standards if the same plant were built in an urban area.

What specific use EPA will make of its newly developed guidelines and standards is still somewhat uncertain at press time; the agency is still formulating policy. But it is more than likely that they will be of use to state

and local pollution control officials. And the effluent reference guides will definitely be useful in administration of the discharge permit program, says Bert Printz of EPA's Permits Program Office. Printz admits that the permit program is complicated, involving as it does, five federal agencies, 50 states, and at least four pieces of legislation. But he is optimistic: He says the permit program is the first new legislative thrust given the EPA with an adequate number of personnel to do the job. The agency's supplemental funding request for 400 people has been approved by Congress, and 150 are on board today.  SSM/DHMB

into five different plant classifications.)

The second group of waste water constituents is more in the nature of a "laundry list," to use La Graff's words. It would contain all the constituents that could conceivably be discharged by the particular industry in question, but no numbers are specified in it. Actual decisions as to permissible discharge levels of these constituents will be made by EPA regional officials with reference to the applicable water quality standard for the receiving stream.

The reference guides for the 20 industries were developed under a series of federal contracts, totaling $1 million in value, that were let earlier this year. La Graff explains that not all water-using industries were covered, some exceptions being the electrical equipment and pharmaceutical industries.

The industries were studied on a manufacturing basis rather than plant by plant. This made it easier to define waste characteristics for the typi-

# Water Quality Improvement Act is now the law

*Reprinted from* ENVIRON. SCI. TECHNOL., **4**, 379 (May 1970)

In conference since last October, the compromise version of the Water Quality Improvement Act is the culmination of efforts begun in 1966 to strengthen the basic 1956 Federal Water Pollution Control Act. Now, P.L. 91-224 provides tighter controls on several water pollutants and sets the stage for control of other harmful ingredients in the water environment.

## Oil

The basic philosophy of P.L. 91-224 is that the oil industry and its shippers, rather than the public, will bear the risk of oil spill cleanup. One legislative delay was the insistence of the Senate version (S. 7) on absolute liability for oil spills from vessels and onshore and offshore facilities.

Now, the federal government is authorized to take immediate action for the cleanup of any spill in the navigable waters of the U.S., adjoining shorelines and waters of the contiguous zone—in essence, any place where there is a threat to aquatic life or of potential shore damage.

The compromise limits of vessel liability on cleanup include the House (H.R. 4148) lower figure of $100 per gross registered ton and the Senate (S. 7) limit of $14 million; the assessment would be based on whichever is the lesser amount. For onshore and offshore facilities the top limit for cleanup is $8 million.

However, if a vessel or a facility spilled oil by willful negligence or misconduct, then the party is liable for whatever cost is incurred by the federal government in its cleanup. Further, failure to notify federal authorities of an oil spill, as provided in the contingency plan, incurs strict criminal penalties. To further ensure that the U.S. will have adequate protection against oil spill threats to the environment (ES&T, February 1970, page 97), the law also contains the following three provisions:

- Revolving cleanup fund of $35 million.
- Requirement for contingency planning.
- Research to develop new cleanup methods.

## Certification

A most important section of the new law requires that applications for federal licenses may not be granted unless water quality standards are protected. For example, federal agencies including the Corps of Engineers, Atomic Energy Commission, Federal Power Commission, and the like must now give consideration to the water quality standards when issuing permits or licenses. Federal facilities would also be required to heed this new requirement. Thermal discharges from power stations may thus be better controlled.

## Vessels

To control sewage discharge from vessels, performance standards for marine sanitation devices will be developed by the federal government. Then, all such devices must be certified by the government before the devices can be sold.

By 1972, all new vessels will be required to have certified devices installed. Older vessels would not be required to install approved devices before 1975, however. For the first time, P.L. 91-224 would require uniform standards for boat owners and operators throughout the U.S.

## Environmental quality

Title II of the new law creates the office of environmental quality, including ten specialists, that will be funded at a level ranging from $500,000 in fiscal 1971 to $1.25 million in fiscal year 1974. This office will provide the staff needed by the three-man Council on Environment Quality (CEQ) to carry out its mandate. One func-

**Offshore rig.** *Spill damages are covered for the first time in act*

tion of the new office would be to monitor all federal pollution control programs. Russell E. Train, CEQ's chairman, will direct the new office.

## Chemicals

In addition to the tighter controls placed on oil, vessel discharges, and thermal releases, the law also specified development of water quality criteria for pesticides in the water environment. Further, federal attention will be given to the designation of hazardous substances and authority for federal cleanup of a wide variety of materials that may find their way to the waters.

Several other less spectacular, but equally important, amendments were retained in the final version of the act including the following:

- Training funds totaling $12 million in fiscal year 1970 and $25 million each in fiscal years 1971 and 1972.
- Acid mine drainage demonstration program totaling $15 million.
- Great Lakes demonstration program totaling $20 million.

# Congress:
# Much ado about water
# pollution amendments

*Reprinted from* ENVIRON. SCI. TECHNOL., **5**, 302 (April 1971)

Senator Muskie

Senator Hart

*Recent water pollution*

*control laws must be wed*

*with the Refuse Act of 1899*

The most pressing environmental matter before this 1971 legislative session of Congress is the messy business of amending and extending U.S. water pollution controls. Obviously, since the present legislation expires this June 30, the goal is to come up with a tough water law, like last session's tough air pollution control law.

One nagging problem in rewriting the law is clarification of some apparently overlapping jurisdictions and the legislative intent of environmental laws already on the books—the Refuse Act of 1899; the Federal Water Pollution Control Acts of 1965, 1966, and 1970; the National Environmental Policy Act of 1969; and the Fish and Wildlife Coordination Act.

The major effort will be to bring the Federal Water Pollution Control Act and its inherent concept of ambient water standards into harmony with the Refuse Act and its inherent concept of emissions standards. Nevertheless, it seems the Corps of Engineers will continue to enforce the Refuse Act and the Environmental Protection Agency's (EPA's) Water Quality Office will enforce the more recent water pollution laws.

## Searching questions

After several rounds of hearings, the consensus maintains that the numerous suggestions already aired, for example, during last year's hearings (ES&T, November 1970, page 888)—effluent standards, streamlining of enforcement proceedings, increase in penalties, extension of federal law to

all navigable waters, removal of distinction between intrastate and interstate water, and the like—should without much further argument be included in this year's amendments. Most probably they will be added. But more evident, as discussions continue, is the need for a complete reshaping of the legislation with a fresh, new, national rationale before the public loses its strong desire for the enhancement of the nation's waters.

While the Senate Subcommittee on Air and Water Pollution explored the status of the EPA program on water quality, joint Senate–House hearings of the Senate Commerce Subcommittee on Environment and the House Government Operations Subcommittee on Conservation and Natural Resources aired the basic 1899 Refuse Act. The joint hearing raised searching and fundamental questions not only on the necessity for the Corps' permit program, which was announced by Executive Order last December 23, but on its very legality.

Thus, while one quarter seeks clarification of recent water pollution control legislation, another quarter is resurrecting the 19th century law with its baroque legal language. So, the race is on to see which of the two can more quickly afford the nation the water quality that has been promised to it ever since passage of the 1965 act.

## Water quality status

In its preliminary oversight hearings on water pollution, Sen. Muskie's air and water pollution subcommittee concentrated on five specific concerns— funding and manpower, regulations to be issued by the EPA, state certification required by the Water Quality Improvement Act (WQIA) of 1970, water quality standards, and effluent guidelines. On another day of hearings, the center of attention was clarification

and interpretation of the 1970 act, specifically the status of vessel pollution, oil pollution, and again, the difficult question of state certification.

In his first official appearance before the subcommittee, the administration's star witness, EPA administrator William D. Ruckelshaus brought the Senators up to date on these items. On financing, Ruckelshaus said that no reimbursables—the monies advanced by states to meet the federal share of construction funds—have been paid to date. By the end of last year, the federal government owed the 22 states which prefinanced the funds a tidy sum of $1.5 billion. But, of the $800 million appropriation for fiscal 1970, $100 million will be available for payment of reimbursables, perhaps as early as mid-May.

The subcommittee chairman referred to the lack of payment as an administrative lag, but the star witness differed with this interpretation and pointed out that the existing allocation formula was not responsive to the needs of some of the heavily populated states, such as New York and Michigan. Ruckelshaus agreed, however, that the federal government needs not only to get the monies to

Congressman Reuss

**Common concern.** *Chairmen of three congressional subcommittees ponder the best way to proceed, on an emergency basis, with the nation's water quality enhancement program*

the states but, in addition, to find cost-effective ways to do so.

Nevertheless, the federal government anticipates reimbursement of fiscal '70 and '71 funds and will probably start payment to the states by the beginning of fiscal year 1972. Already, federal officials have met with Gov. Rockefeller and indicated that New York will be receiving from $100–200 million in reallocated funds.

On effluent discharges, acting commissioner of the Water Quality Office, David D. Dominick indicated to the subcommittee that the purpose of the office's waste water inventory (ES&T, January 1971, page 20) is to canvass 22 basic industries, thereby obtaining discharge data, so that the federal government would be able to define the term "industrial equivalent of secondary treatment."

On the manpower issue, the nation entered 1971 with a manpower shortage, including a backlog of 30,000 persons in the water pollution control field. The precise number and distribution are not available, but a study authorized by the 1970 WQIA presently is looking into the needs for all phases of the water program; a report is promised by mid-year.

The old issue of water quality standards seemingly is becoming more and more of a roadblock to progress and, in particular, with respect to the state certification requirement of section 21(b) of the Water Quality Improvement Act. For example, if a state has given its certification on a particular interstate water and the EPA disagrees, then EPA is the final arbiter on questions on violation of standards.

The problem that becomes difficult to resolve occurs when enforcement actions are considered. Slightly more than half of the states, actually 27, have had federal–state standards for interstate waters fully approved. On the other hand, 44 states have state standards for intrastate waters (no federal jurisdiction); five—Arkansas, Delaware, Kentucky, Tennessee, and Wyoming—do not have standards for intrastate waters, which are not covered under the existing 1965 law and its amendments.

**Permit program**

In a later round of joint House–Senate hearings, the 1899 Refuse Act, which was resurrected in 1969, was the center of discussion because of the new twist of its permit program by the Corps of Engineers. Some circles maintain that the permit program may be the long needed panacea for the nation's water quality ills; others question not only the need, but even the legality of the program. President Nixon's chief environmental adviser, Council on Environmental Quality's Chairman Russell Train, hailed the program as the single most important step this country has ever taken to abate water pollution. So, the race is on to see whether the old law or the recent laws of the past six years are best equipped to proceed, on an emergency basis, with the nation's war on water pollution. Many maintain that the Corps' program may be the nation's best line of defense against polluters.

Although the Corps' guidelines for the issuance of permits had not yet been officially announced at press time, Shiro Kashiwa, assistant attorney general of the Land and Natural Resources Division in the Justice Department, told the joint hearings that the

Justice guidelines of June 13, 1970 were withdrawn. What this means to U.S. attorneys general throughout the country is they are once again authorized to initiate civil or criminal actions referred to them by Corps district engineers or EPA regional representatives (ES&T, January 1971, page 13). Kashiwa indicated that the Justice Department's new instructions are effective when the Corps' guidelines are issued.

Then, ways to strengthen the permit program were discussed; more than likely the following suggestions will be included when the program is officially announced:

• The permit application must be signed by a company executive;

• False statements in the permit application may be prosecuted under 18 USC 1001; violation incurs a penalty of five years imprisonment or a $10,000 fine;

• Citizens will be able to review the permit applications in the Corps' district offices;

• A boilerplate clause in the application, requiring effluent charge considerations.

Whether the new water pollution legislation contains more deadlines than last year's air pollution legislation is of little consequence. What is needed now is a combining of the two water laws—the Refuse Act and the Federal Water Pollution Control Act—and clarification of the judicial test to be used in enforcing the new law. Whether to require an effluent test or an ambient water standards test is a controversial issue that will require much more discussion and interpretations as the hearings continue.

Not until this issue is adequately resolved can the nation make progress in its long overdue program for water quality enhancement. If Congress does not resolve this basic issue in this year's amendments, it will truly return to haunt them at some later date.

What will emerge in the way of new controls for water enhancement this session is far from clear at this time. What Congress has thought it was doing in the last six years for enhancement of water quality is not necessarily what today's status would show. SSM

**Commander William E. Lehr**
*United States Coast Guard, Washington, D.C.*

Besides enforcing federal antipollution laws on U.S. navigable waters, the Coast Guard is also conducting R&D programs

*Reprinted from* ENVIRON. SCI. TECHNOL., **5,** 512 (June 1971)

# Coast Guard activities in pollution control

Polluting inland waters, estuaries, and oceans through deliberate and accidental discharges of foreign materials has been a matter of concern for many years. In the United States, the first water pollution legislation, the New York Harbor Act, was enacted in 1886. However, wide public awareness and concern for environmental pollution in general, and water pollution in particular, are rather recent occurrences. During the past few years events such as the oil platform accident and the various tankship catastrophes have dramatized the detrimental effects of major accidental pollutant releases. Similarly, the discovery of marine animal contamination with DDT and mercury illustrated the dangers inherent in chronic, deliberate discharges of supposedly innocuous foreign materials.

For its part, the U.S. Coast Guard has been involved with federal water pollution activity for over 50 years. This interest is a natural one since several of the Coast Guard's statutory responsibilities bear directly or indirectly on water pollution abatement.

The Coast Guard is the maritime law enforcement arm of the government. As such, it's responsibilities include enforcing federal antipollution laws on the navigable waters of the United States. Typical of such statutes are the Water Quality Improvement Act of 1970 and the Refuse Act of 1899. The Water Quality Improvement Act is of particular interest, giving the Coast Guard additional responsibilities regarding reporting and cleaning up pollutant spills and certifying marine sanitation devices.

The Coast Guard is also responsible for implementing general vessel safety statutes and maintaining the maritime aids to navigation system. While these responsibilities are not directly related to antipollution activity, they do have an impact on pollution prevention. Properly used aids to navigation systems and safely constructed vessels should result in fewer accidents. Fewer accidents, in turn, mean fewer pollutant spill incidents.

From a different point of view, the Coast Guard has a deep internal interest in vessel pollution. As the operator of a major fleet of boats and ships, it also must comply with the various antipollution regulations.

Coast Guard responsibilities therefore cover the full range of prevention, control, and cleanup of water pollution. They are concerned with all potential and (or) actual forms of marine transportation-related pollution. Oil pollution, sewage wastes from vessels, and hazardous pollutants such as bulk chemicals are of particular interest to the Coast Guard.

## Regulations

The best method to control pollutant spills is preventing their occurrence. Activity regarding spill prevention is being carried out in several areas.

One such area is directed toward achieving new regulations to inhibit spills. Activities of the Intergovernmental Maritime Consultative Organization (IMCO) of the United Nations typify this effort. The Coast Guard is an active participant in IMCO as a U.S. representative. IMCO is presently studying the problems of casualty preven-

tion, liability, control of released oil, and pollution from industrial and other waste disposal. IMCO has also been assigned responsibility for revisions to the International Convention for the Prevention of Pollution of the seas by oil.

Other efforts within the Coast Guard have been directed toward analyzing the causes of pollutant spills on navigable waters. From these studies recommendations for new regulations covering personnel training, special transfer equipment, and handling procedures are being developed.

### Vessel safety

Another pollution prevention area is an adjunct result of research supporting vessel safety. While this research is specifically directed toward developing improved ship construction standards and improved ship positioning systems, it furnishes an indirect benefit to pollution prevention. Recently completed research examined the potential bulk release behavior of liquefied natural gas and chlorine. This work has implications for spill control and cleanup as well as providing background data for tank sizing and structural protection for cyrogenic-liquid carrying ships.

A current research project is related directly to pollution prevention in tankship accidents by examining the ability of single and double hull construction to absorb collision energy. When completed, it will provide a tool to evaluate the effectiveness of various types of hull construction to preserve cargo tank integrity and protect against cargo release.

**ADAPTS.** *The ADAPTS diesel engine in a watertight container for airdrop (below) falls during prototype field tests (left). The ship in the background simulates a stricken tanker*

Research and development supporting ship positioning systems has been conducted for many years. The results of this past work are reflected in the existing electronic navigational aids and the channel marking buoyage system. The objective of present research is to improve these systems. New and (or) improved light sources for navigational buoys are under investigation. In addition, an analog model for investigating harbor traffic patterns as a function of channel marker position, ship characteristics such as speed and turning response times, traffic density, and local weather has been developed. The model will determine the need for repositioning harbor navigational aids to assist the safe navigation of modern ships.

A particularly important project is also being conducted in harbor traffic information, demonstrating the feasibility of Harbor Advisory Radar (HAR) services for U.S. ports. Shore-based radars have been utilized in certain European ports since 1946. Experience there has indicated that harbor radars are useful in reducing collisions and assisting in traffic movement during periods of poor visibility.

Assessing the feasibility of harbor advisory radar began in January 1970 with activating an experimental system in San Francisco. The first phase of the experiment, now nearing completion, has four objectives:

• Determining U.S. mariner acceptance of shore-based advisory services;

• Determining manpower and training requirements for HAR operators;

• Providing an economic analysis of HAR concepts in regard to system

costs and benefits resulting from reductions in delays of harbor transits;
• Investigating the capability of HAR services to meet present and future navigational requirements of U.S. ports (in terms of collision and delay avoidance, search and rescue, and law enforcement).

Preliminary phase one results indicate widespread acceptance by mariners. In fact, intraharbor vessels that did not previously report to the Vessel Movement Reporting System are now reporting to the HAR center. Although too soon to report on the economic aspects of HAR and its collision avoidance capability, the preliminary results are, however, encouraging enough to warrant initiating the experiment's second phase.

**Law enforcement**

For many years routine surveillance patrols have been conducted to detect violators of the antioil pollution statutes. These patrols utilized visual sighting to determine pollution incidents. Unfortunately, dependence on visual sightings restricted effective surveillance to clear weather, daylight operations. In order to improve detection capability, a research and development project has been established to produce an all-weather, airborne surveillance system which could detect, map, and measure the film thickness of an oil slick.

Since summer 1968, research has been directed toward defining optical properties of oil slicks and determining the ability of various sensor techniques to detect oil on the ocean surface. During this period, opportune spills such as the *SS Arrow* spill in Nova Scotia and the Chevron platform fire, as well as a series of controlled test spills, have been used to obtain data on sensor capabilities. Typical sensors operating in the ultraviolet, visible, infrared, and microwave regions of the electromagnetic spectrum have been tested. The past work results are now being analyzed. The analysis is considering aircraft capabilities, operational requirements, and sensor capabilities to define the best sensor package to meet objectives. When the analysis is completed, a prototype surveillance system will be constructed.

### R&D Funding in Oil Pollution Abatement
(dollars in thousands)

|  | FY69 | FY70 | FY71 | FY72[a] |
|---|---|---|---|---|
| **Oil pollution control** | | | | |
| ADAPTS | 291 | 579 | ... | ... |
| Containment systems | 84 | 2,011 | 700 | 500 |
| Oil harvesting | ... | 662 | 1,100 | 800 |
| **Law enforcement/surveillance** | 159 | 345 | 1,000 | 1,000 |
| Total | 534 | 3,597 | 2,800 | 2,300 |
| **Total appropriation for maritime pollution R&D** | 804 | 4,288 | 4,000 | 6,700 |

[a] Amount requested in the President's Budget.

As part of sensor system research, a special interim photographic comparison manual has been produced. This manual is designed to assist aircraft crews in identifying violators of the current 100 parts per million and (or) future 60 liters per mile oil discharge limits.

**Shipboard equipment**

All vessels can produce minor pollutant discharges as part of their normal operations. The most common are sanitary waste discharges and excess oil in bilge and ballast water.

Developing seagoing sewage treatment systems is the oldest project in the Coast Guard R&D program. Its overall objective is to provide devices which adequately treat sanitary wastes in Coast Guard vessels. Three sizes of waste treatment systems, identified to meet these needs, include a ship system capable of handling the waste discharges from 50 men; an intermediate, 10 to 20-man system for use in smaller manned vessels; and a treatment device for use onboard intermittently manned patrol boats.

General requirements for all systems include:
• Systems must be self-contained package units which include all special support equipment except electric power.
• The overall size and weight of treatment systems should be minimized.
• Equipment must be simple to operate and not require full time attention by specially trained operators.
• Equipment reliability must be emphasized so that routine maintenance will be at a minimum. Further, individual system components should be sized to facilitate installation through typical existing access doors and hatches.
• Systems must offer complete treatment as required by law.

The R&D program began in the early 1960's with shipboard test and evaluation of several commercially developed ship treatment plants that were marinized versions of land-based package treatment devices.

In 1965, development of a sewage treatment system specifically designed for shipboard use was undertaken. The resulting treatment plant was based on the aerobic digestion process and represented a significant improvement over the previously tested marinized devices. Unfortunately, it did not meet all of the design goals or performance requirements.

At the present time, two new treatment plants are being developed. Both are concentrating on solid/liquid separation techniques. One plant will use a centrifugal separation scheme. The other will utilize ultrafiltration (membrane separation) to effect separation. Both plan to remove solid material and dispose of it by incineration. The relatively pollutant-free liquid which makes up the bulk of sanitary wastes will then be treated in a disinfection process and pumped overboard. A full-size prototype is scheduled to be available for shore test and evaluation within six months.

**Oil spill control**

Notwithstanding the best prevention efforts, occasional massive pollutant spills will still occur. Mechanical failure or human error will continue to cause maritime disasters. Thus, whether due to a ship casualty or oil platform accident, specialized techniques and equipment will be needed to combat massive spills at sea.

The lack of cleanup methods for large oil spills was first demonstrated by the *SS Torey Canyon* accident in 1967 and the breakup of *SS Ocean Eagle* in San Juan Harbor in 1968. In both incidents cleanup could only be effected when the oil came ashore. By that time widespread disfigurement of recreational areas and destruction of marine life and birds had occurred. Additionally, restoration of the shoreline proved to be costly and time consuming. The ability to combat oil spills at their source (at sea) could have eliminated those problems.

If the intact cargo could have been removed, subsequent spills could have been significantly reduced. Unfortunately, most accidents seem to occur in either shoal water with relatively severe weather, or locations remote from empty ships and barges that might be used to receive the intact cargo oil. Development of an emergency system facilitating removal of intact cargo was the first research project undertaken. It has resulted in the Air Delivered Anti Pollution Transfer System (ADAPTS).

The ADAPTS project was initiated in

**Arctic oil spills.** *This oil spread test near Pt. Barrow, Alaska, helps scientists understand oil spread in ice. 90% of the oils was removed*

A high seas cleanup capability has been a major objective of Coast Guard R&D for the past several years. Fund levels for this activity are shown in the table.

Past and current effort is centered on developing seagoing mechanical systems to:
• Reduce oil quantities that may be released during tankship accidents;
• Contain the spread of spilled oil in relatively thick films which will facilitate recovery operations;
• Harvest spilled oil from the ocean surface.

**Reducing releases**

Apparent from past tankship accidents, only a part of the cargo will be lost at the instant an accident occurs.

the spring of 1968. The design goals for the project specified a system to:
• Perform effectively in 40-mph winds and 12-foot seas;
• Be suitable for air delivery at the spill site within four hours of notification of an accident;
• Be suitable for deployment and operation without using surface craft;
• Be complete and not require support from a damaged ship;
• Transfer and store 20,000 tons of crude oil within a 20-hour period.

As developed, ADAPTS consists of three subsystems which are transfer pumping, temporary storage, and air delivery. The transfer pumping subsystem has two major components. One is a two-stage centrifugal pump driven by a close coupled hydraulic

motor. This unit is 12 inches in outside diameter, 9 feet long, and can be lowered through a Butterworth fitting. It weighs 950 lb when packaged for an air drop. Motive power for the transfer pump is provided by a 40-hp lightweight diesel engine driving a hydraulic pump (page 513). The packaged engine unit weighs 1150 lb.

The key components of ADAPTS are collapsible, 500-ton temporary storage containers. Two different styles of containers have been developed for comparison purposes. One is made of rubberized nylon and has overall dimensions of 135 feet long, 35 feet wide, and 6 feet deep. It weighs 13,-000 lb when rigged for air drop. The other container is fabricated of nylon reinforced polyurethane and is roughly cylindrical in shape, approximately 12 feet in diameter, 173 feet long, and 8500 lb in weight. Both containers are interchangeable for use with the other subsystems.

The air delivery subsystem was developed around standard military parachutes and air delivery equipment (page 513). It was designed for use with existing Coast Guard cargo aircraft and helicopters.

### Spill containment

A system to restrict the spread of spilled oil is a practical necessity if effective oil spill cleanup operations are to be carried out. Development of systems to contain the spread of spilled oils was begun in the spring of 1969. Since then, theoretical and laboratory research into the fluid mechanics of oil slick containment, and boom motions and stresses in current and wave fields has been completed. The results of the engineering research were applied by conducting a six-month preliminary design competition for high seas qualified oil containment barriers. The preliminary design competition was completed in June 1970 and provided in-depth proposals for subsequent design and construction of five different barrier concepts. These proposals were then subjected to a detailed technical evaluation to select the concept with the most promising performance characteristics.

At the present time a prototype oil boom is being constructed and will be taken to sea for environmental testing this month. As now designed, the prototype will have a 27-inch draft and a 21-inch freeboard. It will be 1000 feet long and have provisions for anchoring. Buoyancy will be furnished by inflated, horizontal tubes spaced about 60 inches apart. The primary strength member of the barrier is an external tension line. The target performance goals for the prototype include:
• Effectively containing oil slicks in 20-mph winds and 5-foot waves;
• Be capable of surviving 40-mph winds and 10-foot seas;
• Be adaptable for future modification to facilitate air delivery directly to the spill site.

### Spill recovery

R&D for oil harvesting equipment is also well underway. By necessity, this work had to be delayed pending results from the containment barrier project. Advance data concerning barrier retention ability and general physical characteristics were needed to define adequately recovery equipment performance requirements. Preliminary studies were undertaken in December 1969, and in May 1970 engineering feasibility studies of five typical recovery concepts were initiated.

The five concepts investigated were sorbent belt, rotating disk drum, vertical weir skimmer, inverted weir boom, and free-vortex oil concentrator. The feasibility studies utilized model testing in conjunction with a theory to develop parametric equations to predict the capabilities of each concept if developed to full size. The results of the studies indicated the recovery efficiency and recovery rate as well as predictions concerning power and support vessel requirements. Following the engineering feasibility studies, proposals were requested for participation in a preliminary design competition for a high seas oil harvesting system. Proposal evaluation submitted in response to that request was completed in April, and the preliminary design competition is underway. As now envisioned, the desired performance goals for a recovery system include an oil recovery rate of 2000 gallons of oil per minute in 20-mph winds and 5-foot seas.

In addition to the foregoing major efforts, support research in several other areas of oil pollution is also in progress. Defining the physics of oil film spreading has been completed; field tests to determine the problems inherent to Arctic oil spills have been completed (page 515), and follow-on laboratory research has been initiated. Research into special cleanup techniques such as absorbent materials and liquid chemicals for oil slick herding is in progress.

Although much still needs to be done, technical solutions for marine transportation-related water pollution problems are well on the way. In particular, the R&D progress in oil spill control and cleanup equipment, remote surveillance systems for law enforcement, shipboard sewage treatment, and Harbor Advisory Radar is most encouraging.

**William E. Lehr** *is presently assigned at U.S. Coast Guard Headquarters, Washington, D.C., as chief, Pollution Control Branch of the Office of Research and Development. CDR Lehr, an engineering graduate (1953) of the Coast Guard Academy, received his M.S. and a Naval Engineering degree (1961) from the Massachusetts Institute of Technology. After serving aboard Coast Guard cutters in the Pacific, Alaskan waters, and in the North Atlantic, as well as in training positions and combat duty in South Vietnam, CDR Lehr assumed his present post in 1968.*

*The opinions or assertions contained herein are the private ones of the writer and are not to be construed as official or reflecting the views of the Commandant or the Coast Guard at large.*

# Environmental protection:

*Reprinted from* ENVIRON. SCI. TECHNOL., **7**, 26 (January 1973)

**Alexander J. Ogrinz, III** *Washington, D.C. 20007*

Vacationers who enthusiastically enjoy the surf at Virginia Beach, Va., may be nauseated by the prospect of cavorting in the waters of Norfolk Harbor, a scant 20 miles away. This harbor, like most major world ports, has been relegated to the position of a cesspool, receiving the relentless discharges of man and machine since the mid-nineteenth century and the advent of the steamship. Furthermore, the U.S. Navy has established here the world's largest naval complex, which is home to several hundred warships.

From these ships, raw sewage and domestic waste water (galley, laundry, showers, etc.) are being pumped directly overboard daily. Oily wastes—bilge and ballast water—which are not to be pumped in inland waters are occasionally discharged by accidental spills.

### Effects of shipboard pollution

This pollution not only adversely affects the aesthetic value of Norfolk and other U.S. coastal waters, but also poses a threat to public health and endangers the aquatic wildlife indigenous to the area. The detrimental effects of contaminated water in ports have been acknowledged by the Navy, which has followed for years a policy prohibiting the distillation of harbor water for use as potable water. The NAVSHIPS Technical Manual (orders regarding the command of a ship) gives the following definition: "Unless determined otherwise by suitable tests, all water shall be considered contaminated in harbors, rivers, inlets, bays, and land-locked waters, and in the open sea within 10 miles of the entrance to such waters."

The magnitude of the Navy's contribution to harbor pollution can be illustrated by citing the highly concentrated population that inhabits naval ships. The nuclear-powered aircraft carrier USS Enterprise (CVAN 65) with her air group aboard carries over 5000 men, the population of a sizable small town, on a 1123-ft-long floating platform. With about 600 ships carrying an average of 200 men each, the U.S. Navy manages a substantial fleet of ecologically primitive vessels.

However, the degenerate conditions of important U.S. (and international) ports are not the result of naval policy only. On the contrary, the naval contribution to harbor pollution, although significant, is quite small when compared to the large volume of municipal and industrial wastes pumped into these waters. There is virtually no naval presence in New York Harbor; yet it is as polluted, if not more so, than any major U.S. Naval harbor. The U.S. Navy simply followed the practices of the maritime ships of the world, which made no provision for the treatment or containment of shipboard wastes (except petro-leum products, and then only in coastal waters). Due to recent activism among the citizenry and government, this role may be reversed as the Navy seeks to become preeminent in the field of pollution control.

### Naval policy

The effectiveness of a naval environmental protection program can be no better than the official policy of the senior military and civilian managers. In an atmosphere of indifference at the highest levels, the at-sea commanders will be forced to choose expediency and short-term economy in decisions that pit the environment against combat readiness. The task of training men and operating a ship with limited resources and manpower is so dissipating of one's energy that, when upper echelon leaders demand military preparedness and ignore questions of ecology, there is neither energy nor motivation to pursue an individual ecological program. Even if the desire of the individual commander was strong enough to overcome these difficulties, he would be unable to secure the financial resources necessary for such an undertaking without a coordinated policy from the top.

Until recently, generally the only official concern displayed by the Navy was to comply with the Oil Pollution Act of 1961, which prohibited the discharge of oil within 50 miles of shore. Oily wastes were retained on board within the 50-mile limit only to be pumped overboard upon passing the 50-mile line of demarcation. Solid waste—trash and garbage—was retained on board in coastal waters to be carried ashore upon return to port or dumped in the ocean upon reaching the open seas. All other wastes were disposed of by discharge into the waters where the ship lay.

With the passage of the National Environmental Policy Act of 1969 (NEPA) and the ensuing implementing directives, the U.S. Navy has officially recognized the importance of environmental planning in all of its activities. Various congressional legislation and executive orders led the Chief of Naval Operations (CNO) to promulgate an instruction entitled, "Policy and Assignment of Responsibilities for the Environmental Quality Program," which expresses the following policy: "The Navy will actively participate in a program to protect and enhance the quality of the environment, through strict adherence to all applicable regulatory standards, positive planning and programming actions to control pollution caused by installations, ships, aircraft, and other Navy facilities; establishment of methods to monitor the effectiveness of such actions . . ."

Specific responsibilities of the various Naval organiza-

# new Navy duty

Control of shipboard
pollution is a major U.S. Navy
program—older ships are being
retrofitted, and new ships are designed to
eliminate discharges

tions were listed, and a later official order stated, "Addressees are directed to initiate aggressive action to combat environmental pollution in accordance with the responsibilities specified herein . . ."

To encourage individual commanders to take expeditious action at the local level, the Secretary of the Navy announced the establishment of the Annual Environmental Protection Awards "to stimulate outstanding performance in the pursuit of enhancing and protecting the environment."

The Chief of Naval Operations has also issued a separate instruction delineating responsibilities for complying with NEPA, especially in regard to submitting Environmental Impact Statements required by the act. Many impact statements have been filed which deal with shore construction, real estate, new aircraft, target ranges, and target ship sinkings.

To coordinate the efforts of the naval establishment, an Environmental Protection Division was established in the office of the Deputy CNO for Logistics. Captain J. A. D'Emidio, a holder of several degrees in sanitary engineering and a member of the Navy's Civil Engineering Corps, was named the first director of that division. The statements and actions of Capt. D'Emidio indicate that he has a firm grasp on the environmental problems at hand and that he is sincerely pursuing an active policy to minimize any adverse naval effect on the environment. To oversee the Navy's environmental protection program at the upper civilian management level, a new Deputy Undersecretary of the Navy was created in the Fall of 1971.

### Pollution control research

Having only recently acknowledged its ultimate environmental responsibility in the area of ship waste management, the Navy was unprepared to institute an immediate widespread response to the problem. The Navy's fleet of about 600 active warships, originally designed to utilize the most available space efficiently, is under a rigorous schedule of national and international defense commitments. A technological question arose—how to equip these ships with effective pollution control devices in the limited shipboard space without compromising the combat readiness of U.S. Navy ships.

A Navy study, begun in 1966 to meet proposed U.S. Public Health Service (USPHS) sanitation standards, led to the development by Colt Industries' Fairbanks-Morse Research Center (Pittsburgh, Pa.) of an electro-mechanical incinerator system to treat shipboard sewage. An estimated $300 million to equip the fleet with this unit led the Navy to seek assurances that long-term pollution abatement goals would be achieved. Of great concern was the possibility that standards more rigorous than those of the USPHS would be imposed in the future. Several studies were initiated to investigate cost-effectiveness and to determine the optimum system for widespread application among various types of naval vessels.

Besides the Fairbanks-Morse electro-mechanical incinerator system, holding systems and recirculating flush systems were studied in depth. The Fairbanks-Morse system actually treats the sewage aboard the ship by separating liquids from solids, burning the resulting sludge, and chlorinating the liquid, which is pumped overboard. Holding systems basically depend on the transfer of collected wastes to a land-based sewage system or to the open sea after transiting coastal waters. Recirculating flush systems, similar to aircraft recirculating systems, also rely on the holding tank principle that discharges to the open sea and to shore-based systems will prevent the dangerous, unmanageable concentrations of human waste that currently contaminate harbors. Preliminary evaluation of the studies favors the holding tank system once shore-support facilities are completed.

Part of the pollution abatement research included installing operational prototype systems on several active ships. The most notable prototype is the tripartite system in use aboard the submarine tender, USS Fulton (AS 11), in New London, Conn. The first part, the on-board sewage treatment plant, collects and incinerates the Fulton's sewage. The second component, the internal manifold system, collects all other shipboard wastes and transfers them ashore. Finally, a third system collects and transfers ashore the sewage from the submarines when they come alongside for repair. The latter two systems are being installed in the USS Dixon (AS 37) which is now home-ported in San Diego, Calif.

Many other related studies are in progress. An in-depth survey is being conducted to determine the number, types, and causes of accidental oil discharges. The Naval Ship Research and Development Laboratory is currently testing shipboard oil-water separators to determine if available commercial units will be compatible with naval requirements. Investigation into alternative methods of packaging supplies is under consideration as a method of reducing the volume of solid waste. The Naval Facilities Engineering Command has established an automated Environmental Protection Data Base at the Naval Civil Engineering Laboratory, Port Hueneme, Calif., to compile and organize raw information to be used in formulating solutions to naval-related environmental problems.

The Navy is conducting research in the field of oil-spill clean-up, and is developing new techniques for containment and removal of oil spills. On various occasions, U.S. Naval personnel and equipment have assisted in large-scale clean-up operations following civilian oil spills.

### Waste oil control

Recognizing that existing procedures can be ineffective, the Navy is developing new, imaginative procedures rather than simply expanding current programs to meet the present volume of pollutant disposal. When in port, ships are required to pump bilge water, which invariably accumulates and mixes with oily deposits in the lowest parts of the ship, into "doughnuts." These devices are large floating steel rings that work on specific gravity differentiation of oil and water to allow large volumes of slightly contaminated water to be discharged from ships. The oil is contained in the ring, while the water escapes to the harbor. The "doughnuts," however, are often plagued with leaks, sloshing over, unavailability, or improper use, which decreases their effectiveness for pollution control. The Chief of Naval Material is pursuing a program to develop a more thorough oil disposal system for moored ships.

One of the recent substantive steps taken by the Navy to prevent oil spills was to authorize a change in the shipboard fuel oil storage system by issuing a ShipAlt—a mandatory design change which affects all ships of a specific class or type. This ShipAlt reroutes the storage tank overflow lines, which previously discharged to the sea, into an adjacent tank. The chain of tanks ends with a tank in which an overflow alarm is installed to warn of overfilling. This system reduces the number of oil spill opportunities from about 20 (on a destroyer-sized ship) to one per refueling. Owing to the usual time lag involved with modifications, it will take a couple of years for all ships to comply with this ShipAlt.

Prospects for the future should prove to be easier for implementing environmental policy. Warships under design have space and weight allotted for sewage and oil control systems to be installed during initial construction

**Floating cities.** *Existing ships are being retrofitted to hold wastes for adequate treatment prior to their discharge*

phases. The long-range goals envision all wastes—sewage, oil, solids, etc.—be disposed of ashore when in port. When ships are transiting coastal waters, all wastes will be retained on board. At sea, sewage and garbage may be dumped into the ocean where natural biological action will degrade these unconcentrated wastes. Oil-water separators will treat all oil-contaminated water from the bilges and ballast tanks for discharge at sea. The Navy is requesting standards to allow purification of water down to 10 parts per million (ppm) of oil. This standard would not meet the idealistic hopes espoused by Secretary of Transportation John Volpe, who addressed NATO's Committee of Challenges of a Modern Society in 1970: "My government proposes that NATO nations resolve to achieve by mid-decade a complete halt to all intentional discharges of oil and oily wastes into the oceans by tankers and other vessels." However, 10 ppm is a realistic figure that would accomplish Secretary Volpe's ultimate goal, and comply with the present standards of the Water Quality Improvement Act of 1970, which prohibits discharges which cause a sheen on the water. Depending on the type of oil, up to 25 ppm may be present before a sheen is visible.

Rather than equip each ship with a bulky ballast-water oil–water separator, it has been recommended that only replenishment tankers and shore stations carry separators on the supposition that ballast need be pumped only when refueling. A complementary device has been developed for private industry to detect unacceptable concentrations of oil when discharging water from contaminated tanks. When pumping a tank of water supporting a layer of oil, the operator would be notified by an alarm when the oily layer was approaching the pump's intake. With this alarm no special treatment equipment is required on individual ships if adequate opportunity exists to pump the retained oily wastes to stationary or mobile treatment facilities.

The Water Quality Improvement Act requires, that upon the final promulgation of sewage treatment standards by the federal EPA, all new vessels comply within two years and all existing vessels within five years. U.S. Navy spokesmen have asserted that these provisions will be met by Navy ships. As the maritime industry rises to meet water quality standards, shore-based treatment facilities will probably be established to accomodate merchant ships. The Navy's ships would then be able to connect with these facilities, freeing the ships from exclusive dependence on naval facilities at naval bases.

**Other harbor contaminants**

One environmental problem at sea is the precipitation of air pollutants. The Navy is in the process of converting its steam-propelled ships from black oil to a distillate fuel. This conversion was motivated by the economics involved rather than any altruistic drive. Nevertheless, the environment is benefitting by this cleaner fuel, which does not require the frequent "tube blowing" that periodically ejected large sooty clouds of heavy ashes into the air. The fleet's contamination of the seas via air pollution is negligible.

The nuclear-powered Navy has had such an outstanding radiation control program for many years that the recent popularity of environmentalism will elicit no new plans or procedures. Environmental monitoring conducted by the Navy, Atomic Energy Commission, and Public Health Service has led to the conclusion: "No increase of radioactivity above normal background levels has been detected in harbor water where U.S. Naval nuclear-powered ships are based, overhauled, or constructed," which was reported to Congress.

**Funding**

The Navy has met virtually no opposition in obtaining funds for its $1$\frac{1}{2}$ billion environmental program. The Navy has received over $300 million during fiscal years 1968–73 for pollution abatement. Over $31 million in the FY 1973 budget is allocated for ships' waste management programs; this amount will more than double in FY 1974. Funding has been provided over and above regular naval operating appropriations, so that environmental expenditures do not compete with operating expenses to weaken the support of middle- and upper-level commanders.

If the Navy can create and sustain an effective environmental program, many benefits will be forthcoming. Of primary concern is the absolute value of a cleaner environment. Secondly, the Federal Government will be more effective policing private industry, once its own house is in order. Finally, the United States' international image can be improved if "clean" ships visit foreign ports without leaving a dirty souvenir of their visit.

**Additional reading**

"Policy and Assignment of Responsibilities for the Environmental Quality Program," Department of the Navy, Office of the Chief of Naval Operations, Instruction No. 6240.3A, Washington, D.C., Sept. 14, 1971.

"Treatment and Disposal of Vessel Sanitary Wastes," National Academy of Sciences, Maritime Information Committee (Washington, D.C., 1971)

Rear Admiral N. Sonenshein, "What's New in Navy Petroleum," *Nav. Eng. J.,* August 1969, pp 43FF.

Testimony of Vice Admiral H. G. Rickover at the Hearing before the Joint Committee on Atomic Energy, Naval Nuclear Propulsion Program, 92nd Congress, Washington, D.C.

"Policy Regarding and Assignment of Responsibilities for the National Environmental Policy Act and Environmental Impact Statements," Department of the Navy, Office of the Chief of Naval Operations, Instruction No. 6240.2C, Washington, D.C., October 4, 1972.

**Alexander J. Ogrinz, III** *graduated from Duke University with an AB in political science and then served in the U.S. Navy for seven years. As Engineer Officer aboard a destroyer in San Diego, Calif., he experienced the problems of the operating naval forces in minimizing pollution. During a two-year tour of duty in Washington, D.C., he was able to devote his off-duty hours to study of the Navy's environmental policies and procedures. He relinquished his commission to pursue a legal education.*

# U.S. Army modernizes munitions plants

*Headquartered at Joliet, Ill., the Army Ammunition*
*Procurement and Supply Agency aims to mechanize*
*and clean up its manufacturing facilities*

Reprinted from ENVIRON. SCI. TECHNOL., **6**, 986 (November 1972)

Today's Army ammunition requirements can be produced in a total of 26 go-co (government owned–contractor operated) plants in the U.S. Like other industries, the munitions industry is an archaic one which is now under the gun to clean up the air and water pollution burden from the manufacture of the nation's munitions.

The Army has come forth with a modernization program, costing $2.5 billion and spanning 12 years. Pollution abatement is one aspect of the modernization, but mechanization is the other, larger aspect of the overall program. Many old facilities are simply being replaced by modernized ones. All activities of the Army Ammunition Procurement and Supply Agency are headquartered at Joliet, Ill.

It certainly is reassuring to know that manufacturing plants for the production of this nation's munitions are being cleaned up. But one might ask why there are so many such munitions plants. "Just in case," as they say, because since 1915, the U.S. has been engaged in a war, on the average, every 12.5 years.

Progress is being made at these federal facilities to clean up the air and water pollution burdens from their operations,
much as it is with other federal facilities, which are required to have pollution abatement plans complete by the end of this calendar year (ES&T, December 1971, p 1176). Under presidential Executive Order 11507, Joliet is part of the vanguard to implement pollution abatement activities at federal facilities.

In ammunition production jargon there are four types of plants:

• P&E plants (for the manufacture of propellants and explosives); these are the so-called go-co plants, basically large chemical plants

• metal parts plants which make shell casings

• small arms plants

• LAP (load, assemble, and pack) plants.

Any given installation may have both a P&E plant and a LAP plant, another may have just a P&E plant, another just a small arms loading plant, and another just an explosives plant. To what extent of their capacity these plants are operated is, of course, the subject of military security classification, but it is no secret that each of these plants is subject to the same cleanup requirements as are other federal facilities.

Although the total Army moderniza-

> ## Go-co plants involved in the Army's modernization program
>
> • Joliet, a P&E and LAP plant at Joliet, Ill., is operated by Uniroyal, Inc.
>
> • Volunteer, an explosives plant at Chattanooga, Tenn., is operated by Imperial Chemical Industries
>
> • Radford, a P&E plant at Radford, Va., is operated by Hercules
>
> • Holston, an explosives plant at Kingsport, Tenn., is operated by the Holston Defense Corp., a subsidiary of Eastman Kodak
>
> • Indiana, a propellant plant at Charlestown, Ind., is operated by Imperial Chemical Industries
>
> • Badger plant, a propellants plant at Baraboo, Wis., is operated by Olin
>
> • A new explosives plant at Newport, Ind., is under construction and will be operated by Du Pont.
>
> P&E: propellants and explosives
> LAP: loading, assembling, and packing

tion program spans 12 years, certain ongoing pollution abatement activities figure in the overall program. The congressional appropriation to the U.S. Army Corps of Engineers for the Military Construction, Army (MCA) is the appropriation to watch for. Under the MCA planned program, the amount that will have been spent on air pollution controls during the 1969–78 period is $55.4 million; the comparable amount for water pollution control during the same period is $58.4 million. But in the short term (1970–75), $221 million of the overall total is programmed for pollution abatement (box, p 987). Joliet takes $84.5 million of the total.

After 1975, the major cleanup projects will include an air improvement program at the Holston plant totaling $2.28 million during the years 1975–78. During the same period, there are two water improvement projects for a total of $2.37 million, including a $1.7 million project at the Volunteer plant.

## Pollution burden steps

Taking a look at the pollution burdens resulting from the manufacture and production of basic munitions materials

## Typical sources of water pollution

| Sources | Sulfuric acid | Nitric acid | Sulfates | Nitrates | Nitrobodies | Tetryl | Solids | Color | Ammonia | Solutions |
|---|---|---|---|---|---|---|---|---|---|---|
| Acid area | ● | ● | ● | ● | | | | | | Neutralization |
| TNT area & red water plant | ● | ● | ● | ● | ● | | ● | | | Neutralization |
| Tetryl area | ● | ● | ● | ● | | ● | ● | | | Partial neutralization |
| Sellite area | | | ● | ● | | | | | ● | Neutralization aeration |
| LAP area | | | | ● | | | ● | ● | | Carbon |

give some insight into the federal facilities cleanup activity. One of the basic propellants is nitrocellulose, and one of the basic explosives is TNT (trinitrotoluene). Both air and water pollution burdens are associated with the manufacture of these materials.

In a LAP plant, the ammunition to be loaded comes in contact with water. These contaminated waters are passed through filters of diatomaceous earth and then through carbon adsorption columns for cleanup of the water.

Another pollution burden item on a load line is the excess scrap material that becomes waste. The existing operating policy simply states that any material that has been exposed to either propellant or explosive must be destroyed. Realizing that both open burning and landfill operations are slowly coming to an end, the Joliet plant has already installed a prototype incinerator scheduled to become operational before the end of this year, for disposal of such material.

In the manufacture of the basic explosive TNT, the water at the end of the manufacturing process is known as "red water." A red water disposal plant was built at Joliet in 1965 by the Corps of Engineers. It cost $8 million. Basically, the red water results from the purification of TNT and is the largest single water pollution burden at Joliet. The red water is heated and evaporated to a sludge that just barely flows. The sludge is fed into a rotary kiln incinerator where it is heated and mixed until it becomes an ash.

When the plant is operating all areas of production, water usage at Joliet is of the order of 57 mgd, 90% of which is

process cooling water. In the old way of making TNT, the nitro groups were introduced one at a time in separate locations on a plant site. Thus, the chemical process resulting in mononitration was performed in one building; the product was transferred to a second building where a second nitro group was introduced into the toluene molecule, and ultimately a third group was introduced in a third housing installation.

Now, TNT is prepared by a continuous process, which was developed by a combination of Swedish, Norwegian, and Canadian efforts. The process is used in Canada by Canadian Industries Ltd. and was introduced in 1965.

Today, there are three continuous TNT lines at the Radford plant which also has a red water disposal plant, and three such lines are also under construction at Joliet. The construction of the Joliet lines was authorized in March 1970 at an estimated cost of $29.9 million. The lines are expected to be complete in 1973.

A second element of the pollution control aspects under the modernization program at Joliet involves the sulfuric acid regeneration plant. Both air and water pollution controls are necessary in this operation. The regeneration plant is under construction today and is scheduled for completion before the end

of this year at an estimated cost of $15.8 million. When the new plant is completed, Joliet will have no further requirement for oleum (solution of sulfur trioxide in sulfuric acid) produced by the existing acid manufacturing facilities.

The third and fourth elements of the pollution abatement part of the modernization program at Joliet involve the direct strong nitric acid and ammonia oxidation plants, both of which are scheduled for completion by the end of the year. The acid plant was authorized in March 1970 and will cost $11.7 million; the ammonia oxidation plant costs $6.2 million.

In every case, the modernization work at Joliet is performed under Corps of Engineers' contracts with Kaiser Engineers (Oakland, Calif.).

Joliet is currently operating two air monitoring stations to determine ambient air quality. When the ambient air quality exceeds a predetermined concentration, an alarm is sounded and production at the offending facility is curtailed or shut down until the air monitor indicates an acceptable concentration level.

Most of the acid plants that the Army has today are of the World War II variety. Nevertheless, there is a new nitric acid plant going in at Holston, and the construction of an ammonia oxidation plant and a sulfuric acid regeneration plant is under way at Newport (Ind.). But not to be overlooked is the fact that a substantial portion of the funds for pollution control expenditures at existing facilities still goes for the conversion of coal-fired heating operations to gas- or oil-fired ones.

Radford, the first modernized plant in the U.S., has three continuous TNT lines; the authorization for this construction project came on June 23, 1967. Catalytic Construction Co. (Philadelphia, Pa.) won the Corps of Engineers' architectural-engineering bid, and ground-breaking ceremonies for the Radford modernization were held on September 12, 1967. Slightly more than one year later, on October 2, 1968, the plant was dedicated. Each production line has a rated capacity of 50 tons of finished TNT per day.  SSM

## Short-term look at the Army pollution abatement program

| Year | 1970 | 1971 | 1972 | 1973 | 1974 | 1975 |
|---|---|---|---|---|---|---|
| Number of projects | 11 | 4 | 3 | 0 | 0 | 3 |
| Total cost of projects (millions of dollars) | $91.8 | $35.1 | $27.1 | 0 | 0 | $67 |

## Typical sources of air pollution

| Sources | Particulates | Sulfur oxides | Sulfuric acid mist | Nitrogen oxides | Solutions |
|---|---|---|---|---|---|
| Sulfuric acid concentrators | | ● | ● | | Electrostatic precipitators Scrubbers |
| Ammonia oxidation plants | | | | ● | None |
| Nitric acid concentrators | | | | ● | None |
| Oleum plants | | ● | ● | | Demister pads |
| Sellite plants | | ● | ● | | None |
| TNT recovery buildings | | | | ● | None |
| Tetryl recovery building | | | | ● | None |
| Red water incinerators | ● | | | ● | None |
| Open burning | ● | | | ● | None |

# What's the U.S. Army doing in water pollution control?

*For the first time ever, the*

*Corps of Engineers is issuing permits*

*for all discharges*

Reprinted from ENVIRON. SCI. TECHNOL., 4, 1101 (December 1970)

Gone are the days when the Corps of Engineers, in their issuance of permits, are concerned only with the impact a proposed project may have on navigation. The Corps' new regulatory role in pollution control extends their permit-granting authority to include all discharges or deposits into the navigable waters of the U.S. Initially, the Corps will concentrate on major sources of industrial pollution. Although the Corps' announcement of its expanded permit-granting authority came on July 30, official word from Washington has not yet gone out to the Corps' 37 district offices located around the country.

"We expect to issue instructions to the field offices around the first of next year," says Robert E. Jordan III, General Counsel and Special Assistant to the Secretary of the Army for Civil Functions. "The Corps has been issuing permits for structures in the navigable waters for a long time (essentially since the 1899 act). Since 1966, we have had a requirement that applicants seeking permits must consider esthetics, fish, and wildlife effects, water quality, and the like. In general, our long-term permits are reviewed every five years."

## Jurisdiction

Authority for the Corps' expanded program stems from a combination of the Refuse Act of 1899 and the National Environmental Policy Act of 1969 (P.L. 91-190) (ES&T, February 1970, page 103). The latter requires that all federal agencies interpret their authority with environmental impact in mind. Authority for the Corps of Engineers to take action on water pol-

lution is contained in the 1899 Refuse Act. "In our review of the Refuse Act, we realized that here was the authority that should be used," Jordan explains.

Coupled with the Refuse Act, P.L. 91-190 requires federal agencies to step up the government's widespread program on water pollution abatement.

A draft of a proposed interagency agreement between the U.S. Army's Corps of Engineers and the Federal Water Quality Administration (FWQA) identifies the manner in which the two agencies will cooperate to enforce the 1899 Refuse Act. Although the agreement has been discussed with other interested agencies, including the Department of Justice and the Council on

Environmental Quality, the agreement has been delayed, as have other pending environmental matters, in the executive reorganization creating the Environmental Protection Agency (EPA). The recipient of one of this year's Arthur S. Fleming awards for the 10 top men in the federal government, Jordan personally piloted the agreement through the federal establishment. It perhaps will be finalized early next year.

Although the Corps' emphasis to date has been on new constructions in or on the navigable waters, the present federal consensus is that the Refuse Act will be used in the general interest of the public, and the forthcoming Refuse Act permit program attests to this.

For example, actions can be brought under the Refuse Act where the defendant has not met the following conditions: "The defendant does not have a permit from the Secretary of the Army or the Corps of Engineers authorizing such deposit." The Refuse Act (33 U.S.C. 407) can be interpreted to include industrial discharges; the only type of refuse exempted under the act is "refuse flowing from streets and sewers and passing therefrom in a liquid state."

The furor raised by the Corps' July 30 announcement has not subsided yet. The fact that the Corps would be looking at all discharges, including industrial discharges, poses a real concern to industry. Myriad inquiries to local Corps district officials attest to that concern.

## Rationale

"We aren't going into the program with an attitude of arrogance or of

**Robert E. Jordan III**
*Army's General Counsel*

heavy handedness," Jordan explains. "But the Corps and the Federal Water Quality Administration are going to have to work together to come up with sensible solutions to problems.

"The Corps' permit program must be carried out with balance and good sense," he continues. "As the program begins to impact on particular industries, there is going to be some unhappiness on the part of the community to which those industries are important.

"Most of the industries want to do the right thing," Jordan believes. As he testified before the Senate Commerce Subcommittee this July, "You obviously overnight cannot apply the Refuse Act around the country. If you did, you would shut down industry throughout the U.S., and we all recognize that this results in an intolerable economic situation, so we all must exercise good judgment and common sense."

What this recent interpretation of the Corps' role means to industry is that any structure or outfall line leading into navigable waters must be covered by a permit from the Corps, that any discharge from the structure or outfall line must meet all water quality standards applicable to receiving waters, and that state certification must also be included as required by section 21(b) of the Water Quality Improvement Act of 1970, P.L. 91-224 (ES&T, May 1970, page 379). "It's a complicated business," Jordan concedes.

So, the Corps will be issuing permits for industrial discharges, even if the discharge is harmless (it does not violate a water quality standard) or the industry is on schedule in its voluntary compliance scheme established under the administrative procedures of FWQA, in some enforcement conference.

## Permits

"Our permit system would be closely tied with FWQA," Jordan explains. So that industry will have only one form to fill out, the FWQA waste water profile inventory form will be used.

"Submission of the inventory profile is mandatory only in the sense that

**Maj. Gen. Frank P. Koisch**
*Chief, Corps' Civil Works Directorate*

if somebody wants a permit from the Corps, then they must submit the form so that we can pass judgment on their plans," Jordan elaborates. "Nobody is going to be required to furnish the information. But if they don't, then the Corps will not issue the permit."

"In other words, how could the Corps decide whether a plant should be granted a permit for such an industrial discharge unless you know what is being discharged, both in volume and composition?"

In practice, the Corps' group that will be responsible for the permits program is the Office of Chief of Engineers, the Civil Works Directorate, which is headed by Maj. Gen. Frank P. Koisch. The people involved with day-to-day administration of the permits, of course, will be the district engineers and their operational staffs. The operational division of that function is headed by Mark Gurnee.

"The Corps has the machinery (for permits) already in being, and it may as well be used," Gen. Koisch says. "The Corps has a better geographically distributed organization (than EPA) for handling the permits," he adds. "The Corps can perform a real service to the country and materially assist the new EPA."

What is likely to happen in actual operation is that the Corps would send each application for a permit to the FWQA and ask: What do you think? Does it meet the applicable water quality standard? If not, can we grant a permit with certain conditions attached, such as timetables which

would bring them into compliance by a certain date?

Then, the Corps would issue its permit conditioned upon compliance with agreements reached, including specified deadlines for compliance. State certification required under section 21(b) of P.L. 91-224 would, of course, have to be obtained before any Corps' permit could be issued.

## Advice

Many federal regulatory agencies, now perhaps for the first time, are feeling the pressure from the environmentalists and have established citizen advisory boards to provide the agency with more detailed familiarity with some of their problems. One example is the Water Pollution Control Advisory Board (ES&T, June 1970, page 471).

The Corps is no exception. Its six-member advisory board was established in April and represents a broad range of environmental knowledge and experience. Chairman of the Corps' advisory board is Charles H. Stoddard, an environmental consultant from Duluth, Minn.

"I expect that it (the board) will provide not only advice on specific problems, but, perhaps more important, contribute to an enhanced mutual understanding and confidence between the Corps and both the general public and the conservation community," says Lt. Gen. F. J. Clarke, chief of the Army Corps of Engineers.

## Looking ahead

Much remains to be done. Rather than set up a new permit program authority under the aegis of the EPA, the Corps' existing permit-granting authority within the federal government could perform a service in the continued war on water pollution. The Refuse Act permit program is a powerful force; there are tens of thousands of industrial plants discharging into navigable waters and 4000 new ones are being built each year. As the manpower and financial needs are met, the Corps stands ready to perform an invaluable role in the government's crackdown on water pollution.  SSM

# EPA troubleshooters back up enforcement branch

National Field Investigation Centers in
Cincinnati and Denver send out investigative teams
before the agency takes legal action

*Reprinted from* Environ. Sci. Technol., **7**, 296 (April 1973)

**Investigation.** *NFIC technician hand-samples influent as it enters a waste water treatment plant (above). NFIC-Denver recently moved into a new office and laboratory complex (right)*

Ever wonder how EPA obtains evidence for its legal actions and enforcement conferences, or forms a position on emergency situations such as the mercury scare a few years ago, or develops effluent guidelines based on best practicable technology? Well, special teams of EPA water pollution "trouble shooters" have been in action around the country for several years.

These pollution investigative teams are located at two National Field Investigation Centers (NFIC)—one in Cincinnati, Ohio, and the other in Denver, Colo. Each has about 75–80 personnel. The centers direct and conduct studies, surveys, and reports preparatory to enforcement actions taken by EPA; participate in developing effluent limitations; direct and assemble technical support required for legal actions; and provide expert witnesses at hearings in connection with permits.

NFIC-Cincinnati, headed by A. D. Sidio, has been around since 1961. It was formed as a team of investigative experts under the U.S. Public Health Service and was continued under the Federal Water Quality Administration and functioned in response to enforcement proceedings of the governing agency.

In July 1970, NFIC-Denver was created. Tom Gallagher, formerly head of engineering services in EPA's Southeast Water Laboratory, was named director of the Denver operation. Soon after EPA was officially sanctioned in December 1970, Administrator William Ruckelshaus recognized the "need to develop facts as soon as possible if we are to take effective enforcement action against polluters" and relied heavily upon the two centers.

Both centers report directly to Murray Stein, EPA's chief enforcement officer in the enforcement proceedings division. Through Stein and the deputy assistant administrator for water enforcement, the centers are under the auspices of John Quarles, assistant administrator for enforcement and general counsel.

**Official mandates**

Each NFIC has three major functions. First, NFIC personnel are to be available as a major task force as emergency situations arise. For example, in the environmental mercury episode, NFIC-Cincinnati coordinated testing for mercury between regions, cooperated with U.S. Geological Survey scientists working on the same problem, and negotiated controls with the chlor-alkali industries.

Second, NFIC provides large-scale technical support for short-term major problems that are beyond EPA regional resources—requiring excessive amounts of money or large numbers of personnel to handle the situation. At the request of EPA headquarters and its Region VI, NFIC-Denver surveyed the Houston Ship Channel to determine the nature and characteristics of wastes—toxic substances, heavy metals, etc.—discharged into the Channel and Galveston Bay. The final report contained the environmental effects of the discharges and recommendations to meet water quality standards which led to the Galveston Bay Enforcement Conference.

Last, but not least, NFIC provides nonroutine specialty services to EPA headquarters and regions that, by reason of their daily duties, the offices would not be inclined to do themselves. For example, NFIC was

**NFIC's Gallagher**
*"Move in and get out"*

given the task of developing interim effluent guidelines for 25 standard industrial classifications. With sophisticated chemical analysis and industrial waste expertise, NFIC determined the best practicable control technology for each industry. From these interim guidelines, states can issue discharge permits without delay until final effluent standards are adopted. All samples taken during investigations or surveys are held in security storage, and NFIC personnel are on call until all legal action is resolved.

**Expertise in many areas**

Either one or both of the centers have special capabilities in several areas valuable for environmental monitoring. One such area is remote sensing. Through agreements with the U.S. Air Force (using F-11's in their training program) and with photo-interpretation expertise, NFIC-Denver can assess the physical effects of oil spills, algal growths, heat discharges, and septic water discharges on the aquatic environment. This center monitored heat discharges from power plants on Lake Michigan by remote sensing and also documented oil spills in the Houston Ship Channel for referral to the Justice Department.

Another NFIC speciality is industrial waste evaluation. From the NFIC-developed effluent guidelines as well as field investigations of hundreds of industrial sources, the Centers have accumulated a wealth of experience, data, analytical informa-

tion, and education in practical problems of process control.

NFIC personnel are also knowledgeable in municipal waste treatment. NFIC-Cincinnati specializes in biological waste treatment, whereas the Denver office places emphasis on physical-chemical treatment developments and operations. The Centers have undertaken projects involving nitrogen removal at Sioux Falls, S.D., and improving settleability characteristics at Kansas City, Kan., among others.

The Centers use several environmental computer information systems that detail water quality effects from complex waste discharges and provide water quality inventory data. Through remote terminal hookups to computer systems in Washington, D.C., and Philadelphia, Pa., NFIC scientists have access to more than two million pieces of literature. As well as being connected to the STORET system (ES&T, February 1971, p 114), the Denver Center utilized a TOXICON file to support the Houston Ship Channel waste source survey to determine tolerable limits for complex organic discharges.

NFIC laboratories perform complex organic analysis with the best equipment on the market. The laboratories are among the most up-to-date in the country, featuring biology, microbiology, fish bioassay, autoanalysis, wet chemistry, tracer analysis, and quality assurance capabilities. Gas chromatographs–mass spectrophotometers are employed to analyze complex organics in waste discharges with respect to specific compounds in the discharges. The information is used to develop individual effluent limitations for the compounds involved. The organic wastes in the Houston Ship Channel were examined by this means.

In addition, the Denver Center has five mobile labs for extensive testing in the field—one for bioassay, two for chemical analysis, and two for microbiological analysis. NFIC-Denver has moved into a new building, just completed this month, containing 12,000 ft$^2$ of lab space and 18,000 ft$^2$ office space.

Each Center employs 75–80 persons including biologists, microbiologists, engineers, organic and inorganic chemists, hydrologists, lawyers, and administrators. When asked about the small number of people employed to perform this specialized work for EPA headquarters and all regions, Tom Gallagher, NFIC-Denver director, summed up the situation, "When we get a request with well-defined technical objectives, we move in fast, get the job done without sacrificing any quality of work, give them a report, and get out." In spite of being few in number, NFIC personnel complete a number of projects within record time.

**What NFIC does**

At present, NFIC is only mandated to operate in the water pollution control field since the procedures of enforcement in other environmental areas are not finalized. However, if the opportunity arose to investigate other pollutant sources, Gallagher

explains, the matter would not be ignored. In fact, NFIC-Denver has been involved in situations covering both air and water pollution.

There's no real breakdown on the territories covered by the two investigation centers. The Center with particular expertise in a certain area will handle problems in that area. Generally, however, NFIC-Denver handles the five western EPA regions and the southeast region, and the Cincinnati office covers the others.

At press time, the Denver office had 22 ongoing projects; most will be completed in three to six months. Last year, NFIC-Denver undertook a 200-mile survey of the South Platte River and evaluated the waste discharges, climatic conditions, water quality, tributary quality, and recommended controls for waste discharges. Region VII requested the rundown on water quality in July 1971, and NFIC submitted the final report in March 1972.

A report on the Memphis, Tenn., area covering 22 large discharges and evaluating recommendations took only four months. Region VIII requested a report on the Denver, Colo., metropolitan area in preparation for negotiations with the city on what will be needed to meet water quality standards. NFIC covered the status of industry in the area, water sources, waste treatment, sampling, and made recommendations within six months.

When emergency situations arise, the team moves even faster. For instance, EPA wanted to take court action against Rohm & Haas located on the Houston Ship Channel because of the far-reaching effects of their discharges as shown by analysis in the regional lab. The investigators moved in, determined that Rohm & Haas wastes did reach shellfish areas, and were in court within two and one half weeks!

Two years ago, NFIC sampled areas in both Colorado and New Mexico when fish with a high mercury content were found in nearby Navaho Lake. At the requests of Regions VI and VIII, NFIC investigated four possibilities; application of mercury-containing fertilizer, erosion of rocks with a high mercury content, point source discharges, and aerial mercury discharges from the controversial Southwest power plant complex.

One project involves aerial reconnaissance over Lake Superior; here NFIC is working with EPA's Freshwater Laboratory at Duluth, Minn., to show the effects of the taconite tailings that Reserve Mining Co. is presently dumping into the lake. The results will be submitted to Stein and Quarles. CKL

# Uncle Sam's lawyers: a growing force on the pollution scene

*Government attorneys are finding that the law suit—or the threat of one—is a powerful enforcement weapon*

Reprinted from ENVIRON. SCI. TECHNOL., **5**, 994 (October 1971)

To federal government environmental officials, the key word these days is enforcement. And enforcement, it turns out, is practically synonymous with enforcement of water pollution laws. These facts emerge from recent conversations between ES&T and two of the government's top environmental lawyers: Assistant Attorney General Shiro Kashiwa, head of the Justice Department's Division of Land and Natural Resources, and John Quarles, Jr., assistant administrator for Enforcement, and general counsel, Environmental Protection Agency (EPA).

### Emphasizing water

Several factors have combined to make air pollution take the back seat to water pollution as far as enforcement is concerned. First and foremost, use of civil and criminal provisions of the 1899 Rivers and Harbors Act (often called simply "The Refuse Act") has given what all government officials acknowledge as a tremendous boost to water pollution enforcement actions. Second, passage of amendments to the Clean Air Act at the end of last year have, in a sense, given polluters some breathing time; not until fairly lengthy standard-setting and implementation plan procedures have been gone through does the act specify recourse to legal action (see ES&T's "Calendar for air pollution watchers," February 1971, p 107). Nonetheless, the air pollution arena has always been relatively inactive; only one federally prosecuted air pollution case has ever reached the courts. In that one, a chicken processing firm in Maryland was accused—and successfully prosecuted—of causing interstate air pollution. (Ironically enough, the firm is still operating, albeit using some different processes, and has not yet totally escaped the interest of the courts.)

The action on the water pollution scene appears, in contrast, hectic indeed. "What we have done in this field is just tremendous," says Justice's Kashiwa with almost boyish enthusiasm. He cites the vast increase in use of civil (injunctive) procedures under the 1899 act which, prior to 1969, had rarely been used. Criminal provisions of the same act have also been invoked more frequently, says Kashiwa, although he points out that, in fact, they have been used regularly since 1899.

In the decade of the sixties, about 40–50 criminal suits were filed each year by the Justice Department; the number now has risen to a rate equivalent to over 150 annually.

Kashiwa has a large staff to help him enforce the laws—120 attorneys in Washington and one in each of the 90 U.S. attorney's offices around the country. The way they become involved in water pollution cases varies —government attorneys may or may not take the first step toward a suit— but Kashiwa points out that, in general, "We receive the cases. We don't take samples."

Where a discharge is irregular or occasional, such as an oil spill, a U.S. attorney in the field can proceed against the violator under criminal provisions of the 1899 act, assuming, of course, that he has adequate evidence. If, on the other hand, a discharge is continuous, Justice checks with EPA or relies on EPA and the Corps of Engineers to produce the incriminating evidence. Such instances of pollution are then proceeded against under civil provisions of the 1899 act. The object of the action is to enjoin the company doing the polluting from continuing to do so.

Under new guidelines issued by Kashiwa, a U.S. attorney in the field can undertake a civil action without first checking with Justice in Washington, but he must still rely on EPA or the Corps for evidence.

Of course, not all the cases that are filed inevitably end up in court. A preferable alternative, both for the company sued and for Justice, is for the company to file a stipulation that it plans to stop or otherwise control the offending discharge in a manner acceptable to the government.

## Change at EPA

Despite the key role—mandated by federal law—played by the Justice Department in prosecuting water pollution cases under the 1899 act and federal water pollution control law, much of the driving force for enforcement actually rests with the Environmental Protection Agency or, more specifically, with EPA's Office of Enforcement, headed by John Quarles.

"When EPA was formed (in December 1970), we had to take an agency with virtually no experience in litigation and get it ready to go to court," says Quarles, who came to EPA as part of Administrator William Ruckelshaus' team of bright young lawyers. What Quarles found on his arrival in the water pollution area was a rather tentative enforcement agency—FWQA, just

transferred to EPA from the Interior Department. Although numerous enforcement conferences were being held around the country, only one federal water pollution case had ever been brought to court, and that one was in 1960.

"I have focused hardest on individually targeted enforcement actions, because they were totally missing when I arrived," says Quarles. "You cannot have an enforcement agency without a capability to do that." The fact that Quarles has, in fact, been able to single out companies and groups of companies in the way he wants to is in part due to the government's use of the 1899 act, with its provision for civil suits.

At regular intervals, Quarles' office refers cases for action to Kashiwa's office in Justice. Likewise, Kashiwa's people keep EPA informed on criminal cases that may have been instituted by U.S. attorneys and not at EPA's behest, and on Justice-initiated civil cases.

As for the enforcement conference, that much-criticized mechanism for solving water pollution problems, Quarles believes that it still serves a purpose: "Conferences bring attention to a situation with many polluters, and they're also useful in formulating implementation plans," he says.

Likewise, EPA has issued a large number of 180-day notices to municipalities, the first of them—to the cities of Atlanta, Cleveland, and Detroit—within a few days after the agency had been established. The 180-day notice is, at present, the most powerful legal weapon EPA can use against municipal polluters (domestic sewage being specifically exempted under the Refuse Act).

But Quarles leaves no doubt that litigation is where the action is going to be. EPA officials are keeping a count of the actions filed (see table), and each week brings its new list of initiated actions. The potential power and all-around applicability of the Refuse Act is well illustrated by the government's suit against Florida Power and Light Co. over alleged thermal pollution of Florida's Biscayne Bay. The suit has now been settled out of court and the Justice Department has issued a consent decree (see this issue, p 983), representing a considerable victory for the government in this first case filed under civil provisions of the Refuse Act.

Interestingly enough, though, out-

of-court settlement leaves unresolved an intriguing point—whether heat can be regarded as refuse within the meaning of the law. A pretrial hearing judge declined to rule that it was not, leaving the door open for future thermal pollution cases.

If refuse can have a sufficiently broad meaning as to include heated water, as well as logs, oil spills, and industrial waste, then the utility of the 1899 act to the government is obvious indeed.

However, since so few cases have actually come to trial, it is difficult to assess, at present, just what effect the new emphasis on enforcement will have on the overall water pollution picture. Certainly, evidence to date indicates that the mere threat of legal action is a powerful incentive.

Despite the necessarily time-con-

| EPA Enforcement Actions since December 3, 1970 | |
|---|---|
| 180-day notices | 49 |
| Conferences | |
| standards-setting (water quality standards) | 1 |
| new enforcement conferences | 5 |
| reconvened and additional sessions | 7 |
| progress sessions | 1 |
| approval of conference summaries | 7 |
| Refuse Act actions | |
| direct referrals to Justice Dept. | |
| civil actions | 31 |
| criminal actions | 5 |
| Justice-initiated actions with assistance of EPA regional offices | |
| civil actions | 13 |
| criminal actions | 15 |

data as of September 8, 1971; do not include actions involving Justice Dept. alone

Source: EPA Office of Enforcement

suming waits and trials of strength inherent in use of legal machinery, there is no doubt that momentum has picked up. The government is showing, along with its renewed interest in enforcement, a growing confidence that it is on the right track and has the wholehearted support of the public. And when legal pressures against air pollution begin to mount, as Quarles believes they will "about a year from now," the stage will be set for environmental lawyers to start stealing the limelight from "ecologists" who have, so far, had the field pretty much to themselves.     DHMB

**39**

# Water polluters: beware the feds!

*231 criminal cases, 11 injunctive actions, 10*
*180-day notices, 51 enforcement conferences . . .*
*crackdowns increase with no end in sight*

*Reprinted from* ENVIRON. SCI. TECHNOL., 4, 887 (November 1970)

In practice, there are three ways to enforce water pollution regulations at the federal level. The first, the Refuse Act, enacted in 1899, prohibits dumping of unauthorized materials into navigable waters of the U.S. without a permit. The second and third methods are parts of the enforcement procedures which have been built into the present Federal Water Pollution Control Act. The second method involves a 180-day notice to a water polluter for violation of a state's water quality standard, while the third is the now popularized enforcement conference with its three-step procedure—conference, public hearing, and court action.

Each of the three federal enforcement methods has its good and bad points. Despite its 19th century, baroque, legal language, the Refuse Act is the fastest remedy under the law. It is, however, limited in the sense that its use is concentrated on a single water polluter, and it is further limited, in case of criminal actions, by its token penalties. But civil actions can also be taken under the Refuse Act; in such cases, the Justice Department brings suit against the alleged polluter and seeks injunctive relief to stop the polluting, as in the recent 10 mercury cases.

Rather slower in response time is the 180-day notice. Here, the guiding rationale is violation of a water quality standard. In cases where sufficient monitoring evidence has been obtained to show a violation, federal officials seek injunctive relief by filing the notice. If the polluter does not take corrective action within the time limit, he can be hauled into court.

Although the third method is the most tedious, the enforcement conference is used for cases where many polluters (perhaps as many as 50 to 100) in any one geographical area are brought together to abate their water pollution. Even though the present enforcement conference method is the slowest of the three, it is perhaps the best way to bring together the multiplicity of industrial and municipal polluters in a single area.

Together, these procedures in the federal legal arsenal comprise a three-pronged attack on water polluters. To date, all three have been used, and they certainly will be used more in the future.

At last month's 43rd annual conference of the Water Pollution Control Federation (WPCF) in Boston, federal, state, local, territorial, and industrial spokesmen addressed the problem of water pollution enforcement. The consensus:

• Enforcement procedures are cumbersome and time consuming.

• The public must have a swifter response to water quality requirements than existing enforcement machinery permits. "If we don't hurry up and strike while the iron is hot, the public's attention to water pollution may be lost," says the Hon. Kevin White, mayor of Boston.

In the same sense of urgency, the keynote speaker, Sen. Edmund S. Muskie (D.-Me.), urged that rigid legislation be adopted to halt the deterioration of the environment.

## Refuse Act

Under the Refuse Act, actually section 13 of the Rivers and Harbors Act of 1899, several procedures can be used. Under criminal provisions, the polluter is fined from a minimum of

**FWQA's Murray Stein**
*Asst. commissioner for enforcement*

$500 to a maximum of $2500, or imprisoned for 30 days to one year, or both. On the other hand, injunctive action can be requested under the civil provisions, which means that the polluter either reduces or cleans up his effluent to a negligible amount within a certain time. If he fails to do so, he is hauled into court.

In the first eight months of this year, approximately 170 cases involving criminal actions had been filed by various U.S. attorneys throughout the country. Most of these cases had been referred by the U.S. Army Corps of Engineers; most involved oil spills, but some involved acid spills.

The number of criminal cases under the Refuse Act is certainly increasing, in fact by 288% over the last year, according to Justice Department statistics. The statistics: From 37 cases in 1968, to 50 in 1969, to the 144 cases filed in the first eight months of 1970.

Other Refuse Act cases are being terminated more frequently. One example involves the Howat Concrete Company, which recently paid a $2100 fine for polluting the Anacostia River in Washington, D.C. The case was the first successful water pollution case prosecuted in D.C. in the last 45 years, according to one news account.

When a private citizen can supply the Department of Justice attorneys with information which subsequently leads to conviction, he can receive one-half of the fine collected by the court in the criminal action, according to a provision in the Refuse Act.

Thus far, there are two instances in which such awards have been made. The first involved Interlake Steel Co. of Chicago, in which Frank R. Jacklovich provided information to the Coast Guard that led to the conviction against Interlake. Apparently, Jacklovich observed the discharge and reported it to the Coast Guard, the U.S. Army Corps of Engineers, and the Metropolitan Sanitary District. The company paid a $500 fine; Frank Jacklovich collected $250. In October 1969, the Justice Department approved payment to Jacklovich, and Interlake did not appeal the case.

The second citizen's award was made on July 30, 1970. It went to the Hudson River Fishermen's Association. In this instance, the award resulted from four cases against Penn Central Railroad, each for $1000, for dumping oil into the Harmon River in New York. The cases were filed early in 1969, in the U.S. Southern District Court of New York. On November 7, 1969, the company was found guilty and paid the $4000 fine, half of which was paid to the fishermen's association.

**Refuse Act—civil penalties**

As cut and dry as these criminal proceedings may seem, in fact the in-

junctive actions that can be instituted under the Refuse Act put more of a clamp on any gross polluter's operations. It becomes more costly for him to continue operations. In essence, injunctive actions require him to install the necessary equipment to control the discharge of any undesirable material from an industrial or municipal effluent.

Mercury pollution (see this issue, page 890) came into the limelight this April and has been a top priority item at the Federal Water Quality Administration (FWQA) ever since. In 10 civil cases, involving eight industries at 10 locations, Secretary of Interior Walter J. Hickel asked the Justice Department to seek injunctive relief to stop the polluting.

By the first of last month, five companies—Olin at Niagara Falls, N.Y., Allied Chemical at Syracuse, N.Y., IMC Chlor Alkali Co. at Orrington, Me., Diamond Shamrock Corp., and Georgia-Pacific—had agreed to reduce their discharge of mercury to 0.5 lb. per day. Oxford Papers Division of Ethyl Corp. agreed that it would phase out its old plant at Rumford, Me.

At press time, the other four cases were still pending on court dockets. One of the four, Hooker Chemical at

Niagara Falls, N.Y., is coming up with a closed-loop system which looks very promising, according to one federal observer.

In addition to the mercury cases, there is one thermal pollution case involving Florida Power and Light Co. Soon, there may be a ruling on whether heat, as an industrial discharge, can be regarded as a pollutant under the 1899 act.

Another injunctive action under the act is the mandatory injunctive procedure, in which the polluter is enjoined to clean up the pollution from previous discharges. In any event, this procedure would be much more expensive for an industry or municipality than paying a paltry fine of $2500, at most, under the criminal provisions of the act.

**180-day notices**

Before any alleged polluter who has violated a water quality standard can be taken to court, the FWQA must give him a 180-day notice. If, in the interim, the polluter takes positive corrective action, the court case is not decided. Of such notices given, all have taken adequate remedial actions.

Six notices were filed in 1969; another four were filed this year. For example, on August 30, 1969, the FWQA cited six polluters on Lake Erie: a mining company, four steel plants, and the city of Toledo (ES&T, October 1969, page 881). All six indicated that they would comply and none has been taken to court.

This year's notices, along with the dates on which informal hearings and corrective actions were agreed to, include: Penn Central Railroad (June 23); GAF at Linden, N.J. (June 23); Fairfax Drainage District in Wyandotte County, Kan. (July 7); and the city of Fargo, N.D. (July 10). Before the expiration of the 180-day limit, each alleged water polluter corrected his discharge problem. In essence, each agreed to comply with timetables to meet water quality standards requirements, and their actions have been adjudged satisfactory by the federal authorities.

**Enforcement conferences**

There has been a speedup in the number of enforcement conference activities along with the speedup in the number of Refuse Act cases. Under the Nixon administration, six conferences were initiated after January 1, 1969—Lake Superior, Escambia River Basin, Perdido Bay, Mobile Bay, Biscayne Bay, and Dade County. Eight were reconvened, and another three placed renewed emphasis on progress.

To date, 51 enforcement conferences are in some stage of completion. Of the 51, 30 have been taken on federal initiative, 15 on state initiative, and six on both. FWQA's top enforcement man is Assistant Commissioner Murray Stein.

With nine exceptions, all states and U.S. possessions and territories are involved in water pollution enforcement conferences, the first of which was held in January 1957, shortly after the basic 1956 law came into being. But Alaska, Delaware, Hawaii, Kentucky, Montana, North Carolina, Oklahoma, Puerto Rico, and Texas have not been involved in any enforcement conference. On the other hand, some states are involved in several different conferences.

Initiated by the Surgeon General of the Public Health Service, the first conference involved the states of Arkansas and Louisiana over the Corney

Creek Drainage System. Its hearings were held on January 16 and 17, 1957. The most recent conference, the 51st, was convened only last month, for the Dade County area in Florida.

In proceedings established by an enforcement conference, various compliance schedules are involved for the myriad industrial and municipal sources of water pollution. To say that all sources are meeting their respective compliance timetables is gross oversimplification.

FWQA's Northeast and Great Lakes regional offices are heavily involved with enforcement conferences. FWQA recently noted more than 120 water polluters, both industrial and municipal, who are not living up to the compliance dates stipulated in conference proceedings in the Northeast region alone.

Enforcement conference proceedings today are hampered and limited by statutory authority, which ties the federal government's hands for at least one year, and sometimes for one and a half years, before a recalcitrant polluter can be brought to terms. Some deficiencies are now glaringly apparent to the public. At the conference stage, to mention only one problem, there is no federal interaction with the alleged polluter. There is no subpoena power to haul the polluter in so that he can be heard.

**State's view**

At the state level, Thomas C. McMahon, director of water pollution for the Commonwealth of Massachusetts, notes that the state's control effort started in September 1966 with the enactment of four bills:

• $150 million construction grants program.

• New drive for water pollution control with complete authority to implement, administer, and enforce water pollution control in the Commonwealth.

• Local tax incentive systems to industries installing suitable waste treatment facilities.

• Additional $250 million bond issue for construction grants, including prefinancing, mandatory waste treatment plant certification, boat pollution and marina control authority, and streamlined enforcement authority.

It would seem with all these resources—money and laws—Massachusetts would have the cleanest water in the U.S. The fact that it does not can logically be traced to details of enforcement.

State enforcement actions can be filed under criminal or civil provisions of state law. The civil provisions apply in Massachusetts to 130 communities and 400 industries which are currently working on implementation plans approved by the Department of Interior in 1967.

Since 1967, 43 enforcement actions, (21 criminal citations and 22 civil citations) have been taken. "About 75% of the criminal cases received reasonable dispatch and resulted in corrective action," McMahon says.

On the other hand, delays up to $2^1/_2$ years have been incurred under the civil procedures. "At this time, 21 municipalities and industries are on schedules totaling 43 years, or an average of two years per violator," McMahon complains.

Massachusetts needs new authorities (institutional arrangements) to accept wastes and to construct, operate, and maintain treatment plants. Costs for such operations would come from user charges, based on a volume and strength formula. Certainly, this new arrangement is essentially untried and vigorously contested. But three years of delays in Massachusetts attest to the need to change the system. Enforcement procedures at all levels need to be modified.

**Looking ahead**

The present methods of legal action are so cumbersome and time consuming that the public may be losing ground in its war on water pollution.

These deficiencies, plus others, form the basis for further proposed amendments to the Federal Water Pollution Control Act, the statutory authority for which expires next July 1. More than likely, the 91st Congress will have amended both the Clean Air and Solid Waste Disposal Acts before adjourning. To be sure, continuing Congressional emphasis to environmental matters will necessitate further water amendments. Hopefully, these will be a number-one priority for the 92nd Congress.                              SSM

# Federal demonstration projects: What has been achieved?

*Reprinted from* ENVIRON. SCI. TECHNOL., **5**, 498 (June 1971)

Federal support for demonstrating new and untried—yet potentially promising—technological developments for pollution control is a growing force in the battle to stem the tide of environmental degradation. Federal funds have been made available in the past, under a wide variety of granting mechanisms and labels, but the recent establishment of a Division of Grants Administration in the new Environmental Protection Agency (EPA) suggests that a new order will be imposed on what has been in recent years an area beset with complexity.

Until EPA was created last December, grants were separately disbursed by the National Air Pollution Control Administration (NAPCA, now APCO-EPA), the Federal Water Quality Administration (FWQA, now WQO-EPA) and the Bureau of Solid Waste Management (BSWM, now SWMO-EPA). Since the incorporation of the three agencies into EPA, their names have been changed but, so far, the way in which they parcel out money for research, development, and (or) demonstration has not. Earlier this year, however a special EPA task force was organized to take a hard look at the agency's grants operation. Its report, due later this year, will probably include recommendations for change.

In the meantime, Alexander J. Greene, director of the newly established Grants Administration Division of EPA, has stated that the division aims to simplify existing grants procedures, to eliminate anomalous requirements, and to streamline the processing of applications. Greene also wants EPA to develop the flexibility to respond to requests that cut across the "five-media" organizational lines of the agency (separate offices for Water Quality, Air Pollution Control, Solid Wastes Management, Pesticides, and Radiation). For example, Greene cites the need for training grants in the areas of public administration and environmental law; there are at present training grants for water pollution, air

pollution, etc., but no program that cuts across media lines.

Money to be used for the demonstration of new technology to solve environmental problems has been variously labeled (and, of course, has been made available in varying amounts), depending on the federal agency involved.

For instance, demonstration grants for solid waste management have been called just that—demonstration grants —and have been a separate item in the budget of the old BSWM. On the other hand, demonstration projects in the water quality area are not neatly separated for budget purposes, but are included in funds for RD&D (research, development, and demonstration). In the air pollution control field, most of the NAPCA money for demonstration was funded under the program grant or contract.

Now that demonstration grants— under whatever guise—have been made available for several years, and especially in view of the forthcoming EPA assessment of granting procedures, it seems as good a time as any to ask what has been accomplished to date.

Federal officials who are most closely involved have been asking this question themselves. In the past, progress on demo projects was perhaps buried either in technical reports or in papers presented at technical meetings. But federal government people are now intent on communicating with the outside world and have started to brief the potential users of the information already gathered in demo projects. They are getting the word out to state and municipal officials, industry representatives, and to consulting engineers.

Last month, in fact, a solid waste briefing, the first of its kind, was held in Cincinnati. Design seminars aimed at transferring water pollution control technology to consulting engineers are being held monthly by WQO—by invitation only. (Last

month's seminar was in Boston, this month's is in Charlottesville, Va.) To date, no such conscious effort at air pollution control technology transfer based on the demonstration programs has yet been held.

According to EPA's Greene, there are presently 21 grants programs in the agency. They are basically of four types: training grants, RD&D grants, state and local assistance grants, and construction grants. Prospective grant applicants are best advised to contact the EPA regional office interim coordinator in their area (see map).

## Solid waste management

At the beginning of 1971, expenditures of federal funds for solid waste management demo grants had totaled

### Contact for 10 federal regions

| Region | Interim regional coordinator |
|---|---|
| I | L. Klashman (617) 223-7210 |
| II | G. M. Hansler (212) 264-2525 |
| III | L. Gebhard (215) 597-4506 |
| IV | J. R. Thoman (404) 526-5727 |
| V | F. T. Mayo (312) 353-5250 |
| VI | B. V. McFarland (214) 749-2827 |
| VII | J. M. Rademacher (816) 374-5493 |
| VIII | D. P. Dubois (303) 837-3283 |
| IX | P. DeFalco, Jr. (415) 556-4303 |
| X | J. L. Agee (503) 226-3914 |

some $22.4 million, about one quarter of the total federal solid waste program to date. These funds have been used to advance solid waste technology from the research phase to full-scale demonstrations in communities. At the beginning of this calendar year, 128 projects had been initiated with demo grant funds.

"In the past, there were a lot of small projects," says John Talty, director of SWMO's 28-man strong Division of Demonstration Operations, headquartered in Cincinnati, Ohio. "Of course, under the new act—the Resource Recovery Act—the emphasis for demonstrations will be on innovative technology and resource recovery systems. If and when federal appropriations are made, then a large dem-

onstrations program will be initiated," Talty predicts.

One early indication of the success of these projects can be judged from the fact that by the end of March more than 50 letters of intent, in a sense amounting to preproposals, were on file with SWMO. These preproposals included both innovative technology and resource recovery systems. Already, the framework by which the program will be administered has been published in the *Federal Register* (Jan. 23, 1971).

However, Talty points out several distinctions under the new law. First and most important is the fact that the applicant for federal solid waste demonstration funds must be a public body such as a state or municipality. Another fact worth remembering is

that the demo grants program under the new law is not a federal assistance, construction grants program. The money can be used only to demonstrate new and untried systems.

Solid waste demo projects are neatly divided into the following three categories:
• Processing and recovery of solid wastes,
• Collection and disposal,
• Systems management.

How successful or fruitful the expenditures of demo grants funds have been perhaps time will soon tell.

Last month's briefing session in Cincinnati, "Symposium on Solid Waste Demonstration Projects," highlighted the top dozen projects that have made real impact in the three categories.

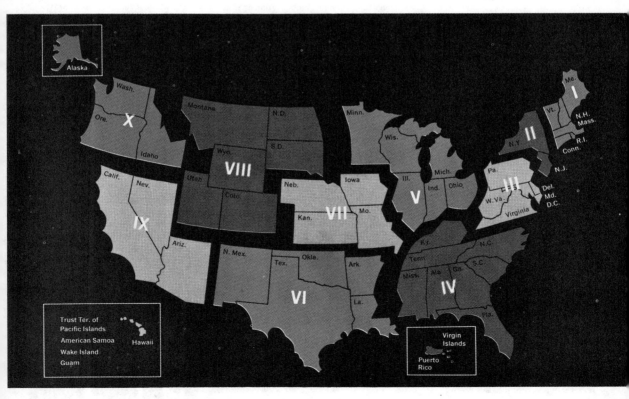

What the more than several hundred attendees heard were presentations by the program managers of the top dozen projects that have been demonstrated to date. A few are new this calendar year, some were completed this year, and others are being continued. Five are noteworthy in the processing and recovery area, four in the collection and disposal area, and three in the systems management area (see box).

Not only is it important to know that these new solid waste management technologies have emerged, but it's reassuring to note that some are being put into practice by state and local agencies.

As one example, the city of Milford, Conn., purchased shredders to grind refuse before landfill operations. The volume reduction achieved by grinding of trash in conjunction with the landfill operation is an advance which was investigated through an ongoing demo project at Madison, Wis. (see box) which has received $641,822 to date.

In another example, the Chilton County, Ala., project established the concept that a central disposal site was acceptable to the citizens and that such an operation is economically and aesthetically feasible.

Then again, one of the 50 preproposals on file is one by the city of Baltimore, Md. Baltimore officials have submitted plans for actual installation of the Monsanto Enviro-Chem pyrolysis system (ES&T, February 1971, page 111); city officials met initially with SWMO representatives late last year to discuss the proposed project.

**Water quality demos**

In the RD&D business at the federal Water Quality Office, no budgetary breakout of funding for the second D, for demonstrations, is made. The RD&D program funding level for fiscal year 1971 reached $60 million. In fiscal year 1968 the funding was slightly more than $43 million; in fiscal 1969 it was $50 million, and in fiscal year 1970, $47 million.

During the past several years, WQO and its predecessor FWQA have held program seminars periodically. But now the express purpose of the design seminars is to transfer existing technology into practice.

"These sessions inform officials from state and local governments, industries, and their consulting engineers about proved technologies that are available

for incorporation into the design of waste treatment facilities that will be built in the near future," says John Convery, special assistant to David Stephan, assistant commissioner for RD&D in WQO.

Progress on water demo projects runs the gamut of the RD&D program and includes advances for municipal, industrial, agricultural, mining, and other pollution control technologies. Technological advances have been made in all categories.

The main emphasis in the pollution control aspects of the program is to develop improved and (or) lower-cost methods of treatment of waste waters. Pollution control accomplishments include the following:

• Demonstration of more than a dozen phosphorus removal techniques, including the use of alum, lime, and iron salts;

• Development and demonstration of several methods of nitrogen removal;

• Demonstration of high efficiency nitrogen removal for use with biological treatment systems which use biological nitrification–denitrification;

• Physical–chemical methods of nitrogen removal which include demonstrations of ammonia stripping in a 3.75 mgd (million gallons per day) plant (Lake Tahoe, Calif.), breakpoint chlorination—in which the ammonia is reduced to elemental nitrogen by chlorine (Blue Plains, Md.), and ion exchange with natural zeolites.

The use of high-purity oxygen and a new filtration process are other noteworthy advances in the municipal category. The experience with high-purity oxygen aeration, a technique which was first tested two years ago in Batavia, N.Y. (ES&T, February 1969, page 109), has been very favorable. Data from this demo project indicate that a potential savings of up to 25% in the total annual treatment costs can be obtained with the system. The cost savings are made possible primarily through the reduction of aeration volume requirements and reduced sludge production.

The other highlight is the moving bed filter process (ES&T, November 1970, page 885). With chemical coagulation, the filter unit is capable of handling variable solids loadings. A 5 mgd moving bed filter plant is currently being operated in the borough of Manville (N.J.) as a tertiary treatment technique to upgrade the quality of the trickling filter effluent. Moving bed filtration, augmented by

the addition of powdered carbon or followed by granular carbon adsorption, is also currently being evaluated as a physical–chemical treatment system for small municipal waste treatment plants.

In the industrial water pollution control technology category, reuse and closed-loop water systems are the watchwords to progress. The project that won the 1970 gold medal award of the American Society of Civil Engineers was Armco Steel Co.'s 2.2 mgd treatment system for acid rinse waters from pickling of steel strips prior to rolling.

Other examples of industrial progress are also available. At the 0.85 mgd treatment plant of the B. F. Goodrich Co. (N.J.), chemical coagulation and activated sludge treatment of a mixed chemical waste from a plastics manufacturing plant were demonstrated. This advanced biochemical treatment system produces an effluent that can meet the Delaware River Basin Commission standard.

A closed-cycle system for treatment of waste waters in the chemical industry was demonstrated at the Dow Chemical Co. plant (Midland, Mich.). The system reclaims phenol and acetate from contaminated brines, and uses the effluent brine for chlorine-caustic production. Another closed-loop system has been demonstrated for the sugar beet industry, where 8.6 mgd of beet transport water is being reclaimed and reused by the Sugar Development Foundation (Colorado).

Another example of the closed-loop system has been demonstrated in Winter Garden, Fla. The waste activated sludge from the biological treatment unit of a juice concentrate plant is being dried, pelletized, and sold as chicken feed supplement for $240 per ton. The revenue from the sale of the by-product is more than adequate to pay for the waste treatment costs for the plant.

The demonstration of recovery and reuse systems, a very worthwhile objective, can be illustrated further with other examples down from the list of WQO-supported demo projects. American Enka Corp. (N.C.) has come up with an original physical–chemical method for removing and recovering zinc so that the water can be reused in the nylon manufacturing process.

If sludge is your problem and your industry is pulp and paper, then the

**45**

project of Crown-Zellerbach Corp. (Wash.) is noteworthy. The sludge is dewatered on a pressure drum filter and used in one of three ways—as supplemental fuel for steam boilers, as a dry mulching material for highway slopes, or as an agricultural soil conditioner.

Oily wastes from refineries can be treated to yield effluent characteristics at least as good as those associated with municipal treatment. The treatment method, which uses API separators, biooxidation, air flotation, and oil sludge return, has been demonstrated at the 30 mgd plant of American Oil Co. (Ind.).

A technique for treating industrial textile finishing wastes has been demonstrated at the 50,000 gpd waste treatment facility of C.H. Masland & Sons (R.I.). The system includes the use of granular activated carbon adsorption with biological regeneration.

Advances in the categories of agricultural, mining, and oil pollution control technology have also been achieved. A synergistic solution to a municipal problem, which also solved an agricultural problem, was the development of denitrification technology for removal of nitrate from irrigation return flows (Firebaugh, Calif.). If this technique were applied, for example, throughout the San Joaquin drainage area, the nitrate load to San Francisco Bay could be reduced by approximately 125,000 pounds/day.

For the oil category, there are some notable advances. A high-capacity centrifuge for separating oil and wa-

ter has been successfully demonstrated in tests off the coast of Santa Barbara, Calif. Another demonstration involved a mechanized labor-saving technique for beach cleanup of spilled oil. The 30-ton per hour system was demonstrated by Aerojet-General Corp. (El Monte, Calif.). The system includes froth flotation to separate oil from the contaminated sand. A smaller unit, an ultrasonic one capable of handling 5 tons per hour, was also successfully demonstrated for cleanup of oil contaminated sand.

**Air pollution demos**

Actual demo projects in the air pollution control program area have been rather few, but what new technology has been demonstrated has, in fact, been done mainly through the contracts or program grants mechanisms. (Demo grants represent only a small portion of the total federal grant funds for air pollution control. The other types of grant are: program, planning, and maintenance.)

"Program grants have many of the aspects of demonstrating technology," says Raymond Smith, acting assistant commissioner for program development in APCO. Part of the funding for a program grant, for example, might be used for monitoring of stack gases, or demonstrating a novel concept.

Nevertheless, on the average, air demo projects have been funded at a level of approximately $1 million annually. Over the past five years, a substantial portion of the actual demo

grants funds has been spent in looking at the availability of low-sulfur fuel.

The closest thing to demo projects in this program area is the demo phase of an R&D project. Approximately $30 million is currently being spent in fiscal 1971 on contracts.

"We are still looking forward to the first commercial demonstration of the feasibility of $SO_2$ control from power plants," Smith says. "It's really part of the research activity." Smelters, the other principal source of $SO_2$ emissions from stationary sources, plus the power plants work received about $15 million in federal funds in fiscal 1971. The goal: proved commercial feasibility for $SO_x$ controls by 1975.

What perhaps is really new with demonstrations of oncoming technologies is the cost-sharing demonstration projects that have been worked out with industry. Over the past two years, a number of these joint federal government–industry demonstrations have been worked out with power generating stations such as Tennessee Valley Authority and Boston Electric, and industrial organizations such as Monsanto and Chemico.

Of course, another category of technology that is being exploited under federal funds goes for air pollution controls from mobile sources. A prime target here is the gas recirculation system for $NO_x$ control from automobiles. Another auto-related problem that is not completely solved is the metals problem for exhaust gas thermal reactors.     **SSM**

# EPA program transfers technology

*Establishing a two-way street between municipal users and federal producers of new technology whereby municipal needs are satisfied by the federal RD&D products*

Reprinted from ENVIRON. SCI. TECHNOL., **6**, 314 (April 1972)

EPA's technology transfer (TT) program is primarily aimed at marketing the products of the federal research, development, and demonstration (RD&D) activities, (ES&T, June 1971, p 498). The products include design seminars, design manuals, and an array of communication items including videotapes, visual displays, and publications.

The TT program has come a long way since it was first announced by David Dominick (then Commissioner of the Water Quality Office and now EPA's assistant administrator for Categorical Programs) at the Boston meeting of the Water Pollution Control Federation in the fall of 1970. Future plans for the program include expansion into other major environmental areas including air, industrial waste water, and solid waste. The expansion of the TT program is in line with current plans to include research, development, and demonstration of both air and solid waste under the aegis of the assistant administrator for research and monitoring, Stanley Greenfield. These plans will be accomplished after this July 1.

The initial goal of the TT program was to make an impact on the construction of municipal waste treatment facilities over the next 3–4 years. The need for the program was first recognized by Eugene Jensen (now head of EPA's water programs office) and David Stephan (now on Greenfield's staff) in the middle of 1970 and before the Boston announcement. At that time, these officials recognized that applications for construction grants were not including new technologies, thus the need for a special TT effort. Within the new organization of EPA, Greenfield appreciated the significance of the program and elevated it to function from his immediate office, in order to allow for its expansion into other environmental problem areas .

### Products

To date, design seminars have focused on the waste water treatment problems in EPA regional areas (see map). For example, phosphorus removal may be a prime target in one region while upgrading existing sewage treatment plants might be the top target in another.

In any event, each of these seminars brings together approximately 100 consulting engineers, sanitary engineers, and municipal engineers. The objective of the program is to ensure that these engineers have all the alternatives and options open to them in designing today's waste water treatment plants.

By the end of last month, a dozen such design seminars had been held in each of the 10 federal regions with the

**EPA's Robert Crowe**

*. . . marketing federal RD&D products*

exception of region VIII (Denver); however, the first seminar for fiscal 1973 is slated for Denver, and the near-term schedule for the remainder of this fiscal year includes seminars in Washington, DC (this month); up-state NY (May); and Pittsburgh, PA (June).

"During the remainder of this calendar year, it is anticipated that at least one design seminar will be held every month at the requests of the regions," says Robert Crowe, special assistant for the TT program, who reports to Greenfield, the assistant administrator for research and monitoring.

"We are not trying to design waste water treatment plants," Crowe says. "We are not in the consulting business, or in the design business, but we want to make sure that both the consulting and design engineers, along with the administrative decision-makers and the

public have all the alternatives and options of new technology available to them," he explains.

"The TT staff has worked closely with the Consulting Engineers Council (CEC) and the Water Pollution Control Federation (WPCF) and continues to seek their guidance and advice in order to maintain relevancy of the TT program," Crowe says.

Process design manuals are the backup material at the seminars; the manuals are loose-leaf volumes and represent all of the available, usable, and practical technology to date. These manuals are prepared by consulting engineers most knowledgeable in the field and the research staff of the Office of Research and Monitoring. They will be updated as the needs arise.

Four manuals have so far been prepared and distributed; they cover the subjects of physical-chemical treatment, upgrading of existing sewage plants, phosphorus removal, and carbon adsorption. At press time, some 20,000 copies of the manuals had been distributed (5000 of each manual) and an additional 6000 individual requests for the complete set of four have been received.

The design manuals are not a duplication of the efforts of the Water Pollution Control Federation (WPCF) but a necessary adjunct to WPCF publications. "We are not in competition with WPCF's manuals; ours are written as a supplemental text to theirs," Crowe says.

Three other design manuals are planned, one on sulfide control in sewage systems, which will get at the odor and corrosion problems in waste water treatment plants, a second one on sludge handling and disposal, and the third on advanced techniques in operation and maintenance.

A videotape on carbon adsorption is another tangible product of the TT program; each EPA regional office has a copy. A second videotape on phosphorus removal is being prepared.

To design a meaningful program to transfer technology to administrative decision-makers, the headquarters team of the TT program is working with a group called Public Technology, Inc. "PTI hopes to function as a focal point for response to the city's need, and we are working with them now to get inputs from the cities and meet with their managers," Crowe says. "We are working primarily with Robert Havlick, a senior vice-president of PTI." PTI is sponsored by a half dozen of the major public groups including the Council of State Governments, the National Association of Counties, the National Governors Conference, the National League of Cities, the U.S. Conference of Mayors, and the International City Managers Association.

"We can see that our (EPA) relationship with the cities will be a two-way street where, in one direction, we will transfer technology to them and, in the other direction, we will know their

needs so that EPA research can work toward solving their problems."

### Staff

In addition to the staff of a half dozen professionals at the headquarters office, there are staff members at the National Environmental Research Center (NERC) in Cincinnati and the regional chairmen for the TT program; they contribute significantly to the program.

In addition to Crowe, members of the headquarters TT team include Robert Mandancy, who is responsible for evaluating new technology for introduction into the program, John Dyer who is responsible for the seminars and is concerned mainly with the technical presentations, Denis Lussier (technical publications), Paul Minor (industrial waste treatment), Kenneth Hay (communications resources), and Bill Herbert

(administrative assistant).

Depending on the specific problem being addressed at any one particular design seminar, a member of the NERC team introduces the subject to the attendees, presents the EPA experiences, and is on hand to answer questions as they develop, since these EPA personnel are the most knowledgeable in the agency on the specific subjects presented. The Cincinnati team members and their specialties include, for example, Ed Barth (nitrogen and phosphorus removal); John Smith (upgrading sewage treatment plants); Carl Brunner (carbon adsorption and phosphorus removal); Sid Hannah (solids removal); Jesse Cohen (physical-chemical treatment); Irwin Kugelman (physical-chemical treatment and phosphorus removal); Art Masse (carbon adsorption); and John English (carbon adsorption).     SSM

## Contact for 10 federal regions

| Region | Technology transfer chairman | Region | Technology transfer chairman |
|---|---|---|---|
| I | Lester Sutton (617) 223-7210 | VI | George Putnicki (214) 749-3842 |
| II | Rocco Ricci (201) 548-3441 | VII | Lynn Harrington (816) 374-2725 |
| III | Warren L. Carter (215) 597-9410 | VIII | Stan Smith (303) 837-3769 |
| IV | Asa B. Foster, Jr. (404) 526-5784 | IX | Irving Terzich (415) 556-7554 |
| V | Clarence Laskowski (312) 353-1056 | X | John E. Osborn (206) 442-1266 |

# EPA's technology transfer: now geared to industry

**Through capsule reports, seminars, and design manuals, EPA publicizes available waste control techniques**

*Reprinted from* ENVIRON. SCI. TECHNOL., **7**, 191 (March 1973)

**Seminar.** *Experts thoroughly explain available pollution control technology*

Over two years ago, EPA began a technology transfer program for marketing the products of federal research, development, and demonstration activities. Technology transfer (TT) was aimed at making an impact on the construction of municipal waste treatment facilities, with plans for expansion into other fields (ES&T, April, 1972, p 314).

EPA activated its industrial technology transfer program to disseminate information to industry on the technology now available for the control and treatment of air, water, and solid waste pollutants. In the past nine months technology transfer staff members have covered a great deal of ground in providing plant management with information of the available pollution control and treatment technology.

The industrial pollution control branch, headed by Paul Minor and assisted by Dennis Cannon for EPA in Washington, D.C., publicizes the latest industrial waste treatment technology in three ways: technical capsule reports, seminars, and design manuals. Capsule reports are concise technical documents, usually 6–8 pages, which describe successful pollution abatement projects proved reliable in either demonstration or full-scale operation. In some cases, they involve non-EPA projects. The purpose of the capsule reports, explains EPA's Minor, is to bring to light the in-plant control and treatment methods available for industrial waste control when the knowledge is first revealed.

Six capsule reports are either distributed or being printed: recycling zinc in vicose rayon plants by two-stage precipitation (American Enka), dry caustic peeling of peaches (Delmonte), hot air blanching of vegetables (National Canners Association), aerated lagoon treatment of sulfite pulping effluents (Crown Zellerbach), color removal from kraft pulping effluent by lime addition (Interstate Paper), and process changes for waste abatement in brass wire mills (Volco Brass & Copper).

Eight other capsule reports are scheduled for completion by the end of June. Potential topics include molecular oxygen used in black liquor oxidation (Owens Illinois), by-products from brewery wastes (Coors), regeneration of sulfuric acid pickle liquor, and total recycle in a coal-fired power plant; others are still under consideration.

**Industrial seminars**

The industrial seminars, one of the most important phases of the technology transfer program, usually consist of a two- to three-day technical meeting. The seminars are geared to provide plant management (usually smaller manufacturers) with a thorough understanding of available pollution control technology. EPA officials as well as pollution control and plant processing experts participate in the seminars. EPA aims to present the best treatment alternatives without emphasizing or recommending any specific method of pollution abatement.

The seminars are pulled together by EPA's Minor and Cannon who draw upon the expertise within EPA as well as industrial experts in that area of process design and pollution control. Seminars are scheduled at the request of any of the 10 federal regions. In this way, EPA can respond to industrial needs and questions.

Eight seminars on five topics—poultry processing, metal finishing, meat packing, dairy products, and seafood processing—are scheduled before the end of fiscal 1973. At press time, six such seminars had been held and enthusiastically received.

A seminar on a particular industry's waste problem consists of several individual sessions. A general session covers regulatory aspects for this industry—the permit program, effluent guidelines, pretreatment requirements, the new water bill—and some basic technology in that area. Then two detailed technical sessions are held, one on in-plant process changes for pollution abatement and the other on waste treatment. Technical handouts explaining in-plant processing control and treatment alternatives are given to each attendee.

The final session includes a management consultant to discuss financial alternatives including any assistance from the Small Business Administration and the Internal Revenue Service. EPA officials encourage questions and discussion (sometimes heated) to become familiar with industry's problems in meeting environmental requirements.

**Design manuals**

Technology transfer's industrial waste group is now preparing design manuals which are developed under contract. A water monitoring manual—a guide for instruments for waste water sampling—will be out by June. An air monitoring manual is scheduled for fiscal 1974. Design manuals for various industrial sectors, aimed at design engineers in one particular industry, are also in the works. The iron and steel and the pulp and paper industries will be covered first. Other areas under consideration for detailed coverage in the manuals include the power and textile industries. The industrial waste program will continue full steam ahead by updating its past works and continually looking into new sources of industrial pollution. CKL

# Industrial waste water: FWQA inventory under way at last

*After years of delay, federal officials are now asking industries*

*to volunteer details of their waterborne discharges*

Reprinted from ENVIRON. SCI. TECHNOL., 5, 20 (January 1971)

A majority of government effort and money has, in the past, gone into the treatment and monitoring of municipal wastes. However, the federal government has known for years that industry discharges the largest volume and the most toxic of pollutants. "The volume of industrial wastes is growing several times as fast as that of sanitary sewage," states the 1969 report, "The Cost of Clean Water and Its Economic Impact." Previously, little detailed information was available for industrial waste water, but the Industrial Waste Inventory, undertaken by the Federal Water Quality Administration (FWQA), will show the effects of industrial discharges on the receptive water.

What is the purpose of this inventory? Jesse L. Lewis of FWQA's technical data information branch, explains that the inventory is intended to reveal whether discharges from industrial plants end up in open waters, public sewers, deep wells, or ground reservoirs (where the waste is stored and does not reach the waterways). Besides the destination of the effluent, according to Lewis, this study will show the liquid's chemical, physical, and biological properties. Incidental information obtained will include the exact location of the discharge and the identification of the plant. The survey is also expected to provide economic data for use by FWQA's manpower training and grant programs in determining the requirements for a water cleanup program. Adds FWQA's George Wirth, "In a broader sense, before we can do anything as far as planning a pollution cleanup program, we have got to know something about the cause and effect relationship of pollution." This inventory will survey rather than solve the problem. Obviously, the polluted state of a body of water can be determined by water quality monitoring or by observing waters' inability to support desired uses. Before the problem can be attacked, however, the relation back

## What FWQA is looking for: Water quality characteristics

**Physical parameters**

flow
pH
temperature
color
specific conductance

**Chemical parameters**

total solids
total volatile solids
total suspended solids
total dissolved solids
acidity
alkalinity
5-day BOD
COD
oil and grease
chloride
sulfate
sulfide
phenols
$NH_3$ (as N)
$NO_3$ (as N)
$NO_2$ (as N)
organic N
Kjeldahl
ortho-phosphorus
total phosphorus
Cr, Cu, Fe, Cd, Pb, Mn, Zn, F, As, Hg

to the cause must be traced—whether industrial, municipal, or other kinds of discharges are going into a stream. "We feel that it is absolutely essential to have this information to provide a cause and effect model which will be the groundwork for planning an attack on pollution," continues Wirth.

### Past efforts

The idea for this inventory project has been around for over seven years. In 1963, the House of Representatives' Subcommittee on Natural Resources and Power, chaired by Rep. Robert E. Jones (D.-Ala.), recommended that such an inventory be undertaken. In 1964, an effort was made to initiate the survey, but the project never got off the ground. A main reason for this was that industry objected. Another was that the Bureau of the Budget (now Office of Management and Budget, OMB) would approve the inventory only on a test basis, meaning that the data had to be taken in only one river basin area and that data obtained would be confidential. Under these conditions, the Public Health Service (PHS), which at that time had all water pollution control functions, did not concur, and the project was never executed.

In 1967, the project was submitted again to OMB. But budget officials again deferred action until two other studies affecting industry were completed. (One study was the feasibility of offering industry incentives to initiate their own abatement actions, and the other was to determine the cost of clean water.) In 1968, Jones' subcommittee issued a report on the project, and it was submitted again. However, industry strongly hinted that the inventory would result in little response unless the data obtained was strictly confidential. OMB deferred action pending resolution of the confidentiality question.

In May 1970, the necessity for conducting the inventory through direct questionnaire to firms was re-evaluated. At this time, the House of Representatives' Subcommittee on Conservation and Natural Resources, successor to Jones' committee and chaired by Rep. Henry D. Reuss (D.-Wis.), pushed for hearings on the inventories. In September, the project was once again submitted for approval. OMB approved a form for use in a test mailing to a representative sample (250 plants) of U.S. manufacturing plants that use water. These forms were mailed in October, and FWQA is currently processing replies.

## Procedure

The Industrial Waste Inventory zeroes in on the major water-using industries in the U.S. Categories include blast furnace and basic steel products, motor vehicles and parts, paper mills, textile mill products, petroleum refining, canned and frozen fruits and vegetables, meat and dairy products, and chemicals. An initial goal of 30 days (which ended December 3) was set for

**FWQA's Dominick**
*Requesting industrial cooperation*

industries' replies to FWQA's Commissioner David D. Dominick's forwarding letter and preliminary survey. At the end of this initial period, Dominick sent follow-up letters and questionnaires to the firms that had not answered, again allowing a 30-day response period.

At the beginning of this month, FWQA's John Dell'Omo started an overall analysis of the response. After determining the percentage of plants that have answered, the percentage of these that have given replies to specific questions, and the exactness of the replies, FWQA will again meet with OMB to review the results of this test survey. FWQA officials are hoping that OMB will approve the questionnaire for a mass mailing to over 10,-000 industrial plants, which they will be able to begin in April or May.

The preliminary survey will give an idea of the percentage of firms that are likely to respond to any mass mail-

ing. (At press time, 34.5% of the 250 industries contacted had responded to the questionnaire.) Of the 350,000 manufacturing plants in the U.S., only 10,000 use 90% of all water intended for industrial purposes. It is the intent of FWQA to cover these 10,000 plants regardless of the number of questionnaires mailed.

## Information requested

The data requested are the characteristics of the effluent just before reaching the receiving water—in other words, at the end of the pipe. For the purpose of this survey, FWQA is interested in the quality of the effluent, not the quality of the stream into which the effluent is discharged. Of course, pollution of federal bodies of water is defined as a discharge into a stream that affects the water quality as established by the standards. The effects can always be observed; but without information on the cause, decisions on the approach to nationwide cleanup cannot be made.

The form consists of a two-page questionnaire. The first page has three parts. Part one is mainly for identification purposes and gives an idea of the plant size, including the number of employees, and whether the plant operates on a seasonal or yearly basis. The second part asks for the name and type of the water source as well as the name and type of the body of water that receives the discharge. Part three concerns economic data—existing abatement costs, number of personnel in en-

vironmental control, and a forecast of funds to be spent in the next five years for abatement facilities and personnel.

The second page asks for data on actual discharges—point of discharge, type of discharge, and various water quality indicators of the intake water and the waste water effluent. The water quality indicators are taken before and after treatment of the intake water and before and after treatment of the waste water effluent (see box). However, industry is asked to furnish only information that they have on hand and can provide. This will give an idea of what is being monitored at the present time.

## Industry response

A big question is how industry is responding to the inventory. "We feel that if a government agency needs information about one of our installations, we supply it," says a spokesman for the Du Pont Co. Generally, there is a cooperative attitude prevalent among the large corporations, according to FWQA's Lewis. "These people want to cooperate and fill out the questionnaires." FWQA officials stress that this trial program is strictly voluntary and is not intended to be used against industry. The matter of self-incrimination has been a touchy point in the inventory. But the survey is being taken primarily to give a picture of what is being discharged to the waterways, summarizes Lewis.

Conversely, a spokesman for Reuss' subcommittee stated that there will be strong pressure from the subcommittee for incrimination and prosecution. But, according to another subcommittee aide, this information may merely be used to determine the cost of the pollution cleanup.

## Data application

What will the analysis of the data received from the inventory reveal? The information will define the problem of pollution in the nation's waterways and perhaps provide a basis for steps in a cleanup campaign. FWQA officials say that the main goal is to enhance the water quality improvement program, but without harming the industries that volunteered the information. Commissioner Dominick has stated that the inventory is "essential to the carrying out of the President's environmental improvement program." Time alone will reveal whether the inventory will fulfill this mission or just stir up a hornet's nest.      CEK

# AWTL key link in water quality effort

*Reprinted from* Environ. Sci. Technol., **4**, 809 (October 1970)

*FWQA lab spreads the gospel that the nation has both technology and money to improve water treatment*

Most observers would agree that a basic problem in achieving any progress in waste treatment is updating the technological tools being applied to the job. What type of approach is needed to manage the development and dissemination of this technology? "We're all missionaries," is the closest analogy that occurs to Frank Middleton, director of the Federal Water Quality Administration's (FWQA) Advanced Waste Treatment Laboratory (AWTL) in Cincinnati, Ohio. The analogy is apt, given Middleton's assessment of the gap that exists in water treatment technology. "There's no doubt about a lag in putting into use some basic and increasingly well-developed principles about water treatment. Our job, simply put, is fighting the technology transfer battle."

Indeed, AWTL does carry the brunt of federal effort to support an updating of the nation's water treatment capability. The lab has developed a significant in-house capacity for process studies; at the same time, AWTL is a key link in formulating FWQA policy on support of innovative water treatment methods. AWTL, a major arm of the Robert A. Taft Water Research Center, has its headquarters at the Taft Center in Cincinnati, where it maintains a full-time staff and laboratory facility. In addition, it has established pilot plants at four other locations:

• **Pomona, Calif.** This plant was built in 1963 with the cooperation of the Los Angeles County Sanitation Districts to study the economics of granular carbon adsorption. Several other process studies have since been undertaken here, including investigations of ion exchange, reverse osmosis, mineral addition, and electrodialysis.

• **Lebanon, Ohio.** The main thrust of this plant, built in 1964, has been the development of processes to improve the quality of effluents from conventional treatment by removal of suspended solids, organic materials, and dissolved salts.

• **Washington, D.C.** Newest (1969) of the AWTL's pilot plants to come on stream, the Blue Plains plant was originally planned to study various tertiary treatment processes, but has developed into a multipurpose facility for large-scale economic studies of various unit processes developed at Blue Plains and elsewhere (ES&T, July 1970, page 550).

• **Manassas, Va.** This AWTL field station was set up under a federal contract to study phosphate removal by biological uptake; it also does specific studies on other means of nutrient removal.

## Funding role

Activities at AWTL's pilot plants tell only part of the story. The full impact of AWTL on the development of advanced treatment must take into account its role in managing development grants to outside contractors. For while all policy decisions on support of advanced waste treatment are made in Washington at the FWQA

**AWTL's Middleton**
*Managing new technology*

Commissioner level, such decisions are strongly influenced by the activities of AWTL. Recommendations for priorities from FWQA's regional staffs are coordinated by AWTL, and all grant applications for advanced waste treatment work are funnelled through AWTL for comment. After advanced waste treatment grants are approved they are administered and closely monitored by project officers assigned from AWTL's staff of specialists.

Middleton estimates that 80% of the funds for advanced waste treatment studies goes to outside contractors in the form of contracts, development grants, and demonstration plant construction. AWTL's current budget for in-house studies is about $2.5 million per year, while at the present time it manages about $20 million in outside grants and contracts.

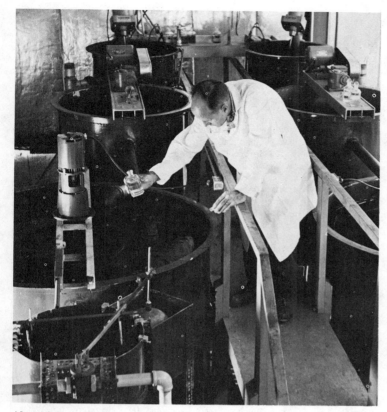

**Absorption.** *Equipment to study carbon absorption process at* AWTL's *Lebanon, Ohio, pilot plant includes two clarifiers in foreground and four slurry tanks*

### Ripe for adoption

Middleton singles out some areas that are high on the list for rapid development and wide-scale adoption. Included are such nonbiological treatment processes as chemical treatment and activated carbon adsorption, although several aspects of conventional processes come under AWTL's concept of advanced treatment. Pure oxygen aeration is one such process that looks promising, according to Middleton, and improved methods of sludge handling and disposal are other areas where he sees a need for more sophisticated, economical methods.

Jesse Cohen, head of AWTL's physical and chemical treatment research programs, outlines the development of AWTL's philosophy in this way: "For many years, the major approach to sewage treatment has been biological, with advanced treatment thought of in terms of a tertiary, or add-on, approach. More recently, especially in the last two or three years, it's been shown that physical and chemical methods—adsorption, membrane processes, and ion exchange—can stand on their own as treatment processes."

But Cohen, and Middleton, too, caution against the expectation of any panacea in the way of a new treatment process that is likely to revolutionize the field and dominate it. Cohen illustrates the point by bringing up the carbon adsorption process now being installed at Rocky River, Ohio, and soon to be installed in Painesville, Ohio. "We won't have a firm assessment of the full potential of this process until after a few years of plant operation. And even then, if the process proves cost competitive, there may be limitations which make it competitive in other situations." Thus, the guiding spirit of AWTL work is the development of an array of feasible processes, from which can be selected those particularly suited to a given situation and set of standards.

As to the effect of advanced treatment on the cost to the taxpayer, Middleton is quite sanguine. He points out that full implementation of many advanced treatment concepts could push treatment costs up to 25 to 30 cents per thousand gallons—about twice the average costs for conventional plants. "But if one makes the usual comparison with what we all pay for electricity, phone bills, and so on, there's no question but that we can afford advanced treatment on a per capita basis." PJP

The $20-odd million funded through AWTL is small compared with the $1 billion in federal funds now authorized for assistance to municipalities for treatment plant construction. But the specific types of grants that AWTL handles are the only ones that have any hope of turning the tide in the technological battle that Middleton alludes to: "As a general rule, local treatment officials, and the consultants they rely on for technical advice, are understandably hesitant about adopting new processes. After all, no one likes to take the risks inherent in pioneering with public funds!" In such a situation, the lure of an advanced treatment demonstration grant may be considerable. Straight construction grants for conventional processes carry a top limit for federal funding of 50%, while advanced treatment grants can go as high as 75%.

Research contracts and demonstration grants are not AWTL's only activities. The laboratory runs a series of seminars for treatment plant engineers and consultants to update them on advanced practices—over a dozen have been conducted so far. AWTL is also planning a series of design manuals of advanced treatment concepts; the initial series of four, now being compiled, include activated carbon adsorption, phosphorus removal by chemical treatment, suspended solids removal, and upgrading the efficiency and capacity of existing treatment facilities.

Aside from AWTL's activities, Middleton feels that the greatest stimulation for adoption of advanced waste treatment practices will be standards enforcement. "There's no doubt that standards have been pushing technology," he says. "Although in most cases, treatment officials have not been told enough about standards, there are no long-term impossible blocks, other than deciding just how to do the job most economically."

# Public advisory group hears water pollution needs around U. S.

*Reprinted from* ENVIRON. SCI. TECHNOL., 4, 471 (June 1970)

The President's Water Pollution Control Advisory Board is a group of nine presidential appointees, each with different professional backgrounds and geographical locations. The board advises the Secretary of the Interior who, in turn, ultimately advises the President on approaches to protect and enhance this nation's water environment. This advisory board meets to review pressing water pollution problems in trouble spots throughout the U.S., usually at the invitation of a governor or the Secretary of the Interior.

Its most recent meeting, actually its 49th since formation of the board in 1956 under authority of P.L. 84–660, was held at the request of board members. The problem concerned pollution of the lower Mississippi River and oil spills—more specifically, Chevron's spill in the Gulf of Mexico. Following a two-day meeting in New Orleans, the board met in the Virgin Islands.

As a result of these meetings, the advisory board passes resolutions and refers them to the higher-ups for ultimate disposition. Obviously, the board has passed many resolutions, some of which have been heeded by the higher-ups and some of which have not. However, the board's batting average seems to have improved over the past year. More importantly, the board serves as an important information source for decision-makers in Interior.

## Explanation

The board's earlier meeting (January 12–14 in Houston, Tex.) illustrates the type of resolutions passed and ultimate disposition of these items. For this meeting, Texas Governor Preston Smith invited the board to comment and review the water pollution problems associated with the Houston Ship Channel and Galveston Bay, a recreational as well as an industrial area. Its January meeting was not the first held in this area on Texas' problems. In 1967, the board considered the problems but failed to implement solutions, hence the revisit.

Following its second visit, the board passed four resolutions; the first is now a *fait accompli;* the second has been incorporated into pending legislation; and the third and fourth items necessitate more study before it can be said that they have been successfully implemented (see scorecard).

Perhaps, other experiences from the Texas meeting also illustrate the functions and operation of the board. "Our findings in Texas revealed that the city of Galveston was polluting its own beach areas by dumping raw sewage into Galveston Bay," says Gordon E. Kerr, executive secretary of the advisory board. At this meeting, chairman Carl L. Klein, Interior's assistant secretary for water quality and research, asked the Texas Water Quality Board (WQB) to take immediate corrective action, by upgrading treatment facilities. "Two weeks after the meeting, corrective action was initiated, continues today, and will be carried forth till complete upgrading is completed," says Gordon Fulcher, the Texas WQB chairman. With a new chairman and emphasis on getting permit holders in compliance, the Texas board—the one and only board responsible for water pollution control within the state—stands ready to make a renewed effort to curb Texan water pollution problems. In fact, the effort has been proceeding since last fall when Fulcher was elected chairman of the Texas board.

Of course, the 1970 problems in Texas are different from those when the board last met there in 1967. Then, no single responsible agency in the Texas legislature could be recognized, or properly and solely charged with the responsibility for water pollution abatement in the state. This is not to say the Texas Water Board was not operating in 1967. But one responsible agency was not properly identified so that the public, industry, and local governments knew to which office they should turn.

Things are different today. Now, the WQB has been singled out and is accountable and directly responsible to the public for water pollution abatement. After the January meeting, the

---

Carl L. Klein

"Under section nine of the Federal Water Pollution Control Act, the board advises, consults with, and makes recommendations to Interior Secretary Walter J. Hickel. I have been meeting with the board in the secretary's stead.

"The board obtains input from as many sources as possible and senses the pulse beat of future water pollution abatement developments. At times, the board has been very critical, but now the board has been coming up with new thoughts and ideas. As a result, the Department of Interior is obtaining a better input on the ways to proceed on water pollution abatement.

"The board can function well. If the individual members function well, then the board functions properly."

### Members of President's Water Pollution Control Advisory Board

Carl L. Klein, Chairman, Interior's Assistant Secretary for Water Quality and Research
Gordon E. Kerr, Executive Secretary
Robert H. Finch, *Ex-Officio* member, Secretary of Health, Education, and Welfare

**Term expires June 30, 1970**

W. James Lopp
  Security analyst

Edward P. Morgan
  Television newscaster

Stephen E. Reynolds
  Engineer

**Term expires June 30, 1971**

Louis S. Clapper
  Conservationist

Ralph W. Kittle
  Businessman

Stuart M. Long
  Newspaperman

**Term expires June 30, 1972**

Melbourne R. Carriker
  Oceanographer

Wallace W. Harvey
  Physician

Parker E. Miller
  Land developer

# Scorecard of advisory board meetings in 1970

| (Problem area, dates, and location) | Resolutions passed by board | Subsequent action by higher-ups |
|---|---|---|
| **Ocean disposal**<br>**May 1 to 2**<br>**Virgin Islands** | • Too early for final comment. | • Too early for implementation. |
| **Oil spills**<br>**April 29 to 30**<br>**New Orleans, La.** | • Too early for final comment. | • Too early for implementation. |
| **Lake eutrophication**<br>**March 2 to 3**<br>**Washington, D. C.** | • Urged board's attendance at the recent FWPCA workshop as observers.<br>• Commended the President on developing and initiating a broad attack on contamination of the environment.<br>• Urged the Secretary of Interior to support legislation which will prohibit the disposal of polluted dredging in the Great Lakes either by public or private agencies and individuals.<br>• Urged FWPCA to investigate the current water pollution problem in the Virgin Islands. | • Included in Nixon's state of the environment message.<br>• Hickel's notice to Reserve Mining Co. (ES&T, April, page 270) and Nixon's message to Congress on April 15 (H. Doc. No. 91-308) prohibiting dumping in Great Lakes dredged spoils.<br>• Board revisited the islands on May 1 to 2 to prod the local government on water pollution abatement action. |
| **Galveston Bay**<br>**January 12–14**<br>**Houston, Tex.** | • Urged spending the $800 million appropriated by Congress to be used immediately.<br>• Urged all waste water being discharged to coastal waters to receive at least secondary treatment.<br>• Urged prohibition of deep well injection of any water that has a chance of getting into aquifers.<br>• Urged spending more money for studies on how to remove viruses from water. Ordinary treatment does not remove them. | • Included in Nixon's state of the environment message of Feb. 10.<br>• Incorporated in proposed legislation which is pending.<br>• Requires more study before any action can be taken.<br>• Requires more study before any action can be taken. |

board and its new chairman, Gordon Fulcher, now look forward to a year of renewed activities.

### Other meetings

Many other advisory board meetings have been held in different locations throughout the U.S. In fact, the board has visited some on more than one occasion. Comments on much earlier meetings would perhaps only serve to bring out old skeletons from the federal closet.

It is not to be inferred that all actions resolved by the board are accepted without question. In some cases, the board made resolutions which were not endorsed by the higher-ups.

For example, at a May 1969 meeting in Washington, D.C., the board passed a resolution urging the financing of waste water treatment facilities. Although the resolution was adopted by the board, the resolution was not adopted by Interior's Secretary Hickel. Perhaps, the timing was premature.

### Criticism

Edward P. Morgan—a Johnson appointee, a television newcaster (Washington, D. C.), and lame duck whose term expires the end of this month—says, "The board is a joke. Under existing circumstances in the present administration, the board has become emasculated, useless, and probably should be removed." With the creation of the Council on Environmental Quality (CEQ), no one knows where this public advisory board fits in with other environmental activities.

"Present assistant secretary Klein is just not with it in the terms of an aggressive policy in terms of solving water pollution problems," Morgan says. "I speak for myself, but I also know that some other members of the board are distressed that the assistant secretary shows no more aggressiveness than bold talk.

"During 1969, the board was dormant, for all intents and purposes nonexistent," the newcaster reports, "We had difficulty trying to set up an executive meeting during the Houston trip (January 1970) to find what the board was being asked to do. Other than listening to what other board members themselves had decided to do, the new members of the board were left completely in the dark on purposes and intent of their involvement.

"The main distinction between administrations is that however ineffective the board may have been under the Johnson administration, at least I personally had the impression that we were being listened to," Morgan says. "Whether the board's resolutions were being accepted or not, I cannot say.

"Since then, I have had the impression and experience that the board is barely being tolerated and furthermore, that we were simply wasting the taxpayer's money, the modesty of the board's budget notwithstanding."

But all is not bleak. The board's public hearings do have a definite positive effect in the local areas. "They stir up interest which gets a flurry of attention in the media," Morgan says. "What lasting effect they have, time will tell."

The one big question still remains. "What are the function, position, and reason for being for this public advisory board in light of the newly created CEQ?" Morgan queries. So far, no clarification has been forthcoming from the higher-ups.

# Water standards near approval

*But continuing federal review is a must*

*Reprinted from* ENVIRON. SCI. TECHNOL., **4**, 199 (March 1970)

Last October, David D. Dominick, commissioner of the Federal Water Pollution Control Administration (FWPCA), stated publicly that he hoped to have all exceptions in water quality standards resolved by the first of January. By mid-January, half of the standards (27) are approved without exceptions, the other half approved with certain exceptions. Standards with approved nondegradation statements number 26—22 states, District of Columbia, and three territories—up from 20 in the last scorecard (ES&T, February 1969, page 120). Of the 27 states with exceptions, 26 have problems in applying recommended dissolved oxygen criteria, and 24 have problems with recommended temperature criteria.

"Statistics alone do not present a complete picture; some major issues remain," says Richard P. Nalesnik, director of FWPCA's water quality standards program.

## In the red

For most states, the federal indicators—red, yellow, or green markings on a standards chart in Nalesnik's office—show that continued federal attention probably will lead to a favorable conclusion. Another 10 sets of standards —Delaware, District of Columbia, Hawaii, Illinois, Kansas, Louisiana, Missouri, New Hampshire, Ohio, and Pennsylvania—for one or more counts are in the red, meaning that federal persistence definitely will be required to resolve exceptions.

Perhaps, Missouri causes the biggest headache, because it is a problem on several counts. Missouri's compliance date of 1982 turns out to be the latest specified in any standard, although five states—California, Illinois, Iowa, Kansas, and Kentucky—have failed to specify compliance dates in their standards. Another problem is the fact that Missouri is upstream, and its downstream neighbors (Arkansas and Louisiana) already have tighter standards. Arkansas has its standard completely approved, and Louisiana's is approved with a dissolved oxygen exception.

What Delaware, Hawaii, and Pennsylvania have in common is the absence of the nondegradation clause. In official correspondence, each has refused to accept a nondegradation policy, stating that existing law is adequate.

Illinois has hedged on its compliance date, conditioning its action on the resolution of bordering states' issues. There is little reason for waiting to see how downstream problems are resolved. Mississippi has dissolved oxygen and temperature exceptions, and Tennessee has a few exceptions.

With New Hampshire, the major problem is its compliance date of 1974 to meet secondary level of treatment. For the remaining states, Kansas' problems are dissolved oxygen, temperature, bacteria, and the compliance date; District of Columbia has a problem with its level of treatment, and Ohio has problems other than those on the Mahoning River.

## Other problems

Upgrading standards presents other problems. Operating on minimal guidelines, the original 10 state standards—Arkansas, Georgia, Idaho, Indiana, Maryland, Massachusetts, New York, North Dakota, Oregon, and South Dakota—were submitted in June 1967. But not until April 1968 did Interior's National Technical Advisory Committee of 86 water experts meet and finalize its recommended guide on water quality criteria (ES&T, September 1968, page 662). This set of criteria was applied uniformly on subsequent state submissions before Interior's approval was granted. In official correspondence, Commissioner Dominick has asked that these 10 states upgrade

**Richard P. Nalesnik**
*Statistics don't present complete picture*

their standards. But the responses have not been uniform nor favorable in all cases. For example, Georgia's official, R. S. Howard, plans to come to Washington to ask why there is need for a nondegradation clause.

On the other hand, another six standards are completely in the green, meaning they are completely approved standards with no exceptions. These include Arizona, Arkansas, Guam, Minnesota, Utah, and the Virgin Islands. "There also is a good possibility that Virginia and West Virginia soon will come forth with a nondegradation statement, bringing that total to 28," Nalesnik says.

FWPCA's transfer of responsibility for water quality standards to its new office of enforcement and standards compliance (ES&T, February 1970, page 91) reflects intensification of federal attention to compliance and implementation.

By the end of January, 49 enforcement conferences have been held nationwide. Each conference produces an updated status report every six months. So, in this regard, the standards program provides vital indicators for locating the action and seeing the progress or lack of it.

# National group looks at water needs and problems

*Study commission listens to the public's concerns*

Reprinted from ENVIRON. SCI. TECHNOL., **4**, 202 (March 1970)

The National Water Commission (NWC), a seven member body selected by the President and not otherwise employed by the federal government, has completed its first of five years of activity. Operating under the legal authority of P.L. 90-515, the NWC has immersed itself in a flood of the public's concern over this nation's future water needs and resources. By 1973, NWC must produce a report to the President and Congress on its findings.

"Water commissions are not something new to the federal establishment," says Theodore M. Schad, NWC's executive director. "Over the last 60 years, more than 20 such commissions have been in existence. The first water resources study commission dates back to the time of George Washington. Then, a governmental commission studied the Potomac River, concluded that no solution to the problems could be found, and recommended further study."

In its first year, NWC met with all of the major federal agencies having responsibilities in the water field, and with the Water Resources Council (WRC). The similar responsibilities of the two agencies suggest the possibility of some confusion and overlap. WRC is composed of the secretaries of the various federal departments with water resources interest—Agriculture; Army; Health, Education, and Welfare; Interior—and chairman of the Federal Power Commission; the secretaries of Commerce and Housing and Urban Development are associate members. So, NWC is to the Water Resources Council what Rockefeller's Citizens Advisory Group is to Nixon's Environmental Quality Council (ES&T, December 1969, page 1249).

One difference between NWC and WRC is that the citizens' group is not required to carry out governmental programs. Thus, NWC can work independently of the large body of federal water resources laws and recommend changes in them.

## Activities

In its first year of operation (late '68-December '69), NWC operated at a budget level of a mere $105,000. Now, NWC has $1.05 million for operations in fiscal year 1970. Its operations include the work of the seven presidential appointees and a full-time professional staff which is expected to reach 35-40. This full-time staff includes three groups, one for engineering and environmental sciences, one for social and behavioral sciences, and the office of the legal counsel.

NWC also set up *ad hoc* panels to advise on ecology and economics, and plans another on behavioral sciences in the hope of bringing the contributions of the behavioral sciences into consideration in the formulation of national water policy. The commission has several studies underway, and more are planned. For example, Resources for the Future, a Washington based organization, is studying ways to forecast the effects of technological developments upon water demands and problems. NWC also is funding a National Academy of Sciences panel to supplement this study by forecasting new technological developments. Other NWC studies now in progress include advanced methods of water pollution abatement, weather modification, and desalting of seawater.

## Public's view

In general, the public rarely grasps the activities of such commissions.

Certainly, it is understandable, since the activities of this commission are nationwide and long-ranged, involving federal as well as other citizens' organizations.

What things are the public saying about present water policies and programs? "Many," says NWC's chairman Charles F. Luce, who is also chairman and chief executive officer of Consolidated Edison Co. of New York. After hearings throughout the U.S. late last year (ES&T, August 1969, page 697) Luce suggested that NWC must consider:

• Means for improving population distribution.

• Replacement of present institutions with tailor-made organizations to meet today's water requirements, since present institutions to develop, use, and conserve these resources are outworn.

• Development of criteria to guide decision-making on proposals for transferring water between basins and between nations.

## Recommendations

The time lag between recommendations and their implementation sometimes is great indeed. "Early recommendations with regard to comprehensive river basin planning go back to the National Waterways Commission which was set up by President Theodore Roosevelt in 1908," Schad explains. "They were reiterated many times and partially implemented in 1927. But it wasn't until July 1965, when the Water Resources Planning Act provided for joint federal-state institutions to accomplish such basin planning, that the idea could be said to have reached fruition." Schad hopes NWC's recommendations will be implemented more quickly. "Obviously, the time is premature for me to suggest areas in which NWC will make its recommendations," Schad concludes.

**Presidential appointees.** *New commission members are (l. to r.) H. Appling, S. S. Baxter, R. C. Ernst, C. F. Luce, J. Wheat, R. K. Linsley, and C. T. Ellis*

# British water pollution control

Regional boards, joint treatment,
and unique inspectorate system are
key features of U.K. program

*Reprinted from* ENVIRON. SCI. TECHNOL., 4, 204 (March 1970)

At a time when most industrial nations are seeking to control the pollution of inland waters by stricter laws or new legislation, Britain can claim that many of its rivers are now in a better condition than they were 20 years ago. Fish are returning to rivers that, not long ago, were devoid of such life; rivers once thought suitable only for use as industrial sewers are showing signs of improvement; and even rivers bearing a high proportion of sewage and industrial effluent now can be regarded as actual or potential sources of public water supply. These results have been achieved by the use of many types of purification plants in order to produce effluents to render them fit for discharge—mostly to public sewers, but also directly to rivers.

It is doubtful whether this improved situation could have been developed during a period when two world wars halted expenditure on waste water

**Samuel H. Jenkins**

*Executive editor*, Water Research
*Oxford, England*

treatment, and at a time of rising population and mounting water consumption by industry, had it not been for a pollution control policy that took the form of a series of parliamentary acts, and the creation of new types of pollution control agencies. An understanding of this legislation and the functions of the agencies is necessary to appreciate why local authorities and manufacturers are pursuing a vigorous policy of waste water treatment.

### Initial legislation

After nearly a century of river pollution, caused to a large extent by industrial effluents, it was realized that such effluents were best dealt with by treatment in admixture with domestic sewage. In 1937, the Public Health (Drainage of Trade Premises) Act gave manufacturers in England and Wales the right to make use of public sewers for this purpose. The municipality owning the sewer was given the power to set conditions under which the discharge could take place, with the proviso that disputes were to be settled by the appropriate government minister. This act brought immediate benefits, but its full effect was not realized until after World War II.

In the River Board Act of 1948, the 1600 or so public authorities in England and Wales that inherited the duty of preventing pollution were replaced by 37 River Boards. These organizations consisted mostly of elected representatives nominated by local authorities in the area of each board. The catchment area of a river or group of rivers was mainly the area of administration, and the purpose of the boards was to implement existing laws on pollution prevention.

The Rivers (Prevention of Pollution) Act (1951) extended the powers of the boards to require them to control all new discharges to rivers and altered outlets. Although boards were given discretionary powers in the standards of quality that could be set, generally, they required effluent standards with a maximum biochemical oxygen demand (BOD) of 20 mg./l. and a suspended solids limit of 30 mg./l. The river authorities have been advised that, whenever they request higher standards, they should show that circumstances justify these demands. Standards not exceeding a BOD of 10 mg./l., suspended solids of 10 mg./l., and ammonia nitrogen of 10 mg./l. are becoming common.

The legal interpretation of a consent to discharge is that compliance with standards must be observed at all times, even when the river authority takes a spot sample without giving notice of its intention. Thus, the average quality of the effluent obtained must always be much higher than the quality permitted in the consent. This differs from the policy adopted by many pollution control agencies where such control policy is based on the percentage of the organic impurity that must be removed by waste water treatment. It may be useful to remember that, in order to attain the BOD and suspended solids standards of 20 and 30 mg./l., local authorities have to remove approximately 96% of the impurity in the sewage they treat.

### Recent requirement

A recent requirement states that the ammonia nitrogen in sewage effluents must not exceed 10 mg./l. This means that a treatment plant must be designed to give complete purification to three times the dry weather sewage flow, and that 60-70% of the ammonia nitrogen must be oxidized at all times. Tanks with short retention periods, even with intense aeration, cannot meet these requirements because the nitrifying bacteria require a longer time to grow—or to oxidize the ammonia—than is provided in tanks of less than two to four hours capacity (calculated on the sewage flow, under ideal conditions of aeration, temperature, and sludge concentration). Consequently, lightly loaded trickling filters are regarded favorably by river authorities seeking to achieve well-oxidized, nitrified effluents that will make the minimum oxygen demand on a river.

Nitrified effluents have been obtained by combination treatment, such as a high-rate activated sludge plant, followed by distribution of the effluent on to a nitrifying trickling filter containing a small grade medium. River control agencies regard nitrification as desirable because of the high oxygen demand of ammonia nitrogen, the toxicity of $NH_4$ to fish under certain circumstances, the reduction in chlorine demand, and the oxygen reserves present in nitrates.

The 1951 Act gave river authorities control over discharges made after 1951, but an Act of 1961 extended this control to pre-1951 discharges. It was administratively impossible for the authorities to deal with the tens of thousands of such discharges within some of the areas, so no time limit was set within which pre-1951 discharges had to be handled.

### Positive approach

To many individuals and organizations, the term "pollution control" or "prevention" strikes a negative note. A more positive attitude toward the subject of water management has been adopted in the Water Resources Act of 1963 for England and Wales. Under this act, newly created river authorities (in place of the river boards) became responsible for water conservation by controlling the pollution of surface and underground waters. By charging a license fee for permission to abstract such water, and an additional sum graded according to the quality of water and its volume and purpose stated in the license, the river authorities are expected to become self-supporting.

Only by detailed control and continuous pressure on discharges to comply with legal requirements given in the consent to discharge, can improvements be made in the quality of river water. It is important to ensure that industrial wastes received by public sewers do not prevent treatment of sewage to whatever standard is required by the river authorities. The fact that the effluents from municipal sewage treatment plants must meet the requirements of the river authorities, and the law that requires municipalities to receive industrial waste, has compelled local authorities to exercise their powers to control the discharge of these wastes to public sewers.

Such control takes several forms. Usually, the total volume discharged in any one day, the rate of discharge, and the composition are controlled. This may call for some pretreatment of the industrial waste before it is allowed to enter the sewer. The simplest form of pretreatment needed may be a balancing tank for equalizing the rate of discharge or its composition. More usual requirements are the removal of excessive amounts of suspended matter or screenable material from abattoir wastes, or vegetable fiber, paperpulp, rags, etc. The removal of oil, down to limits of 10 mg./l., necessitates the use of oil separation tanks, usually after the addition of flocculants at a controlled pH.

### Metal finishing wastes

The metal finishing industry has recognized that treatment of mixtures of effluents containing acids, alkalis, cyanides, and compounds of cadmium, chromium, copper, nickel, and zinc in high dilution is needlessly expensive. The pretreatment tanks required are very large and the cost of the chemicals needed is prohibitive. Much thought has been applied to the development of metal finishing plants that:

• Make economical use of water.

• Minimize the production of unusable effluent.

• Segregate effluents into those that may be treated separately and those that may be discharged without pretreatment.

• Permit the use of plastic pipes carried on trays mounted on the walls for the conveyance of effluents, instead of underground drains that are expensive to lay or replace.

In one plant, overground drainage has saved an estimated $60,000 in effluent treatment plant construction costs. Some three miles of plastic pipework is used at the plant—a Rolls-Royce engine factory in England. The strong wastes are diluted automatically with sufficient weak swill water to allow continuous and automatic treatment; cyanides are treated at a pH of more than 10 with chlorine gas; chromates are treated at pH 2-3 with sulfur dioxide gas; then the two treated liquors mix and are neutralized automatically, either with milk of lime or

**Nottingham.** *Throughput capacity of Stoke Bardolph works now is being doubled*

sulfuric acid. Finally, the effluent is settled for four hours in an upward-flow tank of 48,000 gallons in order to settle out the precipitated metal hydroxides. Only acid effluents which are neutralized may bypass the settling tank. All the sludge is dewatered by filtration through rotary vacuum filters.

Other new ideas on pretreatment that have been put into practice recently at a Rolls-Royce aircraft engine factory in Scotland involve the treatment of spent plating and similar liquors that are so concentrated that the precipitates formed are sufficiently thickened to enable settlement tanks to be bypassed. Plants of this kind are designed with the intention of using water sparingly, partly because a high degree of purification is necessary to meet the standards required to discharge the effluent to public sewers, partly because of the mounting costs of water, and also because payment has to be made for discharges of industrial waste water to sewers.

### Industrial reuse

Water reuse is practiced on an increasing scale in industry. Cooling water either is recycled through lagoons or tanks, or through evaporative coolers. In the steel industry—where,

for example, 180 tons of water are recirculated for every ton of steel produced—the net water consumption is only five tons per ton of steel. In one plant, 25% of the fresh water intake becomes sewage effluent.

Another example of the pretreatment necessary to enable a relatively small sewage works to accept the effluent from a large food factory is the segregation of effluents into strong and weak liquors and the introduction of rigid control measures to reduce water consumption. Suspended matter is removed from strong process liquors by centrifuging and vacuum filtration; all the strong liquors are concentrated in multiple effect evaporators; and solids and evaporated liquors are incinerated in a factory boiler. In fact, incineration is becoming recognized as an ideal solution to the disposal of strong organic wastes of small volume.

In order to ensure that all manufacturers comply with the legal consent they receive from the municipality, local authorities in Britain maintain an inspectorate service. Trained and qualified inspectors have powers of entry to trade premises for the purpose of taking samples of effluent passing to the public sewer. The manufac-

**Wycombe.** *Banks of filter presses at Little Marlow plant are fully mechanized*

turer must provide a suitable sampling chamber for this purpose.

On the basis of the sample analysis, the local authority may calculate the charge to be made for the treatment of the effluent and also determine whether it is of acceptable quality. If it is not up to standard, the manufacturer may have to install a pretreatment plant. When cooperation and/or persuasion fail to bring about improvements, the local authority can take legal action on the basis of a legal sampling procedure. These samples are divided into three parts in the presence of a responsible official of the manufacturer. One portion is to be analyzed, one handed to the manufacturer, and one retained for examination by the court in case the result of analysis is disputed.

The procedure before a magistrates court is simple and effective, since a fine may be imposed for each day on which the offense (failure to comply with the consent condition) is committed. There is little room for procrastination or defense, since the action taken rests only on proving that the effluent comes from the manufacturer's premises and that its composition shows that it is outside the limits of the consent.

Using the consent procedure, it has been practicable, over a period of years, to control the discharge of effluents from 2000 premises in an area with a high concentration of metal finishing plants. About 400 pretreatment plants have been installed for the removal of oils, fats, screenings, suspended solids, alkalis, acids, solvents, nonferrous metals, iron salts, entrails, and organic substances. Without such control measures, it would have proved extremely difficult—perhaps impossible because of the expense—to purify sewage containing 60% of industrial wastes to the standards required by the river authority responsible for water management.

**National policy**

The knowledge that local authorities are prepared to accept industrial effluents allows the river authorities to insist upon manufacturers complying with stringent standards if they wish to discharge to a river. National policy also requires that local authorities make adequate provision for the conveyance and treatment of all industrial effluent in the district at their sewage works when extensions are being planned. If the

scheme does not make adequate provision for such effluents, the omission usually draws attention at the public inquiry held by the appropriate government department to sanction financing for the scheme. Official action even may encourage joint schemes for the treatment of sewage and industrial wastes.

The latest development in encouraging joint treatment of industrial effluents at municipal plants is to ensure that local industry is officially represented at the public inquiries. Industry then has an opportunity of requesting, in public, that adequate provision is made for treating all the industrial waste water in the district.

Industry is required to pay for the treatment of its effluents by the local authority. The general method is a sewage or conveyance charge based upon the volume of the effluent. This principle also is used to assess the charge for sedimentation or for treatment in a plant which is designed on hydraulic considerations. A charge is made for biological oxidation and sludge treatment, depending on the proportion of the plant occupied by the industrial effluent in purifying the mixture of sewage and effluent. The charge may be made on a volumetric

**Scotland.** *Activated sludge unit being built at Kelso*      **Birmingham.** *Aeration tank at Coleshill works*

basis and inclusive of loan repayment and operating costs, or the manufacturer may make an outright capital payment and thereafter pay the operating costs. These charges may be varied from time to time to account for rising costs or charges on new capital expenditures.

### Sewage treatment plants

The emphasis on effluents with a BOD of less than 20 mg./l., and additional standards for metals, cyanides, phenols, and oils, explains why so many plants in Britain appear to be underloaded by comparison with loadings common in the U.S. A further reason for the generous capacity is the insistence that the maximum proportion of storm water also must be purified in order to minimize pollution.

For the past 50 years, trunk sewers in Britain have been designed to convey six times the dry weather flow (d.w.f.). At the sewage treatment plant, screens and grit tanks are designed to handle this flow, and sedimentation tanks, oxidation plant, and secondary tanks are designed to give complete purification to 3 d.w.f. The remaining 3 d.w.f spills over into tanks, usually of at least six hours' d.w.f. capacity, with facilities for sludge removal and for pumping back

for full treatment the aqueous contents as soon as capacity is available to receive it. Overflows of settled storm water sewage to the watercourse are not permitted until full treatment is being given to 3 d.w.f.

In small- or medium-sized works for populations of up to 20,000, the usual method of sewage treatment is sedimentation—biological oxidation in trickling filters, followed by sludge drying on open-air beds with or without sludge digestion. Innovations in sedimentation are in the direction of simpler design and cost saving by the use of tanks with a floor slope of 7.5°, and a single link chain dragged around the floor by the operation of a centrally located electric motor. Sludge with 10% dry solid matter can be moved in this way.

Efficient operation of trickling filters usually is obtained by recirculation or by alternating double filtration, although, if a minimum ammonia nitrogen standard is required, it may be necessary to restrict both the hydraulic and the organic loading. The use of power driven distributors, controlled at a relatively slow rate of revolution, is contrary to the American practice of rapidly moving distributors. But large-scale experiments over many years have proved that excessive surface growth on a filter

can seriously lower its efficiency, and that the way to control such growth is to restrict the rate of travel of the distributor, usually to within 5-20 minutes per revolution.

### Plastic filter media

Plastic filter media of various makes are finding increasing application, mainly for the treatment of industrial wastes. In overloaded sewage treatment plants, such filter media have the advantage of being able to remove substantial amounts of organic matter from poorly settled or even screened sewage. Considerable experimental work is being done on the use of plastics media.

Another simplification in filter design is in the construction of filter walls. For years it was maintained that filters had to be built above ground level so that air could pass upwards from basal ports. Because of this, few filters survived their useful span of life without requiring major repairs to walls, due to the inability of the walls to contract to their original length in winter. Experience in recently constructed plants has shown that, if the base of the filter wall is never waterlogged and the effluent flows freely, the walls can be submerged completely or covered by an earth bank. Waterlogging of

**Samuel H. Jenkins** *is executive editor of* Water Research, *and consultant in water pollution control. Previously, he was president, Institute of Water Pollution Control. He received his B.Sc. and M.Sc. from the University of Manchester (England), and continued studies at London University. Jenkins started his career in the textile industry, spent time in beet sugar and mill processing research, and joined the Birmingham, Tame, and Rea District Drainage Board in 1938.*

filters to control flies is quite unnecessary, since this can be done effectively with DDT or benezene hexachloride.

Trickling filters and secondary tanks cannot always be relied upon to reduce suspended solids to the limit set by a river authority. Since these solids may account for 25-50% of the BOD of an effluent, tertiary treatment to remove them is becoming common. Treatment on specially laid grassland is effective at dosage rates of up to 300,000 gallons per acre per day. Where lack of space makes such methods impossible, treatment in gravity sand filters or microstainers is practiced.

Another, less expensive method is to cause the secondary tank effluent to flow upwards through a 9-12 inch filter of coarse gravel laid on a perforated metal grid, with a covering of finer gravel. Almost all of the upwards flow filter is submerged in the effluent, so that the device may be located in the tank. Much greater attention will be given in the future to the removal of suspended matter from effluents, because recent work has shown that such matter makes a greater demand on the oxygen reserves of a river than was formerly thought to be the case.

In Britain, the activated sludge process was reserved mostly for treatment of large-scale sewage until the introduction of reliable mechanical aerators effective in the small plants. Diffused air, on the whole, is used for large plants, but mechanical aerators are seen in plants treating up to 72 million gallons per day. Air generally is diffused through vitrified domes attached to pipes laid on the aeration tank floor.

Plant-scale experiment has led to the development of a simple activated sludge process. The maximum degree of mixing and the oxygen demand is met at all times by means of a single pass, 60 foot aeration tank, with a submerged inlet the full width of the tank. The air supply is 3.1 cubic feet per gallon, one half of which is provided in the first one third of the tank. One plant of this type is partly in operation for the treatment of 20 million gallons per day and another, under construction, will treat 48 million gallons per day. Both plants are designed to achieve at least 75% nitrification of the ammonia nitrogen.

For the treatment of sewage from small communities, considerable use is being made of extended aeration systems. These include oxidation (Pasveer) ditches, often with a sludge separation tank to allow continuous aeration. The same standards required by river authorities of large plants apply to such small plants, and, consequently, an additional stage of treatment may be necessary to remove excess, well oxidized, suspended solid matter sometimes present in effluents from extended aeration plants. Irrigation over grassland is the preferred method of treatment for this purpose, due to its reliability and simplicity.

Because of the large capital cost of new sewage treatment works, there is constant questioning of cost-benefit relationships, and consideration of alternatives to some of the traditional unit processes that are employed. At present, biological treatment, in comparison with complete treatment by flocculation, ion exchange, sand filtration, and reverse osmosis, remains the cheapest method of removing organic matter in the low and variable concentration in which it occurs in sewage. Thus, the most suitable application for the newer processes is to allow a further stage of purification than can be achieved by the existing unit processes.

### Sludge disposal

Sludge disposal is an acute problem in Britain. With scarcity of land and objections to exposed sludge lagoons, the search for alternate means of sludge disposal has become urgent. A succession of wet summers in Britain has accelerated the move toward mechanical sludge dewatering. Vacuum filtration is on the increase, especially in medium-sized works, and great interest also is being shown in the use of plate pressure filters of new design, mainly because they are compact and produce a thick sludgecake with less than 70% moisture.

Sludge pressing without any coagulant is successful if the sludge first is cooked at a fairly high temperature and put under pressure for a brief period to improve filtration properties. The full potential of this process may be realized when a suitable method has been devised to deal with the organic matter in the sludge liquor.

With growing interest in sludge incineration, there is likely to be a trend toward continuous removal and dewatering of fresh sludge by mechanical methods with the aid of efficient nonbulky flocculants. Such sludge has a high calorific value, and can be incinerated to a slagging ash with only 10% of the volume of the ash in powder form. However, this forecast could be upset easily as the result of some fundamental work being done on the physical properties of sludge.

The present position in Britain can be summarized as one in which rivers are beginning to improve in quality as a result of the extension of sewage treatment facilities. A major improvement is attributable to encouraging the discharge of industrial waste water to municipal sewers. This policy has been possible only by providing municipal authorities with full powers to control all such discharges so that, even when the proportion of industrial effluent is high, the mixture of sewage and industrial effluent still can be treated to the required standards. Cooperation between industry, local authorities, and the river authorities is essential, nevertheless, these authorities still need legal owers to set and enforce their standards.

# Waste control in a fragile environment

*In the wake of a damaging accident, a plant on Canada's rocky eastern coast is trying to demonstrate that industry and environment aren't necessarily incompatible*

Reprinted from ENVIRON. SCI. TECHNOL., **6**, 980 (November 1972)

The term "red herring" has come to mean a specious argument designed to distract attention from real issues. The term originated from the practice of dragging a smoked herring across a game trail to mask scents and confuse hunting dogs.

Environmental issues are often masked by red herrings. But recently, a Canadian chemical company unwittingly created a "red herring" problem with a distinctly environmental meaning. The Electric Reduction Company of Canada, Ltd., (ERCO) put a new elemental phosphorus plant on line at Placentia Bay in Long Harbour, Newfoundland in 1968, and shortly thereafter, numerous fish kills were reported around the bay (ES&T, September 1969, p 811). The fish were herring and they showed a peculiar red discoloration around the gills due to extensive hemorrhaging. Autopsies on some of the fish showed nearly complete disintegration of blood cells. The cause of death was determined to be hemolysis of blood cells arising from contact of the fish with phosphorus plant effluent discharged to the bay. Studies done by the Fisheries Research Board of Canada on cod showed that the fish were extremely sensitive to elemental phosphorus. Static tests proved a concentration of only a few parts per billion of phosphorus in the water was sufficient to be lethal to the fish. In May 1969, ERCO closed the plant down. Other kills had occurred and divers had found vast quantities of dead cod and herring on the bottom of the bay.

After a massive cleanup and public education campaign the plant resumed operation. In more than two years of continuous operation since shutdown, there have been no additional kills attributed to the plant effluent. Placentia Bay is open to fishing—the mainstay of the economy of the villages nestling in the fjord-slashed coast of Newfoundland—and the phosphorus plant and fishermen are once again friends. But the journey to coexistence with a major industry wasn't an easy one.

## What went wrong

The kills in Placentia Bay came as an unpleasant surprise to the management of ERCO's Long Harbour phosphorus plant. When they designed the plant, ERCO engineers were confident that the addition of seawater and dilution by ocean currents in the bay would be more than adequate to prevent any harmful effects of the phosphorus in the effluent. But they overlooked several factors. First of all, currents in the bay were not as effective in dispersing the phosphorus as the engineers thought they would be. Second, the phosphorus tended to settle on the bottom of the bay, just beyond the ocean outfall. Finally, marine life proved to be more sensitive to phosphorus than had been expected and fish migrating through the highly diluted effluents picked up enough phosphorus on the way to kill them.

Toxicity studies were scarce. A literature search turned up only one paper which dealt with the toxicity of elemental phosphorus, and this with freshwater fish only. Kills occurred, therefore, in portions of Placentia Bay which were not actually polluted by ERCO's effluent. Indeed, only a small portion of Long Harbour on the bay was ever polluted with elemental phosphorus. Lobster—although apparently not as sensitive to phosphorus as the fish—were not affected, since they didn't migrate into the vicinity of the effluent.

ERCO makes elemental phosphorus at its Long Harbour plant by electric reduction of fluorapatite ore shipped in from Florida. The ore arrives at the plant in fine granular form and is then crushed to a powder and pressed into pellets. Pellets, coke, and silica are mixed and electrically smelted. The phosphorus is driven off in gaseous form, condensed, and recovered as a liquid

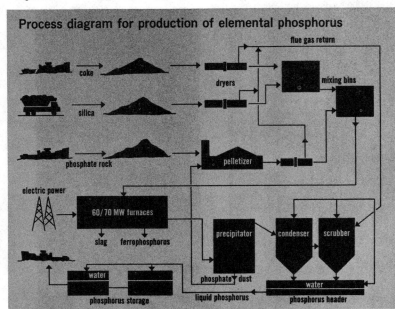

**Process diagram for production of elemental phosphorus**

flue gas return

coke

dryers

mixing bins

silica

phosphate rock

pelletizer

electric power

60/70 MW furnaces

precipitator | condenser | scrubber

slag    ferrophosphorus

water

phosphate dust

water

phosphorus storage

liquid phosphorus

phosphorus header

**Remote.** *ERCO plant sits right on Long Harbour; fishing has traditionally been the local industry*

(flow sheet, p 980). The liquid phosphorus is collected under water since it is combustible when exposed to air.

Plant effluents fall into three categories. Water from the pelletizing plant, water from coke and silica dryers, and furnace process water, called "phossy" water. Water from the pelletizing unit contains most of the fluorides liberated from phosphate rock, water from the dryers contains sulfur dioxide, and phossy water contains phosphorus, cyanide, and ammonium ion.

Prior to shutdown of the plant, the phossy water had been subjected to clarification by hydrocyclones and settling, then all three plant effluents were combined into a single waste stream, diverted to a sump, and diluted with water previously utilized for cooling purposes. The effluent was again diluted with salt water and discharged directly to the ocean water of Long Harbour without further treatment. Fluorides, chiefly in the form of fluosilicic acid can theoretically be neutralized by simple mixing with seawater. Fluorides are reduced by seawater and precipitated

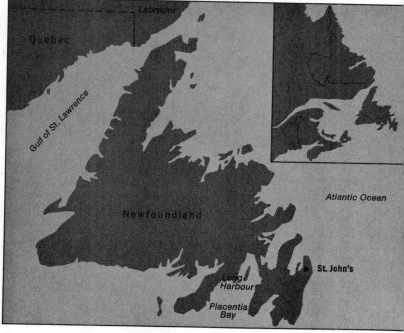

**Dredge.** *Accumulated sediment near the original discharge point was brought up and settled in ponds*

**Tanks.** *Lime is added to effluent to precipitate out dissolved fluorides*

as calcium and magnesium salts. The process requires large volumes of seawater, however. Lack of proper mixing could explain why fish fared so badly with the phosphorus they encountered, says D. R. Idler, the former Atlantic regional director of Canada's Fisheries Research Board now with the Memorial University of Newfoundland. While it was not demonstrated, incomplete neutralization of the fluosilicic acid may have resulted in a pH too low for fish to tolerate. The low pH may have immobilized the fish, thereby making it impossible for them to avoid the lethal concentration of phosphorus. Other explanations could be that fish simply could not detect the phosphorus, were not repelled by the pollutants, or that the fish concentrated the phosphorus in their livers.

Liquid phosphorus from the manufacturing process has certain impurities associated with it, such as silicon dioxide. Such impurities kept some phosphorus in suspended or colloidal forms. The particulate phosphorus in the phossy water discharge settled on a very small area of ocean bottom close to the discharge pipe in much greater concentration than had been expected. Because of poor distribution by ocean currents, settled phosphorus on the bottom was doubly deadly. First there was the problem of the accumulation

itself. Even after discharge of pollutants was halted, accumulations of phosphorus in bottom mud could kill any fish exposed over a period of time. Furthermore, oxygen destroys phosphorus, but with the slower rate of reaction at ambient temperatures on the ocean floor, accumulated phosphorus lasted much longer than it normally would in freely moving ocean currents. Faced with a clearly intolerable situation, ERCO decided to shut down the plant immediately and take remedial steps in its plant design to cope with phosphorus pollution.

**Fishermen unhappy**

One of the biggest headaches ERCO had to face was not an engineering problem at all. The plant had to be shut down and Placentia Bay closed to fishing at the height of the season. Some 400 fishermen were now denied a rich fishing ground. Although the "red herring" were later analyzed by Canada's Federal Food and Drug Agency and found to be safe for human consumption, the fishermen suffered severe economic losses. In most cases, the fishermen had scanty resources and help had to be provided almost immediately.

AN ERCO-Canadian government program was the answer. The government quickly made individual loans to needy fishermen based on records of their past earnings from the catch. ERCO, without admitting any legal liability for the fish kills, set up a $300,000 special relief fund which was distributed to the fishermen affected in amounts equal to twice the loan made by the government.

Although the government had determined that the fish were safe for human consumption, the Department of Fisheries bought all the fish caught during the first month that portions of Placentia Bay were reopened to fishing. No contaminated fish ever reached the market, according to Canada's Fish Inspection Service.

Even with the indemnities offered by ERCO, however, there was strong local opposition to reopening the plant. It took a lot of public education to convince the villagers that the plant could reopen and operate in an environmentally acceptable manner.

**Waste treatment**

ERCO decided to impound all its effluents and treat them on land. In addition, the company initiated a bottom-dredging operation to eliminate

the sediments which could prove dangerous to marine life.

In June of 1969, ERCO constructed a 3.8 million $ft^3$ pond to hold and treat bottom sediments and began dredging. The pond, separated into two smaller ponds by a causeway, promoted aerobic destruction of elemental phosphorus. Sediments and seawater were pumped into the first pond where most of the solids settled out. Clarified seawater was pumped to the second pond where it was exposed to oxygen and lime–alum–charcoal treatment. Effluent was pumped into another excavation after falling over a long drop to give additional aeration, and the treated effluent was further pumped over the land into the harbor.

To minimize damage to fish that might have occurred from the dredging operations, ERCO delayed startup of the project until fish left the area. The company rented and bought nets from local fishermen to seal off the dredging area and keep fish away.

To clean up plant effluents, ERCO initiated on-site treatment. Effluent from the pelletizing operation is now pumped to reaction tanks where fluorides are precipitated out by lime addition. The neutralized effluent is now suitable for reuse in the pelletizing operation and is returned to the plant.

Treatment of the phossy water effluent is somewhat more complicated. Before the plant was shut down, ERCO had installed a battery of hydrocyclones to remove all the phosphorus except that which was in colloidal suspension. The phosphorus thus removed goes to a sump. The remaining phossy water is pumped into a receiving tank and from there into a mixing tank where lime addition removes most of the phosphorus. The effluent—now neutralized and largely free from contaminants—is used as makeup water. In a recycling program, the lime-phosphorus sludge settles out and is later incorporated into pellets.

Marine- and environment-based studies were conducted prior to the building of the plant and repeated on a continuing basis during plant start-up. Benthic analyses were conducted by T. W. Beek consultants, and some station results indicate that marine life in the seabed in the vicinity of the plant has returned to the state it was in prior to the opening of the plant. Other stations indicate a marked improvement, leading to the assumption that they will return to the original ocean-floor environment. HMM

# Canada cracks down on pollution

*Federal and provincial moves parallel U.S.*

*efforts, with some important differences*

Reprinted from ENVIRON. SCI. TECHNOL., 4, 547 (July 1970)

There are both similarities and differences between the U.S. and Canadian approaches to air and water pollution programs. Nevertheless, apparent progress is being made by our northern neighbors.

Some of the similarities are in organization. In both countries, control activities span a multiplicity of levels of governments. In the U.S., federal, state, and local governments are concerned with the problem. In Canada, the levels of concern are federal, provincial, and municipal. Like the U.S., Canada has no one agency responsible for environmental management at the federal level. Air pollution responsibility is vested in the Department of National Health and Welfare; water pollution is in the Department of Energy, Mines, and Resources; motor vehicle emissions control is in the Department of Transport, and so on.

Some of the differences result from the forms of government. In the U.S., competition exists between the legislative and executive branches, often over seemingly trivial points. In Canada, there is no separation of powers. Canada has what is called responsible government. The executive—the government—is responsible to the legislature and is an executive committee of the legislature. Unlike the U.S. government, when the Canadian government introduces a piece of legislation into Parliament in Ottawa, it usually passes. If it fails to achieve passage of important items, the cabinet is replaced by another one. In contrast to U.S. law, which usually spells out the specifics, Canadian law is permissive, enabling, and subject to regulations that the government passes later, from time to time. On the other hand, U.S. law is not usually subject to regulations that are spelled out later.

The consensus of key Canadian officials, at both the federal and provincial levels, is that now is the time to make a renewed thrust on environmental

**Parliament.** *Canadian government legislation usually has easy passage in Ottawa*

programs. Only recently, on the occasion of the 25th anniversary of the Chemical Institute of Canada (CIC), a two-day session on environmental pollution was held at the joint CIC–American Chemical Society meeting (Toronto, Ontario).

## Provincial problems

In Canada, municipal and industrial wastes create serious air and water pollution problems. Water pollution alone costs the Canadian economy $1.2 billion each year, according to the Canadian Council of Resource Ministers. Nearly 70% of this cost is shared by two provinces—Ontario and Quebec. Ontario leads with $437.5 million; Quebec is second with $370.8 million.

Of the ten Canadian provinces, Ontario is the most highly industralized. It is also further along and is setting the pace with its resource management and pollution control programs. Ontario's water program has been underway for more than 15 years. All other provinces now have similar programs,

but they have gotten underway only within the last few years.

Of the ten, six provinces have ongoing air pollution control programs; four do not. Ontario's air program has a staff of nearly 200; but the air management staffs in other provinces do not number more than a dozen each.

The Ontario government is in the throes of requesting $92 million for its provincial air and water pollution control programs in the upcoming year. The provincial regulatory agency is the Ontario Department of Energy and Resources Management (ODERM).

"We want industry in Ontario. Make no mistake about that," says Gordon Hampson, executive assistant to George A. Kerr, the Minister of ODERM. "But we want clean industry."

## Federal–water

The Canada Water Act (CWA), the controversial water legislation which has been gestating for the past two years, was passed by Parliament last month. The main thrust of the act is

the joining of federal and provincial powers to move forward on water management on two paths:
• Comprehensive planning and water development.
• Water quality management agencies.

"The Canada Water Act is enabling legislation," says A. T. Davidson, Assistant Deputy Minister for water in the Canadian Department of Energy, Mines, and Resources. "The act is permissive; many options are open as to how it can be applied."

J. J. Greene, Minister of the Department of Energy, Mines, and Resources —the counterpart of the U.S. Department of Interior—personally piloted the CWA through Parliament. Greene is invested with the power to impose and administer the regulations.

"Primarily, it is not a water quality act, although this is the part that is discussed more than any other today," says Davidson. "It puts emphasis on comprehensive planning for water basin management and implementation of water goals by joint federal–provincial commissions or boards."

"In Canada, we have a bias that is a bit different . . . (from the U.S. on water quality)," Davidson explains. "We think that if you can establish management by basins and provide a large number of tools to those basin authorities as to how to achieve their water quality objectives, then over the long pull we get a better result.

"We do not believe it is sensible to try to establish and set national standards and enforce them by federal law all over the country," Davidson says. "We take the view that water quality management should be part of water management in general, and there should be a basin approach to it.

"Specific details are not spelled out in the act," Davidson reiterates. "Many options are open even now that the law has passed. The most likely option in dealing with the provinces is that where provincial water commissions or agencies have been set up and are operating effectively, then the federal government would only need to reach an agreement with these agencies on timetables for achieving water quality goals.

"Water quality management should be seen in the perspective of regional water management and regional development," Davidson says. "A good deal of the responsibility resides with the provinces, and it should reside there. The federal government does not in-

**Provincial water spokesman Collins**

"The one thing that is unique in Ontario is that every new development—industrial or municipal—has had to meet waste control standards before being allowed to develop."

tend to take over the job of water management across the country."

The major federal concerns for water quality are:
• The Great Lakes, which cause international concern because of treaty obligations.
• The Ottawa River for which the provinces of Ontario and Quebec have failed to reach a water quality management program; the river is interjurisdictional.

Of the 25 amendments to CWA, the one receiving the most attention concerns nutrient additions to receiving waters. The other amendments are largely clarifying in nature. National Energy, Mines, and Resources Minister Greene already has stated publicly that he intends to propose regulations to the cabinet which would have the effect of banning, by August, the manufacture and import of laundry detergent containing more than 20% by weight $P_2O_5$. He intends to work toward a total ban in 1972.

### Provincial–water

The Ontario Water Resources Commission (OWRC), a Crown corporation similar to the Tennessee Valley Authority in approach—was established under the provisions of an act of the Ontario provincial government in 1956. Primarily an engineering corporation, OWRC is in the business of constructing water supply and sewage treatment plants and performing comprehensive planning for development of an area. OWRC serves as the model for other provinces.

D. J. Collins, OWRC's chairman, has 18 years of government services, in-

cluding three deputy assistant ministerial posts and is the full-time chairman of OWRC. "In Ontario, we have $2 billion worth of facilities in the ground—lines and plants; OWRC gets the money from the communities or the federal government," Collins says. "OWRC is a well balanced organization with concern both for water quality and water quantity. Every municipality in Ontario can qualify for full treatment. For the past ten years, every industry has been required to meet OWRC standards.

"The freezing on area development —perhaps, only possible in the Canadian form of government—is done by the Ontario Municipal Board (OMB). If OWRC is not satisfied with the waste control provisions in an area's development plans, the region is not allowed to develop."

On the other hand, Niagara Falls (N.Y.), is one example of an area which has been allowed to grow and develop at the same time that it continues to dump raw sewage into the river. The situation is similar in Quebec, a province in which growth and development have proceeded without proper waste treatment facilities.

Ontario has some 2000 water-using industries, and more than half are tied in with municipal treatment facilities. Some of the other half of the industries have completed their own treatment facilities, but not others.

"The old industries are the big polluters," Collins says. "Pulp and paper and steel are still the major polluters. About half of the mills have treatment underway or planned at this time. Six or seven years ago, muncipalities

### Provincial air spokesman Drowley

"We currently require all new stationary sources to obtain a permit of approval prior to construction and, of course, existing sources are required to conform with the specific regulations."

agreed to put in treatment facilities. Now, industries are in that position."

### Federal–air

At the federal level, Peter M. Bird heads the Environmental Health Directorate in the Department of National Health and Welfare. "The directorate totals about 270 personnel and is divided into four divisions, one each for air pollution, public health engineering, radiation protection, and occupational health," Bird says.

Until a Clean Air Bill comparable to CWA is passed, Canada will not have a major federal–provincial program in air pollution control. However, John Munro, the Minister of National Health and Welfare, has indicated his intention to introduce a clean air bill to Parliament as soon as possible. It may be introduced in the fall session of Parliament.

"One of our first activities involved a joint meeting of federal and provincial air pollution control officials, which was held late last year," Bird elaborates. The following recommendations came from that meeting:

• To develop under federal leadership a comprehensive national air monitoring network.

• To develop national ambient air quality objectives.

• To formalize federal–provincial working level meetings on air pollution.

"We are nearing the final stage of completing a major air pollution study under the auspices of the International Joint Committee (IJC) in collaboration with other responsible agencies on the Windsor-Detroit air pollution

problem." Bird concludes. The report is due later this year.

"The federal air pollution division was added on April 1, operates at a budget level of approximately $600,-000 and is expected to grow to a staff of 29 by the end of the year," says S. O. Winthrop, chief of the division. The new division is focusing attention on an inventory of air pollution sources in federal facilities and advice to the Department of Transport on regulations for motor vehicle emissions controls.

Control of motor vehicle emissions in Canada follows the example of the U.S. federal government. Recently, Canada passed its Motor Vehicle Safety Act. Gordon C. Campbell, director of the motor vehicle safety branch in the Department of Transport, notes that regulations are currently being prepared and are expected to be ready for implementation on 1971 model vehicles. So any new vehicles sold in any province will have to meet these standards, which are equivalent to U.S. federal standards. But neither the Canadian nor, for that matter, the U.S. federal government has the responsibility or authority to ensure that the devices are maintained in good control. This is a job for the provinces and states, respectively.

### Provincial–air

Provincial control of air pollution in Ontario began in January 1968. Prior to this time, the control power rested with the municipalities. Now Ontario's air management program is billed as the best operated and most comprehensive program in Canada.

W. B. Drowley, director of ODERM's air management branch, explains the provincial philosophy of setting emission limitations. "Ontario has set its emission limitations on the basis of maximum permissible one-half hour concentrations at ground level. This requirement means that we must check each individual source.

"Our procedure is to carry out an emission survey in the plant itself and to give the owner a written report with recommendations as to what he must do to obtain a compliance," Drowley elaborates. "This report is followed by a ministerial order—a legal document requiring compliance within a specific time. The normal appeals procedure is also included so that ODERM's air branch does not become too autocratic."

Commenting on air quality standards, Drowley says. "To me a standard is something that is legally enforceable. I have yet to see how you can enforce an air quality number on a multitude of sources unless you have the capability of being able to assess each individual contribution and require its curtailment.

"Recently, an air pollution index was introduced in metropolitan Toronto," the provincial control official explains. "It allows the curtailment of industry when the index reaches certain values." The index is based on a 24-hour running average of two pollutants—$SO_2$ and particulates—to produce a number which is based on data from previous air pollution episodes. The index number 32 signifies a desirable air quality. At values of 50–75, emissions may be curtailed. A value of 100 is equated with the beginning of an episode.

### Looking ahead

In 1972, Canada will be sending its delegation to the United Nations meeting on the environment. The Canadian Department of External Affairs—the counterpart of the U.S. State Department—recently established a new division which has the major responsibility of preparing for the forthcoming Stockholm meeting. W. K. Wardroper, chief of the new division of scientific relations and environmental problems, is in the throes of preparing agenda items for the Canadian input to the conference.

The National Research Council of Canada announced a program which promises environmental solutions.

# Cleveland opts for physical-chemical

Reprinted from ENVIRON. SCI. TECHNOL., **6**, 782 (September 1972)

Several years ago the Cuyahoga River near Cleveland, Ohio, literally caught fire as a result of semitreated and often untreated municipal and industrial wastes pouring into the river. Conditions have improved somewhat since, but Cleveland's existing Westerly Water Pollution Control Facility (one of three waste water treatment plants), located on Lake Erie near the mouth of the Cuyahoga, was built in 1922 and cannot adequately treat its share of the metropolitan area's waste water.

The plant provides primary treatment to approximately 40 million gallons per day (mgd) of municipal and industrial wastes in an Imhoff tank complex. (An Imhoff unit consists of two-story sedimentation tanks with sludge accumulation in the bottom.) At present, the average dry weather flow of waste water through the plant is 50 mgd, and the dry weather peak is 70 mgd, which the outdated plant cannot handle. Nor can the plant produce effluent meeting federal and state water quality standards.

## Criteria

Three years ago, the Cleveland Clean Water Task Force (created to look into water pollution problems and solutions) began looking at two alternatives to the old plant. One was

### U.S. physical-chemical treatment plants

| Plant | Flow, mgd |
|---|---|
| Fitchburg, Mass. | 15 |
| Cortland, N.Y. | 10 |
| LeRoy, N.Y. | 2 |
| Niagara Falls, N.Y. | 60 |
| Alexandria, Va. | 25 |
| Arlington, Va. | 30 |
| Lower Potomac, Va. | 36 |
| Prince William County, Va. | 15 |
| Cleveland (Westerly), Ohio | 50 |
| Rocky River, Ohio | 10 |
| Owasso, Mich. | 6 |
| Rosemount, Minn. | 1 |
| St. Charles, Mo. | 8 |
| Garland, Tex. | 30 |
| Los Angeles, Calif. (Hyperion Plant) | 5 |
| Orange County, Calif. | 17 |
| Selma-Kingsburg-Fowler, Calif. | 10 |
| **Total** | **330** |

Under design or construction.

to construct an island in Lake Erie with waste activated sludge from the plant—this was deemed impractical, ineffective, and expensive. The other—later adopted—was to build a new waste water treatment plant while the old plant continued to operate.

A new plant must meet certain stringent criteria, the Task Force decided. Besides meeting effluent quality objectives, the plant must be capable of handling, without upset, widely varying flows from combined storm and waste water sewers. It must also handle assorted constituents from industrial discharges, which account for nearly 50% of the flow into the plant. The plant site, only eight acres in area, is almost completely built up, but waste water treatment must continue during demolition of the old plant and construction of the new facilities. The design must be flexible for future expansion and addition of new processes. Also, the facility must minimize the effects of toxic materials and provide a potential for removing heavy metals.

What type of plant could handle such a big order? The Task Force knew that biological treatment plants are sensitive to weather changes and that unusual or strong effluents could result in a completely disrupted treatment process. Furthermore, many Task Force mem-

**Transformation.** *The original waste treatment plant for Cleveland, Ohio (left), was torn down in 1922. A 40-mgd Imhoff tank facility (right) was built but is now outdated and unable to handle the amount and variety of wastes passing through to meet water quality standards. It will soon be replaced by a $41 million physical-chemical treatment plant (upper right) capable of handling up to 100 mgd. The new plant, to be completed by 1974, will be the world's largest physical-chemical treatment plant*

*The big Ohio city is turning its
back on biological treatment as it gears
up to modernize its Westerly plant*

bers felt that biological treatment plants cannot provide the high degree of solids and BOD removal required to meet water quality standards.

Physical-chemical treatment is the answer the Task Force decided upon, and the city awarded the design contract to Zurn Environmental Engineers, an affiliate of Enviro-Engineers, Inc. (Washington, D.C.). Physical-chemical treatment is not in widespread use; in fact, only 17 such plants are under design or construction in the U.S. (see box). Nevertheless, it is receiving increasingly serious consideration as effluent restrictions tighten.

### Advantages

The 50-mgd plant will be the largest physical-chemical treatment plant in the world, and will be capable of handling up to 100 mgd hydraulically (see below). Effluent quality will represent 90% BOD removal, 93% suspended solids removal, and 90% phosphorus removal.

One advantage of a physical-chemical treatment plant is that unit processes are based upon hydraulic (rather than biological) requirements. This means that waste water flows above and beyond the design flow capacity do not necessarily mean deterioration in effluent quality. For example, filters can catch solids carried over from the clarifiers. Another benefit is high internal recycle of process waste water. The Cleveland plant will be capable of an instantaneous recycle flow of approximately 30 mgd for backwashing the filters and adsorption columns. The plant is designed to handle recycle flows superimposed upon the peak dry weather flows (70 mgd)—thus total capacity can be 100 mgd for sustained periods.

Not all of the old plant will be demolished. The Imhoff tanks will remain and be modified (for screening and disinfection) to provide storm flows a minimum of treatment. Combined storm and sewer flows up to 100 mgd will be taken into the main treatment plant, and the remainder will be diverted to the storm-water holding Imhoff tanks for later treatment.

### Plant processes

Here's how the new plant will work: Incoming waste water will undergo preliminary screening and settling to remove debris, sand, grit, and other large particles, and to provide some protection for the comminutors which will finely grind the influent. After passing through aerated grit-removal

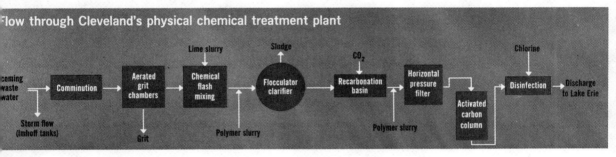

**Flow through Cleveland's physical chemical treatment plant**

Incoming waste water → Comminution → Aerated grit chambers → Chemical flash mixing (Lime slurry) → Flocculator clarifier (Sludge) → Recarbonation basin ($CO_2$) → Horizontal pressure filter → Activated carbon column → Disinfection (Chlorine) → Discharge to Lake Erie

Storm flow (Imhoff tanks) · Grit · Polymer slurry · Polymer slurry

chambers (flow diagram), the waste water will flow into two flash mixing tanks where propeller mixers will stir in a 10% lime slurry. The lime will convert influent phosphorus into insoluble calcium phosphate.

Water from the flash mix tanks will go into two channels, each serving two flocculator-clarifiers. The flow will again be divided and fed to the center flocculating well in each clarifier. Polyelectrolytes can be added at this point as a coagulant aid. Virtually all settleable solids will be removed here as waste particles coagulate and rapidly settle to the bottom of the basins.

The waste water, with a pH of 10.5–11.5 from the lime clarification step, then will pass into the recarbonation basin. There, carbon dioxide will be bubbled through the water to lower the pH.

The dephosphated effluent next will go through a series of pressure filters operating at 10 psi. These 14 horizontal filters will be stacked in two levels of seven each to minimize space. The filter medium has not yet been chosen, but could well be beds of sand or crushed anthracite coal, or a combination of the two. This filtration step will virtually complete removal of the remaining suspended solids in the waste water.

After the filtration step, the semi-treated water will flow into activated carbon sorption columns (30 columns, each 20 ft in diameter, 28 ft high, and having a carbon bed depth of 17 ft) to remove any remaining dissolved organic material. The columns will be stacked in two levels of 15 columns each. Approximately 2000 tons of carbon will be used, and each column will be in service for 43 days between regeneration cycles.

Spent carbon will be regenerated in two multiple-hearth furnaces to burn off the collected organic material. The plant will be able to reuse regenerated carbon again and again. The filtration

and sorption units will be housed in a building equivalent in height to 13 stories.

The treated effluent will finally be disinfected by liquid chlorine. Studies are under way, however, to determine the feasibility of using ozone both for disinfection and as a method to achieve higher BOD removal if it should be required.

**Sludge handling**

Sludge will be pumped from the flocculator-clarifiers to two sludge-holding tanks equipped with sludge-thickener mechanisms. Although sludge concentrations of 10–14% could be achieved in the flocculator-clarifiers, the solids-handling system is conservatively being designed for 5% solids. Centrifuges will further dewater the sludge. Each centrifuge will have a capacity of 15,000 gph, although the design loading is 9850 gph, thus providing capacity for sustained plant operation at 100 mgd. Additional sludge can then be handled easily when waste water flow rate is as high as 100 mgd. Solids capture is expected to be above 90% and sludge cake solids above 30%.

Sludge cake will be fed to the re-

calcining furnaces at approximately 12,500 lb/hr. Zurn Environmental Engineers hopes for a recyclable lime product of 80% purity which can be reused at the flash mixing stage.

Instrumentation, control, and monitoring for the Westerly plant will be automated with economical use of manpower. A programmable computer will monitor all measurements, control positions, and alarms and provide feedback control.

About 60% of the plant design is complete, and bids on major equipment packages will be received this fall. Construction will begin before the year ends, and all phases of the physical-chemical treatment plant are expected to be operating by late 1974. The cost of the entire plant is estimated to be $41 million, and the designers expect to achieve a waste water treatment cost of approximately 26–28¢ per 1000 gal.

This plant represents a significant breakthrough in a waste water treatment alternative not previously explored on a large scale, C. J. Touhill, vice president, Enviro-Engineers, Inc., believes, and will do much for the general acceptance of physical-chemical treatment.　　　　CKL

# Water quality begins at the local level

*Cleveland demonstrates*

*that an aggressive program*

*will find financial support*

Reprinted from ENVIRON. SCI. TECHNOL., 4, 281 (April 1970)

President Nixon's State of the Union pledge that secondary sewage treatment facilities will be provided for every city needing them no doubt came as welcome news to municipal water treatment officials. Yet to be resolved, of course, is the ultimate formula for federal participation in providing the $10 billion necessary to attain that goal. But most proposals advanced thus far still envision a healthy share of local financing and coordination. Thus, the question: Are U.S. cities able to marshal their resources effectively enough to accomplish the immense task outlined by the President, whatever the measure of cooperation forthcoming from Washington?

One hopeful straw in the wind is the water quality program in progress in Cleveland, Ohio. Underway for less than two years, this program already has chalked up impressive gains, leading some observers to comment that, within five years, Cleveland will have the best sewage treatment system in the U.S. And, whatever effect future infusions of federal money may have on the program, Clevelanders can take a certain amount of pride in the fact that the overall program was conceived and established before substantial federal assistance was anywhere near assured.

In its program, Cleveland has taken some approaches that are unique in themselves and that, at the same time, bear all the earmarks of a bootstrap operation. Specifically:

• The city administration was able to get voter approval of a $100 million bond issue for water treatment facilities in November 1968, at a time when any type of municipal bond issues generally had tough going elsewhere.

• The city has set up, within its Department of Public Utilities, a Clean Water Task Force to draw public attention to, and provide the in-depth capability for solving, its problems.

**Mayor.** *Cleveland's Stokes celebrates reopening of beach on city waterfront*

• Last year, the city took the unconventional step of contracting with Dow Chemical Co. to operate one of its treatment plants, with an eye to generating design data for updating and expanding city facilities.

• Construction plans for the city's treatment expansions have taken an aggressive stance on two of the latest concepts in municipal water pollution control—advance waste treatment, and joint municipal-industrial treatment.

**Problems**

Even the most avid Cleveland boosters would have to admit that, a few years ago, only such radical measures could effect any improvement in Cleveland's water quality problems. The city has had secondary treatment facilities for some years. But, for a variety of reasons, the city's program had not kept abreast of advances in sewage treatment technology.

A heavy concentration of industry within the city also has had its impact. Effluents from most of the city's industries drain to the Cuyahoga river, which bisects the city on its way to Lake Erie. The current state of the Cuyahoga has made it the butt of a standard repertory of antipollution jokes: "The only river declared a fire hazard," "Too thin to pave and not thick enough to plow," and so on. And national concern over the imminent death of Lake Erie focused attention on lakefront municipalities, few of which were more visible than Cleveland. Aerial photographs in which the flow of the Cuyahoga can be traced far into the lake have become a classic way of illustrating Lake Erie's plight.

**Reforms**

Against such a background, it is small wonder that, in 1967, incoming Mayor Carl Stokes placed water treatment facilities expansion high on the list of reforms he had in store for the

**Task force's E. J. Martin**

city. Realizing the close link between Cleveland's problems and the fate of Lake Erie, the mayor chose an appropriate setting to unveil his plans, the June 1968 Lake Erie water pollution abatement conference which was held in Cleveland. Armed with a study that showed the need for $200 million as a first step in the city's program, the mayor announced his bond issue proposal and the program it would finance. The proposal came as a surprise to Clevelanders—final details for the program had been ironed out on the eve of the announcement—but the bond issue passed easily in the November 1968 election.

Ben Stefanski, Cleveland's director of public utilities, points out that "At the time the bond issue was passed, we had a master plan, but needed a firmer grip on the technology that would be required." This Stokes and Stefanski got from the Federal Water Pollution Control Administration (FWPCA)— in more than a figurative sense. On one of their many trips to Washington to consult with federal officials on Cleveland's needs, Stokes and Stefanski met with, were impressed by, and eventually hired Edward J. Martin, a former research and development man with FWPCA, who has since become the spark plug for Cleveland's program.

Martin assumed the newly created post of director of Cleveland's clean water task force in January 1969. This department has been superimposed over the former department of water pollution control, in such a way that the task force now has complete jurisdiction over all aspects of the city's water quality program. Martin views the long-range goals of the task force as comprised of three distinct areas: Design and construction of new facilities, construction of a new interceptor sewer system, and a new industrial

waste treatment program.

It is in the area of updating the city's three treatment plants—Easterly, Westerly, and Southerly—that the program has shown the most results to date. At Westerly, for example, the efficiency of secondary treatment—as measured by effluent quality—has been improved substantially. Construction will begin this year on a $38 million expansion project. By taking advantage of new technology in the form of chemical treatment, adsorption, and ion exchange, the city has projected a savings of $15 million on the Westerly project through eliminating construction of a huge artificial island in Lake Erie, once thought necessary.

At Easterly, a $6 million modernization is underway, and contracts will be let soon for advanced waste treatment facilities, mainly to provide for phosphate removal. And, at Southerly, the largest and most complex of the three plants, a $5 million reconstruction is almost completed. Included in this project are improved facilities for treating sludge from all three of the treatment plants. A new treatment unit is planned for Southerly that will incorporate chemical treatment and carbon adsorption; when this plant is completed, Cleveland probably will be the first major U.S. city to adopt physical-chemical treatment as an alternative to biological treatment.

The second major component of the task force program is to provide a new sewage collection and transportation network. Martin says "We're also undertaking this design work in a nonclassical way. We're starting with the problem of water standards for Lake Erie, destination of all our effluents, and working back to the pollutant loads, not hydraulic loads, we'll have to handle." A $2 million design project has just gotten underway, and the whole sanitary waste and storm water collection network probably will account for the major part of the $1.2 billion water quality program Martin thinks Cleveland eventually needs.

The third major segment of the task force program is related to industrial waste problems." Up to 50% of the wastes generated in the city are from industrial sources, and a program that does not integrate treatment of industrial waste is, at best, only 50% effective," notes Martin. Part of Cleveland's current problem with industrial wastes is jurisdictional; most industrial waste effluents are discharged directly into the Cuyahoga, and, thus, are

under state, rather than local, control.

Martin is a firm advocate of joint industrial-municipal treatment schemes. In addition to the usual advantages, economies of scale and opportunities for additional federal financing, there are other technical advantages. Phosphate removal techniques being adopted by the city include provisions for use of acid discharges from steel pickling and coke ovens to precipitate the phosphorus.

**Problems ahead**

Despite his ambitious program, Martin is the first to admit that Cleveland faces difficulties in achieving its goals. The most immediate of his problems is a state deadline, only weeks away, to submit formal proposals for new treatment plant construction. And Stefanski points out the type of enforcement problems that arise from the fact that his public utilities department supplies water services to suburban communities that do not use Cleveland sewage treatment facilities. He argues that one possible interpretation of state enforcement pressure makes his department responsible for those communities whose own treatment facilities do not comply with state standards; such responsibility, of course, implies authority to withhold water service to those communities.

But, despite such short-term problems, both Stefanski and Martin are optimistic about solutions to water quality problems facing most U.S. cities. Stefanski feels that "Cleveland has become a barometer for the country by demonstrating that local commitments must come first; if the commitment is made, financing schemes are available." Martin adds that, although the city's $100 million commitment fills only half of its immediate needs, the city is eligible for an additional $100 million from the Ohio Water Development Authority.

But even this optimism is guarded; speaking of the impact of the federal treatment plant construction program, Martin says "the President is still not talking goals, which ultimately must be set regardless of the cost of attaining them. Estimates of the cost of solving the Lake Erie problem, for instance, include $8 billion to eliminate polluted discharges, plus costs of such things as dredging, aeration, and so on, to restore lake quality. The total cost then looks more like $80 billion, and a $10 billion nationwide program looks like less than a total commitment."

# Policing an industrial river

*Reprinted from* ENVIRON. SCI. TECHNOL., **5**, 996 (October 1971)

For 265 miles it runs, from its source in New York State, through the Catskill and Pocono mountains and the rolling hills above Trenton, down through the coastal plain to the sandy expanses of the Atlantic Ocean. Running into it along the way are tributaries in four states. In all, the river drains some 13,000 square miles in its trip to the sea.

Quite apart from its striking physical setting, the Delaware River is particularly well suited to industrial growth—it is deep enough to accommodate large commercial vessels and its 12,000 ft³/sec average flow rate has, in the past, been considered more than enough to assimilate discharged wastes. And industrial growth has indeed come to the Delaware River Basin, bringing with it people, prosperity—and problems. Many of those problems concern the river itself, which, over the years, increasingly has been used as a repository for all manner of waste as well as an industrial highway and source of usable water.

Recognizing the complex and interstate nature of the Delaware's problems, the U.S. Congress passed (and the estuary states approved) the Delaware River Basin Compact (P.L. 87-328) in 1961 which established the Delaware River Basin Commission (DRBC). The commission is composed of representatives of the federal government and the states of Delaware, New Jersey, New York, and Pennsylvania. This was the first federal-interstate compact given, by law, enough power to implement a river basin development program.

The DRBC, acting as a regional watershed agency, has the power to control water resource development, establish planning standards, and control the operation of all projects and facilities in the basin.

The idea behind such a federal-interstate agreement is undoubtedly a sound one, but with too few staff members and the need for cooperation among four states, the path toward clean water is, in practice, not always a smooth one.

The commission has been able to forecast both short- and long-term water quality (including new or enlarged water supply or waste water treatment facilities) by using a mathematical model of the Delaware estuary based on biological and chemical studies of the water conditions. After compiling this data, DRBC developed water quality standards (which now receive approval by the federal Environmental Protection Agency) and regulations for implementation.

However, some dischargers think that the manual that contains the basin regulations is, in itself, a problem, since basin regulations and water quality are based entirely on stream objectives, not effluent quality. For instance, the manual states: "All effluent shall meet the quality standards of Article 2-1." However, Article 2-1 says of effluents, "All waste shall receive a minimum of secondary treatment, . . . wastes shall be effectively disinfected . . . effluents shall not create a menace to public health, . . ." etc. And those are the discharge standards (although criteria are now under consideration). "In 10 years, they should have come up with something for discharge standards," laments one industrial spokesman. "They have not written a hard set of rules that say you may do this; you may not do that."

There are other areas in the regulations that industrial critics call "fuzzy." In the case of thermal discharges, "objectives are applied to the river; they don't apply to the effluent itself," says another spokesman. The DRBC wants certain temperatures maintained in the river throughout the year (not more than 5°F above the ambient temperature of the water with an 86°F maximum).

The DRBC has said that mixing zones will be established for effluent areas, and from the "mixing zone concept," they will determine what kind of temperature increase should be allowed. "At this point, they haven't defined where the mixing zone is," continues a representative of a large oil refinery in Pennsylvania. In anticipation of criteria, many industries are currently recording temperature profiles of the Delaware River.

In a move to limit pollution, DRBC divided the river estuary into four zones and assigned a maximum allowable oxygen-consuming waste load to each zone. Eighty-one public and industrial dischargers were then assigned a maximum permissible BOD discharge allocation. The allocation system works

this way: DRBC officials decided that the total BOD that the river can effectively assimilate is 322,000 lb/day. They set aside 10% of this total for future expansion and then divided the remainder among the four zones. The allocation system calls for approximately 89% reduction of waste in each zone.

"This was the method determined to be fair and applicable," says James F. Wright, DRBC's executive director. "I think it's a pretty good system," agrees Peter Williamson, Director of Environmental Services for Rollins-Purle which operates a heavy industrial waste disposal plant in Logan Twp., N.J. However, adds Williamson, "the problem is getting the states to enforce with the same degree of muscle." Pennsylvania has about four times as much industry on the river as does New Jersey, so while a New Jersey inspector can easily visit all industries within his district, his counterpart in Pennsylvania may not be able to knock on all doors.

After allocations were assigned to each zone, dischargers submitted abatement schedules to the DRBC for approval, setting definite dates for steps toward compliance. While on an abatement schedule, an industry or municipality may continue to exceed allocations. This action is remedied, in part, by completion schedules containing yearly deadlines leading to progressively more treatment of wastes. "The interim dates are a check as far as the DRBC is concerned," says an oil refinery representative.

The DRBC is also responsible for monitoring the river and its main tributaries and has contracted with the three estuary states (New Jersey, Pennsylvania, and Delaware) to collect and analyze water samples, mainly for dissolved oxygen levels. But even so, manpower is not available to sample the waters adequately. "I wish we could get weekly samples," stresses DRBC's Wright. Presently, many firms are analyzing their own water samples and sending in monthly monitoring reports to the state in which they are situated, as well as to the Corps of Engineers and the DRBC.

## Regional disposal

A proposed high-efficiency waste water treatment plant in southern New Jersey with a 30-mile waste collection network is waiting for DRBC action. The feasibility engineering studies have been completed, and a 50 gal/min pilot plant project has been operating for about a year. The eventual total waste stream to be handled by the proposed facility could reach 120 million gal/day. "The DRBC could perform a real service by acting on this regional plant," says Williamson of Rollins-Purle.

As for power, the DRBC does not have absolute say in all situations, yet does wield a hefty club. Before promulgating regulations, the DRBC holds public hearings (although there has been some complaint concerning the publicity of these) to get testimony pro and con on the issue at hand. "I'm sure that if we have a valid reason for a request, the DRBC would examine it thoroughly and come up with a fair decision," continues the Pa. oil refinery spokesman. However, another industry representative is more critical: "They (DRBC) do have blinders on to some extent, and it takes a great deal of working with

them to swing them around to your point of view or even to see your point of view. The greatest drawback of the DRBC is that they won't say anything hard and fast."

The Commission relies on the individual states for actual enforcement (which, at times, does not lend itself to expediting matters). "They are understaffed like everybody and have trouble with enforcement," says Williamson. If necessary, the DRBC can call in the federal EPA, but thus far, the DRBC has not needed to take such action.

Water-using industries and municipalities realize that the DRBC indeed can get its way, especially through its role in issuing operating permits. "They are hard-nosed, and they have definite ideas of what they want and what they want industry to do," confesses one spokesman. There is general agreement that the staff has good resources and knows the ins and outs of water treatment. "I'm impressed with the DRBC and think their objectives are admirable," comments one spokesman. All dischargers further benefit from the continuing consistency of regulations among the four states since the DRBC came on the scene.

The tale of the Delaware Valley seems to be this: If an industry is on an abatement schedule because it is discharging excessive waste loads, it tends to be reticent about commenting on DRBC and its activities. On the other hand, officials of those industries further ahead in pollution control can afford to be more outspoken about the good and the bad of this interstate agency which some term "bureaucratic."

In spite of it all, things are getting done, even if slowly. "By 1977–80, the backbone of the work will be done," forecasts executive director Wright; at present, "we have over 90% of the waste load under abatement schedules. The 51 dischargers still without approved abatement plans account for only 20% of the allocated waste load to the estuary." CEK

# Hawaii's sugar industry faces tough problems

*Reprinted from* ENVIRON. SCI. TECHNOL., **5**, 1174 (December 1971)

*Roses are red, violets are blue. Sugar is sweet,*

*but mill wastes pollute the ocean*

Hawaii, perhaps more than any other state in the Union, inspires a visceral respect for the warp and woof of the environment. It is a land of primal elements: fire, earth, air, and water. It is a land where the luxuriant is juxtaposed against the sterile, an entire continent in miniature.

Slightly less than 1 million people inhabit the seven westernmost islands of the chain which stretches some 1600 miles from Kure and Midway in the east to Hawaii—or the Big Island, as the natives call it—in the West. Apart from military operations and tourism, the economic mainstay of the islands is agriculture.

By far the most important crop is sugar cane. In 1972, Hawaii will produce some 1.2 million tons of sugar compared with about 1.5 million tons for the mainland's cane producers and 3.4 million tons for beet sugar producers. Cane brings about $1.6 billion to the islands each year. Some 240,000 acres of land are under cane cultivation.

But it's beginning to look like the cane industry on the islands is headed for trouble. One of the state's 26 sugar mills will close this month. Two more will close in 1973. Others will merge to stay alive. A major reason, industry spokesmen say, is that the cost of pollution control facilities will make some of them unprofitable to run.

### EPA cracks down

There is considerable pressure on the industry to abandon waste disposal practices which have been standard for many years. Last September, the Environmental Protection Agency (EPA) asked the Justice Department to take court action against nine sugar mills on the island of Hawaii for causing "severe pollution" off the island's Hamakua coast (see map). The companies cited, EPA Administrator William D. Ruckelshaus said, were discharging "untreated and inadequately treated wastes into coastal waters in violation of the Rivers and Harbors Act of 1899."

The mill managers were miffed, to say the least. All had applied for discharge permits (Ruckelshaus noted, however, that that did not exempt them from prosecution) and were under a state-sanctioned abatement program to clean up operations "as quickly as possible." At a Honolulu press conference, Karl H. Berg, president of the Hawaiian Sugar Planters Association said that the industry would be spending from $10–12 billion in the next five years for pollution control.

Many sugar planters (and state officials as well) privately branded EPA's push as little better than a publicity stunt. Others saw more sinister implications. "Let's face it," one industry spokesman said privately, "EPA is making a whipping boy out of the sugar industry. They wouldn't seriously consider puting Detroit in the profit squeeze they're forcing us into, and the auto industry is a far worse offender. But, by showing the world that they're not squeamish about putting a few thousand people out of work, they hope to bring the big boys to heel."

### Cane processing

Cane processing on the islands is limited to harvesting, milling, and boiling. There is little refining. Most mills ship only crude, impure sugar to the mainland to be refined by California and Hawaiian Sugar Co., an industry-controlled co-op in Crocket, Calif. Only the small amount of sugar needed for local consumption is refined in Hawaii. But at every stage in the operation—from planting to final shipment—the Hawaiian cane sugar industry faces some challenging problems.

Cane is one of the few commercial crops that thrives in the harsh volcanic soils of the islands. It is planted in furrows that follow the terrain of the plantation to minimize soil erosion and conserve water. Because many plantations are located on steep volcanic slopes—quite literally between the devil and the deep blue sea—pollution abatement practices which are technologically simple elsewhere may, on the islands, assume quite different proportions.

Geography is often critical. Rainfall, for example, may vary between 50–200 in./year from the higher elevations to the coastline on a single plantation, often a distance of less than a mile. Consequently, over half of the cane under cultivation in the islands must be irrigated. Since the soil is loose and porous, large volumes of irrigation runoff—or tail water—may find its way into the ocean, carrying with it fine volcanic particles.

An obvious solution would be settling ponds to clarify water and reclaim soil. And some of the plantations in more level spots on the islands do use lagoons. In other locations, notably along the Hamakua coast of the Big Island, settling ponds are not as useful. The tail water problem, like so many problems faced by the industry, is aggravated by the fact that milling operations really haven't changed very much over the years. Only rarely is enough land available—in the right place—to allow sedimentation basins.

Because planting and harvesting of cane was carried out by hand until the labor shortage arising in the aftermath of World War II, mill sites were placed at the lowest elevation on the plantation—frequently poised precariously on the edges of cliffs dropping several hundred feet into the sea below. The design made engineering sense. Cane could be flumed down from the fields in water troughs, and wastes from the milling operations could be conveniently dropped into the ocean. Since

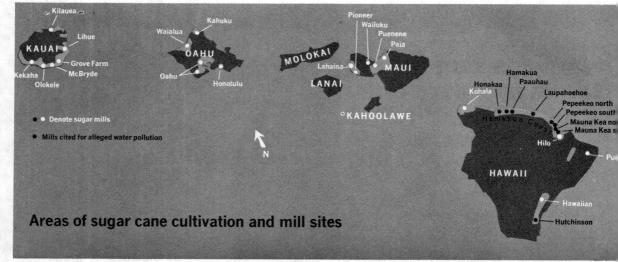

## Areas of sugar cane cultivation and mill sites

Denote sugar mills

● Mills cited for alleged water pollution

the shoreline in this part of the country is remote and inaccessible, few complaints were voiced, except those by occasional fishermen who complained that floating cane trash was fouling nets or interfering with navigation.

Now, although some sugar processors still argue that their effluents are not particularly harmful to the marine environment, they would be willing to keep their wastes onshore—if they could find the room at reasonable prices. Recycling wash water and irrigation tail water would be nice, they say, except that it means pumping it back up the slopes at considerable expense.

### Harvesting problems

Mechanical handling of cane has brought yet other problems. Harvesters are not particularly well suited for the rough terrain on which cane grows. A typical method of harvesting, for example, is one in which cane is pushed into windrows by a plow or push rake. Cranes then load the cane onto trucks to be taken to the mill site. But the machinery gathers, in addition to the cane, huge amounts of mud and rocks which must be removed at the plant by water and mud baths. That means still more soil put into the ocean—soil which allegedly destroys coral and dirties the sparkling water so vital to the tourist industry.

Before cane can be milled, leaves must be removed from the stalks. Traditionally, the best way to clean cane has been to burn the fields before harvesting. The water content of the cane protects it from burning while leafy material—called trash—is quickly and cheaply removed.

Sugar planters will probably have to find another way to remove trash,

however, as prohibitions against open burning gain more support. It is a task which one industry spokesman describes flatly as "impossible."

Once cane is harvested and brought to the mill site, another set of pollution problems must be solved. In addition to removing mud, stones, and the remaining trash, cane must be chopped up, untangled, and pressed between a series of rollers to extract the sugar-laden juice. Water consumption in these steps is high. Average waste water flow rates for a plant which produces about 200 tons of raw sugar per day run as high as 10 million gal/day, according to EPA.

The water is not very clean, either. EPA says that typical wash water contains 1850 lb settleable solids, 1750 lb suspended solids, 650 lb chemical oxygen demand, 12 lb nitrogen, and 5 lb phosphorus for each ton of raw sugar produced. Total coliform counts averaged about 4.85 million/ml in one EPA study, and the concentration of fecal coliforms was around 130,000/ml.

The crushed cane stalk yields a coarse, fibrous material called bagasse. About 80% of the bagasse produced at one mill visited by ES&T was used to fire boilers, and the rest was routinely dumped into the ocean. There is simply too much bagasse produced at most mills to use in-house, mill engineers say, but experiments are under way to determine if it could be used elsewhere.

Burning bagasse presents problems of its own. It's not a particularly clean fuel, and most mills are in the process of installing stack scrubbers to clean up the emissions.

After the juice is pressed from the stalks, it goes to the boiler room where lime is added to precipitate insoluble nonsugars. Clarified juice is thickened in

evaporators, and the resulting syrup is boiled in vacuum pans to form raw sugar crystals. Crystals are separated from molasses by centrifugation, and the molasses may be further evaporated to recover more sugar. The final products—coarse, pebbly raw sugar and molasses for animal feed—are packed for shipment to the mainland.

### Technology available

The EPA says that "facilities for controlling sugar mill wastes are available." And it would seem to be true that control of cane processing wastes requires little in the way of new technology. Old standbys like hydroseparators are used to clarify process water by some mills in the more level plantations, and lagoons for tail water recovery are becoming more common. One mill is experimenting with a "dry cleaner" which uses forced air instead of water to clean cane.

But the problem is more one of economics than technology. Like other industries, Hawaii's sugar industry must reward investors with reasonable dividends. Even with the protection against foreign competition offered by the Sugar Act of 1948, Hawaii faces tough competition from mainland sugar producers.

Because of well-organized unions, the island plantation laborer enjoys a daily wage in excess of $30/day including benefits, according to U.S. Department of Agriculture statistics—twice as much as some of his mainland counterparts. Return on investment, on the other hand, runs between 5–6% after taxes, USDA estimates. "A better investment could be found in bonds," says Frederick Gross, engineer for Castle and Cooke Co.'s Waialua mill. HMM

# Lake Erie group strives for action

*Lake Erie Congress passes resolutions which
it hopes will spur key decision-makers into taking
steps to improve quality of the Lake Erie Basin*

Reprinted from Environ. Sci. Technol., 5, 758 (September 1971)

"There is a burgeoning number of conferences, symposia, meetings, round tables, seminars, conventions, and forums dealing with the Great Lakes," the program introduction began. "Many are of limited value. They retread old ground and recycle old words—while reaching old verdicts."

Not so the Lake Erie Congress, according to its codirectors Harold S. Williams and Robert E. Hansen. The Congress, recently convened under the joint sponsorship of the Institute on Man and Science (Rensselaerville, N.Y.) and the Great Lakes Research Institute (Erie, Pa.), was designed "to make a difference," according to its organizers—"to stretch our minds and lift our horizons."

The mind stretching and horizon lifting was to climax in action. "The program is a Congress with a mission," delegates were told. That mission: "to chart an explicit course of action for dealing with Lake Erie's prospects and problems."

The vehicle for action was a bicameral pseudolegislature—consisting of a House of Context and a House of Interventions. Each house was further split into three committees whose jobs were to draft preliminary resolutions dealing with specific issues in their areas of competence.

The House of Context, for example, had committees on Social Context, Environmental Context, and Economic and Industrial Context. The House of Interventions had committees on Technical, Institutional, and Present Practice Interventions. Each committee was assigned a "rapporteur"—a Harvard, Wellesley, or M.I.T. student—to help polish resolutions, cope with the paperwork, and keep things running smoothly.

As background for their work, delegates were expected to draw upon their particular areas of expertise as well as the International Joint Commission Report on pollution of Lakes Erie and Ontario and the St. Lawrence River (see "Rx for ailing lakes—a low phosphate diet," ES&T, December 1969, page 1243). Additional information was provided by an anthology of interviews, abstracts, and literature summaries prepared for the Congress by Social Technology Systems, Inc. (Newton, Mass.) and distributed to delegates at registration.

## Recommendations

After preliminary resolutions were reported out of committee and passed by their respective houses, joint Context-Intervention committees hammered out the final drafts and presented the distilled resolutions to the entire Congress for ratification. Among the more innovative resolutions passed by the Congress were those calling for:

• A Lake Erie Basin Fund to receive and administer donated money to initiate and implement action programs of benefit to the Lake Erie Basin;

• An environmental trust fund to provide access to capital (not solely on an economic basis) by devices such as low interest or partially guaranteed loans;

• An international Great Lakes Basin Authority (GLBA) which would have power to set standards for water quality, establish and enforce sanctions against polluters, and fund research and development projects. (Such an authority would have a testing laboratory associated with it.);

• A federal-state matching fund program, on a 90-10% basis, to cover capital costs of implementing immediate state-of-the-art technology to reduce pollutants to an environmentally acceptable level;

• A prototype "environmental town" in the Lake Erie Basin for research, development, and demonstration of techniques for pollution abatement, water recycling technology, and relief of urban and suburban sprawl;

• A professional referral service based on a talent bank of qualified consultants.

Now that the Congress has adjourned, was it more than just a bunch of people playing games? Hansen says yes. Proceedings of the Congress will roll off the presses this month and when they do they'll be distributed to legislators and other key decision-makers in industry and pollution control. The format will be "punchy enough," Hansen says, so that "even at the highest desk, the report will have its cover opened and its contents perused." That, the Congress hopes will, at least begin to "make a difference." HMM

**Congress.** *Institutional Interventions committee thrashes out resolutions*

# Recycling sewage biologically

*A novel use of nature's resources may*

*effectively reclaim secondary effluent*

*Reprinted from* ENVIRON. SCI. TECHNOL., **5,** 112 (February 1971)

The typical sewage treatment plant is fairly efficient in removing oxygen-demanding compounds and settleable solids. But the average plant does not remove nutrients, nor does it have a long retention time during the waste treatment process. Although effluent from the sewage treatment plant is an improvement over the untreated influent, this discharge is still rich in chemical nutrients that feed algae, bacteria, and aquatic plants that grow, die, and contribute to the eutrophication of streams, lakes, and ocean estuaries. As a result, the typical sewage treatment plant does not produce an effluent capable of beneficial recycling. Complete recycling requires more time and space, and, generally, neither is available.

In Michigan, however, a 500-acre land and water complex will present the management of waste water from a modern waste disposal plant so that nutrients and contaminants are handled as resources capable of being converted into useful products. Besides this utilization of wastes, the plan includes recreation areas that are so vital in an urban area.

Several interlocking biological recycling systems, designed as an alternative to discharging treated wastes into streams, have been developed by the Institute of Water Research (IWR) at Michigan State University (MSU). This system of recycling and reusing waste water (to be completed in September) is "not just a tertiary step" in an established system, "but is really recycling," says Howard Tanner, assistant director of IWR.

These biological systems, to be established on 500 acres of MSU property, will include a conventional waste treatment plant, an aquatic system of shallow lakes, and a land system of laboratories, wooded areas, and open field plots. The complex will handle 2 million gallons of secondary treated liquid waste per day (equivalent to sewage from 30,000 to 40,000 people). These land and water systems will be combined with a community recreation project to provide complete recycling of the waste water.

### Lake system

Raw sewage will be drawn from a trunk line of the East Lansing (Mich.) sewer system and will be treated by a conventional activated sludge process at the MSU-constructed sewage treatment plant. The waste solids will be returned to the city sewer system; the liquid, secondary-treated effluent, will be piped to the first of three lakes (construction to begin in March) with total surface area of 30 acres.

These three lakes (the first of five that will eventually be built) will be connected by an underground pipe, and the water will flow from lake to lake (by gravity) over a 30-day period. During this movement, the waste water will be stripped of most of its nutritional and polluting characteristics and will eventually, with additional treatment, be used in a swimming pool.

Michigan, like many other parts of the country, has soil that is sandy, which allows water seepage from lakes and ponds into the ground. In the MSU lakes, precautions are being taken to prevent infiltration or water loss into the soil. Many soil sealants are available, but few are effective and most are expensive.

A technique borrowed from agriculture will initially be utilized. An asphalt or clay emulsion will be injected under the ground to seal capillaries, providing greater water-holding capacity.

Rooted aquatic plants will be grown in these 6 to 8-ft deep lakes to maximize the removal of phosphates and nitrates from the secondary effluent. These aquatic plants, which are not algae, are selected for their adaptability to the climate, their high demand for phosphates and nitrates, and their food value. For this, plants have been collected from as far north as the Hudson Bay and as far away as New Zealand. Suitable plants have a growing season

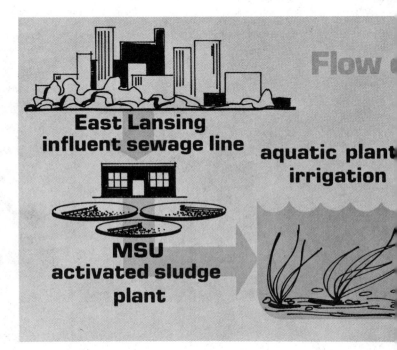

East Lansing influent sewage line

MSU activated sludge plant

aquatic plant irrigation

Flow

from as early as April to as late as October, or until the lakes are covered with ice.

Of course, the first lake will be the richest in nutrient content, but each lake will support underwater plant life that will be harvested three or four times a year. The lakes are designed to facilitate harvesting of foliage.

What the harvested corp will be used for remains a question. One answer is to feed it to livestock. The nutritional value of these aquatic plants is equivalent to that of alfalfa, perhaps even higher. But they will be more costly to harvest since they have to be cut underwater. However, the cost can be partially offset by its contribution to the reclamation of the waste water. Laboratory testing has shown that the plants can be dehydrated and pelletized for future use or fed directly to livestock. Another possibility is the extraction of proteins from this material for feeding the increasing human population.

## Water usage

Fish will be grown in the last two lakes, and possibly in the first lake. Forage fish that are potentially marketable as bait will be stocked in the first lake, with their use dependent upon the amount of BOD present. The last two lakes will each contain two fish populations: catfish, a high-protein source, could provide as much as 500 to 1000 lb of fish per acre per year; bluegill-bass will be a source of recreation for the sporting fisherman.

A recreational area will be an additional contribution from this reclaimed water supply. Grounds surrounding the lakes will be landscaped for public recreation. As the waste water is being purified, it will be odorless and esthetically pleasing to the eye, says Frank D'Itri, staff chemist, IWR. Thus, the area can be used for walking, boating, fishing, and picnicking.

Spray irrigation on adjacent lands will also be used to strip nutrients from the water. This water will be distributed to forage crops, selected row crops, coniferous tree plantations, woodlands, and old farm lands. The increased plant growth and the effects on the soil and water will be measured after nutrient removal by the interaction of soil, sun, and vegetation. Irrigation water will be available from any of the lakes and will be used on almost every conceivable type of plant that lives in the Michigan climate. The land will be irrigated year round, perhaps with increases during the winter, when biological activity in the lakes slows down.

Another use for this "biologically cleaned" water relates to the geography of the area. Lansing and the surrounding region are above a large underground aquifer—i.e., on a large bowl of water. In 1935, before the region's population grew significantly, this underground water supply was full and often overflowed as springs. Excess water from the aquifer fed the streams and rivers. With the growth of industry and population, the large

wells in this region draw heavily on the groundwater, causing a drop in the water table. The more shallow wells are now located above the table. Because of this change in the water table, not only are the people with shallow wells deprived of water, but waters from streams are now returning to the aquifer rather than being further supplied by it. Eventually, the water will be poorer in quality yet more costly to bring to the surface.

Solving such problems will take time, but this water reclamation plan offers the possibility of spraying the cleaned water over the surrounding land, where it will eventually return to the ground aquifer, thus recharging the larger underground water source. This project offers a potential closed hydrologic cycle where water is drawn from the wells, used and dirtied, cleaned, and returned to the ground to be used again after an elapsed period. Careful monitoring will assure a clean water supply for future generations.

## Pioneer

When the first three lakes are completed, the final two will be added along with a swimming pool, which will make use of water from the system after additional treatment. This project, if successful, could serve as a pilot plan for communities with inadequate water treatment facilities to overcome their deficiencies. The entire country will be on watch for the outcome of this MSU "two for the price of one" system.                    CEK

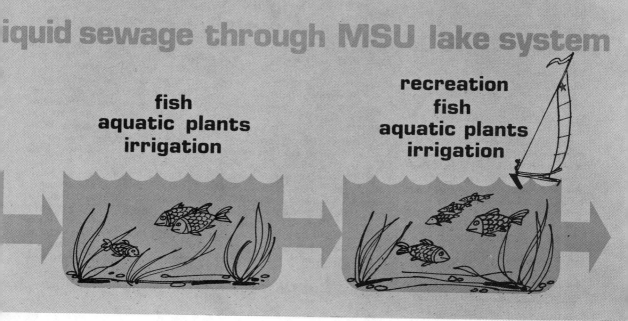

iquid sewage through MSU lake system

**fish
aquatic plants
irrigation**

**recreation
fish
aquatic plants
irrigation**

# Water quality in industrial areas: profile of a river

The lower Hudson, like most tidal areas, must
serve many needs, and its biological status
is a direct measure of its ability to do so

*Reprinted from* ENVIRON. SCI. TECHNOL., **4**, 26 (January 1970)

The Hudson River drainage area,
one of the major watersheds of eastern North America, encompasses
about 12,400-14,500 square miles, and
supports a human population of about
10.5 million. The flow in the southerly portion is controlled by a dam at
Troy, about 155 miles north of the
river mouth. The channel of the
river can be traced well out into the
Atlantic, and, at the northern limit of
the tide at Troy, the river bottom is
still four feet below sea level. The
physical characteristics of the river
bed indicate that it is a "drowned
river."

The lower Hudson is a major transportation artery and, at the same time,
a source of water for industrial and
domestic purposes, an unsurpassed recreational resource in an area of great
natural beauty and important historical associations, and serves as a drain
for industrial and domestic wastes.
The economic, industrial, and recreational potential of the river can
hardly be overemphasized, yet, regret-
tably, little is known about its hydrology, biology, or chemistry. At a time
when there is a need to predict the
impact of projected developments,
much basic data still are unavailable
or controversial.

Part or all of the lower Hudson
can be defined as estuary. The more
seaward stretches are mesohaline
(from two thirds to one third seawater), an intermediate zone is oligohaline (one third to one tenth seawater), and there is a final limnetic
(or fresh water) stretch. The extent
of these zones in the lower Hudson
is closely dependent on fresh water
runoff from the drainage area, and,
consequently, the saltfront—or the intrusion of seawater—varies from year
to year, and with the seasons.

## Hydrology

The average annual fresh water inflow into the lower Hudson at Green
Island has been estimated at between
12,000-23,000 cubic feet per second
(c.f.s.). This flow is uneven, with
spring maxima as high as 40,000 c.f.s.
(as in April 1967), and summer min-
ima as low as 2000 c.f.s. The summer
low flow period lasts from four to
seven months each year. As the fresh
water flow diminishes, the salt water

front pushes upstream from its spring
limit near Tappan Zee (30 miles
from Battery Park, Manhattan),
reaching Newburgh (59 miles), Poughkeepsie (74 miles), or even Kingston
(88 miles) in the summer and fall,
depending on the annual rainfall. At
any station on the river, the salinity
increases progressively from May to
November. There is generally a minor
additional flow of fresh water before
the cold temperatures of mid-winter
reduce the flow. Although salinity is
reduced somewhat (and the salt front
pushed southwards), it remains relatively high throughout the winter,
until the spring thaw brings about a
massive dilution in March or April.

As seawater pushes up a river at
flood tide, it tends to form a wedge of
more dense saline water at the bottom
of the river bed. Since the Hudson
River is more than 100 feet deep in
places, this saline water might be expected to remain near the bottom. In
fact, only relatively slight differences
have been seen between the surface
and the bottom, indicating that the
water gets fairly well mixed, at least
north of the Tappan Zee.

The northward motion of the tides
at some periods apparently reverses
the river flow; since the tidal flow
is massive—approximately 300,000
c.f.s.—it dwarfs the summer fresh
water flows. Upstream flows of 19,-
000 c.f.s. have been measured at Governor's Island, 5000 c.f.s. at Riverdale,
and even 2100 c.f.s. at West Point.
The Federal Water Pollution Control
Administration's Hudson-Champlain
project, using dye marking studies,
showed virtually no flow during the
summer of 1965, when dyes added to
the river at different sites were followed for 14 tidal cycles. At the head

**Gwyneth Parry Howells and Theo. J. Kneipe**

*Institute for Environmental Medicine, New York, N.Y.*

**Merril Eisenbud**

*Environmental Protection Administration of New York, N.Y.*

of the tidal influence at Troy, the tidal excursion was only three miles, but the velocity of net movement of the dye mass downstream was a mile and a half for each tidal cycle, or about 20 miles in a week. At Kingston and further south, while tidal excursions were greater, there was no net movement of the dye downstream.

The overwhelming effect of the tidal flux appears to produce a seiche-like movement of the brackish water, but little effective exchange. The implication for pollution and eutrophication effects are clear. The large volume of water (150 miles long with an average cross section of 150,000 feet) behaves as a brackish lake rocked north and south by the tide. The inflow is sufficient to exchange only 0.3-2% each day, assuming the simplest model of a single compartment, no evaporation or withdrawal, and no rain or additional inflow. The mean life of pollutants that remain in solution or suspension thus ranges from 40 days to more than 300 days, depending on the flow. Effluents and nutrients discharged into this brackish lake will, as in a true lake, be recirculated during dry summers between water, sediments, and biota. Only the high spring flows provide a flushing volume of water necessary to prevent an accumulation of pollutants and eventual eutrophication.

Too little is known about the physical hydrology of estuaries generally, and of the complex Hudson River system in particular. Without detailed knowledge of the self-purifying capacity of a river or an estuary, we cannot make a reasonable forecast of the effects that will follow a polluting load. A British study of pollution in the River Thames demonstrated that the narrow parts of estuaries have a comparatively limited capacity to purify polluting material. The water flowing to and fro is substantially the same water from one day to the next, and, at times of low fresh water flow, pollutants may remain within the system for several months, building up in concentration. The narrow tidal stretches of the Hudson, then, are most vulnerable to pollution, with a limited surface area and mud-water interface available as either a source or a sink for pollutants.

### Hudson ecology

The Laboratory for Environmental Studies, part of New York University Medical Center's Institute of Environmental Medicine, has been studying aspects of Hudson River biology and chemistry since 1963. The study, financially supported in part by the New York State Health Department and the Consolidated Edison Co., developed initially from an interest in environmental radioactivity problems, and has broadened into a wider ecological study of the effects of various pollutants on the biota. We are accumulating survey data from the river, and trying to relate our information to that from other studies, especially those of water flow and industrial use of the river.

Those who have been actively engaged in this study (aside from the authors) are A. Perlmutter and H. Hirshfield of the biology department, and A. McCrone of the geology department at the Washington Square

Campus of New York University, and Dale Bath at the Lanza Laboratory.

Investigators pursued several lines of study:

• Considerable knowledge was acquired about the abundance, distribution, and variety of animals and plants in the river, from microscopic forms to fish.

• Seasonal changes in the nutrient anions as well as trace cations were followed.

• A long, continuing study was made of levels of radioactivity—both natural and man-made—in the water and biota.

• A study was made of levels of organo-chlorine pesticides in the river water and mud, and their accumulation in selected species of the biota.

• Present and future effects of heat additions to the river from industrial cooling and processes are being evaluated.

We are trying to see how the effects of varying fresh water flow and tidal cycle influence the distribution of inorganic pollutants, nutrients, pesticides, and heat, and how these affect the biota of the river. In short, we are trying to predict the future of the river in terms of eutrophication, in the face of increasing industrial and domestic utilization.

## Nutrient loads

The sparsely populated agricultural Mohawk watershed, stocked with farm animals and supplied with fertilizers, provides significant nutrient input. This is seen in the relatively high nutrient levels of Mohawk river water (0.85 mg. nitrogen/l.) compared with other northeastern American rivers. About 17% of this could be attributed to natural runoff, and the remainder to artificial sources.

The 10.5 million population of the Hudson watershed produces 61 million kg. of nitrogen and 5.5 million kg. of phosphorus as domestic wast in a year, much of which ultimatel will be carried seawards by the river Other wastes such as domestic de tergents (contributing two thirds o total phosphate in one municipal dis charge), and wastes from meat an dairy industries, yeast production and paper manufacture, also contribut to the nutrient load.

Present municipal waste discharge north of Yonkers provide 1.6% of th spring volume of river flow, and 16% in the summer. At the Verrazan Narrows (Brooklyn), the proportion are much greater, because of th population density of the New Yor City area. The use of some rivers a sewage conduits has led to deoxygena tion and conditions impossible for fis life, especially when sewage disposa is coupled with other industrial uses.

The nutrient content of the rive water reflects its use. At Indian Poin (43 miles from the Battery), the phos

**Distribution of common

Mesohaline zone, 1/3-2/3 seawater

Oligohaline zone, 1/10-1/3 seawater

Anemones, jellyfish

Barnacles

Copepods

Shrimps, prawns

Crayfish

Crabs

Oysters, clams

−10    0    20    40    60

**Miles from Battery**

Statue of Liberty    Battery Park    George Washington Bridge    Tappan Zee Bridge    Newburgh    Poughkeepsie

phorus content of the water ranges from 9.5-2.5 $\mu$g.-atoms/l. and, at the southern tip of Manhattan, a maximum of 12 $\mu$g.-atoms/l. was recorded. (Seawater has about 2 $\mu$g.-atoms/l.) Most of the phosphorus (70%) is present as inorganic phosphate, available for immediate plant assimilation and growth, while the remainder is dissolved organic phosphorus or particulate material. In general, phosphorus in river water usually is higher than in lakes, but these values are high by any measure—2.8 $\mu$g.-atom/l. is the approximate upper limit of unpolluted water. Lake Washington (Seattle, Wash.), where eutrophication is slowly being reversed by a costly sewage diversion, has a phosphorus concentration of 7.5 $\mu$g.-atoms/l.

Ketchum deduced a theoretical relationship that oxygen demand is equivalent to the oxygen supply from photosynthesis at concentrations of about 2 $\mu$g.-atoms of phosphorus/l.

At higher phosphorus concentrations, the net oxygen demand during darkness will deplete the dissolved oxygen levels in the water. On this basis, we might expect that south of Albany, and around Manhattan, there will be net oxygen depletion. This has been shown to be true by monitoring studies of dissolved oxygen and biological oxygen demand levels. However, at many sites on the river, the dissolved oxygen level is adequate to maintain a healthy fauna in spite of the high phosphorus content.

Nitrate nitrogen concentrations in the mid-Hudson range from 0.2 mg. to as much as 1 mg. N/l., and are somewhat related to the tidal incursion. Inflowing water from the upper Mohawk watershed has a concentration of about 0.85 mg. N/l. These levels may be compared with values of 10 mg. total N/l. in the Thames River in England, and values around 1 mg. or less in relatively un-

polluted lakes. A seasonal fall in nitrate concentrations, such as that seen in the late spring of 1968 at Indian Point, could be a limiting factor in phytoplankton growth in this region, although many other factors may be implicated. Nitrates can act as an important reserve of oxygen, even with appreciable concentrations of dissolved oxygen, but, in anaerobic conditions, nitrogen compounds are reduced to nitrogen or even to ammonia. The nitrogen cycle is related to dilution, oxygen, and temperature, and complicated by the effects of nitrogen fixation or denitrification by bacteria or plants.

Preliminary data on sulfate levels indicate about 25 mg. $SO_4$/l. at Nyack, fairly closely related to salinity. As with nitrate, sulfate can be reduced to sulfide by bacteria in anaerobic or near anaerobic conditions.

The availability of these important anions—nitrate, phosphate and sul-

## species in the Hudson River

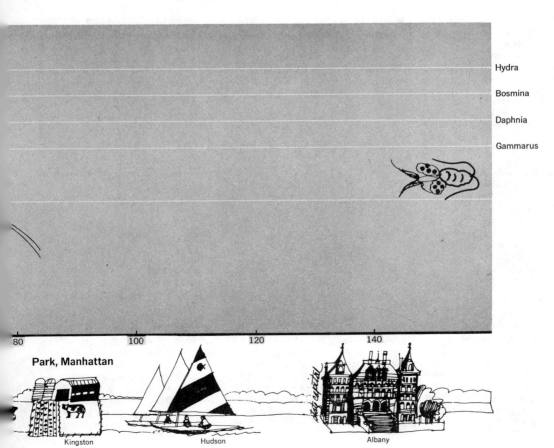

Limnetic zone, fresh water

Hydra

Bosmina

Daphnia

Gammarus

80    100    120    140

Park, Manhattan

Kingston

Hudson

Albany

## Common invertebrates of the lower Hudson River

| | Mesohaline zone | Oligohaline zone | Limnetic zone |
|---|---|---|---|
| Protozoa | Dinoflagellates | *Difflugia*<br>*Arcella*<br>*Ochromonas*<br>*Polytomella* | *Volvox*<br>*Synura*<br>*Paramecium*<br>Amoebae |
| Coelenterates | Sea anemone (*Sagartia*)<br>Jelly fish<br>Ctenophores | *Hydra oligactis*<br>Occasional jelly fish | *Hydra oligactis* |
| Rotifers | *Trichocerca* sp. | *Keratella cochlearis* | *Philodina* |
| Worms | Polychaetes | Polychaetes | *Tubifex* sp.<br>*Aeolosoma*<br>Nematodes |
| Crustacea | Harris crab (*Rhithropanopeus*)<br>Blue crab (*Callinectes*)<br>Prawn (*Palaemonetes*)<br>Shrimp (*Crangon*)<br>*Bosmina longirostris*<br>Barnacle larvae<br>Copepods: *Microarthridion*<br>    *Cyclops bicuspedatus*<br>    *Acartia tonsa* | *Gammarus fasciatus*<br>Prawns<br>Shrimps<br>*Bosmina longirostris*<br>Barnacle larvae<br>Copepods: *Microarthridion*<br>    *Cyclops bicuspedatus*<br>    *Eurytemora hirundoides*<br>    *Ectinosoma eurticorne* | *Gammarus fasciatus*<br>Crayfish (*Orconectes*)<br><br>*Bosmina longirostris*<br>*Daphnia pulex*<br>Copepods: *Microarthridion*<br>    *Cyclops bicuspedatus*<br>    *Diaptomus pallidus* |
| Mollusca | Snails (*Physa*)<br>Clam (*Mya*)<br>Oyster (*Crassostrea*) | *Congeria leucophaeata*<br>*Sphaerium* | *Elliptio complanatus* |
| Insecta | None | *Chaoborus albipes*<br><br>*Tendipes* | Chironomid larvae<br>Dragon fly larvae<br>Stone fly larvae<br>May fly larvae |

fate—is important in a consideration of possible eutrophication of the lower Hudson. The present move to treat sewage so as to provide a liquid effluent may do little to reduce the nutrient levels of Hudson River water. In fact, it may worsen the present situation, since the treated effluent will provide nutrients in a readily available soluble form. Nitrate and phosphate levels in the Hudson are more than sufficient to develop algal blooms. Profuse blooms would result in a net oxygen depletion, with the subsequent reduction of nitrate and sulfate to noxious gases. In a flowing aquatic system, the interaction of nutrient concentrations and water temperature may be of major importance in controlling algal growth rates. If algal production exceeds the rate of removal by downstream flow and predation, nuisance blooms could result.

### Heat additions

A further problem of growing importance is the use of water for industrial cooling. At the present rate of increase of water use in U.S., about 20% of the total fresh water runoff in the next 10 years will be required for cooling. Considering seasonal variations, about half of total runoff will be required for two thirds of the year. In the highly developed northeastern states, there are a number of rivers whose total flow is utilized, sometimes more than once in the passage downstream. Excessive water use will build up the heat load of fresh waters and estuaries; fortunately, heat is not conserved and can be dissipated to the atmosphere, provided the volume of water is adequate. Here is a situation where good management of water resources is essential to maintain water quality. The siting and design of power stations must be considered from the point of view of biological effects, so that overuse of the water resource is prevented. The difficulty lies in determining at what level of heat addition the biological effects are significant.

Long-term heat changes have been well documented for the Thames River, where heat released to the river rose from 555 mW in 1930 to about 3700 mW in 1950. The total volume of cooling water was approximately 2000 c.f.s. with a summer minimum of only 260 c.f.s. Of the total heat load in 1950, 75% was contributed by fossil fuel power stations, 6% by industrial effluents, 9% by sewage effluents, 6% by fresh water discharges, and 4% by biochemical activity. The yearly average temperature in the river over these 20 years rose from about 53° F. to about 60° F. (an estimate corrected for changing meteorological conditions). On the basis of the proportions of contributors above, three quarters of the 7° F. rise (or about 5.3° F.) is due to about 3000 mW of power. The biological effects of such a temperature rise are not known, but the additional temperature in the Thames increased the oxygen deficit by about 4%. This effect is not large, and the estuary would have remained anaerobic even if its water had not been used for cooling. However, it was calculated that the reducing conditions, together with the temperature, increased the evolution of hydrogen sulfide by 20%. Following the introduction of strict legislation to control the quality of industrial discharges other than heat, hitherto fishless zones of the metropolitan Thames once again support fish life, albeit of species tolerant of the low oxygen.

The flow of the Hudson is about 15 times greater than that of the Thames. The thermal capacity of the four power stations on the river is only 3000 mW, but projected development within the next few years includes an additional 10,000 mW. While the flow of the Hudson is not so intensively utilized as the Thames, and ambient air temperatures in the two countries are not the same, the parallel is interesting, indicating the sort of temperature change, if not the magnitude, we might expect.

## Biology of the Hudson

The variety of plants and animals in the Hudson is known only imperfectly. A study by the New York State Conservation Department some 30 years ago gives a good account of fish and rooted vegetation, but little information about the plankton or benthos (bottom living fauna) of the river. To evaluate the effects of such pollutants as excessive nutrients, trace metals, pesticides, and heat, we have been inventorying the kinds of animals and plants in the river, their relative abundance, and their distribution with respect to salinity. With so little baseline information, it is difficult to draw conclusions about changes in the biota that may have occurred. Even for fish studies, the methods of investigation have not been repeated. Hopefully, our present knowledge will be adequate to allow us to predict future changes in the river.

Our studies indicate there are two rather distinct environments in the Hudson River—the main channel and the shore. The main channel has abundant plankton, a rich variety of species of protozoa, diatoms, algae, rotifers, and the small crustacea which are important as food for larval fish. The shore has a rather different fauna and flora, partly because of different physical conditions, but also perhaps

because pollutants from shore-sited industry sweep along the shores before they are diluted by the main body of the river.

In the oligohaline stretch of the river, the dominant phytoplankter during the spring and summer is the *Melosira ambigua,* but, during the fall, this species gives way to others. At the same time, salinity rises with the summer seawater intrusion, allowing survival of a more marine-type biota at this site. When the estuary water is diluted by early winter or spring rains, a return of *Melosira* is

seen, together with a variety of other fresh water forms.

At a more southerly station on the river, a similar microflora to that at Indian Point was found during the summer. The limnetic zone in the summer is characterized by other species of *Melosira* and by other fresh water species. At all stations along the river, a number of species are found which might be considered as indicators of eutrophication. None of these were found as blooms, but the biological potential for nuisance algal growths (eutrophic species and more than adequate nutrient levels) undoubtedly is there.

The zooplankton characteristically is dominated by the microcrustacea, largely a flourishing copepod fauna. A number of crustaceans appear to be ubiquitous throughout many miles of the Hudson. Barnacle larvae are common in the plankton of mesohaline and oligohaline zones, derived from a benthic adult population extending to Peekskill. (Some of the copepod species show a transition with changing salinity.) An important zooplankter for fish nutrition, *Mysis oculata,* was found at Indian Point only in the fall of 1968, when salinity reached about 25% of the seawater levels.

Another group common in the zooplankton is the rotifers, sometimes frequent enough to be termed blooms, again with a succession of species inhabiting different zones. The rotifer fauna appears to be rich and varied.

**Navigational aid.** *Esopus Meadows Lighthouse illustrates transportation, recreation, and aesthetic value of Hudson River*

## Fish seined on lower Hudson shores, July—August 1965-68

| | Mean number/100,000 ft² |
|---|---|
| Blueback herring | 598 |
| Freshwater killifish | 512 |
| Spot-tail shiner | 333 |
| Northern silverside | 317 |
| Johnny darter | 200 |
| White perch | 180 |
| Common sunfish | 164 |
| Alewife | 102 |
| Goldfish | 93 |
| Bay anchovy (absent 1967/68) | 43 |
| Golden shiner | 42 |
| Striped bass | 30 |
| Carp | 10 |
| Shad | 9 |

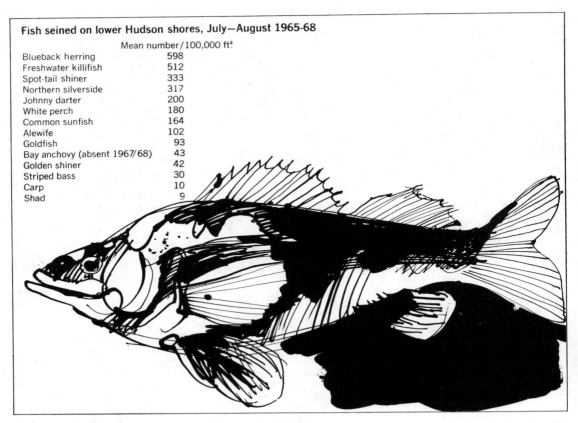

The protozoan fauna is also rich in species in all parts of the river, but we do not have quantitative assessments of the relative dominance of species. Ciliates (25 genera) and flagellates (15 genera) are common, as are shelled amoebae (15 genera), in shore collections. Of the larger invertebrate inhabitants of the river bottom and of the shores, we know little about relative abundance, but something about distribution. Larger crustacea are represented by the Harris crab and blue crab in the more seaward stretches of the estuary, and by shrimps and prawns in less saline water. Replacing these decapods in the limnetic zone are crayfishes. The presence of the crabs upstream of Tappan Zee seems to depend on the seawater intrusion during the late summer. The apparent scarcity of crabs in very recent years may then not be due to pollution, but rather to freshening of the river after the drought years of 1964-66.

Mollusks also show a salinity related distribution. One species of oyster has been found as far north as the Tappan Zee; in the fresh water, gastropods are common. Insects are scarcely present in the more saline reaches, only the larval *Chaoborus* and *Tendipes* extend into the oligohaline zone. Other insect larvae are found in the limnetic zone, but are restricted to the shore, except for chironomids. River mud provides habitat for worms.

The river has a large population of endemic fishes. Some 70 species recorded in 1936 include a tremendous migrant population of diadromous fish, many of which migrate into the Hudson or its tributaries to spawn. The shore environment serves as a nursery ground for the juvenile fish who feed on the snails, shrimp, and insects available there. This population has been sampled at selected stations along the river for the past five years by shore seining, and about 35 species have been recorded. The fish sampled by this technique are mostly in their first or second year, and some fish known to be present in the river—such as the hogchoker and the sturgeons—have not been seen. The shore sampling thus does not represent the true population, but only juvenile and small species which seek sheltered and shallow inshore areas. However, the consistent annual sampling can be used to indicate water quality changes. Most of the fish show little distribution related to salinity, since they are euryhaline, and able to live in waters of a wide range of salinity. The fresh water variety of killifish is dominant in the upper reaches of the river, while the euryhaline type is more common further south, and their distribution reflects the seawater intrusion. It is difficult to make valid comparison of the fish fauna of the river now and in 1936, but it is clear that the Hudson still presents an environment rich in variety and quantity of fish, even though commercial fishing in the river has severely declined.

### Radionuclides

The lower Hudson River receives direct industrial waste discharges, as well as material from the watershed area, and the levels of some waste products in the river and its biota are of interest. We followed radionuclide and pesticide concentrations in the water and mud and their accumulation in the biota of the river.

The radionuclides are derived from three sources:

- **Natural products, such as radium and potassium-40.** This group reflects the geological character of the watershed and the degree of salt water intrusion. Potassium-40 is the major contributor of radioactivity in estuarine water and its biota. South of West Point, the mean potassium-40 concentration during the past five years was 22 picocurie per liter (pCi/l.), less than 10% that of seawater, while, in the limnetic zone, it is only 1.5 pCi/1. Natural radium-226 and radium-228 derived from soils each contribute only about one tenth of a pCi/1.
- **Fission products, derived from fallout from weapons testing.** The nuclides derived from fallout are cesium-137, strontium-90, cerium-144, and ruthenium-106. Levels of these nuclides in the water or biota are low, and together contribute only about 1 pCi/1.
- **Activation products released as a result of nuclear production.** This group is the most interesting, even though levels are very low in water. Among the nuclides seen are cobalt-60 and manganese-54. Although their levels are so low that accurate estimates are difficult to make, their accumulation in plants or animals leads to the use of selected species as natural monitors. The situation is illustrated best by manganese-54, which is found in river muds, especially downstream of the nuclear power station at Indian Point, where physical and chemical conditions of seawater promote the salting out of manganese compounds.

Both stable manganese and its radioactive analog are accumulated by some water plants, especially those in the genera *Chara*, *Potamogeton*, *Valisneria*, and *Myriophyllum*, which are found along the shores of the Hudson. Those plants growing closest to a reactor effluent site naturally show the highest levels of activity. The natural soluble manganese concentration in Hudson River water ranges from less than 0.5 to 12 $\mu$g./1., and total manganese (mostly particulate) about 1 mg./1. In *Potamogeton crispus*, it is about 2.3 mg./g. wet weight, indicating a concentration factor of about 190,000, with regard to the soluble nuclide concentration in the water, or 2300 to the total nuclide concentration. The total radioactive manganese levels in river water during 1966-68 indicate concentration factors of about 10,000 (range 3000-18,000) for this and similar species, in the same order of

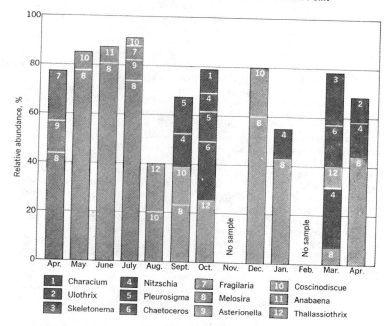

Phytoplankton species in Hudson River at Indian Point

| | | | | | | |
|---|---|---|---|---|---|---|
| 1 | Characium | 4 | Nitzschia | 7 | Fragilaria | 10 Coscinodiscue |
| 2 | Ulothrix | 5 | Pleurosigma | 8 | Melosira | 11 Anabaena |
| 3 | Skeletonema | 6 | Chaetoceros | 9 | Asterionella | 12 Thallassiothrix |

magnitude as total manganese. Plants of other genera, or phytoplankton samples, showed much lower levels of manganese or manganese-54.

Fish with a stable manganese content of only about 6 $\mu$g./g. wet weight do not show appreciable accumulation of radioactive manganese, even though exposed to the same concentrations in the ambient water. Even if species of fish which are known consumers of plants are considered, there is no appreciable accumulation of manganese-54 observed. Hence, although these fish may ingest food high in manganese-54, low absorption by the gut probably limits its uptake in fish. Among invertebrates in the

river, blue crabs did not show any manganese-54, and only traces of two fission products, cerium-144 and cesium-137 (in 1964). Other crustaceans—crayfish, prawns, and shrimp—have similarly low concentrations. On the other hand, fresh water clams, which feed by filtering plankton and sediment from the water, showed a larger accumulation of radionuclides, including manganese-54, though not so much as plants. Other filter feeders—barnacles, oysters, etc.—were not sampled, but might be expected to have similar values. The relative accumulation from the two routes—direct uptake from the aquatic environment and uptake from ingested food—is not known.

#### Pesticides

While the distribution of radionuclides in the river water, mud, and biota reflects the distribution of the stable elements, as well as chemical interactions and the physiology of the biota, the distribution of pesticide residues reflects more clearly accumulation through trophic levels of the biota. Hudson River water contains numerous pesticide residues of the chlorinated hydrocarbon type, but only during the spring runoff are concentrations as high as 0.25 $\mu$g./1. At other times of the year, the levels are too low for effective quantitation, generally less than 0.010 $\mu$g./1., but the residues can be more readily identified in biological samples.

Like other major northeastern rivers, the Hudson River water appears to have little DDT, although its metabolites are present. Dieldrin is present in about the same levels of concentration as the DDT metabolites. This pattern of occurrence is reflected in the biota, where accumulations of dieldrin and DDT metabolites are seen, rather than DDT itself. The muds take up pesticides from the water or retain them in settled sediments after the spring runoff. Muds contain the pesticides at 0.01-0.05 $\mu$g./g. dry weight, generally several thousand times higher than the water. Plankton in the water, perhaps feeding on both microforms and suspended particles, build up concentrations of pesticides of about 0.02-0.06 $\mu$g./g. wet weight (0.1-0.3 $\mu$g./g. dry weight). This is about five to ten times higher than the sediments and about 20,000 times higher than water.

Clams feeding on the plankton would be expected to accumulate the residues to an even higher degree, but, in fact, pesticide concentrations in the fresh water clam are of the same order, 0.03-0.07 $\mu$g./g. wet weight. Various fish species in the river have pesticide concentrations ranging from 0.05-0.8 $\mu$g./g. wet weight. Local birds, such as heron and killdeer, thought to be consuming fish or invertebrates from the river, have 0.3-3 $\mu$g./g. wet weight on a whole body basis. The pesticide residues are particularly concentrated in fat. Thus, the continuing concentration through the food chain results in relatively high pesticide concentrations at the top of the food chain even though the base levels in the water are relatively undetectable. This has been observed in other environments.

Trace element concentrations in the river water show great variability with sampling station, state of the tide, and temporal differences which, perhaps, reflect intermittent discharge of effluents from industries. Generally, in the limnetic zone, levels of trace metals are much less than those permitted by state or federal drinking water standards and do not appear to cause nuisance. In the more estuarine reaches of the river, high concentrations of iron, copper, and cadmium have been seen. Extensive studies would be required to locate the sources of such sporadic trace metal concentrations. Evaluation of potential biological effects would require application of sophisticated multivariate analysis.

Muds in the river may play an important role in the sequestration or regeneration of toxic material derived from pollutants, in addition to their role in the oxygen cycle. Our survey of radionuclides and pesticides has indicated that muds always are much higher in concentrations of these materials than the overlying water. There are other indications from gas chromatograph tracings that they are accumulators of diverse organic residues as well.

Most mud samples from the lower river are clay silts with about 5-6% of easily oxidizable humic material, 50% silt, and 20-45% clay. The cation exchange capacity of the muds is considerable, even in the brackish reaches where cations are readily available in the water. Consequently, most of the exchange sites on the muds are occupied by hydrogen under predominantly reducing conditions.

The muds contain a substantial amount of manganese, (0.72 $\mu$g./g. wet) compared with the ambient water, (0.01 $\mu$g./ml.). Hence, there is a sedimentary reservoir of manganese which could be available if reducing conditions prevail. Appreciable amounts of iron in the ferrous state also are present. Organic matter in the muds, oxidizable by hydrogen peroxide, accounts for more than 65% of the cation exchange capacity. Together with the unsaturated nature of the exchange sites, this indicates that, in relatively well oxygenated

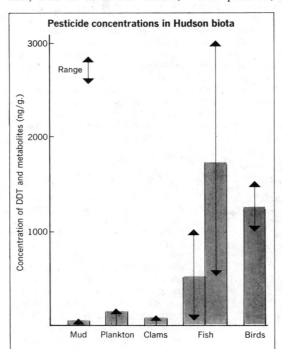

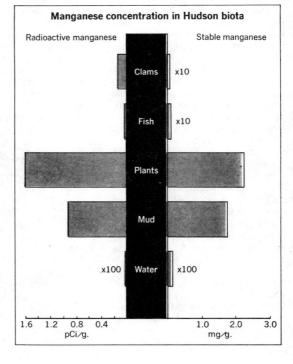

conditions, the muds may have considerable capacity to absorb chemical pollutants, including radionuclides. More research is needed to evaluate the role of the muds in this capacity.

## Conclusions

It is not difficult to predict an increased use of the Hudson River for the disposal of sewage wastes, industrial effluents, and cooling water, and for augmenting existing water supplies. At the same time, the burgeoning population of the area has a need and a right to use the river for recreation. How can further development be controlled, so that the water resources can be exploited, but still provide for that refreshment of the spirit so necessary for the urban inhabitant? Changes that might be expected from increased use are:

• First, an increasing nutrient load from domestic sewage and some industrial processes.

• Second, an increasing heat load.

• Third, an increased demand for industrial and domestic water.

Our studies have made clear that potential eutrophic nuisance species of algae are present in the river, and that the shores are populated by animals indicative of sewage pollution. Yet, serious fouling and deoxygenation have so far been avoided for most of the river. From this it could be deduced that the present situation need cause no concern; however, examples of other eutrophic water bodies give us warning of the potential rapidity of changes, and should encourage effective sewage treatment and the control of nutrient sources.

Heat additions to the aquatic environment are a major concern. In the Hudson, the volume of tidal flow can be utilized to disperse such heat; at the same time, it is clear that the capacity of the river as a heat sink is severely limited during the peak summer demand by a low net flow and high ambient air temperatures. If significant overall temperature rise in the river were allowed, it is highly probable that the species composition of the fauna and flora would be unbalanced. This interacting with the high nutrient levels in the river could easily tip the balance between nuisance conditions and the relatively healthy biological situation seen today. The effects of a temperature rise restricted to localized sites on the river have yet to be evaluated. There is a great need for more detailed hydrological and thermal studies of the river to evaluate the capacity of the Hudson to receive heat addition. There is also need for studies of species endemic to the river to determine their response, singly and together, to changes in temperatures.

The extraction of additional water for any purpose—pumped storage schemes, industrial use, or domestic use—is also of great importance, and is closely related to the other uses of the river. Almost any increased water extraction, except industrial cooling intakes, will make a volume of water unavailable at least for limited periods or limited stretches of the river. The effects of this on the present hydrological pattern in the river remain largely unknown. It seems probable, however, that the extent and duration of salt water intrusion up the river will increase. This will limit the sites for drinking water extraction and reduce the capacity for exchange of effluent discharges with the ocean which depends largely on the net fresh water flow.

How can the situation be controlled? We need more information about all aspects of the hydrology of the estuary, about the fauna and flora of the river and their response to existing and predicted conditions. We need strictly controlled use of the river for all purposes and at all levels. And, finally, we need to know how to alleviate pollution problems when they have arisen and how to channel waste materials, including heat, to other outlets.

## ADDITIONAL READING

W. T. Edmondson, "Water Quality Management and Lake Eutrophication: The Lake Washington Case," in "Water Resources Management and Public Policy," edited by T.H. Campbell and R.O. Sylvester, University of Washington Press, 139-178 (1968).
"Effects of Polluting Discharges on the Thames Estuary," Water Pollution Research Technical Paper No. 11, Department of Scientific and Industrial Research, Her Majesty's Stationery Office, London (1964).
B.M. Ketchum, "The Flushing of Tidal Estuaries," Sewage Ind. Wastes 23, 198-209 (1951).
G.G. Polikarpov, "Radioecology of Aquatic Organisms, Reinhold, New York (1966).
D.W. Pritchard, "What is an Estuary: Physical Viewpoint," in "Estuaries," edited by G.H. Lauff, American Association for the Advancement of Science, 3-5 (1967).
U.S. Department of Health, Education, and Welfare, "Report on Pollution of the Hudson River and its Tributaries," U.S. Government Printing Office, Washington, D.C. (Sept. 1965).

**Gwyneth Parry Howells** *is senior research scientist and director of Hudson River ecology studies, New York University Medical Center Institute of Environmental Medicine, a position she has held since 1967. Previously (1962–67), she was senior scientist, Medical Research Council, London. Dr. Howells received her B.Sc. (1946) and M.Sc. (1948) from the University of New Zealand, and her Ph.D. (1953) from Cambridge University (England).*

**Merril Eisenbud** *is administrator of the Environmental Protection Administration of New York City, a position he has held since 1968. Previously (1959), he was professor of environmental medicine and director of the laboratory for environmental studies, NYU Medical Center Institute of Environmental Medicine. He received his B.S. from New York University (1936), and D.Sc. from Fairleigh Dickinson (1960).*

**Theo. J. Kneip** *is acting director of the laboratory for environmental studies, NYU Medical Center Institute for Environmental Medicine, a position he has held since 1968. Prior to joining the institute (1967), he was with Mallinckrodt Chemical Works (1954-63). He received his B.S. from the University of Minnesota (1950), and his M.S. (1952) and Ph.D. (1954) from the University of Illinois.*

# ORSANCO: Pioneer with a new mission

*New roles for 22-year-old water*

*pollution agency add to an impressive*

*list of past achievements*

Reprinted from ENVIRON. SCI. TECHNOL., 5, 22 (January 1971)

The term "regional approach" has become somewhat of a cliche, so much so that mere utterance of the words tends to conjure visions of the addition of yet another layer of bureaucracy onto a field already long on programs but short on results. This is unfortunate, for although full implementation of regional action is a concept perhaps more honored in the breach than in the observance, the basic concept is a sound one. That it can work has been amply demonstrated by the Ohio River Valley Water Sanitation Com-

**Executive director Horton**
*Not dominated by feds*

mission (ORSANCO), the 22-year-old pioneer in interstate cooperation on water pollution problems. ORSANCO is alive and well, and shows no signs of flagging from its original mission—restoration and continued protection of 981 heavily industrialized miles of the Ohio River.

By mere coincidence, the ORSANCO organization dates exactly as far back as the federal government's nationwide water pollution control effort. P.L. 80-845, the first major federal involvement in water pollution abatement, was signed by President Truman on June 30, 1948, the same date that the ORSANCO compact became effective. Because of the coincidence, the temptation to compare ORSANCO's rate of progress with that of the fed-

eral program is strong. However, OR-SANCO's track record is strong enough not to require comparisons.

## Monitoring

The most well-known and concrete of the Commission's achievements is the monitoring network that has been established on the Ohio River. What began as a rudimentary manual sampling grew rapidly with the development of automatic monitoring equipment, and by the early 1960's had become a showcase for demonstrating remote monitoring and data acquisition and handling concepts.

But the Ohio River's quality, and not monitoring techniques, is the Commission's primary concern, and the trends charted in the *ORSANCO Quality Monitor*, a monthly summary of data generated by the network of sampling stations, shows the slow but inexorable trend of improvement in the quality of the Ohio River. Early in its development, the Commission had established a set of water quality standards for the portion of the river that flows through the boundaries of the eight signatory states—Illinois, Indiana, Kentucky, New York, Ohio, Pennsylvania, Virginia, and West Virginia. Of the 21 chemical and biological criteria that have been adopted for appraising river quality, fully two-thirds are met routinely 100% of the time. To be sure, some of the deficiencies are troubling, including such vital parameters as dissolved oxygen, coliform density, and pH. But even on these criteria, considerable progress has been charted over the years, and will continue as the compact states continue to implement their abatement plans.

## Achievements

ORSANCO is a regulatory agency, not a treatment authority, but the pace of treatment plant construction along the Ohio is a direct result of judicious use of the Commission's enforcement powers and coordinating efforts. Twenty-five years ago, prior to the establishment of the Commission,

less than 1% of the population in the Ohio basin was provided with sewage treatment facilities. But today, treatment facilities serve more than 95% of the population with at least primary or intermediate treatment. In its last annual report, the Commission notes that about 30% of the basin's treatment plants require upgrading, mainly as a result of upgraded state water quality standards. But most of these municipalities, particularly the larger cities on the Ohio—Pittsburgh, Pa., Cincinnati, Ohio, and Louisville, Ky.—are well along on the design of improved facilities.

Furthermore, about 87% of the industrial effluents in the Ohio basin are in compliance with the ORSANCO standards, and all but a handful of those industries not yet complying have abatement programs in various stages of design and construction.

With such a record of achievements, even the most cynical critic of the "regional approach" would have to concede that ORSANCO must be doing something right. Just what it has been doing right has been documented by Edward J. Cleary in "The ORSANCO Story," published a few years ago by Resources for the Future. Cleary's book presents considerable detail on the ORSANCO compact: its genesis in the concern of a group of Cincinnati citizens in 1936 over the condition of the Ohio River; the tortuous route of negotiation of the compact, both among the states and with the federal government; and, finally, its rapid growth and implementation. The thoroughness of Cleary's account is overshadowed only by the objectivity of the author who, as ORSANCO's first executive director and chief engineer, was deeply committed to ORSANCO's goals.

## Future directions

To assess ORSANCO's future role in the light of a growing federal involvement in water pollution control—three new water quality acts have been passed since the Commission was activated—ES&T visited with Robert K.

Horton, who has taken over as executive director (Cleary still serves the Commission occasionally as a consultant). Horton agrees that ORSANCO's *modus operandi* must change, and in fact has been changing subtly, with changing public attitudes.

Horton points out that the strength of ORSANCO lies in the fact that the original compact, stripped of its organizational provisions, is basically a pledge of "faithful cooperation" by the signatory states. He explains that many people are misled by the fact that the Commission was formed under Congressional authorization, and has three members from the federal government. "We are in no way dominated by the federal government," he says. "ORSANCO simply consists of a mechanism whereby the states can devise compatible actions."

Perhaps because of this "self-policing," the Commission has had to use the enforcement powers delegated to it by the compact only sparingly, and only as a "last resort," according to Horton. For example, when a member state itself has been unable to get satisfactory enforcement from local courts, the Commission, acting as an interstate agency, can bring the ac-

tion to a federal district court. Even so, enforcement can only be brought on agreement by both a majority of the signatory states and a majority of the three members of the delegation from the state involved. In no case, however, has this requirement hindered initiation of interstate agency intervention.

The Federal Water Quality Act of 1965, with its standards-setting provisions, did not create any confusion in ORSANCO's mission, according to Horton. "At the time of the act, the Commission was getting ready to update its standards anyway," says Horton, and the only confusion was over whether the Commission should hold joint hearings on standards. Although the Department of Interior did sanction joint hearings, the Commission opted for unilateral action and submission by separate states. "We lost a little time in ironing out the inconsistencies," says Horton, but he points out that uniform standards have recently been approved by the Commission.

What is ORSANCO's future direction? "A most important thing we are doing, and will continue to do, is our river monitoring effort," says Hor-

ton. "We will continue to become a service agency, coordinating what the states can't do for themselves." With new and upgraded standards to be enforced soon, this mission implies surveying the operation of treatment plants now abuilding and determining where new facilities will be needed; ORSANCO's monitoring network will be the backbone of this operation.

In fact, its river-wide monitoring capability will play a major role in most of ORSANCO's future plans, according to Horton. The Commission has just received a federal research and development grant to define river quality forecasting procedures. Such a capability would offer several advantages.

ORSANCO will soon have a role in long-range planning in the Ohio River basin. The Commission will be a member of the recently formed Ohio River Basin Planning Commission, established under the Federal Water Resources Planning Act. Says Horton: "ORSANCO will be the agency for reviewing water quality aspects of long-range planning in the area. Input from our monitoring system will be a major contribution to the planning commission."                                    PJP

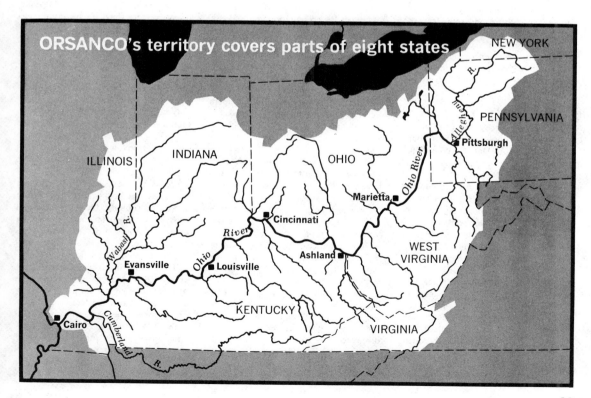

ORSANCO's territory covers parts of eight states

# Purdue conference keeps rolling along

**Ringmaster.** *Don E. Bloodgood is the driving force behind Purdue Industrial Wastes Conference*

*Reprinted from* ENVIRON. SCI. TECHNOL., 4, 552 (July 1970)

The 25th annual Purdue Industrial Wastes Conference, which has been attacking the subject of industrial pollution since 1944, was held at Purdue University in May. The conference has grown in reputation as well as in size—from 200 attendees in 1944 to over 800 at this year's gathering—yet in format and function it has remained essentially unchanged over the years.

Originally designed as a service to Indiana agriculture and industry (the rationale for its status as a Purdue extension program), the conference nevertheless has broadened sufficiently to be considered at least a Midwest regional affair, if not perhaps a truly national one.

## Function defined early

The driving force behind the organization of the conferences, and chairman of all 25, is Don E. Bloodgood, professor of civil engineering at Purdue. Reflecting on the early years of the conference's history, Bloodgood recalls that it was initially rather difficult to get papers from industry: "In those days, people were not so willing to talk about their problems as they are today." Yet representatives of the dairy, brewing, and canning industries in Indiana did indeed discuss their problems at the 1944 conference, and in the presence of many Indiana government officials, thus setting a precedent for the open exchange of ideas and information that has persisted as the main function of all subsequent conferences. Although other industries —petroleum refining, meat packing, and paper manufacturing, for instance

—now are on the programs, the agriculture-related industries of Indiana and neighboring states still represent the core of each conference.

It would possibly be unfair to Bloodgood's hard-working associates on the conference organizing staff to describe him as the ringmaster, but nevertheless that seems an apt description. There is no program committee; the final decisions on the suitability of any paper for presentation are made by Bloodgood himself. Working through what he describes as "an informal, worldwide network of 50–60 people," Bloodgood keeps on top of advances in industrial waste treatment and each year invites participants to the conference. In 1970, 150 unsolicited papers were submitted but not accepted for presentation.

At any one time during the conference (which lasts for three days) there are six concurrent sessions, a matter of some aggravation to those who want to hear everything. But Bloodgood is adamant: he wants to cover all of the very broad field of industrial wastes and realizes that to do so he needs a large number of papers (103, at the 1970 conference). An attempt is made to reduce potential conflict between simultaneous papers but, naturally, some tends to occur.

Another matter on which the ubiquitous chairman is adamant is the length of each presentation—30 minutes for the paper and 30 minutes for discussion. Session chairmen who do not abide by the rules are likely to be chided by Bloodgood who has been sitting quiet and unnoticed in the back

of the room. Discussion at the conference is lively and stimulating; any discusser who doesn't maintain audibility, however, runs the risk of being sternly reprimanded: "There's a private discussion going on down there, and Bloodgood doesn't like private discussions!"

## Recipe for success

The success of the Purdue conferences—and they are by most measures very successful—is probably due to the subject matter, which is almost wholly technical and is treated in an unabashedly pragmatic way. The attendees are, by and large, technical men rather than managers. The practical approach, coupled with the enjoyably human aspects of the conference, makes for a most educational three days. Where else can you hear, on the same program, a federal official ruefully confessing that he has been banned from local fishing contests because of his access to 7-lb. black bass killed by feedlot runoff, a chicken farmer attributing the lack of solid waste disposal problems in his hen houses to "the will of God," and a refinery engineer describing new techniques for waste water treatment?

There are lingering problems, of course. Attendees annually curse the unavailability of papers at the conference and bemoan the fact that conference proceedings are rarely published within 18 months. But these seem small prices to pay for the privilege of attending a conference that really has a function, and actually comes quite near in fulfilling it.

*In Texas, the Gulf Coast Waste Disposal Authority, a new governmental entity and political subdivision of the state, is now . . .*

# Managing regional water treatment systems

*Reprinted from* ENVIRON. SCI. TECHNOL., **6**, 402 (May 1972)

One item in the clean water amendments for 1972 is the requirement of regional planning for waste treatment plants. If the regional officials do not develop such plans for regional treatment of their wastes, then they will not be eligible for federal funds authorized by the nation's new water law.

But Texas legislators are ahead of their federal counterparts. The Gulf Coast Waste Disposal Authority (GCWDA) is one of the newer, if not the newest, regional waste disposal authorities in the U.S. The authority is a political subdivision of the state and it has both the powers to plan and to implement regional waste treatment systems. In addition, GCWDA is eligible to receive construction grant funds from the federal government.

The seed for the authority was planted in 1965 when Texas State Senator Criss Cole, now a judge for the Texas juvenile courts, recognized the need for some sort of super waste treatment agency that would not have to recognize county boundaries. The result, five years later, was GCWDA with boundaries of the authority being the three Texas counties encircling Galveston Bay (see art).

Legislation creating the authority was enacted by the Texas legislature in 1969, and although its general manager, Jack Davis, was not hired till June 8, 1970, now it has both a staff and a board of directors. In addition to Davis, the present staff of 10 includes Mike Eastland, assistant manager for the authority; Joe Watts, director of finances; Charles Hayes, research coordinator; Donald Vacker, chemical engineer; and Larry Crow, planning coordinator. The immediate goal is a central staff of 15–20. Depending on the number and different types of facilities that GCWDA would own and operate, the authority could eventually increase to more than a 100 operating staff in several years.

In addition, the authority has a nine-member board of directors. Three mem-

**GCWDA's Mike Eastland**
*. . .with industrial projects in sight*

bers are appointed by the Governor, three by the Commissioner court in each county, and the remaining three by the mayors of each county.

By statute (Texas Civil Statutes Article 7621d-2), the authority was empowered with the ability to levy a tax. It was also empowered with the authority to issue municipal bonds. Although the voters defeated a referendum to permit the institution of this taxing power in the three counties in the general election of 1970, the margin of defeat was slight—54% against, 46% for. Considering the fact that GCWDA was a new governmental entity, the outcome was not all that bad.

"We don't have to have these tax revenues to continue operating," says Mike Eastland, assistant manager for the authority. "We can issue revenue

bonds anytime after legal notice is given. In fact, the authority's first bond issue for $10.5 million was underwritten last December 28 by Eastman Dillion Securities of New York, N.Y. The 25-year bond issue was sold at a 5.48% interest rate. Earlier, (July 22, 1971), the authority issued $3.6 million worth of revenue bonds for the city of Houston to assist them in upgrading some of Houston's waste treatment plants.

**What it does**

Under contract with either industry or a state governmental agency, GCWDA operates, owns, and constructs waste treatment facilities. Thus far, the authority has three industrial projects in sight. The first is under way; it's the one for which the municipal revenue bond was sold. A second is in the contract negotiation stage; signing appears imminent. At press time, the third had suffered a setback because one industry, after considering the pipeline transportation expenses of the venture, had a change of heart.

GCWDA's first project involves treating waste waters of Union Carbide. The contract was signed last October and the municipal bonds now have been sold for this project. The construction will be done in two phases; a primary phase to be completed by July 1973 and the second phase by July 1976. Union Carbide's BOD load was well over 1000 ppm. Union Carbide was passing their wastes through settling ponds, providing some aeration and then discharging them to Galveston Bay.

A primary plant will reduce the burden of pollution to 150 ppm of BOD by the 1973 date. Then, a secondary plus polishing stage in the second construction phase will reduce the BOD burden to less than 50 ppm by the 1976 date.

There is the probability that Monsanto will come into the second phase of this project. Monsanto would provide treatment of its wastes in its own primary plant but would then send their primary effluent to a secondary treatment facility that is owned by GCWDA. At press time, Monsanto wastes are being analyzed from a treatability and a compatibility standpoint to see if it would be feasible for joint treatment.

In the second project, for which signing of a contract appears imminent, five industrial companies—U.S. Plywood-Champion Paper Co., Atlantic Ritchfield Co. (a refinery), Crown Central Petroleum Corp., Goodyear Tire and Rubber Co., and Petro-Tex Corp. (a petrochemical company)—are planning to combine their waste streams and let GCWDA provide the regional treatment service for their combined wastes. For this project, the authority is negotiating to purchase Champion's existing treatment plant which already provides secondary treatment and good BOD and suspended solids removals. The Champion plant went on line in 1969.

With project two, the authority plans to change the existing plant, to increase the treatment capacity by 10–25%. "There will be some construction," Eastland says. "We will add more clarifiers and aeration basins to increase the capacity of the plant. We will also add sludge-handling facilities including a filter press and an incinerator. Of course, the cost will be prorated among the five companies. The total cost for the second project is $20–25 million; this figure includes $10 million for the purchase of the existing plant," Eastland explains.

Earlier this year, the third project was in precontract negotiations but now has been scrapped. This project involved combining waste streams of Ethyl Corp. and Tenneco Hydrocarbons, Inc., and treating their wastes in one plant. Due to the cost of laying a pipeline between the two plants, the industries decided that it would be more feasible for them to perform their own construction and modification to their existing waste treatment plants, rather than to tie in with the authority.

## Why go GCWDA?

"When a waste treatment plant is built with municipal revenue bonds, the industry gets a financial break because of the difference in interest rate paid on a municipal bond vs. an industrial bond. Also the treatment facility is taken off the tax rolls," Eastland explains. "What is more, the facility pays no sales tax on the equipment that goes into the plants."

"Because financial advantages accrue to the industry, we must own and control operations at the facility," Eastland continues. "Industry pays the debt service on the bonds that are issued; they also pay the operating cost, and on top of that the authority charges about a 7.5% figure on the average annual debt service as a management fee."

"We don't offer as many benefits to a municipality as we do to an industry because municipalities can issue the same type of revenue bond as the authority. Unless, of course, you combine treatment facilities for municipal wastes, the authority cannot offer a significant financial advantage to a municipality," Eastland elaborates.

"We are trying to follow an evolutionary course here. We start off with small subregional plants and build up to a regional plant. What we are doing is keeping one plant from being built by combining it with another plant. The authority is trying to halt the proliferation of small waste treatment plants." GCWDA has three ongoing municipal projects in which it eliminates the need for three proposed plants by combining the three with existing facilities.

## What's ahead?

The reasons for considering regional waste treatment are many and varied. Economies of scale and eligibility for federal funds are only two arguments, but certainly good ones. Permitting one plant to treat wastes from several locations in any one geographic region makes sound economic sense.

Recently, the Texas Water Quality Board, after public hearings on February 7, ordered industries which discharge their wastes to the Houston Ship Channel to reduce drastically their existing BOD discharges. Presently, there is a collective existing discharge of approximately 100,000 pounds of BOD per day which probably will be lowered to a range of 35,000–50,000 pounds of BOD per day.

The Texas Water Quality Board issues permits. Then, industries of course, had to and still have to file permit applications with the Corps of Engineers. For the Union Carbide project, for example, both sets of permits are being transferred from the industry to the authority.

What is encouraging, at least to the new authority, is that permits issued by the Texas Water Quality Board, its so-called waste control orders, could all one day be turned over to the regional authority—the GCWDA. Then, the authority would get all of the industrial permits issued for the three-county area. The authority believes that should this happen, it will be easier for the authority to institute a monitoring and water-quality management program that would assist in the cleanup of the Ship Channel, Galveston Bay, and the Gulf of Mexico. SSM

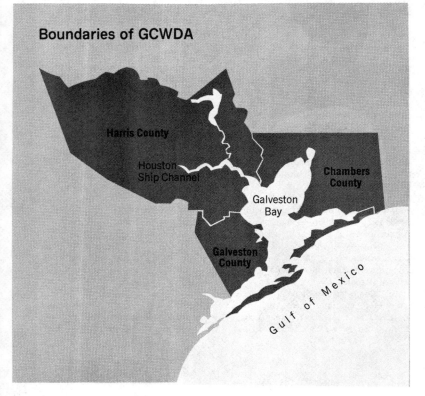

**Boundaries of GCWDA**

Harris County

Houston Ship Channel

Chambers County

Galveston Bay

Galveston County

Gulf of Mexico

*The Interstate Commission on the Potomac River Basin, hampered by a lack of enforcement authority, tries to help formulate the master plan for . . .*

*Reprinted from* ENVIRON. SCI. TECHNOL., **6**, 1074 (December 1972)

# Cleaning up the Nation's River

According to usually reliable sources in the Nation's Capitol, George Washington once threw a silver dollar across the Potomac River. Today, there are places where the Potomac is so thick, he might as easily have walked across with it.

Silt, agricultural runoff, sewage, acid mine drainage, and inadequately treated industrial effluents continue to pollute the Potomac.

Yet some progress is being made. Parts of the Potomac are relatively clean. There is hope, at least, that beefed-up federal laws and new agreements between the jurisdictions that make up the Interstate Commission on the Potomac River Basin (ICPRB) will provide the muscle needed to clean up the river.

The ICPRB was established by an act of Congress in 1940 with passage of the Potomac River Sanitation Compact and charged in a general way with helping to abate pollution along the Potomac. Membership in the ICPRB is made up of delegations from Virginia, West Virginia Pennsylvania, Maryland, the District of Columbia, and the federal government (see box, p 1075).

The depth of the commitment to pollution abatement during the 1940's and 50's is perhaps best judged by the fact that during those years the ICPRB had no full-time staff. The sole employee of the commission was a part-time executive director.

In fact, the Commission did not have a full-time executive director until 1967 when Carl J. Johnson, a former Director of Conservation with the state of West Virginia, was named to the post. Johnson held the job until his death early this year and is generally credited with greatly enhancing the effectiveness of the ICPRB.

The present Executive Director, Paul Eastman, was named to his post in July 1972.

In the early days of the ICPRB, virtually all of the Commission's efforts were directed toward generation of public support for pollution abatement. Even today, much of the ICPRB's activity

**Drainage area served by Interstate Commission on the Potomac River Basin**

centers around grass-roots education and the marshalling of public support for basinwide programs the Commission sees as necessary to carry out its mandate.

### Scope of compact

The Commission in its present form began to take shape in 1970 following a 10-year tug-of-war among the signatory powers which came to an end late in 1970 when the federal government inked an amended Potomac River Basin Compact.

The new Compact widened the sphere of influence of the ICPRB to include water quality and supply management and land use planning for the basin as a whole. Prior to the new compact, the Commission's authority had been limited solely to questions of water pollution.

Even newer proposed amendments to the Compact—formulated by a Potomac

**97**

River Basin Advisory Committee—would once again broaden the scope of ICPRB's activities to include enforcement powers and financing authority. The proposed amendments have been ratified by two states—Virginia and Maryland—and require signatures from each of the remaining four jurisdictional bodies before they become effective.

Under P.L. 91-407, approved by the Congress in September 1970, several broad powers are given to the ICPRB. Briefly, they include instructions:

• to collect, analyze, interpret, and otherwise prepare data bearing on the problems of pollution or water quality in the Potomac Valley Conservancy District—the area designated by Congress to be under ICPRB's jurisdiction

• to cooperate with state agencies and other interested groups or persons in promoting uniform laws for abatement of pollution or development of water resources

• to conduct public information and education campaigns, provide liaison with agencies dealing with pollution and related water problems in the Potomac River Basin

• to review or comment on proposed plans by any agency within the Conservancy District relating to water pollution or related water quality and supply matters.

Among the goals of the ICPRB are full compliance with currently approved water quality standards, in accordance with the implementation plans of the signatory agencies; a Basinwide program

**ICPRB's Eastman**
*reasonably optimistic*

for water management (to be prepared by the end of 1973) which would assure adequate water supply through the year 2000; and development of comprehensive land use plans by the end of 1974 for areas adjacent to the major streams draining the basin.

### Lacks enforcement

ICPRB, unlike several other river basin commissions of which the Ohio River Valley Sanitation Commission (ORSANCO) is more or less typical, does not have even limited enforcement authority ("ORSANCO: Pioneer with a new mission," ES&T, January 1971, p 22) according to Eastman. Under the new amendments to the compact yet to be ratified,

ICPRB would finally have some power to enforce its recommendations and long-range plans.

The new amendments to the compact would permit ICPRB to hold hearings, issue subpoenas and compel testimony under oath. ICPRB could sue and be sued in its own name and could impose penal sanctions up to $500 for "petty" offenses.

Perhaps as important as the power to set and enforce regulations would be the power granted under the new amendments to enter into financing agreements for projects in the Potomac Basin Conservancy District. The Commission could negotiate for loans, grants, or other types of revenues to carry out any of the projects necessary for pollution abatement or water resource management. A special article in the new Compact would establish procedures for issuing bonds or other instruments of long-term financing.

The Commission could condemn property through eminent domain and could establish the amounts payable by each member government for projects to upgrade water quality in the basin.

The new amendments to the compact would place responsibility for the affairs of the ICPRB closer to the top of state governmental affairs. The new compact would make the Governors of Virginia, West Virginia, Maryland, and Pennsylvania, and a presidential appointee representing the federal government voting Commission members.

Eastman feels reasonably optimistic that the states which have not yet ratified the amendments to the compact will do so in the future, perhaps even in the coming year.

He is not so optimistic about the recently enacted water legislation however. "The widely advertised promise of over $24 billion of federal funds and uniform nationwide federal controls to eliminate all pollutants by 1983 will bear scrutiny and perhaps some skepticism by the Potomac Basin states as well as other areas of the nation," Eastman comments.

Eastman terms the new legislation "complex beyond belief," and questions the wisdom of massive federal control over state and local pollution abatement efforts. "Geographical and other differences between the upper Potomac mountain creeks and the giant Mississippi in the lowlands of the south have been eliminated by federal statute," Eastman charges. And he sums up his misgivings by quoting T. S. Eliot: "Between the idea and reality, between the motion and the act, falls the shadow." HMM

---

### Commissioners and officers of ICPRB

**Commissioners**

**District of Columbia**
Lee F. Dante
Paul V. Freese
Malcolm Hope

**Virginia**
James J. Corbalis, Jr.
John A. K. Donovan

**Maryland**
Oscar W. Meier
Marvin Mandel
A. T. Brust, Jr.

**U.S. government**
Lawrence M. Fisher
Eugene T. Jensen
Arnold Sternberg

**Pennsylvania**
Maurice K. Goddard

**West Virginia**
N. H. Dyer
Cliff Umscheid
William A. Powers

**Officers**
A. T. Brust, Jr., Chairman
Maurice Goddard, Vice Chairman
Lee F. Dante, General Counsel

Oscar Meier, Treasurer
Paul W. Eastman, Executive Director
G. L. Carley, Jr., Asst. Executive Director

*Charges to industry for municipal sewer use, are becoming more widespread as the Environmental Protection Agency ties construction grants to cost recovery systems*

Reprinted from ENVIRON. SCI. TECHNOL., 5, 1000 (October 1971)

# Cities treat industrial process wastes

Phineas Fictitious, president and chairman of the board of Federated Fictitious Firms, is a man with a problem. His biggest company, Fictitious Fortified Foods, had to stop dumping processing wastes into the Slough River after the Army Corps of Engineers denied him a discharge permit. He could have built a treatment plant, but profits were squeezed from every side. Besides, he said, he was in the food business, not the waste business.

Undaunted, Fictitious tied into the Megalopolitan Municipal Sewer System. That might have been the end of it, except that a short time ago, Megalopolis passed an ordinance which gave the city power to assess a surcharge to recover operating costs for treating industrial wastes.

"It's only fair," said City Hall. "Private citizens are complaining that by treating such wastes free of charge, the city is subsidizing industry. Besides, federal law requires that we establish an equitable industrial waste cost recovery system, or we don't get any federal construction grants for new plants that we desperately need."

"Fiddlesticks," fumed Phineas. "Fictitious Firms pay the city for water, and we pay enormous property taxes. Besides, the law is highly discriminatory because it doesn't clearly define 'industry.' How about my commercial laundry, Fabulous Fabric Finishers, or my Fast-as-a-Flash carwash?"

"Fabric Finishers and Fast-as-a-Flash are services, not industries," Megalopolis maintained, "and are therefore exempt. Fine Foods has to pay."

"Bureaucratic balderdash," thundered the tycoon, "I'll see you in court."

**No case**

Until recently, Fictitious might have carried the day. Today it is unlikely that he would. Sewer surcharges are rapidly being accepted by industries as part of the cost of doing business. Recent court decisions have bolstered the legal position of surcharges, and

municipalities across the country are being prodded into adopting them. "A few years ago, such a surcharge was rare," says Earl W. Knight, chief pollution control officer for the Metropolitan Sanitary District of Greater Chicago. "Now there are at least 36 of them that we know about."

Undoubtedly, more are on the way. Public reluctance to foot the bill for cleaning up the waste of private industry is certainly a factor, but a more

**Stickney Plant.** *Chicago's giant West-Southwest sewage treatment plant (850 million gpd, dry weather flow) treats both municipal and industrial wastes*

powerful incentive is provided by the federal government. Part 601.34 (c) of the Construction Grant Regulations published by the Environmental Protection Agency in the Federal Register last July required a fair-cost recovery system for treating industrial wastes: No surcharge system—no more construction funds. It's not exactly a secret, according to Knight, that Chicago's recent decision to implement the surcharge was heavily influenced by the EPA ruling.

Cities ranging in size from Warren, Pa. (population 10,000) to New York City (population about 8 million) have adopted surcharge ordinances, as have major cities including St. Louis, Houston, Denver, Cincinnati, Pittsburgh, and Kansas City.

The surcharge formula spelled out in the New York ordinance is fairly

representative in that it requires payment for waste concentrations that exceed a fixed limit for suspended solids (ss) and biochemical oxygen demand (BOD). Chicago, which does not consider pollutant concentrations in its formula, pegs its treatment costs to the volume of water discharged above a maximum deductible allowance.

**New York surcharge**

The New York ordinance has its roots in a 1938 codification of regulations for city sewer use begun during the first term of Mayor Fiorello La Guardia. It soon became apparent, however, that better control over discharge of grease, toxic metals, acids, and other deleterious wastes was necessary. In 1957, new regulations were published, but they still fell short of the mark. Violations of the sewer code were punishable only by civil penalties, and there was no machinery for recovering costs of treating exceptionally strong wastes.

In 1961, Local Law No. 2 finally established the procedure for collecting the surcharge and beefed up enforcement. Further reworking of the regulations in 1963 gave the laws under which the city operates today. Although some changes are being contemplated, the city's Commissioner of

Water Resources, Martin Lang, says the present code is working well.

The New York law provides that the city has the power to levy the surcharge and to prescribe what wastes are acceptable in the city sewers. The surcharge constitutes a lien on real property, and its collection can be accomplished in the same manner as property taxes. Violation of provisions of the code are criminal offenses with penalties as stiff as three months imprisonment and/or $50 maximum fine. Each day a plant is in violation counts as a separate offense. For continuous "flagrant violation," the city has yet another remedy: "When they balk at our admonitions and exhortations," Commissioner Lang says, "we can shut their water off—and I have."

The surcharge is based on the formula $D_s = CFV[(ss - 350) + (\text{BOD} - 300)]$ where $D_s$ is the amount of surcharge in dollars, $C$ is the actual cost per pound of removing pollutants, $F$ is a conversion factor for changing mg/l. to lb/million ft³, $V$ is the volume of waste water discharged into the sewer in ft³, ss is concentration of suspended solids in mg/l., and BOD is biochemical oxygen demand in mg/l. The figures 350 and 300 are, respectively, concentrations of suspended solids and BOD below which there is no charge to industry. The city finds that typical domestic wastes have ss content of about 155 mg/l. and a BOD of about 125 mg/l. "We double the average for industry," Lang says, "and give them a substantial free ride. We think it's equitable to both the city and industry."

Such a "free ride" exempts many smaller industries which would vastly complicate the collection and inspection machinery and would eat up more in administrative costs than they would bring to the city's coffers.

The cost factor, $C$, is recalculated each year and is currently $0.025/lb. The volume, $V$, is determined by the amount of water delivered to the industry, corrected by a retention factor. Where water consumption (as opposed to discharge) is high—say in an

ice manufacturing plant—the correction factor may reach 85%. Where retention is low—in launderettes, for example—only 5% will be allowed.

Measurement of ss and BOD provides something of a problem in New York. "New York City is not a big wet-industry town," Lang notes. Indeed, its small-to-medium-size industrial plants seldom cover large areas of land, and may be located in several nonconnected buildings or even lofts. Space for manhole-type sampling stations complete with flumes and weirs may not be available, and the prospect of maintaining a crew of municipal samplers to monitor the strength of process effluents stuns the imagination. Instead, the city prefers to rely on industrywide averages obtained by representative samples of operations categorized by standard industrial classification numbers. Recently, a New York state appellate court upheld the legality of the city's practice of using industry averages to compute ss and BOD.

Chicago's experience with a surcharge-type system dates back to a 1935 Illinois statute prohibiting indus-

try from dumping more than 10,000 gallons of waste per day into municipal sewers. However, the law provided that the city could take more than 10,000 gpd if it chose and was compensated for its service.

Several attempts were made to implement the power granted by the statute, but for one reason or another, they were not successful, according to the Sanitary District's Knight. Renewed attempts during 1969 and 1970 were more successful, he says, and the Sanitary District Board of Trustees adopted the surcharge ordinance in December of 1970. The Board's action was promptly met by a lawsuit filed by several area industries which appears, at press time, to be settled in favor of the District.

The Chicago ordinance provides for recovery of capital costs incurred by the Sanitary District for the industrial waste portion of the flow to its facilities, operation and maintenance costs, and additional costs as may be necessary to assure adequate waste treatment on a continuous basis.

The ordinance allows a 10,000 gpd (3,650,000 gpy) exemption from the

## N.Y. City treats industrial wastes of various strengths

| Industry | Firms sampled | Average concentration (mg/l) | |
| --- | --- | --- | --- |
| | | BOD | SS |
| Slaughterhouse | 2 | 1,785 | 1,920 |
| Meat processor—no pretreatment | 5 | 762 | 414 |
| Meat processor—with pretreatment | 7 | 485 | 255 |
| Frozen desserts—open refrigeration | 3 | 537 | 157 |
| Frozen desserts—closed refrigeration | 3 | 982 | 440 |
| Milk processor | 6 | 964 | 266 |
| Juice & fruit drink bottler | 5 | 1,813 | 116 |
| Brewers' spent grain processor | 1 | 23,762 | 4,848 |
| Commercial baker (sweet products) | 7 | 3,745 | 1,530 |
| Candy manufacturer | 7 | 1,635 | 121 |
| Brewery | 2 | 1,090 | 422 |
| Soft drink bottler | 3 | 488 | 38 |
| Potato chip manufacturer | 4 | 1,110 | 1,657 |
| Commissary | 9 | 586 | 236 |
| Instant coffee manufacturer | 2 | 594 | 237 |
| Semi-industrial laundry | 6 | 1,045 | 601 |
| Pharmaceuticals | 2 | 957 | 137 |
| Fur dresser or dyer | 8 | 791 | 962 |
| Laundry (family) | 10 | 535 | 237 |
| Laundry (linen) | 12 | 695 | 256 |
| Diaper service | 4 | 270 | 231 |
| Industrial laundry (wiping cloths) | 2 | 2,506 | 1,661 |
| Rug cleaning | 7 | 121 | 74 |
| Prepared salads (potato salad, cole slaw) | 6 | 4,386 | 5,041 |

Source: WPCF Journal, December 1968

surcharge, regardless of the concentration of waste products in the effluent. For amounts in excess of allowable volumes, a flat rate of $0.021 per thousand gallons of liquid flow is assessed, together with $0.014/lb of $BOD_5$ and $0.024/lb$ of ss. The unit costs are computed with reference to capital costs, depreciation, maintenance, and operation costs.

The exemption, Knight says, excludes small industries and saves administrative costs. The estimated value of one exemption, Chicago figures, is only about $200–300/year, depending on the strength of the waste discharged. The estimated total surcharge collection for one year is about $15 million, Knight adds, and the total exemptions granted only amount to about 5% of that total. Knight emphasizes that even exempt industries pay something for sewer services through ad valorem taxes. "It is likely that the exemption is more than balanced out by taxes collected," he says.

The Chicago ordinance makes sampling of process effluents the responsibility of industry. The ordinance prescribes that for effluents of constant strength, sampling once a month is sufficient. Where there is wide fluctuation in waste concentration, it may be necessary to sample every day. Analysis must be confirmed by an independent laboratory, and the Sanitary District also makes spot checks.

The district uses automatic sampling equipment and trailer sampling crews to obtain 24-hr composite samples proportional to flow. About 3% of the companies liable for surcharge are subjected to these 7–10-day formal studies each year. Another 10% are subject to less intensive, 2–3-day, 8-hr composite sampling.

The surcharge is figured according to a schedule provided by the Sanitary District (see sample). Credits are allowed for any tax actually paid to the Sanitary District from ad valorem assessments. A correction factor for sewage from plant employees is included in figuring the surcharge.

The payments are sent directly to a bank, and the returns are verified by computer. Figures are cross-checked against county tax bills, water use bills, and previous data filed by the company. Penalties carried by the ordinance are 5% of the surcharge due for failure to file and 2% for late filing.

Since companies must list their standard industrial classification numbers, as in New York City, the District is able to establish a set of average values for industries of any particular type. These average values are compared with data submitted by individual industries, and if they differ by more than a few percentage points, the industry is subject to audit. "We call the company in for a hearing," Knight says, "tell them we got different readouts, and ask for an explanation."

In some cases, a company may not be discharging nearly as much water as it bought because of losses in the product or by evaporation. In other cases, pollutant concentrations in effluent will be considerably lower than industry averages because the company may have installed pretreatment or reclamation facilities in the plant.

### Surcharge fair

Both New York's Lang and Chicago's Knight agree that industry, by and large, thinks the surcharge is fair. "All the harbingers of doom who said it would die out were wrong," says Lang.

The reason Chicago didn't adopt the surcharge years ago, Knight says, was that "unless everybody does it, you'll lose industry to the surrounding area." Lang concurs: "In the future, there won't be any haven for polluters like the Passaic Valley," he says.

Most cities will have to adopt surcharge ordinances, Knight thinks, because they all need federal funds. Industries won't move out, he adds, because they could buy time only for a year or two at the most. "Industry is smart enough to see that municipalities will all adopt the surcharge sooner or later," Knight says, "Besides, most industries wouldn't consider moving into an area just because they might be free to pollute—it's not good for their images."

Knight is "amazed" that industries are going along with the Chicago program as well as they are. "The ones that are not going along are the marginal guys—the small businessman who has maybe 15–20 employees and is making sausage in a little factory that couldn't stand up well under federal inspection anyway." Such an operator doesn't want to call attention to himself, and therefore doesn't file or doesn't remit payment. "We're going after them," Knight says, "and we'll catch them."                    HMM

# Water money needs require more than promises

*Officials from large city and small municipality agree that changes are necessary in the federal role of assistance*

Reprinted from ENVIRON. SCI. TECHNOL., **4,** 278 (April 1970)

Money—which may or may not be the root of all evil—certainly is a necessity for any progress in pollution abatement in the U.S. The Nixon administration's program for water pollution cleanup is awaiting Congressional action. S. 3472 calls for $10 billion for construction of municipal waste treatment plants and seeks an authorization of a full $4 million federal share in fiscal year 1971. Under this proposal, the Secretary of the Interior will enter into contracts with municipalities for the construction of municipal waste treatment plants at the rate of $1 billion for the next four years.

The remaining $6 billion would be financed by state and local authorities or by the newly proposed Environmental Financing Authority (EFA). Faced with the inability to borrow necessary funds in the market at a reasonable interest rate, state and local authorities under the proposed S. 3468 will be able to sell their bonds to the new EFA. High interest rates have hindered the sale of many state and local bonds for such purposes. After purchasing these bonds, the EFA would sell them in the capital bond market under presumably more favorable conditions.

Also proposed is reform in the present allocation formula for granting funds to state and local municipalities. The proposal states that:

• 60% of the total federal money for a given fiscal year will be allocated to states in accordance with the existing allocation formula.

• 20% of the federal funds will be allocated to those states or localities with matching funds programs, thus insuring a positive incentive to states to contribute financing.

• 20% of the federal funds will be allocated by the Secretary of the Interior to those areas of greatest need and from which the greatest water pollution control benefits can be realized.

Under the provisions of proposed legislation, states which have prefinanced construction facilities, includ-ing New York, Maryland, Pennsylvania, Wisconsin, Michigan, and others, will be refinanced the $320 million that was spent from 1966 to the present. Proposed S. 3472 specifically protects the eligibility of those states which prefinanced the cost of treatment facilities. However, no new reimbursables would be . authorized after 1973. Presumably, the entire water pollution control effort would be reviewed in 1974.

## Views

A fundamental point made time and again on water cleanup throughout the U.S. is that the federal promise must be backed with a financial commitment from the federal government if the U.S. is to make progress on this pressing domestic problem.

At the recent 4th Annual Legislative Seminar (Washington, D.C.), sponsored by the Water Pollution Control Federation, the viewpoints of spokesmen from both large cities and small communities were aired in continuing cleanup dialogue.

The needs of a large metropolitan area were described by James R. Ellis, legal counsel for the municipality of metropolitan Seattle (Wash.). He noted that cities are in a catch-up situation and that the surface of the problem in large cities has just been scratched. During the period 1965-69, 35% of the sewered population in the U.S. received only 5% of the federal grant dollar. In the same period, small municipalities (with less than 5000 population) received 74% of the federal dollar, according to Ellis.

Another factor Ellis considers significant is that, in the past few years, the cities have grown considerably while small municipalities essentially have remained static. He finds that the cost for all municipal services is 60% higher in cities than in small municipalities, but that the per capita income is only 20% higher.

In summary, Ellis makes six suggestions that would help cities, including the following:

**James R. Ellis**

• Amendment of the federal water pollution legislation to remove the funding restriction on construction of waste water treatment facilities. His suggestion would make such water pollution abatement works as combined storm and sanitary sewer systems eligible for federal funds.

• Assistance in the marketing of local bond issues. In this regard, Ellis considers the prospects for the new EFA favorable.

• Encouragement of sewage service charges. Wherever possible, local construction should use the element of sewer charges to finance their plants.

Sounding the concern of small municipalities, John L. Salisbury, executive secretary of the Maine Municipal Association (Hallowell) restated the need for adequate funding on federal promises already made. The need for secondary treatment (required by Federal Water Pollution Control Administration in doling out funds), is being challenged, according to the Maine spokesman. Projects upgrading facilities from primary to secondary treatment must be stopped, as these upgrading projects only produce marginal effects. What is needed is a more flexible regulatory approach, Salisbury notes. The water pollution abatement needs of the small municipality should be considered on a river basin management basis. The needs should be considered with regard to the assimulative capacity of the streams.

Salisbury also was critical of the greater than 250,000 people criteria for obtaining funds on federal projects. It is not a true measure of U.S. needs, he says. Rather, he prefers using a regional river basin planning approach to a small municipality's water pollution abatement plan.

*Reprinted from* ENVIRON. SCI. TECHNOL., **6**, 868 (October 1972)

# Thomas O'Boyle

*Thomas O'Boyle is president of Ecodyne Corp., a subsidiary of Trans Union Corp. Incorporated in 1970, Ecodyne is running hard to be the No. 1 water company in the U.S. O'Boyle tells ES&T's Stan Miller that Ecodyne is the largest diversified company devoted exclusively to water treatment, water cooling, and waste treatment. The company does about everything one can do to water—cools it, cleans it, softens it, conserves it, treats it, recycles it, uses it, and protects it. Ecodyne sales hit $75 million last year (calendar year), out of $255 million for Trans Union, up from $65 million and $200 million, respectively, in 1970.*

## Rationale

**Is Ecodyne ready for passage of a new water pollution control bill?**

That question should really be answered in two parts. Do we have the technology to handle the demands which will arise under the new water bill? And, do we have the manpower and physical capacity? Frankly, there is no question but that from a technology standpoint Ecodyne can easily handle any demands for advanced waste water treatment. As I have said many times before, the processes and techniques for effective control of water pollution have been available for years and our expertise in this field has solid depth. From a manpower and capacity standpoint, in many of our divisions we have already anticipated a growing market potential and have geared up our operations to cope with it.

**When you say "geared up," how much expansion do you anticipate?**

We could probably double the municipal waste water treatment output of our Smith & Loveless Division without any problem whatsoever. We can handle our share of biologically treated industrial waste. Our Graver Division which handles both physical-chemical municipal waste treatment and chemical treatment of large industrial wastes, is an engineering company subcontracting to independent fabricators. There is no reason why we cannot substantially expand Graver's output since we have no plant restraints. Frankly, the only problem we would have at Graver is the ability to recruit competent engineering personnel to handle increased volume. This is always a problem in a high-technology, custom engineering company. We have learned to live with this problem over the years. As a matter of fact, engineering talent seems to be more available today than it was a few years ago, so I see no problems here.

On the cooling side, we have grown considerably over the past few years and have already expanded our facilities several-fold. I might note that because of the long lead time required on major cooling tower jobs, our cooling group can book con-

siderably more business and still have ample time to add to its physical staff and capacity on a planned schedule.

**Where should the national focus in water pollution cleanup be at this time?**

It's not glamorous but it's absolutely essential that we, as a nation, upgrade a vast number of municipal treatment plants that are currently operating with inadequate primary and secondary treatment facilities. In industry's case, there must be a real attempt to formulate consistent standards throughout the U.S. so that an industry in one state does not have a competitive advantage over a similar industry in another state. It's better, in the long run, to pursue a methodical solution to the municipal and industrial problems. The recent actions against phosphate detergents and DDT are examples of crash programs instituted by emotional issues rather than logical ones.

**To what extent is there a delay in the nation's water cleanup activity?**

Well, it is rather difficult to quantify. As a nation we have lost a couple of years because of a lot of talk and little action. On the funding side, the federal government has consistently under-obligated appropriated funds. We can catch up, but it is becoming more expensive to do so. Still, we have the ability to catch up, and I am sure we will once meaningful federal legislation is enacted.

## Position

**If Ecodyne is the No. 1 water company, what company is No. 2?**

Thanks for your vote of confidence. Seriously, we have taken the position that we are the largest diversified company exclusively in the water field. Our broad product line and our substantial sales volume, again exclusively in water, certainly supports our claim. When you ask what company is No. 2, what you are asking is "who is our main competitor across the board?" Unfortunately, there is no simple answer to that question. The fact is there is no single competitor that we meet head to head in every line of our business. There is no single company that does all the things—water treatment, water cooling, and waste water treatment that Ecodyne does. Certainly, there are companies that are larger in specialized segments of our market but the real point is that Ecodyne has achieved a commanding position in the most important areas of the water field.

**Does Ecodyne anticipate any move into other pollution control areas?**

At the moment, no. We feel strongly that our future in water is so challenging that water can absorb all of our energy over the next few years. If a particular and unusual opportunity presents itself, then we might reconsider, but at the moment we have no plan to go into other areas. The Ecodyne philosophy is to establish a dominant leadership position within a given industry. It would be inconsistent with this philosophy to go into another industry such as air, for example, unless there was the potential to achieve a commanding role. There doesn't seem to be any opportunities like this at the present time.

**I understand that Ecodyne is only two years old. Don't your roots go back farther than that?**

We've been in the water field since we acquired Graver in 1957 and Graver's roots go back to the 1940's. In our later acquisitions, we've stayed in the water field, and the water field alone. So, over the years, we've built our reputation in water. We acquired Lindsay and Smith & Loveless in 1959, and formed Unitech in 1967. In 1968 all of these divisions, each of which was operating successfully, were brought together into the water and waste treatment group of Union Tank Car Co. which subsequently became Trans Union Corp. So Ecodyne was operating as a single company, in a sense, as the water and waste treatment group of Trans Union in 1968 with the history of its division going back some decades. In 1969, we purchased Fluor Cooling Products, now Ecodyne Cooling Products Co., a company which dates back to the early 1920's. This was our entry into the evaporative cooling tower business. And in 1971, we acquired McKenzie-Ris which is in the air-cooling field.

### Products and services

**What is the breakdown for Ecodyne's sales?**

Ecodyne's total sales in 1971 were divided among the following four markets: Industrial, $33 million; Electric utility, $17 million; Municipal, $14 million; and Residential. $11 million.

**What percentage of the sales is equipment; what percentage is engineering services?**

It's all equipment sales. The engineering Ecodyne does is always in connection with the sales of its products. We do not charge separately for engineering. We may have a specialized charge for field services, repair work, that sort of thing, but we have no separate charge for basic engineering. The engineering costs associated with a particular installation are included in the overall price of the product along with other material and labor costs.

**Does Ecodyne retain the services of consulting engineering groups? Does the work that you do come in the door by the consulting engineer, from the municipality, or does Ecodyne have a marketing group that helps?**

Let's break those questions down. On the municipal side, consulting engineers are practically always involved. The work comes to us through our sales offices or representatives who keep in touch with potential clients and consulting engineers to discover what jobs are coming up. On the other hand, industrial customers often utilize their own engineering staffs to perform studies and make recommendations on treatment needs.

The one exception seems to be in the utility field where large consultants are often used. Depending on the size of these consultants' staffs, industrial customers may never deal with a consulting engineering firm. In this case, our field sales force deals directly with the industrial customer involved. Many inquiries come directly to us because Ecodyne is well-known in its fields. But there is always a lot of bird-dogging that goes on to see that sales offices and representatives do not miss an opportunity.

### Turn-key

**What is the present company position on turn-key construction, both municipal and industrial?**

We have no inflexible position on it. We are capable of doing turn-key jobs although we have not done any great number of them up to the present time. In the municipal field we have done none, but in the industrial field, we have been involved in several turn-key projects. As a matter of fact we have recently bid on a number of such projects in Canada. There has been a lot of comment lately that more and more industrial companies are turning to turn-key projects. It is true that there has been more interest in this type of approach on industry's part, but

---

## Ecodyne milestones in water and waste water treatment

**1972:** Contract awarded for first U.S. physical-chemical municipal sewage treatment plant to Graver

**1971:** Introduction of the concrete mechanical draft cooling tower

**1969:** Tertiary filters for reducing municipal sewage effluent to potable water quality levels following secondary activated sludge treatment installed by Smith & Loveless

**1969:** First "PFR" concentrator producing 63% solids black liquor from southern pine for direct firing into odor-free recovery boiler designed and furnished by The Unitech Co.

**1966:** Introduction of PVC fill in cooling towers to replace wood by Ecodyne Cooling Products

**1962:** Powdex(R)—High-quality condensate polishing for electric utilities

**1956:** First use of sewage plant effluent for boiler feedwater after treatment by Graver clarification, filtration, and ion exchange

**1946:** Factory-built sewage pumping stations pioneered by Smith & Loveless

**1946:** Phosphates reduced to negligible level by first Graver tertiary treatment plant designed to recover industrial process water from sewage effluent to supply cooling water for oil company

**1921:** First cooling tower for an oil refinery built by Ecodyne cooling products predecessor

*"We feel strongly that our future in water is so challenging that water can absorb all of our energy over the next few years."*

**Ecodyne's O'Boyle**

I really can't say that we've seen any evidence of a trend. In the municipal field we just don't think the votes are in yet. I can't really say whether municipalities will move in this direction or not, but I doubt it. Ecodyne can handle any turn-key opportunities which might arise. If the market demands turn-key, Ecodyne can be counted on to provide it.

**What are the problems of industrial users of municipal waste treatment plants?**

There may be a very severe economic problem for industrial users with a surcharge program. Once the industry is liable for surcharge there seems to be a very automatic way that the additional costs are passed along to industry rather than absorbed by the municipality. Surcharges, of course, are not applicable in all waste water treatment agencies. The Metropolitan Sanitary District in Chicago has a surcharge program. Some other municipalities do also but, at the moment, only a minority of agencies have surcharges. You would expect to see more and more programs like these because one of the provisions of the new water bill requires that the industrial user pay his prorated share of the treatment cost of the municipal plant. Of course, industry can always make other arrangements to handle its waste. By bringing wastes down to the acceptable municipal standard, industry can eliminate not all, but probably most, of a surcharge.

## Marketing

**How does Ecodyne resolve in-house squabbles between competing technologies, such as physical-chemical treatment vs. biological treatment of municipal wastes?**

Fortunately, to date, any squabbles we might have within the company are theoretical rather than commercial. Naturally we have strong advocates of both biological and physical-chemical systems within our divisions. I think this is healthy. There is nothing like a vigorous competition of ideas to promote the development of the art.

On the practical side, almost invariably the process that is going to be used in a particular application has already been decided on by the customer and his consulting engineer when the job comes to Ecodyne. In other words, it has already been specified whether a biological or physical-chemical process will be used. The same is true in the cooling field. The customer has decided that he needs an evaporative cooling tower or an air-cooled heat exchanger by the time he comes to Ecodyne. Up to now the marketplace has not generated any real squabble between competing technologies. As an aside, it gives me some reassurance that whatever the technology chosen, Ecodyne can provide it.

**Is there a marketing function in Ecodyne which resolves differences? Who figures the best economical and financial fix for the use of any particular technology?**

We don't have a corporate staff marketing function, if that's what you mean. We haven't adopted that kind of an organization structure. Ecodyne is organized to stress the autonomy of the divisions, but it is also flexible enough to handle a joint approach to a specific problem. There is constant communication between all levels of our divisions: marketing, research and development, and product planning. This relationship is not as difficult to foster as you might think. We also have regular meetings of the division presidents to examine product strategies, research and development activities, and to concentrate our marketing efforts more effectively. Solid decisions come out of these meetings. Positions are set and product and process responsibilities are allocated. No, we don't have any trouble resolving differences. It works very well.

## Technological innovations

**It has been said that all of the technology is available today to solve all the problems? If Ecodyne is innovative, then must the innovations of necessity be more economical?**

Not necessarily. The basic technology is available today but we need continuing innovations and refinements, not only to improve the effectiveness of that basic technology, but also to broaden its applications. For example, the principle of the cooling tower is the same as it was 50 years ago. Ecodyne has made substantial strides in improving the efficiency and design of these towers but not all of these improvements have reduced the cost of the towers: for instance, the PVC fill material which replaces wood. The PVC fill is not less costly, but other advantages, such as fire resistance and improved thermal performance, accrue from its use. So, innovations do not necessarily have to be more economical.

In the physical-chemical area, the Rosemount, Minn., job is a refinement of technology that has been available for years. We are just designing it into a highly efficient system. The same is true of biological treatment plants where we are continuing to up-grade, refine, and shorten the treatment cycle. So not all innovations reduce the cost; they improve the product. Basically, we are still using a time-tested technology. There are no black boxes that have appeared and there are not likely to be any.

**What are the major innovations introduced by Ecodyne for better water and waste treatment?**

Ecodyne has a history in the water treatment field running back to the 1920's. We introduced significant innovations even in those days. As you can see (box, p 104), we have continued to innovate up to the present date.

# Camp Dresser & McKee's Joseph Lawler

*Reprinted from* ENVIRON. SCI. TECHNOL., **7**, 489 (June 1973)

**What are the basic groups within CDM?** Each CDM project is handled by a specially organized team. The firm is project-team oriented, though it does have special support groups —for example, a structural group, an electrical group. A senior officer of the company has responsibility for every design job; he deals directly with the client. A waste water treatment plant, for example, for a city might be a $20–30 million job. In this case, an officer of CDM at the vice-president level would be in direct charge. Of course, CDM handles all sizes of projects. The firm handles jobs as small as $5000 in construction costs for many of its clients. Now it wouldn't make sense for the firm to go all the way to Texas for a $5000 job for a new client, but it makes very much sense for CDM to handle one that size for, say, the town of Andover, 20 miles from Boston, since we have been dealing with them for many years.

**How many subsidiaries does the firm have and does the firm have a marketing function?** As far as geographic distribution goes, CDM has four offices in the U.S. and seven overseas. CDM also has a partnership. In some states, for example, New York and Virginia, CDM needs a partnership because a corporation cannot operate as effectively there. CDM owns 50% of Camp Scott Furphy, a subsidiary in Australia, and acquired the New York firm of Alexander Potter Associates last year. Potter is a fairly sizable firm of about 60 people and billings of about $2.5 million. Potter covers the geographical area from New York down to the Washington, D.C. area.

CDM does not have a marketing function, as such. The firm does not have anybody beating the bushes, so to speak, not even one person and never has had. I don't mean to imply that there is anything unprofessional or unethical or anything of the sort; CDM just hasn't found it necessary to do it. Maybe some day we will.

The CDM parent corporation was established in 1970, so we are on our third year as a corporation now. The international subsidiary was set up in 1968 as a corporation, prior to the incorporation of the parent, for various reasons including tax purposes.

**Do CDM personnel refer to themselves as environmental engineers, as opposed to sanitary engineers?** Yes, in all of our professional cards CDM uses the term environmental engineers. But we still have some clients with whom we use the term sanitary engineer; they prefer the latter terminology.

## BUSINESS

**What is the current number of active projects and the short-term projection for new construction design under the water law?** Current engineering contracts, including reports, total 190; in aggregate they have an estimated construction cost on the order of $2.0 billion. Including our subsidiaries, CDM has about $800 million worth of work under design or construction at this time. Incidentally, the Potter acquisition took place in September 1972, so only the last quarter figures show up in the total figure for 1972 billings.

**Is the CDM business 100% in the municipal area, or are there some examples of industrial projects?** Private industrial clients have never been a big proportion of the CDM total, although we do have a number of private industrial clients, perhaps representing 5% of our total volume. However, a lot of our municipal plants receive major contributions of industrial wastes and require much engineering study for the industrial wastes alone. Take CDM's job at Niagara Falls, for example. This is a large municipal job; construction costs run over $50 million. But it is largely an industrial waste problem. About 70% of the waste water which will enter the Niagara Falls plant will

*At the helm of one of the largest U.S. consulting engineering firms, Joseph Lawler is president and chairman of the board of Camp Dresser & McKee, a corporation specializing exclusively in sanitary engineering and related sciences, which more appropriately these days has been termed environmental engineering. Lawler tells ES&T's Stan Miller that over the past 20 years, the firm has completed more than 1700 engineering projects and has a current staff of more than 600. With offices in Boston, Washington (D.C.), New York City, Bangkok, Bogota, Dacca, Singapore, Bermuda, and Pasadena, CDM's billings hit $13 million in fiscal 1972 (calendar year).*

be industrial waste. This plant will not use the conventional biological process; it's a physical-chemical treatment plant which will go under construction this year. In fact, it's one of the largest physical-chemical plants—50-mgd—that will be built.

**How many advanced waste treatment (AWT) plants does the firm have under design or actual construction?** CDM has a lot of them. Major projects where the design has been completed by CDM or its subsidiaries include: a $41-million plant for Arlington County, Va.; a $40-million plant for Fairfax County, Va. (both using a combination of biological and physical-chemical processes); two plants for Fitchburg, Mass., totaling $29 million in cost (one biological-chemical and the other physical-chemical); a $43-million biological-chemical plant for Chicago, Ill.; and, of course, the physical-chemical treatment plant for Niagara Falls, N.Y. In addition there are a couple of smaller AWT plants on which we are working.

The Salt Creek Plant for Chicago is a CDM design; the plant is under construction now and is of particular interest because it was the first major plant designed utilizing the two-stage aeration process for nitrification with denitrification in a final stage of filtration.

Some of the plants referred to are presently under construction. All of them should be under construction before the end of the calendar year.

**What is the current thinking on going public? Any announcement?** CDM is definitely "going public." Depending on stock market conditions, the public stock issue will probably take place in the first quarter of 1974, for which we will probably issue about 20% new stock, the remainder of the stock being held within the company.

The firm of White Weld & Co., one of the top underwriters (investment bankers) in the U.S., will handle the operation.

**How does CDM go about getting new accounts—by word of mouth, marketing activity, too much business to go around, or what?** Most of the business in any good firm comes by word of mouth. When we get a job in a new area that we haven't been in before, and if we do it well, then in due course somebody else comes along and pretty soon CDM may have a number of new clients in an area where we have never worked before.

Obviously, a firm needs to keep its name before the public. One of the best ways to do this is to write technical papers and present them at various professional and technical meetings around the country. Also, if CDM undertakes to submit a proposal for a new project, we make strong efforts to do a really good job on the proposal.

**Does the firm have activities or work in most states?** At one time or another CDM has worked in the majority of the states, but not all at one time. CDM doesn't make an effort to work in any particular area; it goes primarily by the job. If there is a project that CDM is interested in, and there is a chance that CDM will be considered, then we write letters, send brochures, and if possible meet the key people there so that our firm will be among those considered for the project. Of course, the primary reason for making acquisitions is to get the firm into new geographical areas.

## WATER CLEANUP FUNDS

**What are your comments on the announced controlled spending of cleanup funds? Will there be enough**

funds to go around the sanitary engineering firms? As far as commenting on whether or not the President made the right decision, he is in a better position to judge spending priorities in this country than we are. After all, the funds being made available for water pollution control are still much larger than we have ever had before. There is plenty to do for all of the consulting firms. I'm not personally concerned about this so-called "slow down"; it's all relative to what some people thought was going to happen.

As far as CDM is individually concerned, the firm has a certain rate of growth which we believe we can satisfactorily manage; and if there were 10 times as many jobs available, CDM just wouldn't take them. The firm is not going to grow at a faster rate than it can handle.

## TURN-KEY

**Does the EPA turn-key proposal have any chance for success?** I do not think that turn-key operations for municipal waste water treatment plants are going to work. I want to differentiate between turn-key for municipal waste water treatment and turn-key for waste water treatment for private industry, because there is a vast difference. The turn-key approach for private industry can be quite sound. There are several reasons for this. First, the public interest is not involved. If mistakes are made or if a company makes a bad decision, then no one is hurt but that company. Second, most industrial companies of any size have sophisticated staffs that can properly specify and monitor the turn-key work. And the third and most important reason is that industry doesn't have to take bids for turn-key work, and in most cases they don't. The company can pick from among the top and most reputable

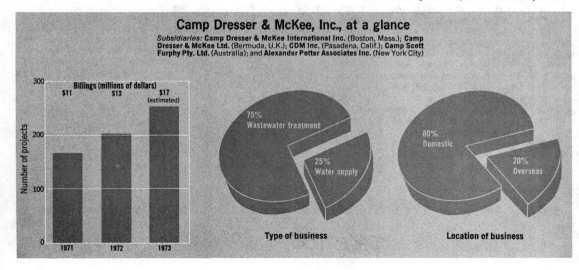

### Camp Dresser & McKee, Inc., at a glance

*Subsidiaries:* Camp Dresser & McKee International Inc. (Boston, Mass.); Camp Dresser & McKee Ltd. (Bermuda, U.K.); CDM Inc. (Pasadena, Calif.); Camp Scott Furphy Pty. Ltd. (Australia); and Alexander Potter Associates Inc. (New York City)

Billings (millions of dollars)
$11   $13   $17 (estimated)
1971   1972   1973

75% Wastewater treatment
25% Water supply
**Type of business**

80% Domestic
20% Overseas
**Location of business**

*"CDM has had a steady annual growth of 15–20% for about 15 years."*

Camp Dresser & McKee's Lawler

design construction firms and work something out or negotiate. For all these reasons, industry can usually get a good turn-key job. Industry is not confined in the same way as is a municipal turn-key operation.

**But what about turn-key in the municipal area?** These same advantages don't work at the municipal end because the municipality will have to go through a bidding situation based on performance standards. In municipal contract work, it is very difficult, and in some cases impossible, to throw out the bids of firms not adequately qualified to do the work; this is where the turn-key approach to municipal waste water treatment plant construction will fail. The municipality may end up with a contractor who may not be really capable of doing the job.

You may want 90% BOD removal, but this performance requirement doesn't enable you to write a specification on every single item that goes into that plant. On a major plant there are literally hundreds of such items as electric motors, pumps, chemical feeders, metering devices, and control instruments. If one could write detailed specifications and make detailed designs on all of these items, then of course you are right back to the conventional approach. The turn-key contractor may get the 90% BOD out but the question is whether the motors, pumps, and ancillary equipment will last for 20 or 25 years. The municipal turn-key operator will install the minimum to last past the guarantee period, three years or whatever is in the contract. With turn-key and public bidding you will not get a long-term low-cost operation. The total life cycle costs will be higher.

**Will turn-key actually happen?** Turn-key construction of municipal waste water treatment plants is probably going to happen in spite of it all because EPA has obviously made the judgment that it will work. I say that it won't work, but it will probably be several years before the problems show up. First, some municipal plants must be built by the turn-key method and operated for awhile; then the problems will become evident. Ten years from now there won't be any more waste water turn-key work in municipalities—maybe even sooner than that. In the meantime, the mistakes are going to have to be made.

When turn-key becomes part of the EPA operational policy, I predict that there will be a tremendous tendency on the part of many municipalities to go this route because it appears on the surface to be simpler. Many of the smaller municipalities don't have the professional staffs or the degree of technical sophistication to know what they really want, and then to follow up and be sure that they are getting the best. Some do, of course. Many of the larger cities, such as New York and Chicago, are first rate in the waste water treatment field, but there are those that just don't have that kind of expertise.

**What can you tell ES&T readers about CDM's new construction management service?** Construction management goes far beyond the usual job supervision. Let me give an example. CDM is providing construction management services on both the Fairfax County and Arlington County waste water treatment projects in Virginia. On those projects, which are both large ones, CDM has packaged the work into a number of smaller construction contracts to obtain more competition in bidding. Equipment, piping, pumps, and the like are being purchased ahead of time. Experience on these projects has shown that lower costs are being obtained by using several smaller contractors than by using one large one.

The construction manager is responsible for making sure that all of the equipment gets to the job in the right place at the right time, that the work of the many contractors is coordinated, that the mountain of paper work and project administration is carried out in timely fashion, and that the work is being coordinated with federal, state, and local government officials. One important aspect is that the construction manager be a part of the project from the beginning. A number of other civil engineering consulting firms are now providing construction management services.

**From your first-hand experience as president of the international subsidiary, how did the company get into this facet of the business?** CDM started in the overseas work in 1964. There were a number of reasons for moving in this direction. We felt that there was going to be a big future in the waste water and water supply fields in many of the developing areas of the world. We started responding to overseas invitations. Our first job was in East Pakistan (now Bangladesh), and, by the way, we are still there. CDM has also worked in many other countries including Australia, Brazil, Singapore, Thailand, Canada, Ireland, Taiwan, Turkey, Colombia, Puerto Rico, and the Virgin Islands.

The overseas work averages about 20% of our total volume; half is in the waste water treatment field, and the other half is in water supply. The two biggest jobs that CDM has overseas right now are both water supply—a $100 million job in Bangkok and a very large water filtration plant (840 mgd) for Sydney, Australia, which will be the third largest in the world. (The two largest are in Chicago.)

**Basically, what type of company does CDM look for in acquisitions? What elements of their operations would be an expansion for CDM?** CDM is looking for firms that are exclusively in our field; this is one policy that we adhere to strictly. We don't want to diversify into other fields and therefore we aren't interested in acquiring any firm that's even partly outside the environmental field.

The primary reason CDM is interested in the acquisition of other firms is to move into new geographical areas. There are two basic ways to move into new areas. One way is to develop an office from scratch, as we did in California. This is a long process. The other way is the acquisition route.

Geographically, CDM has divided the country into seven principal areas. We have worked in all those areas in one way or another, but it is difficult to develop much ongoing work without an office in the area. We therefore have long-range plans to establish branch or subsidiary offices in each of these areas. Most of the proposed new offices will involve acquisitions, which, in turn, are tied in with our plans to go public.

Our most recent new office is in Washington, D.C. CDM has recently agreed in principle to acquire a firm in Singapore and one in Saigon; CDM would like to participate in the new redevelopment/reconstruction program in South Vietnam.

We also might be interested in acquiring companies which would complement our technical strengths—companies which specialize, for example, in groundwater, irrigation, and R&D. The primary reason for CDM's going public is to obtain the additional capital and public stock to enable us to move in these directions.

## NEW TECHNOLOGY

**To what extent are new processes being incorporated into the CDM design projects?** There has been much talk, particularly from EPA, that consultants are too much inclined to use the old "tried and true" methods, and that by going to the turn-key approach, more "innovative" designs will be forthcoming. The idea that turn-key will suddenly generate exotic new designs that consulting engineering firms cannot themselves produce is, in my opinion, sheer nonsense. Sure, consulting firms failed to turn out very many advanced designs in the past, but that was because there was little or no expressed need for them—water quality standards, by and large, didn't require such advanced designs in the past. But the better consulting firms are certainly turning out such designs today.

In the case of CDM, we like to think that our designs are as "inno-vative" as any being turned out today. We have a design review committee comprised of some of the top professionals in our firm. This committee keeps on top of all of the latest developments in the field.

As to some examples of new processes being incorporated into CDM projects, I have already mentioned the two-stage aeration process for the Salt Creek plant being constructed in Chicago, in which nitrification takes place in the second aeration stage. In a third stage, filtration, the nitrates will be reduced to nitrogen, passing off as nitrogen gas. This was the first major plant in the country to be designed using this process. One of the plants we have since designed for Fitchburg, Mass., utilizes a similar two-stage aeration process.

**To what extent is CDM involved in air pollution control and solid waste disposal projects?** On the whole, CDM is not heavily involved in air pollution control. But the firm does a fair amount of work in the solid waste field. For example, CDM recently completed a major study for Erie-Niagara Counties, which includes the city of Buffalo and hundreds of industries in the area. The study featured a computer program for collecting and transporting the wastes and for locating transfer stations.

CDM is currently serving as a consultant to a new company which has developed a very promising method of disposing of solid waste. The process involves shredding, baling, and encapsulating. The material is highly compressed and encased in heavy sheet plastic which is shrunk-fit. The blocks of compressed material have a density greater than that of water, and can be buried in landfills or in water and collected for reprocessing at some later time. Additionally, these blocks can be used to reclaim certain islands that are washing away, provided that proper perimeter protection is provided against wave action.

## PERSONAL ASIDE

**What was the most satisfying assignment in your long and varied professional career?** It has not been any single assignment. Rather it has been the satisfaction of working on many projects over the years, starting with modest jobs and in later years being associated with some of the very largest projects of their type. The Sydney water treatment project, which will be one of the world's largest, is an example. We were able to come up with a fairly unique plan, which is about to be carried out.

Starting way back in 1940, I was the firm's first cooperative-work student, while an undergraduate civil engineering student at Northeastern University. You might say that I have had an opportunity to grow with the firm, because we had less than a dozen people in those days.

Starting out in the planning stages of some of these more complex projects and seeing them develop is really a lot of fun. I'm one of those guys who regularly work extra hours evenings and weekends. But then I play hard too; last month I went big game hunting for ibex in the mountains of Iran. People sometimes ask me why I work so hard. I tell them, "Don't feel sorry for me; I love it. You should be so lucky!"

# Peabody Galion's John McConnaughy

*Reprinted from* ENVIRON. SCI. TECHNOL., **7**, 97 (February 1973)

**Who is responsible for the concept of Peabody Galion as a total environmental company?** By analyzing the product line of the then-existing company, Hercules-Galion, we (mainly McConnaughy and A. J. Giglio, vice-president of corporate development) chose the environmental business as our base for growth. In terms of future development for the company, two product categories were singled out for growth potential, profitability, and a number of other factors. One was the refuse equipment business; the other was the fast food business. The refuse equipment business was used as the basis for future growth. Since refuse is, in essence, land pollution we decided not to look just at refuse equipment but to look at air, water, noise, and so on.

The first implementation of the concept was the acquisition of Peabody Engineering. The Peabody name was well recognized in the air pollution scrubbing business. In order to indicate the new direction, the company name was changed from Hercules-Galion, which was primarily known for garbage truck equipment. The name change was an external way of pointing out the expanded direction of the corporation. After getting into the air field, we moved into waste water treatment in July 1, 1970 and into the environmental services area last year.

Before coming to the company in 1969, I was with Westinghouse for seven years and then with Singer for about another seven in several positions, the first as controller of U.S. operations, then marketing director for the eastern half of the U.S., and before leaving, running Singer's European consumer business which hit about $400 million.

**Do the four groups act autonomously?** There is a president of each group, and although each group operates independently, they do some business with the other groups; but their number one assignment is to make their profit objectives individually as groups.

For example, when Peabody Galion was bidding on the Stamford, Conn., solid waste job which included a difficult incineration problem, the air group bid together with the solid wastes group. At the top level, the presidents of the groups get together to coordinate on problems, but it is important to note that the company does not force cooperation of its groups. Cooperation either happens naturally or it doesn't happen at all.

Recently, Philip Van Huffle became president of the air group; he replaced John Dunn who elected to retire but will do consulting work for the company. Within this air organization, there are three separate entities—Peabody Canada, Peabody U.S., and Peabody, London. Until recently, the three had been relatively separate. Each has its own leader, who in turn reports to Van Huffle, the new president who does the coordination and liaison within the group.

Varying degrees of cooperation are found within the groups. Again in the air group, for example, there is a great deal of cooperation between Peabody GCI in Paris with Peabody Roberg and Peabody Ltd. Each of the three reports to Peabody Ltd. in London; they have a lot of problems in common. Roberg and GCI distribute Peabody Ltd. products in France, Italy, and Scandinavia. In return, Peabody Ltd. will pick up their products and bid them in Canada, South Africa, and so on. It's a built-in working relationship.

## ACQUISITIONS

**What can you tell ES&T readers about acquisition strategy and organizational pains, if any?** There are five things that Peabody Galion looks at. First, the acquisition must make conceptual sense, either from a product, marketing, or manufacturing standpoint. In this way, the acquisition ties in with something Peabody is trying to do. Let me give you an example. In the solid wastes management group, a company called DeWald was acquired on the West Coast. The acquisition was essentially a marketing strategy because Peabody was weak in terms of marketing its products there. The acquisition gave us immediate marketing capabilities plus two plants, one in Los Angeles, the other in San Francisco.

Second, the company has to be relatively well managed and doing well. Third, Peabody only believes in the friendly type of acquisition, where they want us and we want them. Four, we won't dilute our earnings for any reason. Last, Peabody takes a very hard look at the balance sheet.

**What can you tell us about the more recent acquisition now known as the services group?** The services group

*At ease in his Manhattan office, John McConnaughy, Jr., is president of Peabody Galion Corp., a total environmental company that derives more than 70% of its revenue from balanced positions in all segments of pollution control—air, water, and solid waste, and testing services. The 42-year-old executive tells ES&T's Stan Miller how the individual groups were pulled together to make the corporate entity that the company is today. Since he joined the firm on May 1, 1969, Peabody Galion has showed that there is room for total pollution control companies in the U.S. clean-up operation; sales hit $134 million in 1972 as reported (year ending September 30), up from $90 million in 1971.*

is regarded as an excellent opportunity for growth and profit. In the area of stack gas testing and water pollution testing it is very important to know what pollutants are going through a system before corrective control technologies are applied. One of our early mistakes on a pollution job was taking the customer's word for what was coming out of the plant. The reason that the controls were not working well was that what was going through the system was not well known. Terrible mistakes can be made by taking the customer's word and not knowing the parameters of moisture and temperature, for example, on air emissions. To make sure that a customer's pollution control problem is corrected, it is all-important to do the testing beforehand and make sure you know what to correct.

**ES&T did an earlier interview with Stewart Udall and noted that he served on the board of directors for two years and came off in 1972. What influence did he exert?** Most of the Peabody Galion directors, whether internal or external, were primarily businessmen and not environmentalists, if you will. The directors felt that it would be very helpful to have a conceptual input and to give the company an overview in the area of environment. Mr. Udall gave us his ideas in terms of what really was happening and what he felt was going to happen. While he was here, he was well worth having on the board.

**What else does the service group do?** The basic business of our Commercial Testing subsidiary is oriented toward the testing of coal as much as anything else. The fuel which goes into a combustion system is normally the source of most of the air pollutants. Knowing what is going in, one can design and fabricate efficient control equipment to handle it. Peabody has performed stack testing for Commonwealth Edison in Chicago and also has a long-term contract with Ontario Hydro. Our X-ray Engineering subsidiary performs nondestructive testing for nuclear-fueled utilities. They perform nondestructive testing to ensure that the equipment has no leaks and measures up to the requirements of the Atomic Energy Commission. Our Industrial Leasing subsidiary provides a capability for the other groups to provide financial backing arrangements for their customers. In many cases, Peabody is finding that the customer would rather pay so much each month or each year, than a lump sum.

Peabody Cactus, a recent acquisition, is our approach toward the petroleum industry. Cactus' major activity is the capping of abandoned wells and the recycling of used well tubing and casing for resale to the petroleum industry. Cactus charges a certain fee for the proper recapping of an abandoned well and the handling of the water and oils so that there is no pollution damage to groundwater and surface soils.

The newest acquisitions are Leonard S. Wegman Co.—the well-known consulting engineering firm in the solid waste area—and the Boston-based firm—Hayden, Harding & Buchanan—in the waste water treatment field, and the John L. Doré Co., a construction firm. The Wegman operation is included in the services group despite the fact that it concentrates on solid waste; we feel that its operation should not be tied in with solid waste equipment manufacture.

BUSINESS SPLIT

**What is the breakdown of sales figures into products and services?** Hardware equipment items account for 70% of total sales, services the remaining 30%. (The market splits within the various groups as shown in the box.) The industrial segment is normally a bit more profitable, particularly in the solid wastes area. The company plans to enlarge its solid waste activity in the municipal market as part of its natural growth.

**What parcel of the recent water law does Peabody intend to carve out for itself?** Under the law, water pollution controls will affect industry as well as municipalities. As such, it will impact on all entities in the water group plus some in the services group. Specifically, Peabody expects to provide consulting engineering services, water testing and analysis, hardware, and construction. A lot of testing will be required to prove that a certain effluent limitation has been achieved. Whatever the total expenditure of federal funds in the water area, the expenditure should bring about dramatic growth for the water group.

ACROSS-THE-BOARD

**Why does a customer come to Peabody rather than to a company that specializes solely in air or water pollution controls or solid waste management? What attraction does the company offer the potential customer?** We are not going to be all things to all people by any means. The attraction mainly is the customer knowing that he is dealing with someone he respects and who is

"*If the administration really spends all the money in the new water law, we could be swamped with work.*"

Peabody Galion's McConnaughy

working on his particular problem. Many times, the problem is much wider than the specific one the customer brings to Peabody. So the Peabody contact man might say, for example, you need to correct these related pollution control problems as well; let me bring some other groups in for a total answer to the problem. This is where the total environmental capability plays a very important role.

**What is it that makes Peabody unique?** The company is unique in that it is relatively more balanced than its competitors. Peabody has a broader scope and feels that there is more safety in balance, including a number of different types of balance —balance sheet vs. profit and loss; balanced positions in air, water, and land; and a balance between hardware (capital equipment) and consumables (services, fabric filter replacement bags, and chemicals). Overall, if one area is down, another is up.

MARKETING

**How does the company attract potential customers to the individual segments of its business? Is there an overall marketing function?** Peabody has no formal marketing organization on a company-wide basis; generally, the marketing is done within an individual group. Each handles its own marketing activity; they vary in size, and it is difficult to say just what is marketing and what is engineering, particularly in the air group. Many of our engineers are, in essence, marketing experts who go out and sell the project while they are doing the engineering and so on. There is no standard approach. Quite often the customer brings the job directly to the competence within the company.

In the solid wastes area, the company recently introduced a new rear loader which is primarily aimed at municipal markets. The company is also in the business of transfer stations and has supplied equipment for 35 stations in different communities. The sales of these new product lines—stationary compactors and transfer stations—require different marketing direction from the sales of mobile types of equipment that have been historical. The marketing of two large SO₂ removal jobs in the air field is also quite different from the sale of normal combustion systems or off-the-shelf equipment products. In the latter case, we have moved away from agents and toward factory branches and in-house experts.

**Where does the company concentrate its sales effort? In jargon, what piece of the cleanup action does Peabody hope to get?** It varies with each group. The marketing must be cost effective for each individual segment or each individual business. Peabody is active in cost reduction projects; we are concerned with administrative costs of one kind or another, but when it comes to marketing organization, we feel that the local management has a much better idea of how to manage their products than we would here in New York.

Take the municipal business of Peabody Petersen, in the water group, which builds waste water treatment plants. They know where the jobs have been given out, and their marketing effort goes about getting the jobs. The president of the water group does not monitor the market in that particular case. Peabody Barnes, which manufactures pumps, has an entire marketing organization with regional managers, myriad agents of all kinds, factory representatives, and the lot. With Peabody Welles, which supplies specialty hardware including clarifiers, aerators, and filtration systems, again in the water group, sales are more oriented toward the engineer.

### INNOVATION

**What control technologies is Peabody backing that may well pay off in the future?** Certainly, stationary compactors and transfer stations in the solid waste area. There is a line of air products aimed primarily at control of internal air in a plant. The device is relatively new in the industry and can be used to control solid particles such as smoke from welding fumes or aerosol particles such as those from lubricating liquids used on machine tools. The interest here is both from an air pollution control standpoint as well as an occupational health standpoint. The interest stems from the fact that if you can return clean air in the plant, you do not have to make up air from the outside. Peabody also expects to play a role in SO₂ absorption technology. We are now building a commercial plant (Detroit Edison) using limestone, and have test projects in magnesium oxide and citrate absorption.

When you ask about new products we can go on and on and on. A lot of them have fantastic potential; whether the potential is realized depends on many things in the marketplace, but Peabody Galion does not invest in an idea unless there is a very high potential that the idea will succeed. Peabody has an exclusive license for the sale of the Lugar fabric filters in the U.S. The filters are used in the glass and aluminum industries, for example. We also have a license for sale of the Lurgi radial flow wet scrubbers. Other items include an inerting system for oil tankers which snuffs out any spark that might cause a fire or explosion in the tanker, methanol combustion systems, and liquid waste incinerators. Last year, Peabody introduced 22 new products in the solid wastes area.

Each of the groups handles its own research and development to varying degrees. The air group has a separate research and development organization in Stamford, Conn.; the Ven-Kinetic air scrubber was a direct result of its research.

## Peabody Galion, the environment company, at a glance
### (all figures in millions of dollars)

| | 1972 sales | 1971 sales | |
|---|---|---|---|
| **Air pollution control group** including Engineering (U.S. and Canada), Limited (London), Compower, Engineering (Power and Combustion), GCI (France), Roberg (Sweden), Gordon-Piatt and American Brattice Cloth | $33.0 | $26.5 | industry 68% / government 2% / utilities 30% |
| **Water resources group** including Barnes, Welles, Hart, and Petersen | 24.3 | 15.7 | industry 55% / municipal 45% |
| **Environmental services group** including Commercial Testing and Engineering, X-ray Engineering, Ryan Instruments, Industrial Leasing, Cactus-Pipe,* Leonard S. Wegman Co.,* and Hayden, Harding & Buchanan* | 24.8 | 7.4 | industry 70% / municipal 5% / utilities 25% |
| **Solid wastes management group** including DeWald and Rudco | 22.0 | 16.6 | industry 70% / municipal 30% |
| **Truck equipment** | 17.5 | 14.9 | |
| **Others** | 12.4 | 9.2 | |
| **Totals** | **$134.0** | **$90.3** | |

*Recent acquisitions not included in 1972 sales figures
Figures do not include restatement

# Frank Zurn

*Reprinted from* ENVIRON. SCI. TECHNOL., **6**, 495 (June 1972)

*An across-the-table discussion with a third-generation Zurn, a 45-year executive whose company is setting a fast business pace in the total environmental field*

*Frank W. Zurn is president of Zurn Industries, Inc., a company firmly dedicated to the concept of environmentalism. Sales totaled an estimated $185 million last year (year ending March 31, 1972). Frank Zurn tells* ES&T's *Stan Miller that the company was founded in 1900 by his grandfather, John A. Zurn, and grew into a multimillion dollar, family-owned corporation during the 30's–50's under the leadership of his father, Melvin A. Zurn, (who died in 1970) and his uncle, Everett F. Zurn, the present 64-year-old chairman of the board. The corporation went public in 1961, has acquired 18 companies during the past five years, and now has products and services in all four major environmental pollution control areas—air, land (solid waste), water, and noise.*

## Environmentalism

**When did Zurn management direct itself to total environmental pollution control?**

In the mid-sixties, we made a decision that if we were going to be in the business of pollution control, we were going to be involved on a total environmental control basis—air, land, water, and noise—because of the close interrelationships existing among these areas, both from a causative standpoint and the problem-solving mechanisms utilized. We set out to organize the growth program from within. We invested substantial monies in developing certain components in the air pollution control field, for example, but we concluded that the most efficient and economical method to growth within our time constraints was to initiate an acquisition program concentrated in the pollution control areas.

Our first major acquisition, in 1966, was a company known as Erie City Iron Works, one of the oldest corporations in Erie, Pa., and the third oldest corporation in Pennsylvania. The expertise that they brought into Zurn Industries was that of incineration, solid waste disposal, as well as a certain amount of air pollution control technology, primarily in the waste heat energy recovery field with emphasis on cooling and cleaning hot, noxious gases—all of which is tied in with generating steam from incinerated wastes.

**What are some of the reasons for the changeover to a total environmental company?**

Zurn customers were running into a problem with the disposal of solid waste which we entrapped in our water pollution control system. We had solved the water pollution control problem but ended up with a solid waste problem, and usually there were one of two answers—incineration or landfill. We realized that if we were to incinerate, then we had to watch out for air pollution emissions.

All three control areas—air, land, and water—had to be done on a total basis or we were not doing the cleanup job thoroughly and properly. Before the move to environmentalism, Zurn was basically a hardware-oriented company; we dealt in water pollution control systems only.

After we had acquired various air, land, and water pollution control equipment companies, we recognized that we had the hardware, but we did not have the software which, in our parlance, consists of research, analysis, and design feasibility—the beginning of solving a pollution control problem; neither did we have the construction capabilities. We set about to find a company or companies that had both software and construction capabilities. After considerable searching, we came up with Ludwig Engineering and Science in late 1968. With that acquisition, Zurn then was ready to begin handling total turn-key projects because then we could start essentially at ground zero on a pollution control problem and take it to the ultimate solution, including training of operators, monitoring of plant effluents, and assuming total responsibility for round-the-clock performance of the plant.

**What is the breakdown of industrial and municipal clients in the four categories of pollution control?**

As a general statement, about 75% of the business is industrial, 25% municipal. However, this has been changing a percentage point or two each year in favor of the municipal. Starting this new year (April 1, 1972), we will probably be 70% industrial and 30% municipal. In water pollution control, municipalities are heavily involved with water and waste water treatment plants, but in air pollution controls the problems are with industries, power generation, and municipal incineration.

Noise pollution control is one of our newest directions; thermal pollution control is the other and is classified under the water pollution control activity of the company. Noise, our smallest business category, does about $3 million in volume—$2 million in hardware and $1 million in software. Zurn has been involved with the hardware side of the noise business primarily through the development of sound attenuation, mechanical, electrical, and hydraulic devices for nuclear submarines. Zurn's Mechanical Drive Division supplies noise-attenuating drives for industrial machinery applications of the type developed for submarines. Zurn has also developed Accumultrol and Shoktrol products for attenuating or eliminating noise and shock from pipeline flow which are caused by quick-closing, actuating solenoid valves. In the absence of these shock arrestors, severe damage can be done to piping systems, not to mention shock noises reverberating through "quiet" buildings such as hospitals, schools, motels, and institutions. We don't see a huge market in noise pollution control in the immediate future, even in the equipment end of that business. Our work in the noise area, as far as the software is concerned is heavily oriented to municipalities—studies on airports, subways, and rapid transit.

On the other hand, thermal pollution control is our newest growth field. We regard this area as one of substantial growth with a very substantial market in dollar volume and potential profits. Balcke cooling towers, which Zurn can provide, are normal hyperbolics of the natural draft type as well as the induced draft type. The induced draft towers have a much lower silhouette than the natural draft. They are about one third the size of the natural draft towers in physical size. Of course, the advantage of the induced draft towers is that the flow rate across the water we are trying to cool can be varied regardless of the humidity of atmospheric conditions, and achieve extremely efficient cooling capability.

## Products and services

**Does Zurn have a complete line of products for air, water, solid waste, and noise?**

We think we have a complete line capability in-house. However, we do specify and use other types of equipment that we do not manufacture. A good example is electrostatic precipitators. Zurn has investigated this market thoroughly, not only from design capability in this country but also from foreign licensing capabilities. We came to the conclusion that there is nothing really new and unique to offer in the electrostatic field. We are able to buy such equipment from a half dozen sources. The electrostatics market has peaked out; we feel that high-velocity wet-type scrubbers and bag-type filters are going to be taking over the market. Zurn has capabilities in both high-velocity wet scrubbers and bag filters; our bag filter work is coming on very strong.

**Are the hardware items fabricated by Zurn?**

We take responsibility for the equipment which we may specify and purchase which goes into the total pollution control system. It may be our design or it may be someone else's design and manufacture. Of all the products listed in various Zurn brochures, 99% of them are manufactured in-house and 95% of the 99% are proprietary. About 45% of the total Zurn manufacturing capacity is in the city of Erie, Pa., with more than a million square feet of manufacturing space. The other 55% of the capacity is spread all over the U.S. For example, scrubbers, mechanical dust collectors, and bag filters are manufactured in Birmingham, Ala.; metal bellows which go into our new instrumentation line of air-monitoring pumps are manufactured in Sharon, Mass., and Chatsworth, Calif.

## Turn-key

**Can Zurn perform turn-key operations in the four categories for both industrial and municipal clients?**

We have been very successful in selling the turn-key concept to industries and are becoming successful in selling this concept to municipalities. With the economies, the quality of

*"There is no reason why turn-key cannot be applied to the municipal market. We are 100% in the direction of turn-key."*

installations, and performance and the efficiency of such installations that turn-key offers, there is no reason why turn-key cannot be applied to the municipal market.

Zurn Industries is 100% in the direction of turn-key. We think it's the most economical, efficient way to solve the massive pollution control problems. We think also in turn-key that there is plenty of business available for all available consulting engineering firms, for hardware equipment manufacturers, and for construction companies. We take total responsibility for the study analysis, for the design of the installations and the systems. We take responsibility for the equipment which we may specify or purchase. It may be our own design or it may be someone else's design and manufacture. It is true that Zurn does not produce every nut and bolt that goes into a pollution control plant. Take shredders as an example. We market a shredder under the Zurn name, but it's not produced by us. Electrostatic precipitators are another example; we do not produce these.

**Are there examples of Zurn turn-key projects in the water category?**

Although the project at the City of Cleveland (Ohio) which is being carried out by our affiliate—Zurn Environmental Engineers—is not specifically a turn-key contract, it contains the critical element of one. There, the Westerly plant project involves a departure in terms of liquid waste treatment; the untreated waste coming into the plant is a combined municipal and industrial waste that is quite complex and not suitable for treatment by conventional biological treatment means (activated sludge). Zurn submitted its proposal to the City of Cleveland three years ago in concert with Battelle-Northwest for conducting a feasibility study which included: • laboratory analysis of the incoming wastes and surrounding receiving waters • design, construction, and operation of a pilot plant facility at the plant site • analysis of pilot plant data • preparation of initial cost estimates for the full-scale facility. Now, Zurn Environmental Engineers is proceeding with preparation of final designs, plans, and specifications for the full-scale plant.

Initially, the plant will have a capacity of 50 million gal./day with provisions for handling storm water overflows up to a peak loading of 1600 ft/sec. Later, the plant can be expanded to 100 million gal./day. It will be located on approximately

8 acres of land adjacent to Edgewater Beach. When completed, the Westerly plant will be the world's largest municipal waste water reclamation facility utilizing physical-chemical treatment concepts. Construction cost alone approximates $35 million.

A good deal of the work that Zurn does in construction is more appropriately termed construction management. We do not necessarily use our own actual labor in all of these projects. We have made a point to use local available subcontracting construction companies in various locations. Of course, we would have the overall management responsibility for their proper performance, but we have made a point to use local available firms or the more talented construction engineering firms in joint ventures with Zurn where we are the lead responsible company. We use them on a subcontract basis and this has worked extremely well.

### Technology markets

**How many business agreements are being negotiated between foreign technology firms and Zurn for application of foreign technology to U.S. markets?**

It's a two-way street. It's an international business challenge of tremendous magnitude. Our philosophy has been one of licensing our know-how and patents to overseas firms; Zurn has some 50–60 licensees to foreign companies. Then, through a cross-licensing arrangement Zurn brings their technology to U.S. markets. Balcke is a good example here. We brought over the thermal pollution control technology of Balcke (an engineering-manufacturing firm based in West Germany), and we have been negotiating with them to take on certain of Zurn's water pollution control technology. In the area of the Dusseldorf incinerator system, Zurn introduced this to the U.S. market. The patent for the system, as you know, is built around the cylindrical rotating inclined-plane grate for greater agitation of the solids and for cleaner and more efficient combustion of solid waste. In turn, we are negotiating with Deutsche B&W, the parent firm, to market our water pollution control products in the international market. In Japan, Mitsubishi has our water pollution control technology. Hawker Siddley, in England, also has this water technology. In South America, we have done feasibility studies of watershed developments and waste treatment plants for Sao Paulo and Rio de Janeiro in Brazil as well as many comparable projects in other parts of the world.

Zurn also is doing construction work in many major free-world nations. Zurn developed certain techniques in slit forming concrete foundations that speed up the construction of waste treatment plants and similar waste control structures such as waterways, sluice ways, and the like.

**Is there any reluctance by foreign countries to accept U.S. technology and vice-versa?**

No, we have found that foreign countries eagerly look to the U.S. because they realize that the U.S. has the biggest industrial pollution problem. Japan, too, has a major pollution problem as well as certain European nations, but nothing of the magnitude that the U.S. has generated. Therefore, foreign countries eagerly look to us for the most economical solution to their problems. They eagerly await U.S. technological leadership in this area.

We know of some areas in the world where international state planners want nothing but U.S. technology in the particular area of pollution control. These are the new emerging nations such as Saudi Arabia. The whole group of Arab nations have discovered substantial oil reserves and are now building new communities literally from the 15th to the 20th century overnight. They want the latest, best, and most sophisticated approach to the problem, and they are coming to the U.S. for that pollution control technology.

### Recent acquisitions

**What can you tell ES&T about acquisitions?**

Since 1966, Zurn has acquired 18 companies. We had a problem with three of the smaller acquisitions which resulted in a one-year moratorium on acquisitions (April 1, 1970– March 31, 1971) until we got our house in order. Zurn has an internal audit committee that travels to each of our operations on a regularly scheduled basis and makes a very thorough investigation of how things are coming along from a financial control standpoint. Because of the rapid rate that we were acquiring companies—one almost every two or three months—the audit committee did not get to a couple of the companies in sufficient time to initiate and implement the proper corporate controls. During the moratorium we continued to investigate potential acquisitions but we had a policy of no consummation. We felt that any company that wanted to come with us during the moratorium was worth waiting for.

Post moratorium, Zurn acquired the Robert Irsay Co. in June 1971. Their speciality is the design, engineering, and construction of air pollution control systems. Previous year sales were $13 million but sales for the current year will be close to $18 million. Robert Irsay does not produce any of the auxiliary equipment for these systems, such as induced draft fans, forced draft fans, and air pollution collecting equipment. These units are furnished through other Zurn subsidiaries and this again gives Zurn a total turn-key capability and responsibility in the design of air pollution control systems.

In the fall of 1971, we acquired Vulcan Manufacturing Co. (Cincinnati, Ohio). Their main expertise is in working with exotic metals, which is very important when air pollution controls are fabricated for highly corrosive liquids or gases such as in the continuous process industries—including petroleum refining and petrochemical manufacture. Vulcan will do about $8 million in volume this year.

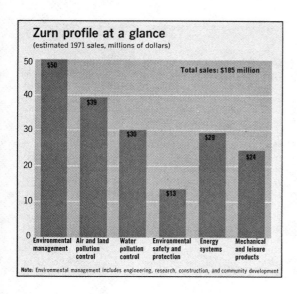

## Zurn profile at a glance

(estimated 1971 sales, millions of dollars)

Total sales: $185 million

| Category | Sales |
|---|---|
| Environmental management | $50 |
| Air and land pollution control | $39 |
| Water pollution control | $30 |
| Environmental safety and protection | $13 |
| Energy systems | $29 |
| Mechanical and leisure products | $24 |

Note: Environmental management includes engineering, research, construction, and community development

Wilkins Regulator Co. (suburban Los Angeles) was another acquisition. Last year, sales were $1.5 million; this year they will probably hit $2 million. Wilkins has developed and patented a pressure-regulating valve used in interior fire protection systems. In high-rise buildings, regulating the pressure flow of a water supply system in a building of this type is very difficult and costly, but the Wilkins pressure regulator does this in a very economical way. For example, in a 40-story high-rise building, these regulators might cut the cost of such fire protection equipment by as much as 50%, and it is all done automatically.

In December 1971, Zurn acquired Crowell Designs, Inc., a company in the business of designing and producing bilge pumps for all types of pleasure boats and recreational vehicles. In the next several years, Zurn may broaden into two additional pollution control areas—instrumentation and chemicals. We will probably get into the instrumentation field by acquiring a company in the $10–20 million volume range. With chemicals, Zurn will not manufacture the chemical itself. We will buy these in bulk and do the mixing and compounding and then merchandise them under the Zurn name. We see a big market coming in this expendable products area as we as a nation move more and more to the physical-chemical type treatment of waste waters. A company in each of these two areas will probably be acquired within the next six to 12 months.

### Jack-of-all-trades

**By virtue of its activities in all areas, is Zurn the jack-of-all-trades yet the master of none in the environmental area?**

I never thought that Zurn wanted to dispel the notion that it did not have specialization in certain areas. Zurn is probably the most renowned solids interception company in the world with all types of interception devices and controls—screening, straining, and the like. The same holds true of our leadership in various other speciality fields within the pollution control market. I do not think any other company can touch us in these areas of specialization. A company with a different speciality might win out; they might manufacture a pump, for example, or they might make valves or pipes. But Zurn as a total environmental company already manufactures many of these so-called specialties and could produce many others.

**Can you cite certain recent contracts in the total environmental area?**

We, of course, have a large number of on-going and new contracts too numerous to mention here. But of the more interesting I would name the following: on the industrial side, U.S. Steel (Gary, Ind.) has contracted Zurn for elimination of air pollution and recovery of waste heat, a project of about $4 million. On the municipal side, I would mention various feasibility and engineering studies, the most recent being the James River study for the Commonwealth of Virginia. A main objective here will be the involvement of a study leading to the preparation of a total water management plan for the James River, which will not only consider appropriate waste treatment criteria, but also recommend parameters for proper water utilization from the river.

In solid waste, Zurn constructed the new incinerator system for Disney World in Florida for $1.1 million. In air pollution control, we are about to "start-up" a breakthrough approach for controlling sulfur dioxide gas at Key West, Fla., utilizing a unique filtering medium for capturing the sulfur dioxide. But municipalwise, the largest proposal outstanding to date is for incinerators for New York City, totaling about $45 million.

# Turn-key turns on construction firms

*Construction of municipal waste treatment plants is a rapidly growing business; it merits a test of the turn-key option to get on with cleanup of the Nation's waters*

Reprinted from ENVIRON. SCI. TECHNOL., 6, 20 (January 1972)

Now that banks have changed their business operations, for the first time in 34 years, to a floating prime rate from the prime lending rate, how much longer can it be before the U.S. government allows turn-key construction of federally funded municipal waste treatment plants? Under the traditional method, consulting engineers prepare a plant design for a municipality. Then, contractors bid competitively for construction based on that design.

The furor touched off by the turn-key proposal that was announced by the EPA administrator last September 15 stirred the hottest environmental conflict in years. The issue, simply, is whether or not to permit turn-key operations as an alternative option that would be open to municipalities. The EPA proposal would open the option.

Whether turn-key operations are in the public interest only tests and time can tell. Judging from the sharp comments made on the EPA proposals, it may be months before the regulations are finalized, if indeed they are finalized at all. At press time, the deadline for comments (November 29) had passed, but it was too soon to know the fate of the proposal.

According to Webster's, a turn-key job is a job or contract in which the contractor agrees to complete the work of building and installation to the point of readiness or occupancy. (For a commonly accepted definition, see box.) In any event, turn-key involves three elements—engineering, procurement, and construction.

The turn-key approach has been around for years; it's been used in the industrial sector with much success. Although turn-key is a radical departure from the traditional way of performing municipal waste treatment plant construction in this country, it's common practice in foreign countries.

It is used in just about every other country of the world except the U.S. on public works projects involving water and water treatment.

In the past, because of state regulatory bodies, there has been neither competition in municipal treatment construction nor advances in process design or technology. Nor has there been a single source of responsibility to ensure that a municipal plant employing traditional treatment methods will even operate as designed. This, at any rate, is what the turn-key proponents say.

The advantage most often cited as the prime consideration for turn-key is the fact that plants are built at a lower cost and in a shorter time. Savings accrue because one or more of the three elements of turn-key operations are provided concurrently. The turn-key contractor has complete control of the work because he normally employs mechanical and other workers directly, rather than through subcontractors. Other advantages, according to proponents, are that turn-key operations:

- are used in many foreign countries
- promote technological competition
- pinpoint one responsible management
- guarantee performance after construction.

Why EPA made the turn-key proposal in the first place is not certain. One school of thought maintains that pres-

## Turn-key proponents

The 29 regular members of the National Constructors Association (NCA) have provided extensive turn-key operations in the industrial sector and two thirds of them have capabilities in all or some phases of waste water treatment, according to a 1971 NCA survey.

The Austin Co., Cleveland, Ohio
The Badger Co., Inc., Cambridge, Mass.
Bechtel Corp., San Francisco, Calif.
Blaw-Knox Chemical Plants, Inc., Pittsburgh, Pa.
C. F. Braun & Co., Alhambra, Calif.
Burns and Roe Construction Corp., Paramus, N.J.
Catalytic, Inc., Philadelphia, Pa.
Chemical Construction Corp., New York, N.Y.
Crawford & Russell Inc., Stamford, Conn.
Dravo Corp., Engineering Construction Div., Pittsburgh, Pa.
Ebasco Services, Inc., New York, N.Y.
The H. K. Ferguson Co., Cleveland, Ohio
Fluor Corp., Los Angeles, Calif.
Ford, Bacon & Davis Construction Corp., Monroe, La.
Foster Wheeler Corp., Livingston, N.J.
Kaiser Engineers, Oakland, Calif.
The M. W. Kellogg Co., Houston, Tex.
Koppers Co., Inc., Pittsburgh, Pa.
The Litwin Corp., Wichita, Kan.
The Lummus Co., Bloomfield, N.J.
Arthur G. McKee & Co., Independence, Ohio
The Ralph M. Parsons Co., Los Angeles, Calif.
J. F. Pritchard & Co., Kansas City, Mo.
Procon Inc., Des Plaines, Ill.
The Rust Engineering Co., Pittsburgh, Pa.
Sanderson & Porter, Inc., New York, N.Y.
Stearns-Roger Corp., Denver, Colo.
Stone & Webster Engineering Corp., Boston, Mass.
United Engineers & Constructor Inc., Philadelphia, Pa.

EPA Administrator Ruckelshaus

*"What I am asking is not that you take a calculated risk on unproved methods, but that you take . . . advantage of systems that already have been shown to work."*

sure for the EPA announcement came from the top—the administration. The story goes that construction firms, contributors to the coffers of political candidates, see business opportunities in construction of waste treatment facilities and want their share of the action. What truth, if any, there is in this line of reasoning is left to the environmental speculator.

More importantly, turn-key proponents say they want to be given the opportunity of having one, two, three, or dozens of tests cases—the exact number is unimportant—to prove that the alternative choice of construction method can work. Turn-key, then, is a basis issue.

### Pro arguments

R. M. Santaniello, vice-president and general manager of Gulf Degremont, Inc., says, "The turn-key approach has been and is being used in just about every other country of the world, except the U.S., on public works projects involving water and waste treatment—with a high degree of success. A form of turn-key bidding is now being utilized on planned construction of a 30-million gpd waste water treatment facility for Garland, Tex. It is proving to be successful in reducing costs and developing new technology, all with true guaranteed performance of proposed facilities."

The pressure is on to make turn-key a test case for a construction grant from the federal government. Many municipal plants treat industrial wastes and need more technology than that offered by run-of-the-mill plants.

Harvey A. Stephenson, mayor of the city of Keokuk, Iowa (a city that plans to build a secondary sewage treatment plant), says, "Keokuk requires more than a conventional sewage plant design. Three manufacturers in this city of 14,500 people contribute waste components sufficient to produce a waste strength approximately 10 times that of a normal city of this size. In order to obtain the best design at the most satisfactory capital and operating costs, we wish to obtain the benefits of design competition. We could do this, as I see it, by inviting bids against performance specifications if we were certain that we would not sacrifice our eligibility for a federal construction grant."

Monsanto Enviro-Chem, the catalyst for the Keokuk case, started talking about 2–3 years ago with the responsible federal, state, and local officials on ways to introduce alternative business views in the waste facilities construction field.

C. C. Kemp of Monsanto Envirochem Systems, Inc. says, "We have long advocated use of the turn-key construction method as a means of obtaining competition at the technical level in municipal treatment plant construction. We believe that the proposed change will be especially important in ensuring effective design for the increasing number of joint industrial and municipal treatment plants."

C. C. Pascal, president of Zurn Engineers (Upland, Calif.), notes that "the turn-key concept adds one of the more important functions of engineering design—economics. The incentive to design economically is practically nonexistent with the separation of builder and engineer. Indeed, the incentive is to design uneconomically—i.e., the greater the construction cost, the greater the engineering fee, generally."

R. E. Siegfried, president of The Badger Co., notes, "We strongly feel that the same progressive chemical engineering expertise which we have provided for years to the petroleum and chemical process industries can also be applied to the field of municipal waste treatment. . . .We expect to function in the waste treatment field in much the same way as we do in the process field, namely, to develop and offer a more economical competitive technology than our competition."

Another turn-key supporter is John B.

Dwyer, vice-president for research and engineering of The M. W. Kellogg Co. He notes that, "This (turn-key) undoubtedly reduces the burden of costs which will have to be borne by the public, by permitting utilization of best available technology. Furthermore, it will encourage private companies to invest in research and development of improved processes, since business will be available to them provided they guarantee performance. This move is certainly in the public interest."

John H. Robertson, manager of the environmental systems division of Catalytic, Inc., notes that "We (Catalytic) would strongly favor a proposal which permitted a design and construct approach to municipal projects." Robertson goes further and offers suggested language on turn-key: "It is recommended that a qualified environmental engineer certify, in such cases (turn-key), that alternatives have been considered, and the solution proposed is economically feasible as well as technically approved."

J. D. Spink of The Rust Engineering Co. notes that "The turn-key contracted method will be available to municipalities whose waste treatment facilities must not only treat the domestic waste in that municipality, but also, in many cases, various industrial wastes. This situation alone in the past has been detrimental to some form of traditional treatment processes which municipalities have had to use because of the state regulatory bodies."

Additional support for the turn-key proposal was received from Westinghouse Electric Corp. (Pittsburgh, Pa.); R. B. Humphreys Construction Co., Inc. (Indianapolis, Ind.); and Fluor Utah Engineers & Constructors, Inc. (Los Angeles, Calif.)

Organizations, too, are behind the turn-key proposal. The National Constructors Association (Washington, D.C.) is an association of engineering-construction firms (see box) that have provided turn-key operations in the industrial sector for years. The firms have regularly provided the engineering, procurement, and construction of plants to deliver performance in accordance with the design. Under the right set of circumstances, these firms do not see why they would not be interested in the construction of municipal waste treatment facilities.

## Con arguments

On the other side of the proposal is the consulting engineering profession.

So too are state water pollution control officials. From coast to coast and in every state, consulting engineers are not happy with the proposal, which is putting it quite mildly. In fact, Wesley Gilbertson, deputy secretary of the Pennsylvania Department of Environmental Resources, says, "We have talked with no one who has indicated they favor the proposal."

The fear of competition is never so evident than in considering their comments. "My experience in foreign turn-key projects indicates that we would be taking a step backward if such a proposal is adopted," says Philip Abrams of Consulting Engineers,

<hr>

### Turn-key: how it works

A contracting method under which all major activities—engineering-design, procurement of plant equipment, and construction—are included in a single contract, with the contractor having a professional relationship with the owner or client to produce optimum end results as to the quality of work, lowest possible cost, and earliest possible completion. Turn-key contract terms vary but may include performance guarantees and such price guarantees as lump sum or guaranteed maximums. For example, if a contract were negotiated for $10 million, and the actual cost exceeded that figure, then the contractor would absorb the overrun up to an agreed amount such as all or part of his fee.

<hr>

Inc. (Palm Springs, Calif.). "Quality of construction and contract negotiations are subject to irregularities, which are difficult to resolve."

R. H. Albanese, a consulting civil and sanitary engineer of Port Washington, N.Y., points out, "It is our opinion, as practicing professional sanitary engineers, that such approval would eventually lead to contractors dictating the designs of treatment works to professional engineers (retained by the contractor) and that the engineer would no longer be protecting the interests of the municipality, but instead his client, namely the contractor."

"Not only will the effectiveness of the projects suffer," he continues, "but also the well-established competitive bidding on a well-defined design will be lost. Such a loss will be detrimental to the sanitary engineering profession as it may lead to the substitution of sub-

standard equipment once a contractor (not a design) is selected."

One consulting engineer points out a few examples of turn-key projects which are in operation today, and that they are in trouble. "City of Omaha is currently undergoing problems with two waste treatment and paunch manure plants that are of the turn-key variety," says Raymond G. Alvine of the Nebraska-based R. G. Alvine & Associates. He comments, "I know from personal observation and experience that turn-key contractors will cut every corner possible on a project to increase their profit."

But one real concern regarding turn-key that is not easily dismissed is that public bodies will be held accountable for their choices. "The tragedy of the turn-key proposal is that public bodies will necessarily be required to make decisions which they are not competent to make without experience, re: costs, the adequacy and fairness of which they are incompetent to judge, and under a plan where the acceptance of the finished work is by the same party that performs the work," says R. Howson of Alvord, Burdick & Howson Engineers (Chicago, Ill.).

A similar theme is found in the comments of Thomas McMahon, director of water pollution for the Commonwealth of Massachusetts, who objects to the EPA proposal, saying that the turn-key approach might result in "marginal" construction and greatly increase responsibility of state and federal agencies.

### Looking ahead

Perhaps many of the apparently conflicting views on the merits of turn-key stem from the fact that the September 15 announcement was not specific enough. Presumably, the language will be more specific if and when it is finalized.

Given the opportunity of test cases in the waste water construction field, turn-key operations may pave the way for construction in the future. But it is important to note that instant answers are not possible. For example, even if a turn-key operation were approved for municipal construction tomorrow, only after several years— when the construction would be turned over—would the results be available for a final decision on the merits of the approach. But how much longer can we continue to try and solve pressing problems with yesterday's, if not yesteryear's, procedures? SSM

# Central waste disposal: New service looks for some action

*Disposing of other people's wastes*
*may become one of pollution control's*
*most profitable business ventures*

Reprinted from ENVIRON. SCI. TECHNOL., 4, 195 (March 1970)

Ask anyone with an industrial waste disposal problem what he would like to do about it if given the choice, and chances are that he would just as soon hand the problem on to someone else. The trouble with this wishful thinking is, of course, that there is generally no one to pass the buck to. Even though a few enterprising garbage collectors have ventured into the business of carting away the wastes that industry cannot or does not wish to treat, evidence is mounting that such small operators do not have the knowledge (or, unfortunately, in many cases, the wish) to dispose of the wastes in a pollution-free manner. The story is told of the large oil refinery which contracted with a small local entrepreneur for the latter to take away large quantities of a particularly noxious waste. According to the contractor, the waste was being dumped into a deep well approved by local health authorities. In fact, the contractor merely was pouring the liquid into a lagoon he had scooped out of nearby land. When the waste started seeping through the ground and into local waterways, both the contractor and the oil refinery were severely embarrassed.

## Central disposal makes sense

Two facts have emerged in recent years concerning the treatment of industrial wastes:

• It is the responsibility of every industry to ensure that its wastes are properly treated.

• The processes needed for the destruction of most industrial wastes are relatively sophisticated and not easily undertaken by those unfamiliar with a whole range of waste treatment technology.

For many small firms which have waste streams unacceptable to the local municipal treatment plant and which now are restrained from polluting local bodies of water, the prospects of having to install in-plant treatment equipment are not pleasant. Small firms often are ignorant of the technology required (if, indeed it is available) or, perhaps just as important, cannot justify the often large capital investment. Larger firms, too, may balk at having to tie up a large amount of capital in treatment plants when it does not contribute one cent to earnings. This is particularly true now, when the price of borrowing money is so high.

This background may help to explain why the concept of contract

**Optimism.** *Rollins-Purle officials say centralized waste disposal will pay*

waste disposal—paying someone else to treat one's wastes—seems to be finding increasing favor with industrial executives.

For instance, Dow Chemical Co. president Herbert D. Doan says that "it is entirely likely that pollution needs will establish a major new service business" (see this issue, page 179). One of the reasons Doan can speak authoritatively on the matter is that his own company has shown that centralized waste disposal makes sense both technically and economically. Dow does not treat anyone else's wastes, of course, but it has so much of its own at its huge Midland (Mich.) manufacturing facility, that it has built what many industry observers call a model integrated waste treatment plant.

The plant covers 50 acres and serves more than 500 processing units in the adjacent manufacturing area. Incoming wastes are segregated by chemical nature: Strong phenolic liquid wastes are blended with cooling water and the phenol removed in trickling filters; general organic wastes are neutralized with lime, sent to settling basins and then subjected to activated sludge treatment; burnable solid wastes and certain tarry liquids are incinerated in a fashion designed to eliminate air pollution. Solids from the treatment plant either are burned, sent to a sanitary landfill, or disposed of in underground caverns (where certain brine wastes also are pumped).

## Separate treatment steps

An essential feature of Dow's plant, and one which appears to be neces-

## Which treatment process R-P uses depends on nature of waste

- **Acids.** Neutralized with lime or other alkali. Calcium sulfate precipitate can be landfilled, soluble neutral salts are disposed of in ocean. Ocean disposal also used for fluffy iron precipitate from spent pickle liquors, where precipitate is difficult to dewater.

- **Alkaline wastes.** Often are alkaline detergents contaminated with organics from equipment washing. Alkali is first neutralized and neutral solution is treated biologically.

- **Water soluble organics.** Biologically treated. Contaminants such as trace metals must be oxidized or precipitated out of solution.

- **Insoluble organics.** Incinerated. When present in aqueous emulsion, emulsion is first broken physically or chemically, organic phase is skimmed off and burned. Aqueous phase receives appropriate treatment (generally biological). Heavy emulsions can be burned directly.

- **Metals in solution.** Chemical treatment used to precipitate salts or oxides.

Sludge from biological treatment processes is surface landfilled. Ash from incinerator is washed to dissolve water soluble materials, then landfilled.

---

sary for the optimum operation of any central waste treatment facility, is the segregation of incoming waste streams. If the wastes from a variety of chemical manufacturing steps were allowed to mix in one central sewer line, then the total volume of liquid would have to undergo all the treatment steps needed to remove every contaminant. Quite apart from the obviously uneconomical aspects of such a procedure, there is also a real risk of explosions in the transfer line as chemical reactions proceeded unchecked.

Despite the disadvantages inherent in having to treat a mixed liquid waste containing many different, and probably nonbiodegradable, components, several municipalities have shown interest in treating industrial wastes as well as domestic sewage (ES&T, October 1969, page 887). Although they generally do not have the technological capability to treat industrial wastes, many municipalities feel that the sheer volume of these wastes will enable them to build very much bigger treatment plants than they otherwise would need to treat only domestic sewage. They hope to achieve the economies that a large plant could make possible.

## Profit in waste disposal?

A growing number of companies in the last year have come to the conclusion that there just may be profits in waste. E.S. "Bud" Shannon, waste control manager for Dow at Midland, thinks that a well designed and integrated plant could charge customers prices they could afford to pay to get rid of their liquid and solid wastes and make money in the bargain. Plants of this type are being built by Rollins-Purle, Inc. (Lansdowne, Pa.).

Shannon's views are not shared by everyone who has looked into the possibilities, however. Spokesmen for Hytek, Inc. (Cleveland, Ohio), say that, at one time, they were interested in setting up a pilot plant to treat a variety of wastes from chemical and other plants in the Cleveland area. But the company ran right into the plans of the City of Cleveland, whose municipal treatment plant presently accepts and treats (quite inadequately, according to many) wastes from any source whatsoever. "How can you compete with the city?" asks Hytek. Cleveland authorities apparently are determined to have one of the largest and most efficient municipal plants in the world. Nevertheless, R. C. Sargent, Hytek's vice president for engineering, says that his company has Ohio Board of Health approval and still intends to go ahead with a central disposal project "when the time is ripe."

**Complex.** *Dow plant spans 50 acres*

The ripe time is not now, according to COPE, Inc. (Houston, Tex.), another company hoping to make an entry into the disposal business. COPE (Consolidated Oxidation Process Enterprises) has done considerable groundwork to discover just how profitable contract waste disposal would be for them. For instance, the company has calculated that customers would have to pay anywhere from $8-14 per ton of liquid or solid waste treated. COPE has looked extensively at prospects in the heavily industrialized Houston area, and also at industrial locations "in the northeast." The company has concluded, however, that the climate is not favorable at present, and gives two main reasons:

- Pollution laws on the books are not being enforced to the utmost, so that it may be cheaper to pollute than to find cleanup methods.

- Prospective customers are not willing to sign up with COPE for long-term (five years or more) contracts, because many firms believe they can engineer around pollution problems and get rid of their wastes at the source within a few years.

### Active plants

The number of operational central waste treatment plants in North America is, by all indications, very small. There are probably many plants that are able to treat a limited number of types of waste from multiple sources—such as the Friendswood Development Co. facility (Bayport, Tex.) described at last month's ACS meeting in Houston. The Friendswood plant is selective in accepting wastes from the chemical plants it serves, however. Its activated sludge unit can treat only biodegradable wastes and sets acceptability criteria in terms of biological oxygen demand, pH, etc., even for those.

One of the few waste treatment complexes in the continent is that of Goodfellow Enterprises (Corunna, Ont.) which has been in service since 1957 and modernized in 1968 at the urging of the Ontario government. The Goodfellow plant accepts liquid and solid wastes from the numerous chemical plants in the Sarnia (Ont.) area, and disposes of them by burning, burial, or deep-well injection. The plant is highly regarded in Canada, and Goodfellow reportedly is planning to build another similar facility in the Toronto area, although company spokesmen are loath to discuss

**Sludge gulper.** *Special tank truck vacuums sludge from industrial waste lagoon*

plans while negotiations with Toronto industries and government officials are going on.

U.S. critics, nevertheless, point out that the Goodfellow plant in Corunna is not a particularly good model for the pollution control service industry to follow. They cite such things as the plant's lack of any biological treatment processes and the relative unsophistication of its incinerators (which have few or no controls to prevent air pollution).

**Optimistic entrant**

Among all the hesitant entrants in the field of central waste treatment, it is somewhat surprising to find a company that is decidedly optimistic about prospects for making money in the business. John W. Rollins, Sr., chairman of Rollins International, Inc. (Wilmington, Del.), stated in March 1969 that he expected his company's subsidiary, Rollins-Purle, Inc. (R-P), to be doing $50 million worth of business providing pollution abatement services within five years. R-P, said Rollins, would build 20-25 waste treatment plants across the U.S. "in the next five years." Since Rollins first made these projections, he has upped his estimate of construction to 100 plants within the same period.

R-P presently is just completing its first central treatment plant (Gloucester County, N.J.) to treat wastes from an area 50 miles around Philadelphia, and expects to start operation of another plant (Baton Rouge, La.) in mid-summer. R-P's marketing director, James J. McLaughlin, indicates that the company has about 50 industrial firms lined up (under contract or negotiating) to have their wastes treated either at the New Jersey plant or at a smaller facility (Wilmington, Del.) where acids and alkalis are neutralized and the resultant salts shipped for ocean disposal. McLaughlin says that R-P has had little trouble in signing up firms for 2-3 years. Although many of the firms who have contracted with R-P are big names in the chemical industry, they are not willing, at the moment, for their identities to be revealed—apparently out of a fear that they would be admitting that they had pollution problems.

A trump card in R-P's hand is its access to the large fleet of tank trucks operated by Matlack Bulk Distribution Services, another Rollins International subsidiary. These trucks, supplemented by about 30 specially designed "Sludge Gulper" vacuum tankers, are the main means by which customers can send their wastes to an R-P plant. R-P also can arrange for wastes to be piped into a plant if the client is located nearby. McLaughlin further indicates that R-P is willing to operate existing pollution control facilities or to build and operate a plant on the customer's site.

For the moment, however, the main thrust of R-P's business is in relieving a company of its wastes at the plant site. This is to the client's advantage, since contracts are so written as to remove his responsibility for the waste once it is off his property.

R-P will accept all types of liquid and solid wastes except those that are radioactive or explosive. All the company's centralized plants will treat wastes by one or more chemical, biological, or physical processes; which one is used depends on the chemical nature of the waste (see box). Ability to segregate wastes is vital, says R-P's technical manager, Peter Williamson; only by separating incoming streams is it possible to ensure that plant operations are safe, and to provide complete waste deactivation at a price reasonable to the client.

An important feature of all R-P plants will be an insistence that they do not contribute to any form of pollution. Thus, incinerators will be equipped with scrubbers, and the large amounts of insoluble wastes produced in a plant will be disposed of in landfills with leachate collection systems.

Naturally, control of the wastes coming into a central plant is vital: R-P emphasizes that it must know the composition of a waste before it can quote prices on cost of disposal. So emphatic is R-P on this point that it is reluctant to accept wastes from independent contractors who may misrepresent or may not know the chemical nature of the wastes that they pick up from their industrial customers.

**Costs**

The cost to a customer for having wastes treated at a central plant obviously depends on many factors in addition to the nature of the wastes: The volume to be picked up, the distance of the customer from the plant, etc. But Williamson does give some general guidelines:

• Wastes that require just one-step treatment might be charged as little as 3 cents per gallon

• Wastes requiring two treatment steps might cost from 4.5-6 cents per gallon

• Difficult wastes such as mixed chlorinated hydrocarbons—which have too low a caloric value to support combustion and from which the chlorine must be removed—could cost as much as 30 cents per gallon.

• Baled trash that can be directly landfilled might cost $2-4 per ton. These costs are very similar to estimates made by COPE in Houston.

It remains to be seen whether R-P plants will be able to generate the large dollar volume in business projected for them by company chairman Rollins. What is certain is that if R-P appears to be making a going proposition of its stake in the new pollution control service business, it will not be without competition for long. Apart from the firms mentioned here, there is good evidence that comparative heavyweights such as Zurn Industries (Erie, Pa.) and Petrolite Corp. (St. Louis, Mo.) are watching with interest on the sidelines.

The closed-loop cycle for waste water
reuse is focused on minimal or zero
discharge units

**George Rey**
**William J. Lacy**
**Allen Cywin**

*Environmental Protection Agency*
*Washington, D.C. 20242*

*Reprinted from* Environ. Sci. Technol., **5**, 760 (September 1971)

# Industrial water reuse: future pollution solution

The Water Quality Office of the Environmental Protection Agency (EPA) has in progress a research and development program for industrial water pollution control which touches on industrial waste water reuse as a tool for pollution control and abatement. Many others have also discussed the subject of water and waste water reuse, recycle, and even the zero discharge industrial plant. Man must turn from a "throwaway" economy to a "recycling" economy!

### Industrial water uses

To reduce waste water discharges from an industrial plant by implementing reuse techniques, the first step is determining the major water use requirements (water use volume balance) for the plant. The Department of Commerce "Census of Manufacturers" is a source of general water use information and distribution for industrial groups. In addition to water volume needs, water quality requirements must also be established. At the other end of the process spectrum, the quantity of effluent water to be treated and the level of treatment must be taken into account for the plant "water balance and cost picture." These starting criteria are necessary to plan for the future.

In general, water use within a self-supporting industrial complex is divided among three major functions: process use, cooling purposes, and steam production. The quality requirement for each function is different.

For the year 1964, intake water requirements by selected industries were distributed by function (see below). These are the makeup requirements for each function and do not reflect the extent of internal water reuse which occurs within any one function. Obviously, cooling water makeup needs are greatest, followed by process demands, and the requirements for steam generation. In current practice, steam condensates are nearly always recycled, spent cooling waters are becoming more frequently recycled, and process waste waters are seldom recycled or reused.

To reduce intake demands, and consequently waste water discharge volume, it is necessary to further recycle spent water from each function. Or, in addition to internal recycle, to reuse the spent water from one function as makeup for another function. Thus, recycle and reuse reduce and regulate water intake needs.

### Reuse–recycle

Before proceeding further, multiple use, reuse, and/or recycle need(s) to be defined. Multiple use of water—a method of reuse—implies its use more than once, but each time for a different purpose; for example, the countercurrent use of water for successively dirtier applications, but never for the same application, until it is no longer needed. This in contrast to a once-through use of water (used only once in any application).

Recycle (also a method of reuse) implies using water over and over again for the same identical application from which it came. By this method, the total water intake of a plant, where reuse is practiced extensively, can be substantially less than a similar plant using water on a once-through basis (see page 762).

In the plant illustrated, the three major water use functions are:
• Process use—contacts products and raw materials in the manufacturing process;
• Cooling—for process operations and power production;
• Steam production—for process use and power production.

Also assumed, in the flow plans of this hypothetical plant, is the water quality acceptability of process waste water for cooling functions and likewise cooling water blowdown as acceptable to make boiler feed. In these figures the average gross water usage is identical (i.e., 15 units of water). In the once-through system, the water intake requirement is the same as gross water use. For the flow scheme of multiple use and reuse–recycle, gross water use is the same 15 units as in the once-through system, but intake and discharge requirements are substantially less.

### Cooling Water Makeup Is Greatest Need in Industry

| Industry group | Water Intake Use by Function, % | | |
| --- | --- | --- | --- |
| | Process | Cooling | Steam |
| All (U.S. industries) | 26 | 67 | 7 |
| Chemicals and allied products | 15 | 80 | 5 |
| Primary metals industries | 22 | 74 | 4 |
| Petroleum and coal products | 6 | 87 | 7 |
| Paper and allied products | 64 | 30 | 6 |

Source: 1963 Census of Manufacturers.

# Industrial water use and pollution control projects

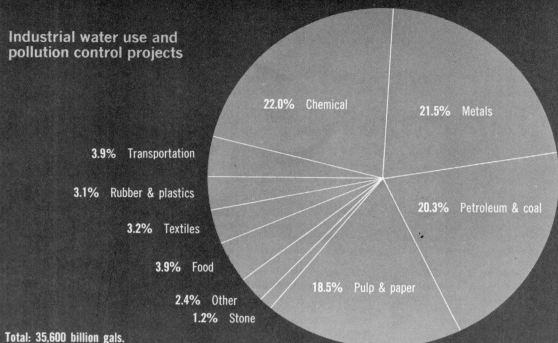

22.0% Chemical

21.5% Metals

3.9% Transportation

3.1% Rubber & plastics

3.2% Textiles

3.9% Food

2.4% Other

1.2% Stone

20.3% Petroleum & coal

18.5% Pulp & paper

**Total: 35,600 billion gals.**

| Industry[a] | Untreated discharge, %[b] | Industry[a] | Untreated discharge, %[b] | Industry[a] | Untreated discharge, %[b] |
|---|---|---|---|---|---|
| **Chemical** | 85 | **Pulp and Paper** | 56 | **Machinery and transportation** | 90 |
| Dow Chemical Co. | | Georgia-Kraft Co. | | | |
| Mineral Pigments Co. | | Green Bay Pkg. Co. | | **Textiles** | 76 |
| State of Florida | | International Paper Co. | | American Enka Corp. | |
| State of Louisiana | | Pulp Mfg. Research Inst. | | Fiber Ind. | |
| | | S. D. Warren Co. | | | |
| **Metals** | 69 | St. Regis Paper Co. | | **Rubber and plastics** | 79 |
| Beaton and Corbin Mfg. Co. | | Weyerhauser Co. | | **Other** | 60 |
| Fitzsimons Steel Co. | | **Food** | 76 | Leather goods | |
| Interlake Steel Corp. | | Archer Daniels Midland Co. | | Blueside Tanning Co. | |
| S. K. Williams | | Beefland Int. Inc. | | S. B. Foote | |
| Volco Brass and Copper | | Beet Sugar Dev. Foundation | | Water supply | |
| | | Central Soya | | Gainesville, Fla. | |
| **Petroleum and coal** | 25 | Crowley's Milk Co., Inc. | | | |
| | | Green Giant Co. | | **Stone, clay, glass** | 83 |
| | | Maryland Dept. of Health | | Johns-Manville Prod. Corp. | |
| | | National Canners Assoc. | | State of Vermont | |
| | | North Star R&D Inst. | | | |
| | | State of Vermont | | | |
| | | Winter Garden Citrus Prod. Coop. | | | |

a Industries listed are involved in projects on closed-loop systems, water reuse, or product recovery.
b Source: 1967 Census of Manufacturers.

# Water use methods

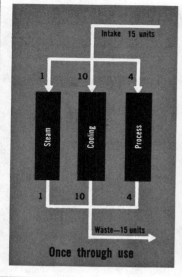

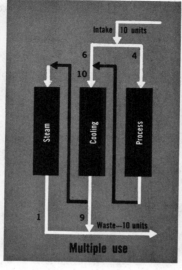

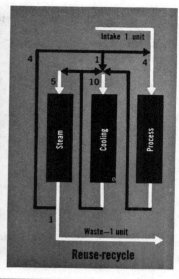

To reduce waste water discharge 93+% by reuse–recycle, the waters used must be continually purged of waste materials which accumulate (build up in concentration) with usage. It is this buildup which affects water quality and limits the extent of water reuse achievable. The concentration of salinity, hardness, alkalinity, organic matter, and suspended solids must be controlled in such a system by their selective removal at appropriate locations and in appropriate amounts. In this manner water quality may be controlled to meet reuse or recycle requirements.

A hypothetical plan for purging limiting waste materials from the reuse–recycle system is described above. It is assumed that cooling, process, and boiler feed makeups are produced by a multieffect evaporation method; that alkalinity is removed by chemical methods; and that organic and suspended solids are removed by conventional techniques. In addition to providing high quality makeup water, a multieffect evaporator controls salinity and hardness buildup in the system. To accomplish this, the evaporator blowdown would need to be taken to dryness, or near dryness, by some appropriate method which prevents uncontrolled release of the materials in the blowdown to the environment. In addition to salinity and hardness, this single blowdown effluent would be expected to contain many other accumulated nonvolatile (possibly toxic) substances from each water use function. Toxic substances could include heavy metals from process and corrosion products and toxic organic and inorganic inhibitors and catalysts.

If the ratio of gross water to water intake is an index of reuse, the once-through system would have a reuse ratio of 1.00. The multiple use system would be 1.50, and the reuse–recycle system would be 15.00. In a total reuse system the ratio would be infinite. In practice, this can never be achieved due to water consumption (evaporation and drift losses, leaks, etc.) and the final fluid blowdown requirements imposed by salinity buildup and sludge removals. Some water intake will always be necessary.

## Industry trends

Water reuse in industry is increasing. The trends of two major industry groups since 1954 are shown on page 126. The reuse ratio shown is the effective number of times water is reused by the industries. Although reuse has been increasing, a much higher reuse ratio is potentially possible, under appropriate circumstances. The past trends have been motivated by the pressures of limited water supply, poor water supply quality, and, more recently, environmental considerations. The latter should accelerate the trend in the near future.

As practiced today, cooling waters and condensed recoverable steam are largely reused waters. Reusing steam condensates for direct boiler feed and new steam production, practiced since the industrial revolution, was motivated by engineering factors related to boiler operations and construction. Waste waters are and can be used, directly or indirectly, as feed waters for boiler feed makeup. Similarly, process waste waters have also been utilized as cooling waters and for cooling water makeup. Accordingly, a plan for a waste water recycle system is not a pipe dream and is within the realm of engineering reality.

Water quality standards established by states should sharply increase industrial efforts to reduce water pollution discharges within the next few years. However, even if discharges per capita were held constant, increasing population and consumption would dictate increasingly stringent standards since the assimilation capacities of the receiving environment are rather constant.

Therefore, farsighted industrial management should consider a water use plan which results in the zero discharge, or as close to this goal as is possible. Engineers have pointed out that a closed cycle for water use in which no outfall returns effluent either to local supplies or groundwater can eliminate the cost of continuous mon-

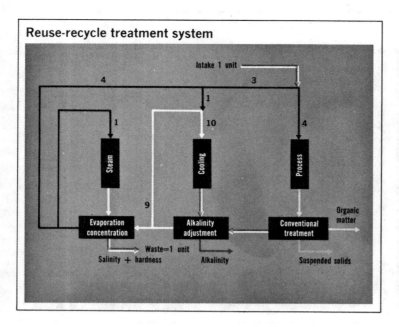

## Reuse-recycle treatment system

Intake 1 unit

4 — 3

1

Steam — Cooling — Process

1 — 10 — 4

9

Evaporation concentration — Alkalinity adjustment — Conventional treatment — Organic matter

Waste=1 unit

Salinity + hardness — Alkalinity — Suspended solids

itoring, which may, in addition to being costly, be required within a few years. Add to these advantages cost reductions from eliminating outfalls and lesser intake water supply facilities for new plants. Now, the dream starts to look like a reality.

In view of current and prospective environmental standards, each industrial plant should ask itself which way is best to proceed to comply with regulatory demands. In considering the alternatives, there are really only two: no discharge, or discharge complying with current and future requirements of the regulatory agency.

Where conditions may initially dictate treatment for discharges, the resulting effluent might be suitable for reuse within an industrial plant —either as cooling water and/or boiler feed makeup. So why throw it away? Especially when permits may be needed to do so.

Recent developments indicate waste waters of very high salinity, hardness, and containing organic matter can be successfully utilized directly as boiler feed without prior treatment. This is particularly significant since boiler feed water normally requires high quality water. In some cases boiler feed water is indirectly produced by an evaporator generating a distillate for boilers. Under this condition evaporators may be considered as waste water concentrators, which have a wide operating latitude of ac-

ceptable feed quality. In fact, evaporators operating on low quality water can still produce a high quality distillate suitable for boiler or cooling tower makeup.

### R&D—the key

The Water Quality Office program for industrial water pollution control views closed-cycle water use as an ultimate goal for industrial plants to control water pollution. Accordingly, water pollution control reuse–recycle projects are high priority areas for R&D support.

Basic technology for technically and economically feasible closed-cycles is generally accepted as available. More often than not it is practiced in piecemeal fashion in specific applications. Now the time has arrived to consider putting the pieces together to de-

velop totally engineered systems suitable for typical industrial plants, and to research water quality parameters which are controlling factors in water reuse and the affected economies. How can this be done?

A planning guide for an industrial water reuse R&D program (page 127) was prepared by using three major industrial water use functions as a basis for the closed-cycle system which must effectively minimize water intake, waste water discharge, and treatment (water supply as well as waste water) costs. Interconnecting functions (hydraulically) and more fully utilizing capabilities of waste water treatment facilities of each water use function (normally expected to be required) follow these guidelines. Also, expanding each operation's capabilities to serve a multiple purpose would further take into account the economies involved. But it is not that simple since operation and maintenance of cooling towers and boiler systems are very sensitive to water quality. Furthermore, experience with total reuse–recycle systems in industrial complexes is rare, which indicates R&D efforts are needed in this area. As a consequence, future R&D emphasis should establish more firmly engineering feasibility. This particular concept is predicated on maximizing waste water reuse–recycle, and the water reuse plan is essentially the same as shown previously (pages 125–6) in more general terms.

Even the more detailed planning guide must be considered as general with respect to specific needs of typical industries with different operating characteristics. Therefore, actual plans which will evolve in industry can only be expected to be "of the kind" shown in the general scheme. However, the time is now to plan for the future.

## Water Reuse–Recycle Trends

| Industry | Year | Gross water use, bgy | Water intake, bgy | Reuse ratio |
|---|---|---|---|---|
| Chemicals | 1954 | 4290 | 2690 | 1.59 |
| and | 1959 | 5230 | 3240 | 1.61 |
| allied | 1964 | 7670 | 3900 | 1.96 |
| products | 1968 | 9460 | 4510 | 2.10 |
| Petroleum | 1954 | 4150 | 1250 | 3.22 |
| and | 1959 | 5780 | 1320 | 4.38 |
| coal | 1964 | 6160 | 1400 | 4.40 |
| products | 1968 | 7220 | 1370 | 5.25 |

Source: 1954, '58, '63, '67 Census of Manufacturers.

How can industry start? Before a plant considers how to move toward the no-discharge system, it must first make a water use and materials (affecting water quality) balance. For new plant designs, industry should locate water use and distribution systems to the best advantage for a closed-cycle system. For old plants, obvious reuse possibilities should be implemented along with other capital plant improvement projects. For either type plant, existing supply water quality and receiving water standards should be evaluated to identify the water quality parameters which limit existing systems from achieving greater reuse and/or pollution control. Finally, plans should be developed to use the minimum number of treatment operations and to handle the least amount of pollutants inherent to the industrial operation and treatment processes applicable for use.

This system (below) also includes considerations for environmental factors other than water pollution control. These factors, also to be considered in the future, concern use of waste waters for air and thermal pollution control functions and waste residue usage for heat and power production and/or water treatment (ash). However, even an ideal plant system will have a net discharge of waste material. In the system discussed, this includes excess ash from thermal power production and a waste water blowdown of high salinity, hardness,

and toxicity. These wastes may be blended for some beneficial purpose, to ease proper handling, or for controlled assimilation by the environment in appropriate disposal sites. These areas have to be given consideration as man proceeds to meet future environmental demands.

**Economics**

Numerous unit operations and processes may be applied to waste water treatment. Several publications have listed the various methods and treatment costs by respective unit processes or operations employed. Such information can be very useful, but also misleading because it does not present less obvious factors which can influence costs considerably (for example, the concentration of pollutants, scale of operation, etc.). Rather than discuss economics with typical costs based on amount of water treated (i.e., cents per 1000 gallons) or on specific processes employed, it may be more beneficial, at least initially, to look at total cost potential in a simpler, more general way. Water treatment costs are primarily related to volume of water treated; amount of suspended solids, alkalinity, hardness, organic matter, and dissolved salts removed.

Barring any unusual waste water ingredient, constituents that are controlling factors in industrial water usage and waste water treatment fall into the above major categories. These

controlling factors and their approximate cost of removal from hypothetical waste water system are shown on page 128 (on the basis of amount removed). To obtain water pollution control costs for an industrial plant, the amount of these materials to be removed in various reuse schemes and water volume to be handled should be calculated.

Employing this approach should increase understanding of what each industry faces as potential water pollution cost, and alternatives of by-product recovery vs. disposal or treatment can be approximated more easily. Total water reuse schemes may also be evaluated on a more manageable basis. When making complete material and cost balances, the removal cost of the same components from intake water supplies as well as waste waters should be included. In poor water quality areas the total water recycle method may provide minimum cost for pollution control since these water supplies bring into a plant several of the same materials whose removal is required when the water is treated for use or discharge. The net effect is that more suspended solids, organic materials, and dissolved salts would be handled on a once-through basis than if only net increases due to internal (closed-cycle) plant functions were to be handled. This point should be kept in mind when industrial water pollution control plans are being made.

# Industrial waste water reuse scheme

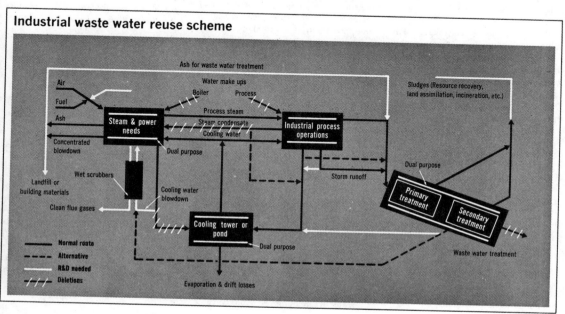

## Water Treatment Costs

| Undesirable component | Assumed method of removal | Range of cost, ¢/lb removed |
|---|---|---|
| Suspended solids[a] | Sedimentation | 2 |
| Organic matter[b] (as BOD) | Biological oxidation | 2 |
| Total dissolved salts[c] (including hardness) | Multieffect evaporation and solar evaporation | 3.5 |
| Alkalinity (as CaCO₃)[d] | Acid addition | 2 |

NOTES:
(a) Primary treatment at 2.5¢/1000 gal with 200 ppm removal and sludge disposal at 0.5 ¢/lb.
(b) Secondary treatment at 5¢/1000 gal with 400 ppm removal and sludge disposal at 1¢/lb.
(c) Feed at 3500 ppm, hardness 250 ppm. One MGD plant with total costs at $1.00/1000 gal and no credit for product water produced.
(d) Total cost at two times chemical cost. Sulfuric acid at 1¢/lb.

Costs of using evaporation techniques for pollution control can be misleading, at first glance, when considered in terms of cost per volume unit directly treated. However, the gross water quantities in a plant need not be entirely treated by this method, but rather only a small fraction of the total. If this cost is spread over overall water and waste water management costs, it then does not appear unreasonable—particularly if using the resulting water lowers other water operations' costs. Toxic, hard-to-treat, nonvolatile pollutants may still be present in process effluents and cooling tower blowdowns even after some conventional treatment; therefore, the effectiveness of an evaporation–concentration method to accumulate and concentrate these contaminants should be considered. The value of high quality water produced for reuse should be subtracted from the cost of its production. Using "new" water can maximize the water reuse ratio within an industrial complex.

### Future demands

Closed-loop industrial waste water and water systems are vitally necessary to maintain continuity in future industrial expansion. The huge water demands and high water usage growth rate of American industry cannot continue to rely solely on traditional water supply sources. Even in water-abundant areas, intake water supplies for industrial use are fast becoming restrictive. Trends toward water reuse have already been started and must be accelerated now—if an adequate base for future industrial expansion is to be provided.

Current and future environmental standards for waste water discharges are expected to increase the pressure on industry to reduce both the pollution discharge loads and the magnitude of effluent volumes in order to minimize environmental impact. Industrial water quality requirements for reuse are less demanding as a general rule, than for municipal supplies. Accordingly, industrial water reuse should be technically and economically achievable earlier than comparable municipal water reuse systems.

Waste water reuse is not only a resource conservation measure, but also a method of pollution control—a step in tune with future demands. Adequate R&D activity in this area is the key to accelerating implementation of extensive waste water reuse systems and, eventually, the total closed-loop cycle. The latter, with no effluent discharge, would comply with any quality standards now or in the future.

### Additional Reading

Brooke, Maxey, "Water," Chem. Eng., 77, Dec. 14, 1970.
Cywin, Allen, "Engineering Water Resources of the Future," ASME 70-WA PID-8, Nov. 29–Dec. 3, 1970, New York, N.Y.
Cywin, Allen, "The Urban Hydrological Cycle," EASCON 1970, American Conference of Professional Group on Aerospace and Electronics Systems (IEEE,) October 26–28, Washington, D.C.
Dykes, D. R., Bry, T. S., Kline, C. H., "Water Management a Fashionable Topic," Environ. Sci. Technol., 1 (10), 1967.
Lacy, William J., "The Industrial Water Pollution Control R&D Program," National Association of Corrosion Engineers, 26th National Conference, Mar. 2–6, 1970, Philadelphia, Pa.
Symons, George E.,"Industrial Wastewater Management—A Look Ahead," Water Wastes Eng., 6, January 1969.

**George Rey** is acting chief, Industrial Pollution Control Branch, EPA, Office of Research and Monitoring. He received his BS and MS in chemical engineering (University of Idaho). Prior to his EPA position, Mr. Rey worked with air and water pollution from hazardous chemical and nuclear materials and atomic waste handling, treatment, and disposal. Address inquiries to Mr. Rey.

**William J. Lacy,** a graduate of the University of Connecticut (BS in chemistry), is presently acting director of EPA's Division of Applied Science and Technology. Mr. Lacy has worked in the area of waste treatment and water decontamination since 1951. He joined the Federal Water Quality Administration in 1967 and assumed his present post in May.

**Allen Cywin** is a mechanical engineering graduate of Rensselaer Polytechnic Institute. As acting chief of Water Quality Research in EPA, Mr. Cywin plans, directs, and implements R&D programs in preventing, controlling, and abating pollution from storm and sewer overflows, industrial, agricultural, mining, and other sources of pollution.

# WWEMA shows industries its wares

At its first industrial conference and exposition,
the Waste Water Equipment Manufacturers Association
drew 128 exhibitors of water pollution control equipment

*Reprinted from* ENVIRON. SCI. TECHNOL., **7**, 404 (May 1973)

**Manufacturers and
their speciality**[a]

**Allied Chemical Corp.**
  chemicals for waste water treatment
**BIF Purifax Inc.**
  chemical oxidation systems for paper
  manufacturing
**Calgon Corp.**
  activated carbon, treatment and service
**Carus Chemical Co.**
  waste treatment
**Chemical Separations Corp.**
  ion exchange technology for waste
  water treatment
**The Dow Chemical Co.**
  chemicals for treatment
**Ecodyne Corp.**
  total water treatment
**General Signal**
  Mixing Equipment Co., Inc.
  DeZurik
**Houdaille Ind. Inc.**
  jet aeration system (gas liquid mass
  transfer process)
**Komline-Sanderson Engineering Corp.**
  air flotation in waste water treatment
**Met-Pro Systems Division**
  industrial waste water systems
**Nalco Chemical Co.**
  chemicals for waste water treatment
**Peabody Barnes & Peabody Wells**
  pumps and equipment including clarifiers, aerators, and filtration systems
**Rexnord, Inc.**
  air flotation for waste water treatment
**Swift & Co.**
  electrocoagulation for treatment
**Union Carbide Corp.**
  oxygen systems for waste water
  treatment

**Instrument companies**

  Calibrated Instruments, Inc.
  EMR Schlumberger
  Ethyl Corp.
  The Foxboro Co.
  Hach Chemical Co.
  Honeywell, Inc.
  Technicon Industrial Systems

List not exhaustive.

Industry has a cleanup date. By July 1, 1977, all industrial water dischargers must have the best practicable control technology, according to the new national water cleanup game plan, P.L. 92-500. What's more, industries must have permits that require this best practicable technology. These permits contain target dates of construction and compliance, which will be the conditions of the permit.

Where will industry obtain all the equipment and controls for its cleanup activity? From the more than 300 equipment companies who are members of the trade association, WWEMA. In order to get the word out to industry leaders, WWEMA held its first industrial pollution control conference and exposition (Chicago, Ill.) in March. It was attended by more than 3500 persons, and representatives from 27 industries.

Alphabetically, from Airvac, Allied Chemical Corp., through Westinghouse Electric Corp. and Zurn Industries, Inc., all told 128 exhibitors attended the first WWEMA industrial show.

Dale Bryson, spokesman from the EPA Region V office (which includes Chicago), indicated that 23,000 industrial applications for permits have already been filed formally with EPA on a nationwide basis and that more were coming. Every point source of water discharge needs a permit by the 1977 deadline date. Bryson indicated that EPA is concentrating on 2700 permits, without mentioning specific industries to which they apply, and that 2000 of them were in draft form ready for approval.

Industry, of course, is not the only polluter under the gun to clean up. Municipal waste water treatment plants must achieve secondary treatment by the same date. And by July 1, 1983, publicly owned treatment works are required to incorporate best practicable technology.

But industry leaders, aware of the fact that the money for municipal waste water construction is being funded at less than half the con-gressionally authorized funding, wonder if EPA regulators will present a hard or soft enforcement stance when enforcement time comes around. In the absence of full federal funding for municipal construction, industry leaders are beginning to question if they must be the scapegoat for cleanup.

Congressman William Harsha (R.-Ohio), ranking minority member of the House Public Works Committee, in his remarks at the general session of the conference said, "While new source standards set a rigid requirement for industries, there is a significant benefit accruing to a source that meets a new source standard." He continued, "Once a facility has complied by meeting a new source standard, that facility cannot be subjected to a more stringent standard for the subsequent 10 years or, at the option of the industry, the five-year depreciation period authorized for pollution control investments."

**Who they are**

Many of the papers at the conference dealt with process equipment that is being used today and will be used in the future. If vacuum transport of sewage may solve your problem, then Airvac Construction/National Homes Corp. may have your answer. There are a half dozen installations of this system in the U.S. today. One industrial application at the Scott Paper Co. (Mobile, Ala.) started operation last October. Airvac claims that the vacuum system is particularly advantageous in areas with high water table, rock conditions, existing underground utilities, and in the separation of sanitary and process waste in existing industrial mixed waste systems.

Autotrol Corp. offers the BIO-DISC process, which is analogous to the trickling filter, the exception being that a rotating disc provides an abundant population of aerobic microorganisms and aeration to make contact with the waste water. Over the past years, pilot plant studies have been conducted with the BIO-DISC process and food processing wastes. Lower power consumption, space requirements, and resistance to upset from fluctuations in hydraulic or organic leading are the attributes claimed for BIO-DISC.

A Crane Co. spokesman told of the

**129**

use of reverse osmosis, using the hollow fiber polyamide material of Du Pont, in cleaning up plating waste effluent. Crane also offers ozone generating units with capacities from 1–200 lb/day of ozone.

Diamond Shamrock Corp., a major chlorine producer, offers the SANI-LEC system for on-site generation of sodium hypochlorite. The only service connections are water and electricity. In cost comparison with gaseous systems and hypochlorite solutions, the SANILEC systems would pay for itself in 4.4 years of operation.

FMC Corp. has a basket-type sonic strainer that can be used to reduce the concentration of suspended matter in industrial waste water. The strainer can be used on waste waters with solid particles that are only a few microns in diameter and with suspended solids concentrations as low as 10 ppm and as high as 30,000 ppm. FMC also offers hydrogen per-•oxide, which has commercial significance in the oxidation of sulfides in industrial wastes.

Avco Corp. spokesmen told of their use of a direct contact freezing process as an alternative method to more conventional methods of concentrating wastes containing dissolved solids.

A Nichols Engineering and Research Corp. spokesman summarized filter press applications as a means for dewatering. Nichols said that high solids content in the filter cake and low solids in the filtrate are obtained. Their Vertical Tray Filter is an automatic shifting device which effectively eliminates previous labor requirements for his type of filter.

Neptune MicroFloc Inc. told of their experience with direct filtration of process waters. Raw waters with from 50–500 JTU turbidity are first treated with various coagulants and then filtered to obtain boiler water and general process water.

A Permutit representative noted that high rate softeners can eliminate the problem of sludge handling. Permutit claims that the system removes substantially all hardness in the form of a free draining granular mass.

Although air floating was used as far back as the 50's in the treatment of petroleum refineries waste, a Rexnord, Inc. spokesman explained how the technique is now being used to clean up waste waters from railroad and aircraft maintenance shops, from meat and poultry processing plants, as well as waste from tank truck washing operation. The Rexnord speaker said that reverse osmosis units can be used to remove both soluble organic and inorganic metals from an aqueous solution. Units are being used today to purify metal plating rinse waters as well as boiler feed waters.

Pennwalt Corp. people told of their experience with centrifuges for dewatering pulp and paper waste streams. Centrifuges can handle both primary, various ratios of primary and secondary, and secondary effluents, and lime-treated mill waste waters wherein color removal is being accomplished.

Kenics Corp. (Danvers, Mass.) has a subsurface aeration system which consists of helical elements inserted in a pipe and no moving parts. Its STATIC AERATOR has been used to treat the waste from Chemetron Corp.'s plant at Holland, Mich. And Union Carbide Corp. was there with its UNOX process for the secondary treatment of industrial waste.

Apparently plastic media add new life to trickling filters. Mass Transfer Ltd. (Eng.) uses Filterpak and Filter-stak plastic media to overcome the disadvantages (low BOD removal) of conventional trickling filters.

Westinghouse Electric Corp. exhibited the capabilities of capillary action in dewatering certain water and waste sludges. The capillary extracting force is exerted by a synthetic woven belt that also serves to convey the sludge from feed to discharge.

Zurn's Enviro-System Division, an equipment and control design/fabrication group, supplies special systems to industry on a subturnkey basis. Zurn interfaces with the consultant, client, and contractor by supplying only that amount of system engineering, construction, and erection services necessary to meet its responsibilities as a subsystem supplier. This division supplies five categories of subsystems: two-stage activated sludge units, Verti-Matic upflow sand filters, upflow-downflow carbon adsorption systems, physical-chemical sewage treatment systems and a sludge heat treatment system. The Verti-Matic upflow sand filtration system has been used by Marathon Oil Co's refinery at Robinson, Ill.

Operating since 1908, WWEMA is the trade association of approximately 300 equipment manufacturers and suppliers of waste water equipment in the U.S. A representative list of the types of equipment that member companies supply is found in last year's ES&T special report, "The Business of Water Pollution" (Nov. 1972, pages 974–9). For a list of individual companies that supply any one or more of the specific 21 equipment items, the reader is referred to the 1972–73 ES&T Pollution Control Directory, published last October and available from ACS Special Issue Sales.

SSM

*With increased emphasis on water conservation and
thermal pollution control, more and more industries are
turning to cooling towers for recycling hot water*

Reprinted from ENVIRON. SCI. TECHNOL., 5, 204 (March 1971)

# Cooling towers boost water reuse

Water, the colorless, transparent liquid in rivers, lakes, and oceans, is a major environmental concern. Water use and pollution increase as population and industry grow, until man is made aware of the necessity for conserving this precious natural resource. U.S. industrial manufacturers require 42 billion gallons of water per day (gpd); agriculture's demand is even greater, 140 billion gpd. Utilities require 95 billion, and household use of water totals 36 billion gpd. Although the water consumption figure will jump to 500 billion gpd by 1980, there will be "no water famine if it is intelligently used," says Bob Cunningham of Calgon Corp.

One major but relatively unpublicized means to conserve water—through recycling as well as reducing the incidence of chemical and thermal pollution—is through the use of cooling towers. Cooling towers—their uses, problems, and the growth potential of the industry—were discussed at the annual meeting of the Cooling Tower Institute in January held in Houston, Tex. Since effluent volumes and subsequent treatment costs can be reduced by use of cooling towers, this means of increasing the cycles of water use can "save industry millions of dollars per year," Jim Axsom of Sun Oil Co. emphasizes. A major Gulf Coast chemical processor further asserts that cooling towers are generally more economical than a dry air cooling system.

## Design

A cooling tower is a component of an open recirculating cooling system which is generally used for cooling water that has been heated by passage through process heat-exchange equipment. (A heat exchanger is a metal device consisting of a large cylindrical shell with tubes inside the shell. As a hot process fluid passes through the shell and cool water flows through the tubes, or vice versa, heat exchange takes place across the tube wall.)

Hot water, pumped to the top of the tower, trickles through the fill—redwood boards or polyvinyl chloride placed in a crisscross pattern in the tower—which in turn spreads the water uniformly and assists the cooling process. Air is pulled into the base or sides of the tower and exhausted through the fan stack. As air and hot water mix, the water is evaporating and condensing thus causing cooling. Finally, the droplets fall into the sump (concrete collection basin at the bottom of the tower), and the cooled water is ready for recycling through the heat-exchange equipment. Chemicals are added to the water to prevent scaling or corrosion of the cooling system components due to various problems aggravated by the chemicals in the water and the high temperatures.

The open recirculation system is compared with two other types of water systems—the once-through system and the closed-recirculating system. The once-through system merely borrows water, usually from a surface stream, warms it a few degrees, and discharges it further downstream. Of course, chemical treatment is required to prevent pollution, and these costs can become excessive due to the large volumes of water treated.

The closed-recirculating system is used, for example, in office building heating and cooling, refineries, and chemical plants. There is no intentional water loss; therefore, little makeup water is required, allowing more exotic chemical treatment than would be feasible in one of the other systems. Typical chemical treatments in this system are usually very high in chromates (up to 2000 ppm) and borax-nitrite mixtures (2000 ppm).

## Prototype

The forerunner of the modern cooling tower was a spray pond where water was sprayed into the air to be cooled. However, high drift rates (water droplets carried by the air) hindered this early method, so the first

towers were built. These original "atmospheric towers" were narrow and tall (56-ft high and 12 to 13-ft wide), depended on the wind velocity moving through the tower to cool the water trickling down, and were usually inefficient. Fans were later mounted at the bottom of these towers to blow the air up through the structure (forced draft). However, the air velocity coming out of the tower could be so low that wind sometimes forced the air back into the tower, cutting efficiency by as much as 50%.

Then fans were mounted on top of the tower to pull the air through (induced draft). This operation is the most widely used in the U.S. today. "The forced draft tower is only built periodically—just for special orders. No manufacturer has a forced draft as a standard model," says Jim Willa, vice president of Lilie-Hoffman Cooling Towers, Inc. However, under certain installation parameters, the forced draft tower is the most feasible model to install, according to a spokesman for the Marley Co.

There are two types of induced-draft cooling towers:

• In the counterflow tower, water enters as a spray at the top of the tower and trickles through the fill down to the sump. Air is sucked through inlet air louvers in both sides of the tower, travels up through the tower, and flows out of the top. The air and water mixtures travel counter to each other—the water down, the air up.

• The induced-draft crossflow tower still has water entering the top, but only at each side. The center of the tower, where the fan is located, is left open. As the water falls through the fill on each side of the tower, the air is pulled through the fill across the water. The air moves horizontally—through the side louvers, the fill, and then to the center of the tower. Here, it makes a 90° turn and flows out the top. The air moving horizontally across the water moving vertically creates a crossflow.

**131**

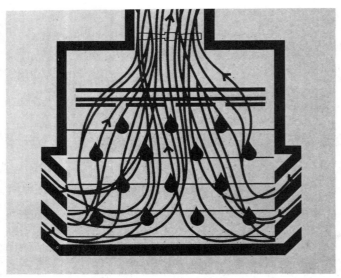

**Tower types.** *Counterflow cooling towers (above) have water moving down through the fill, while the air flows up. In the crossflow tower (right) the water again is moving down, but the air is flowing horizontally across the fill. Each of these units has a fan system in the top of the tower to induce or draw air through the fill. However, the natural draft or hyperbolic cooling tower (below) functions by air flow caused by the density differences between the moist air inside the shell and the denser air outside*

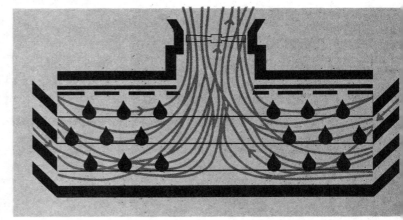

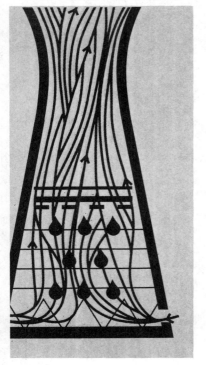

## Tower costs

The typical image but recent example of a cooling tower is the natural draft or hyperbolic tower which is, though tall and graceful, expensive and perhaps unreliable in this country. These towers have a large chimney rather than a fan to force or induce the air. Air densities inside and outside the chimney create a differential in pressure that causes a draft to be sucked through the cooling area.

The only justification for building this tower, Willa contends, is its low power requirement for a 20-year period or more. The hyperbolic tower costs 5 to 10 times more to build than the crossflow or counterflow unit. However, other major companies contend that the ratio rarely goes higher than 2:1 or 3:1. Presently, the major market for these towers is the eastern electric utility plants that previously used rivers and lakes to cool their hot water. Not one client in the U.S. who has ever owned a standard cooling tower has purchased a hyperbolic tower, says Willa. Besides the prohibiting difference in cost, the hyperbolic tower is not efficient functionally in the U.S. Designed, introduced, and built in parts of Europe where latitude is equivalent to that of Labrador, the hyperbolic tower (in the U.S.) may yield the least performance at the peak period of demand—during the hottest days of the summer months (because of the small pressure differential), according to Willa. (Willa's company, Lilie-Hoffman, does sell and build hyperbolic cooling towers, incidentally). On the other hand, Marley Co. customers are ordering additional hyperbolic towers as well as the mechanical draft models.

Cooling tower prices range from cheap models (untreated wood or plywood fill, galvanized steel hard-

**Concrete cigar.** *The hyperbolic (natural draft) cooling tower, measuring as high as 400 ft, functions most efficiently in locations with high humidity conditions*

ware), whose cost is estimated at $5 per gallon of water it will eventually recycle, to the more expensive ones (pressure-pretreated redwood fill, and yellow brass, silica bronze, 304 or 316 stainless steel hardware), which will cost $12 to $13 per gallon of recycled water. Performance requirements also affect the cost considerably. Short range, long approach (large temperature difference between the cold water leaving the tower and the wet bulb temperature of the air), easy-duty cooling towers are naturally less expensive than the long range, close approach, heavy-duty tower. For example, petroleum and petrochemical industries capitalize only on a five-year basis. If the tower cost cannot be justified in five years, the system stands a good chance of being obsolete.

## Water treatment

Cooling towers have four common operational problems: corrosion, scale, deposition fouling, and microbiological attack. In cooling tower systems, the problems are related to cooling tower blowdown (the intentional removal of a portion of the recirculated water in the cooling system to limit the buildup of dissolved solids beyond a certain concentration), high temperatures, and the air scrubbing action of the cooling tower itself. The untreated makeup water added after blowdown to maintain constant volume may also cause these problems.

Corrosion is an electrochemical phenomenon occurring in the system's piping, heat-exchange equipment, and other metallic components. Water soluble corrosion inhibitors, added to the water to reduce corrosion, form a monomolecular film at the metal–water interface. These compounds are:

• Inorganic polyphosphates.
• Inorganic polyphosphates plus zinc.
• Chromate-zinc and chromate-zinc-phosphate.
• Nonchromate inhibitors, which include amino-methylene-phosphonate (AMP) plus zinc, polyol-ester phosphate with or without zinc, and polyacrylamide-silica polymers.

The choice of treatment depends upon the existing water pollution control laws, since some cooling water will reach receiving streams during blowdown. A high-phosphate chemical also contributes to pollution problems. Usually, corrosion is controlled with a mixture of chromate and zinc, with or without additives.

Scale formation is a result of precipitation of limited solubility salts. Essentially, during continuous recycling, a salt reaches a concentration that exceeds it's solubility product, and it is deposited as scale (usually $CaCO_3$ or $CaSO_4$). Scale formation can be prevented by pH control. If the cycles of concentration cannot be economically reduced in a system, various phosphates are used to alter the crystalline structure of the precipitate and to prevent deposition, or to cause the deposit to form a soft sludge which can be easily washed away. However, using phosphates in cooling towers with high temperatures and long residence times may cause reversion. (The metaphosphate that controls scale can revert to

orthophosphate, forming an insoluble salt with calcium that will deposit in the system.) Sulfuric acid addition will lower the pH to prevent scale, but this will accelerate corrosion. Using a phosphate–zinc combination lowers the amount of potentially revertible phosphate; however, much emphasis today is placed on AMP, which is resistant to reversion and can therefore be used with little or no pH control.

Fouling occurs when silt, mud, and debris accumulate in the cooling tower system. Silt is treated with a synthetic polymer to "fluff up" the material, thus creating a larger surface area. The particle size expands allowing the force of the flowing water to scrub effectively the lower velocity areas. Dispersants have also been used to prevent deposition. "A major breakthrough for pollution control, performance, and ease of control," explains Paul Puckorius of W. E. Zimmie, Inc., "is obtained with a combination of organic polymers with inorganic polymers. These treatments with effective scale inhibitors, give a complete scale, fouling, and corrosion control system."

Microbiological attackers—fungi, bacteria, and slime—get into the cooling system through the makeup water or air. These organisms can foul lines as well as damage wood fill. The most effective and least expensive biocide available is chlorine. With proper pH control and correct dosages, chlorine will not cause wood deterioration and will prevent pollution in the blowdown effluent. Over 90% of industrial cooling tower users favor chlorine treatment. Many nonoxidizing microbiocides, chlorophenols, or organo metallics, can also be used for good microorganism control without significant contribution to wood deterioration.

## Market growth

The major cooling tower manufacturers (Marley, Lilie-Hoffman, and Fluor) forecast rapid growth for the industry. Air-conditioning uses contribute to this growth. Cooling towers are now widely used in the petroleum and petrochemical industries, especially with the present drive for pollution control. The largest market for cooling towers is electrical plants. Electrical power use doubles every 10 years, and the number of cooling towers required by the electrical industry will grow proportionally. Cooling towers and their contribution to environmental control will become more evident as time passes.          CEK

**133**

# P–C treatment gets industrial trial

## Activated carbon cleans up effluent to meet state standards at reasonable capital and operating costs

*Reprinted from* ENVIRON. SCI. TECHNOL., **7**, 200 (March 1973)

Physical/chemical treatment using large-scale activated carbon adsorption technology is getting a major industrial test at the Tuscaloosa, Ala., plant of Reichhold Chemicals, Inc. Just how well it does may determine, to a large extent, the future of physical/chemical plants in industrial waste treatment and what part activated carbon will play in industrial waste water cleanup.

Reichhold's Tuscaloosa plant is located on the Black Warrior River, part of the Warrior-Tombigbee River system which enters the Gulf of Mexico some 358 river miles south. The plant, along with several other manufacturing operations, including paper making, foundry, by-product coking, asphalt, and chemicals, is on the Warrior Pool, an eight-mile section between Oliver and Holt Lock and Dams.

The Tuscaloosa plant, which began operations in 1943 with synthetic phenol production, makes sufuric acid, formaldehyde, pentaerythritol, sodium sulfite, sodium sulfate, orthophenylphenol, and a number of synthetic resin and plastics in its production units. Its effluent streams, therefore, represent a diverse combination of types and poses a complex.

In 1966, in conjunction with the Alabama Water Improvement Commission, Reichhold determined that its effluent loading 16,444 lb/day BOD, 26,718 lb/day COD, 1540 lb/day phenols and had an average pH or 9.8 with a range of 5.4–12.3. Effluent volume from all plant sources ranged from 10–15 million gpd.

In accordance with an agreement between the company and the Alabama Water Improvement Commission, a target reduction of 90% was agreed upon. (Although legislation and regulations applicable under current law required a 75% reduction, the industries in the Tuscaloosa area were requested to achieve an 85% minimum reduction.) In addition, because of special conditions surrounding operation of a peak-load hydroelectric generating plant immediately

upriver, Reichhold, along with all other industries on the Warrior Pool, was required to install capacity for holding its entire discharge for five days.

### Getting started

Reichhold then started work to find out how the proposed reduction in discharge could be met, according to its regional vice-president T. P. Shumaker who is in charge of the Tuscaloosa plant. The program attacked the problem from three directions: reducing the effluent loading by inplant process and/or equipment improvements, reducing the hydraulic loading by segregation and separation of effluent-bearing streams from once-through streams, and seeking the most economical and efficient means for effluent treatment.

In-plant improvements included redesign and replacement of caustic concentrators where entrainment had added 4000 lb/day of caustic soda to the effluent stream. The loss has been reduced to fewer than 40 lb/day. Sodium sulfite, previously washed from returning rail cars in a

cleaning process, now has been returned to the process water rather than to the effluent.

The program to reduce hydraulic loading included a complete rebuilding of the entire plant sewer system where once-through cooling waters were separated from effluent streams. The volume of effluents to be treated was reduced from 15 million gpd to half a million gpd.

Looking for an answer to its treatment problems, Reichhold found there was no "off-the-shelf" technology then available which would fill the bill. The company looked at several alternatives, Shumaker notes.

Municipal treatment by the City of Tuscaloosa was impossible since the city did not have secondary treatment facilities, nor would secondary treatment be available in time for compliance deadlines to be met.

Reichhold studied the possibility of concentration and incineration of the wastes but found that incineration could not handle the entire problem. Certain process effluent streams had relatively high levels of alkali metals (such as sodium formate and sodium

**Flow diagram of Reichhold's physical-chemical waste**

acetate), and concern for refractory life led Reichhold to drop evaporation/concentration plans.

### Deep well injection

A preliminary feasibility study of deep well injection was done and looked pretty good, according to Shumaker. So a program was established with the Environmental Protection Agency, the Geological Survey of Alabama, and Alabama Water Improvement Commission to drill, monitor, and conduct an extended research program on applicability of deep well injection. A research test well, drilled to a depth of 8097 ft, has been completed at the Tuscaloosa location and continues to be studied.

Although deep well injection holds potential for future application and possibly total elimination of discharge, it was not considered a part of the primary process for meeting the current discharge standards, according to Shumaker.

Reichhold mounted a major effort in process design based on biological treatment with research done by both commercial and university-related firms. The bio-oxidation route looked promising, according to Shumaker, and after two years of laboratory work, the company built a nominal 25,000-gpd pilot plant at the Tuscaloosa site.

After nine months' operation, however, the results were less than spectacular. Six separate waste streams were not amenable to biological treatment and would have needed incineration. The company estimated the capital costs of the combined treatment scheme to be in excess of $2 million with direct operating costs of an additional $280,500 annually.

While Reichhold had been investigating biological treatment, the company, with the help of the Calgon Corp., had also been looking at carbon adsorption as a polishing process. "At first look, the operating costs of once-through carbon were prohibitive, although the results achieved were very interesting," Shumaker says.

But with the idea of carbon regeneration and the costs savings that could be realized, Reichhold took a second look—this time to evaluate carbon adsorption as the principal method of treatment.

Batch clarification and adsorption studies were run on grab and composite samples of plant waste water to establish the feasibility of physical/chemical treatment. Once the feasibility was established, dynamic carbon column tests were run to establish design parameters.

The batch clarification tests showed that a combination of neutralization to pH 6.5–8.5 followed by a 2-mg/l. dose of WT-2690, a nonionic water-soluble polymer manufactured by Calgon, would reduce the suspended solids to less than 20 mg/l.

Batch adsorption isotherm studies demonstrated that the organic content of the waste could be reduced to an acceptable level via adsorption.

These tests, together with the column tests, brought about an adsorption system which Reichhold thought would reduce BOD to 1650 lb/day, and COD to 2675 lb/day (about a 90% reduction). Phenolics would be cut to fewer than 27 lb/day and the pH of the final effluent would range from 6.5–8.5.

Total capital costs were pegged at $1.3 million for a plant capable of treating 500,000 gpd, including the cost of a pretreatment system, holding basin, and adsorption system. Direct operating costs—including labor, fuel, power, makeup carbon, and maintenance—were projected to be about $320,000 annually.

### Full steam on PCT

On the basis of those projections, Reichhold decided to build a full-scale physical/chemical plant. The process flow looks something like this:

The process waste waters flow to a 1,250,000-gal earthen equalization basin with a residence time of 2.5 days which serves the triple function of equalizing flow, pH, and organic content. Four turbine-type agitators provide complete mixing and prevent solids settling.

Waste water is then pumped to an acid-mixing chamber—concrete basin with a rapid mix turbine-type agitator—where concentrated sulfuric acid is added to maintain the pH in the 6.5–8.5 range. Although the pH of the raw waste water can be as low as 5, a caustic soda feed system is not required since the periodic acid slugs can be handled by equalization.

Waste water flows from the acid-mixing chamber to a flocculation basin. The nonionic polymer is added to the neutralized waste water as it leaves the acid mix chamber and the waste is gently agitated in the flocculator to enhance the formation of large floc particles.

The flocculated waste water then flows to a clarifier—a 40-ft square concrete basin with a circular clarifier mechanism—where essentially all suspended solids and floating material are removed.

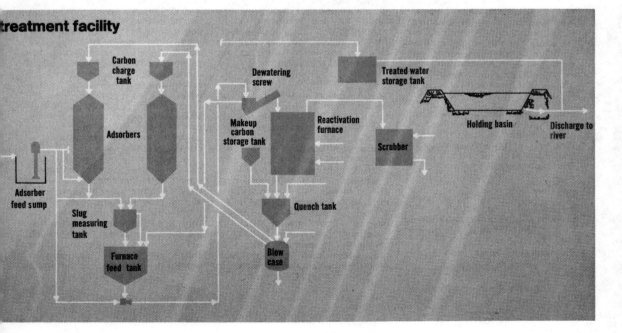

**treatment facility**

Carbon charge tank

Dewatering screw

Treated water storage tank

Adsorbers

Makeup carbon storage tank

Reactivation furnace

Holding basin

Discharge to river

Scrubber

Adsorber feed sump

Slug measuring tank

Quench tank

Furnace feed tank

Blow case

Settled solids are periodically withdrawn from the clarifier sump and pumped to a thickener. The supernatant from the thickener is recycled to the equalization basin. The thickened solids are presently hauled to a land disposal site, Shumaker says, although dewatering studies are now under way.

### Carbon adsorbers

After pretreatment, the clarified waste water flows to the adsorber feed sump. The waste water enters the bottom of two parallel moving-bed adsorbers through a circular pipe header. Each lined carbon steel adsorber is 12 ft in diameter with a 36-ft sidewall height and a cone-shaped bottom and top. Each adsorber contains 124,000 lb of Calgon's granular activated carbon, Filtrasorb 300. At the design flow of 175 gpm per adsorber, the superficial or empty bed contact time is 173 min.

The waste water flows upward through the carbon beds, and dissolved organics are physically adsorbed. When the carbon at the bottom of the bed becomes exhausted, the spent carbon is removed from the bottom of the bed in batches or slugs. The volume and frequency of slugging is dependent on the rate of carbon exhaustion, according to Calgon, and at the Reichhold plant, each moving bed adsorber is slugged once per shift. (One slug volume is equivalent to approximately 5200 lb of carbon or about 5% of the bed.) During the slugging operation, a valve at the bottom of the adsorber is opened allowing a slurry of spent carbon to flow to the slug-measuring tank. Simultaneously, a valve at the top of the adsorber is opened allowing reactivated carbon to flow from the surge tank to the top of the ad-sorber. When the slug-measuring tank is full, both valves are closed, completing the slugging operation. The contents of the slug-measuring tank are transferred to the furnace feed tank for reactivation.

### Carbon regeneration

Spent carbon is transferred in slurry forms from the furnace feed tank to an inclined screw conveyor. To minimize utility water requirements, clarified waste water is used to transfer spent carbon.

In the conveyor, the spent carbon is dewatered by gravity and the water overflows to the furnace feed tank. The dewatered carbon is discharged into the reactivation furnace, a 13' 6'' o.d. x 5 hearth furnace with an integral or "zero" hearth afterburner capable of processing 42,000 lb of carbon/day. A central rotating shaft with rabble arms at each hearth moves the carbon across the hearths and through drop holes causing it to pass through the furnace.

The furnace is normally fired with natural gas with LPG used on a standby basis. Supplemental steam is added to the furnace to control the reactivation atmosphere. In the furnace, the damp carbon is dried and heated to 1700–1800°F. The spent carbon is reactivated by controlled oxidation of the impurities in the pores of the carbon granules.

The off-gases are vented to an integral afterburner to ensure complete combustion of the organics and to prevent odors. The afterburner exhaust gases exit to a wet scrubbing system where the hot gases are cooled and any particles are removed. Treated effluent is used for scrubbing.

The reactivated carbon is discharged from the furnace by gravity into a quench tank. There the hot reactivated carbon is cooled and wetted. The quenched carbon drops to a blowcase and is transferred pneumatically to the two adsorber charge tanks. A time sequence programmer automatically controls the transfer of reactivated carbon.

During the transport and reactivation of spent carbon, carbon losses occur which make periodic addition of make-up carbon necessary. An accurate make-up carbon requirement has not been established for this plant to date since it only recently began operation, according to Shumaker, but make-up will probably be 3–5% by weight. Make-up carbon is added from a storage tank.

After carbon treatment, process water is collected in a trough at the top of each adsorber and sent to a 5500-gal FRP surge tank from which a portion of treated water is pumped for various uses throughout the waste treatment plant. When the treated water is used for scrubbing, polymer dilution, virgin carbon unloading, and quenching, the utility water requirement is minimized. Approximately 70 gpm of treated water is reused on a continuous basis.

The overflow from the treated water tank flows by gravity to the Black Warrior River. During periods of low flow, treated water is diverted to a five-day holding basin as required by the Alabama Water Improvement Commission.

With its new plant, Reichhold now has the "largest and most advanced" waste water treatment facility for chemical waste water effluents in Alabama, Shumaker says. Time will tell if the plant lives up to its expectations, but Reichhold is confident that the physical/chemical treatment unit will allow the company to exceed the state's requirements for industrial waste water discharge.          HMM

# Putting the closed loop into practice

*Dow Chemical Co.'s latex plant in Dalton, Ga., was built with complete water reuse in mind; for the past five years it has completely recycled its process water*

*Reprinted from* ENVIRON. SCI. TECHNOL., **6**, 1072 (December 1972)

Waste water discharges are becoming more and more limited by federal law (see story on the new water bill, p 1068), and increasing emphasis is being placed on a closed-loop system of water recycling both to conserve water and to prevent pollution. Few manufacturing plants can yet boast of complete water reuse. However, a Dow Chemical Co. plant in Dalton, Ga., is successfully operating such a system.

The Dow plant manufactures styrene–butadiene latex for carpet backing and carpet construction and latex foam for carpet backing. The plant is ideally located, since at least 240 carpet manufacturers producing tufted carpeting (only 5% of the market accounts for woven carpet) are located in the Dalton, Ga., area. In fact, Dalton is known as "the carpet capital of the world."

Carpet manufacturing itself explains the need for latex. Carpetmaking machines with up to 12 needles per inch continuously punch loops (or tufts) of yarn into a jute or synthetic scrim (backing). Since there is no "lock" stitch, any tufted carpet could be unraveled by pulling one of the yarn loops. Latex applied to the back of the scrim locks the tuft. When water evaporates from latex—an industrial adhesive—a synthetic rubber film is formed that keeps the carpet from unraveling. Latex also affixes another scrim to the carpet as secondary backing.

## The Dow plant

In view of the importance of latex to the carpet industry, Dow officials decided to construct a latex plant in Dalton rather than ship large amounts of latex to its carpet customers. The plant site was chosen in 1965, explains Jim Kaye, plant manager at Dalton.

In the early planning stages, Dow officials considered an intake water station at the Conasauga River which flows along one side of Dow's 420 acres of property. However, the river water would need to be treated and cleaned to remove clay silt before use in the plant. Also, effluent discharged back into the river would have to be treated in this once-through system.

Water from the municipal system was available, but would have been too expensive to buy for once-through use. Also, the tremendous volume of water needed for latex production could seriously affect local supplies in times of drought.

Dow officials then decided, with the approval of the Georgia Water Quality Control Board, to use municipal water for makeup and sanitary water, and to provide a closed-loop system for reusing process, wash, and coolant waters. The goal was to eliminate discharge to the river and to benefit economically by recycling water. The plant was then built in 1966 and began operation in 1967.

Latex produced at the Dow plant is a water suspension of very minute synthetic rubber particles formed by a proprietary process for emulsion polymerization of styrene and butadiene. An initiator starts the reaction in a water phase, with soaps added to emulsify the oil–water interface. The latex particles produced are very tiny (1500–2500Å) but with a tremendous surface area per unit volume. The final latex product is approximately at a 50% solids level.

The large surface area of the latex particles accounts for the fact that only trace quantities of them will give water a cloudy appearance.

Besides visual pollution, effluent from the manufacture of latex contains bits of latex, trace quantities of unreacted monomers, and ammonia. The effluent has a high chemical oxygen demand (COD), but

## Dow's water recycle system

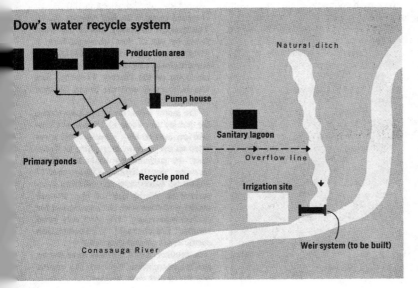

Production area

Pump house

Natural ditch

Sanitary lagoon

Overflow line

Primary ponds

Recycle pond

Irrigation site

Weir system (to be built)

Conasauga River

is "essentially nil in BOD," emphasizes plant manager Kaye. Latex particles are chemically stable and will remain in suspension. Coagulation and removal are essential for COD reduction as well as for appearance.

Large amounts of wash water are used in the Dow plant. Latex is easy to wash away while it's still wet, but after it dries, it forms a skin that is difficult to remove. Thus, equipment, tank cars, and tank trucks are repeatedly washed to prevent buildup of dried layers of latex.

### Closed loop

Here's what Dow does with its effluent: About a million gallons of water are used daily for processing, cooling, and washing. All drips, spills, wash water,

**Reuse.** *Another latex plant discharges wastes directly into a stream (below); however, Dow's latex effluent is treated in a trough (above) and pond system and then is ready for recycling (above left)*

and process water end up in the outflow stream from the plant. The plant's effluent stream flows through a series of concrete trenches leading to the primary ponds. Metered amounts of alum (aluminum sulfate) are added to the waste stream to destabilize latex and to coagulate it. The coagulating effluent is then neutralized by either sulfuric acid or caustic soda to maintain a narrow pH range, preferably around 7. Optimum coagulation occurs below pH 8. Also, neutrality protects the process and fire system piping when the water is later reused. Alum, and acid or caustic when needed, are added in the trench to promote mixing by natural turbulence.

The trench system leads to four primary ponds. Each one and one-half

million gallon pond is used alternately. Any three can be shut off while the remaining one is filling with effluent. Under optimum conditions, floccing occurs in the pond within 2–3 hr; however, each primary pond of effluent usually undergoes a settling period of one to one and one-half days. About a million gallons of water daily flow to the primary ponds for settling; each pond is thus filled once every four to six days.

After settling, the clarified water drains by gravity into the $10^1/_2$ million gallon "recycle" pond. On the edge of this large pond is a pump house where vertical turbine pumps recycle the clean water continuously back to the plant through buried pipelines. Also, two 2500 gpm fire pumps, one diesel and one electric, feed the sprinkler, deluge, and hydrant system. The recycle reservoir serves as a fire protection reserve since both major raw materials (styrene and butadiene) are very flammable, and the plant is located in a rural area. "We keep a minimum of four million gallons in the pond," says Kaye, "for fire protection purposes."

The Dow plant also has two other water systems. In one, a cooling tower recirculates two million gallons of cooling water each day. In the other, sanitary wastes from the plant are treated in a state-approved biological oxidation pond. In periods of flood or heavy rains, the sanitary pond does sometimes overflow; however, the capacity of this lagoon is such that the discharge is of acceptable quality.

### Overflow

In past years during the rainy winter months, some overflow from the recycle pond flowed through a drainage ditch which eventually led to the Conasauga River. Last year, Dow installed a system that it hopes will prevent any overflow from entering the river.

Land around the plant and pond system was graded so that rainwater would not run into the ponds. Rain runoff, in the winter months, has in the past contributed to filling the ponds to capacity and causing overflow. To boost capacity, the recycle pond was dredged from its original nine million gallon limit to its present $10^1/_2$ million gallons.

An irrigation system was set up to handle surpluses and permit pond-level management in anticipation of the wet winter season. Underground pipes carry water to an adjacent area for spray irrigation. The high nitrogen content of the recycled water (400–500 ppm) contributes to vegetative growth, and this system avoids direct discharge to the river.

A single, natural low-line leading to the river takes any outflow from the active plant site, the ponding systems, and rainfall runoff from the irrigation area. Dow also plans to install a weir dam where this drainage line meets the river. A V-notch weir will measure the flow of water. Water samples taken at the dam site will be regularly tested for quality. "The real test will come this winter during the rainy season," says Kaye, but he is confident that "we'll have no discharge all the way around."

The ponds are monitored at least once per shift (the plant operates 24 hr/day, 365 days/year) and up to five or six times per shift. The primary ponds are monitored for pH and clarity, and a log is kept to record whether each pond is filling, settling, draining, or empty.

Last year, two of the primary ponds were dredged to remove the accumulation of sludge which had reduced pond capacities to about 750,000 gallons. The latex sludge presents some problems in landfilling, largely connected with drying. CKL

# Industry looks at water reuse

## National Conference on Complete WateReuse outlines problems and prospects for industry

*Reprinted from* ENVIRON. SCI. TECHNOL., **7,** 500 (June 1973)

Look for reuse of waste water to become more commonplace in industry as discharge limits move inexorably toward the zero mark by 1985. That was the consensus of opinion at the recently convened National Conference on WateReuse, sponsored jointly by the Environmental Protection Agency and the American Institute of Chemical.Engineers.

Public Law 92-500, better known as the Federal Water Pollution Control Act Amendments of 1972, is putting the screws to polluters. The national goal of achieving zero discharge of water pollutants by 1985 is making industry take a fresh look at reuse. More often than not, the conclusion will be that it's cheaper to treat for reuse than to treat for discharge.

### Costs and benefits

The decision on treat or reuse, like so many others in pollution control, must be made with regard to both the costs and benefits of the proposed plan of action, according to Lawrence Cecil, a Tucson, Ariz., chemical engineering consultant who chaired the conference. While many industries "use" large quantities of water, the basic concept of water use is that water is "borrowed" instead of "consumed," pointed out E. D. Dyke, of E. D. Dyke and Sons, Ltd., Ashford, England.

Only a few industries incorporate sizable quantities of water into their products—industries such as food processing, ice manufacture, and the like—while most others simply use the properties of water to attain their purposes. "Complete cycle stream to stream is the ideal; a closed cycle cannot be closed absolutely," Dyke pointed out. He told conference attendees that consideration must be given to the enlarged cycles, taking into account oceans and atmosphere, and said that due weight must be given to "rights and liabilities at point of access," to water supplies.

In addition to economic discussions of the costs and benefits of

zero discharge and water reuse, there were discussions of the various technologies available for reusing water and case histories of successful operations. In the forefront of available technologies are reverse osmosis, activated carbon adsorption, continuous ion exchange, and eutectic freezing processes. Such techniques, while they may not remove every last trace of pollutants, produce water of sufficiently high quality that it can be recycled or reused.

Several participants pointed out the distinction between recycling and reuse, where the former refers to using treated water in the same application for which it was previously used, while reuse could include other applications where water quality was less critical. Still other discussions centered around reclaiming incidental contaminants, including low-level waste heat, for productive use.

Medium sized lumber mills, for example, have a tough time disposing of wood wastes, Don White of the

**Lawrence K. Cecil**
*Chairs meeting*

University of Arizona, Tucson, told the conferees. A conceptual study shows that a possible solution would be to hydrolyze the wastes with acids and ferment the sugars, turning the waste into a high-protein feed. Water would be reused in the system so that no liquid wastes would leave the process stream, White reported.

### Using heat

The use of low-level waste heat has been receiving considerable attention from the Tennessee Valley Authority, according to TVA's C. D. Madewell. Projects under investigation include using waste heat from steam plants to heat and cool greenhouses, livestock, and poultry housing; subsoil heating to increase the growing seasons of certain crops; and enhancing catfish production in raceways.

Several symposium participants pointed out that secondary municipal sewage plant effluent could be used to augment cooling water make-up requirements for power-generating plants. R. A. Sierka of the University of Arizona went the other way in describing uses of waste heat from nuclear power plants to augment a physical-chemical sewage treatment system. Such a combination power plant–sewage plant would use each other's wastes in a form of industrial symbiosis.

L. J. Boler of Cherne Industrial, Inc. (Edina, Minn.), proposed a similar solution using waste heat from a 1000-MW generating plant to enhance operation of an extended aeration biological sewage treatment plant. The two plants in tandem could take care of the power and sewage treatment requirements of a city of 1 million people. The benefits: reducing need for fresh water to cool the power plant by using sewage as a coolant, trapping waste heat for useful purposes rather than merely dissipating it, and reducing the horsepower necessary for handling both heat rejection from the power plant and introduction of oxygen into the sewage treatment plant.     HMM

# Unox gets first industrial test

*Lederle Laboratories is the first industrial concern to use Union Carbide's pure oxygen aeration package, and both companies are happy with the results*

Reprinted from ENVIRON. SCI. TECHNOL., **6,** 878 (October 1972)

Lederle Laboratories, the nation's leading maker of broad-spectrum antibiotics is also on the leading edge of industrial waste treatment technology. Last July, the company dedicated a $4 million waste treatment plant at its Pearl River, N.Y., manufacturing complex, and became the first industrial user of Union Carbide Corp.'s proprietary Unox pure oxygen aeration process.

Lederle, a division of American Cyanamid, manufactures pharmaceuticals at a 500-acre plant site in Pearl River. Most of the plant's waste is spent fermentation broths produced by the microbes Lederle uses to make antibiotics. It is highly proteinaceous and contains some fats, oils, antifoaming agents, and salts. The plant effluent—some 1.1 million gpd—contains 15,000–20,000 lb BOD and 10,000 lb suspended solids (SS). Prior to construction of its Unox plant, Lederle had been treating the wastes with high-rate trickling filters before discharging to the Orangetown, N.Y., municipal sewage system.

Lederle's major problem was odor. The plant sits in a small valley, and during weather extremes, Lederle "had its own private little inversion problem," according to Charles Isberg, Lederle's Director of Community Relations. With homes less than 200 ft from the plant's perimeter, Lederle was looking for some way to control odors and treat its waste adequately and cheaply.

## Enter Unox

Lederle's answer was Unox, Isberg says. The Unox process takes place in covered reactor trains, thus alleviating the odor problem. Furthermore, Unox promised improved treatment results over conventional air units with considerably less capital investment and greater control over the process. Lederle's plant has been operating for several months now, Isberg says, and

the company is "very pleased" with the results.

Unox got its first full-scale test in 1970 at a 3-million gpd (mgd) municipal treatment plant in Batavia, N.Y. The test, sponsored by the Federal Water Quality Administration (FWQA) "triggered an extremely high level of interest in the Unox system," according to Union Carbide, and today there are four Unox systems onstream, five plants under construction, and 13 more for which contracts have been signed. Some 40–50 additional contracts are under negotiation, according to Carbide's Jack W. McWhirter, Unox's developer. The company is certain that it has only scratched the surface of the market.

McWhirter takes pride in the speed with which Unox was commercialized. "I don't think there's been any industrial program in the history of man that's been brought into existence and advanced as rapidly as Unox," he says with characteristic confidence. "People have been jawing about physical-chemical treatment for 10 years, and I still can't identify a single physical-chemical treatment plant that's more than a peanut curiosity," he adds.

## Process technology

Unox is a fortuitous marriage of "recent advances in gas-liquid contacting and fluid control systems with well-established air separation and waste water treatment technology," according to Carbide.

Apart from the oxygen supply system, the workhorse of the process is the reactor—a multichambered aeration tank fitted with a gas-tight precast concrete lid. Although requirements differ for various applications, a typical reactor might have three chambers or stages—with depths ranging anywhere from 10–30 ft—separated by baffles. Waste water and oxygen are introduced into

the first chamber—the oxygen filling the space between the liquid level and the tank lid—and an agitator whips up the waste to contact the oxygen. Oxygen and waste water flow concurrently through each of the three contacting stages, biological oxidation occurs, and the effluent from the third stage is settled in conventional ponds. Settled activated sludge may be returned to the first stage to be blended with raw or settled influent.

Oxygen to the system is supplied essentially on demand. As demand increases, oxygen pressure in the reactor decreases, activating a flow controller which provides more oxygen. Small amounts of surplus oxygen and waste gases such as $CO_2$, produced by respiring bacteria, escape to the atmosphere through a vent located above the third stage.

To a large extent, the economics of Unox hinges on an inexpensive source of oxygen (although the increased efficiency of pure oxygen compared with air allows considerable savings in land and capital construction costs). Liquid oxygen, while useful as a backup supply or in peak shaving, is too expensive to

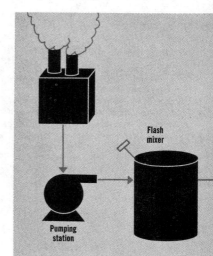

Flash mixer

Pumping station

## Cutaway diagram of Unox reactor with surface aerators

**Control valve**
**Aeration tank cover**
**Oxygen gas feed**
**Agitator**
**Exhaust gas**
**Waste liquor feed**
**Mixed liquor effluent to clarifier**
**Stage baffle**
**Sludge recycle**

use on a large scale. Plants with upward of 75-mgd nominal flow can use conventional cryogenic air separation plants. For small- or medium-volume users, Carbide's Pressure Swing Adsorption generator (PSA) fills the bill.

PSA works like this: Atmospheric air is compressed and fed to one of four adsorption columns packed with a granular material which Carbide describes only as a "molecular sieve." Air enters at the bottom of the unit and the packing preferentially adsorbs nitrogen, water, and carbon dioxide, leaving behind a nearly pure oxygen stream. The process is sequentially repeated in each of the four units. While one unit is producing oxygen, the other units are in various stages of regeneration.

The system's "pressure-swing" name comes from the fact that the adsorption process takes place in a pressurized column, and regeneration takes place when the column is depressurized to atmospheric pressure.

### Lederle's system

At Lederle, waste from the manufacturing operation is piped to a pumping station which feeds a lime-alum flash mixer and clarifier-flocculator. Clarifier effluent, with about 50% of the BOD and 85% of SS removed, then goes to the Unox reactor for oxygen treatment. The effluent stream from the primary clarifier is split and enters two parallel Unox reactor trains. Each train consists of three contacting chambers with appropriate baffles and weirs. Each of the three chambers has a surface agitator to promote oxygen–waste water contact.

Oxygen to the system—supplied by a PSA unit with liquid oxygen backup— enters the reactor and fills the 4-ft head

space above the 16 ft deep liquid. An oxygen bleed valve on the opposite end of the reactor assures the continuous flow of oxygen through the reactor trains, at pressures slightly above atmospheric. Of the oxygen that goes in, says Donald Reinhard, the microbiologist in charge of Lederle's treatment plant, about 5% escapes through the bleed as oxygen, another 5% becomes $CO_2$ which also escapes from the bleed, and the remaining 90% goes to make up the cells of the microbial population comprising the activated sludge. The reactor trains also feature a sludge return to recycle activated sludge to the system. Surplus sludge is dried on a Komline-Sanderson Coilfilter and composted.

Effluent from the reactor passes to a series of parallel settling basins. Clarified effluent is dumped to the sewer, virtually free of suspended solids and with a BOD reduced by about 90%, according to Reinhard.

Lederle's system features a number of extras, including provisions for ozonation and chlorination for further odor control and disinfection. "Frankly, I doubt that we will ever use them," Reinhard says. In the event its PAS oxygen generator malfunctions, a liquid oxygen supply sufficient to treat 48 hr worth of waste is on hand. A combustible gas analyzer and total organic carbon analyzer check effluent from the clarifier-flocculator before it goes to the Unox reactor, to avoid introduction of solvents used in the pharmaceutical extraction processes into the reactor trains. Effluent can be diverted to holding tanks before it enters the reactor to help shave off peaks and avoid shock loadings or poisoning of the sludge through accidental discharge of waste chemicals from the plant.                                   HMM

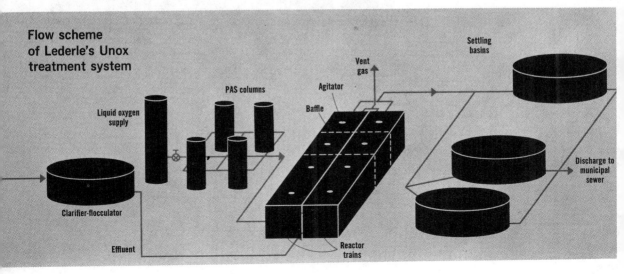

## Flow scheme of Lederle's Unox treatment system

**Settling basins**
**Vent gas**
**PAS columns**
**Agitator**
**Baffle**
**Liquid oxygen supply**
**Discharge to municipal sewer**
**Clarifier-flocculator**
**Effluent**
**Reactor trains**

*A specialized chemical conversion industry takes wastes from petroleum refineries and manufactures marketable products*

Reprinted from Environ. Sci. Technol., **5**, 306 (April 1971)

# Reclaiming industrial waste water

Today, recycle is *the* word in environmental circles, especially among businessmen who foresee reclamation as a profitable step in industry. While the treatment of gas, liquid, or solid wastes is more or less already required, recycling has yet to become a completely adopted practice.

Reclaiming waste streams from petroleum refineries is an example of potential pollutants becoming a successful chemical business. Merichem Co., based in Houston, Tex., represents one of the first such recycling ventures. "All of the materials that we bring in are totally recycled back into the economy," stresses A. Roy Price, executive assistant to Merichem's president. Many of a refinery's highly toxic waste streams contain valuable chemical compounds which can be recovered and marketed.

Waste caustic, produced in manufacturing gasoline, contains on the average 20–30% alkalinity (used to reduce the sulfur content of fuel). The caustic treating process, widely employed for many years, is still used in the newest refineries and is one of the most economical ways to sweeten gasoline. Besides removing sulfur compounds, caustics also extract organic acid materials. Although not true acids chemically, these organic materials, classified generically as phenolics, are commonly called cresylic acids.

The caustic solution, containing 20–30% dissolved organic materials after its reaction with gasoline, used to be dumped into the refinery effluent and discharged into the plant's receiving body of water. Pollution became a serious problem for two reasons: the alkalinity of the discharge was too high, and the phenolic materials (cresylic acids) are highly toxic, especially to fish and vegetation. Furthermore, drinking water supplies could become contaminated. Toxicity is not the problem here (dilution reduces toxicity); taste is. Phenolics have some of the lowest taste thresholds of any organic material. When a drinking water supply receives the usual chlorine treatment, phenolics become chlorinated. This chlorinated form has a low taste threshold of 2–5 ppb. Therefore, no phenolic material should be allowed in any surface water that will eventually be used for drinking.

## Process

In the late forties, John Files, president of Merichem, developed a process that recovered cresylic acids from waste caustic solutions. This process was developed commercially before the present push on pollution control. Presently, 700 million pounds of waste material are processed and marketed annually.

Here's how Merichem recovers cresylic acids from refinery waste streams. The process begins (see flow sheet) with a purification step that separates the organic sulfur compounds and hydrocarbons from the cresylic-rich caustic mixture. The mixture is neutralized which causes the dissolved cresylic acids to separate out and float to the surface. This upper organic layer, further purified and separated by fractionation, is divided into phenol; *ortho-, meta-,* and *para-*cresol; xylenols; and alkylated phenols. All the components of this organic layer are marketed to industry, except for a heavy-bottom pitch.

The heavy pitch, collected during the cresylic recovery process, is burned in an incinerator. One common problem encountered in burning pitch is difficulty in sustaining combustion. The secret is atomization or breaking up the heavy residue into small droplets. Each particle is then surrounded by enough oxygen for complete combustion. If the residue is not atomized, incomplete combustion takes place with a resulting heavy black smoke.

The system Merichem adopted uses sonic energy—a high velocity steam jet that shears the pitch particles. The pitch, kept in a molten state at temperatures above 300° F, and process air are injected with the jet stream into the incinerator. The high velocity steam shears the molten pitch into finely divided droplets.

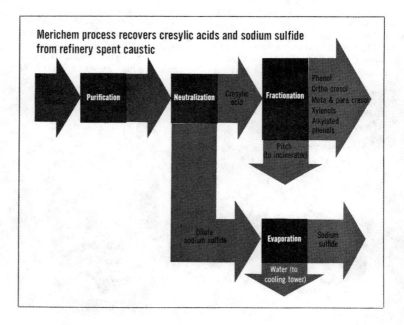

Merichem process recovers cresylic acids and sodium sulfide from refinery spent caustic

The bottom, aqueous layer (after neutralization) contains a dilute sodium sulfide solution ($Na_2S$ or $NaHS$). This is concentrated and evaporated before being marketed and sold. Inorganic salt solutions form the second category in Merichem's product line, the first being organic compounds.

## Pollution control

Besides being part of the pollution abatement program for some 100 different refineries, Merichem has developed two internal waste disposal systems (for air and water) for its recovery process. In purifying organic materials, large quantities of air are needed. Also, the organic compounds contain large amounts of sulfur picked up during gasoline purification steps. These organic sulfur compounds (the mercaptans) may be quite odoriferous.

Steps are taken to prevent air pollution from the recovery process. If an air stream is vented to the atmosphere, it must pass through a caustic scrubbing unit that removes soluble organic sulfur compounds. Since no unit is foolproof or can operate 100% of the time, a back-up system collects the waste gas streams after passage through the scrubbers and feeds them to an incinerator. Any remaining gases are burned odorlessly and vented to the atmosphere as sulfur dioxide ($SO_2$), carbon dioxide, and water. The $SO_2$ occurs in such small amounts that no air pollution is created.

The waste caustics from gasoline refineries contain about 20–30% sodium hydroxide and organic matter chemically combined with sodium hydroxide. In fact, the term for this material is sodium cresylate. The rest of the solution is water. Some of this water leaves the plant in the 30–40% sodium sulfide products. However, when the sodium cresylate enters the process separation step, the aqueous layer is very dilute—only 15% sodium sulfide. Large quantities of water are removed in the evaporator. This water, even though distilled, may still contain trace quantities of sulfides or phenolics and, consequently, cannot be discharged to the receiving body of water (in this case, the Houston Ship Channel).

Because of the water's low mineral content after distillation, it is used internally as cooling tower makeup water. The Merichem plant has no process water discharge into the Houston Ship Channel. The only effluent is a small amount of blowdown from the boilers, this is treated and discharged. Merichem has a state-issued permit to make the discharge.

The water control system also consists of dikes and drains. The entire plant process area is diked on concrete slabs with drain lines and sumps that collect all rainwater. To eliminate the risk of ground contamination from the runoff, this water is pumped into the process system. In this way, rainwater which would ordinarily wash the area and run off into the bayou will not pollute the receiving water.

## Markets

Merichem produces 70 million pounds of cresylic acids annually for use as plasticizers for plastics, as synthetic lube oils, gasoline additives, or hydraulic fluid. Also, cresylics are sources of phenols, cresols, and xylenols.

The pulp and paper industry is the main consumer of sodium sulfide (a replacement for salt cake or sodium sulfate in the digestive processes). A pulp mill, before using sodium sulfate, has to reduce it to sodium sulfide. Merichem's material, already in the form of aqueous sodium sulfide, can be used directly in the digesting system.

Full-scale commercial shipments of sodium sulfide are going directly into pollution abatement. Added to effluent streams, it will precipitate heavy metallic metals (lead, mercury, zinc) out of solution. For example, mercury, in the presence of a sulfide ion, forms a mercurous sulfide precipitate that is highly insoluble. (See "Mercury in the environment," ES&T, November 1970, page 890). This precipitate settles and can be removed from the effluent stream before it is discharged.

## Economics

In the U.S., about half of the cresylics in gasoline is presently being recovered by caustic treating. Nevertheless, most of the other 50% is also recoverable. The technology is available for almost 100% recovery of the cresylics in gasoline, according to Merichem.

The entire concept of making money from wastes sounds ideal, but this is far from the full story. This industry, emphasizes Merichem's Price, "is in the chemical business, and the markets for our products go up and down with the fortunes and failures of the chemical industry." First of all, Merichem pays for the wastes it receives. It's a waste material to the refiner, but is purchased, transported, processed, and then marketed.

Freight is a large expense, even though Merichem has its own barge system. Hauling materials that are 80% water results in small payloads. Also, markets have to be developed—sodium sulfide has been used by the pulp and paper industry only for the past few years. Pollution control within the plant itself must be considered as an expensive (although, of course, necessary) item.

Lastly, the materials marketed often have to compete with synthetics. The cresylics compete with organic synthetic cresylics and coal tar-derived cresylics. The sulfide competes with salt cake and sulfide manufactured from other processes. In short, these products are in constant competition. Revenues realized are also subject to the cycles of the chemical industry.

Some chemical producers are saying that the supply of waste material for cresylic manufacture is drying up. However, Merichem does not foresee any drastic changes. Hydrogenation is sometimes used as a method for sulfur removal rather than the caustic treatment, but while it more thoroughly removes sulfur, hydrogenation is more expensive. On the whole, refineries are not using this instead of caustic treatment; in fact, some refineries use both processes. An added incentive is the revenue benefit from selling caustic solutions to such plants as Merichem. Says Price, "You'll find some refineries abandoning caustic treatment in favor of hydrogenation, but this is not an overall trend. For every one that does, there is another going the other way."

## New ventures

With its already developed collection and transportation system established east of the Rocky Mountains, Merichem stands ready to apply its waste-to-marketable-product concept to other industrial wastes. In fact, the company is mounting a major diversification program toward recovering additional salable materials. "We're a specialized chemical conversion system," explains Price; "besides being in the chemical business, we are in the pollution abatement industry in the sense of recovering economic material from industrial wastes."   CEK

# Reuse, recovery lower pollution from brewery

*Reprinted from* ENVIRON. SCI. TECHNOL., **6**, 504 (June 1972)

An environmental science writer's dream come true: the chance to do a story on pollution control in the brewing industry. Writers and Brew, after all, get on famously—the one extolling the virtues of the other; the second aiding and abetting the vice of the first.

Americans like beer. Last year, they drank more than 127 million barrels of it, according to U.S. Treasury Department statistics (one barrel contains 31.5 gal. of beer). And while Americans consume only about half as much beer per capita as northern Europeans—the world's most enthusiastic tipplers—the U.S. brews twice as much beer as the first runner up, West Germany.

The potential for pollution is high. Brewing consumes large amounts of water. Beer itself is about 93% water and the average brewery uses about 10 gallons of process water for each gallon of beer it makes. Effluent from even a small brewery—one which produces half a million barrels of beer annually—averages 700,000–800,000 gpd. BOD is in the neighborhood of 4000 mg/l. and total solids are in the 1000-ppm range. Solid wastes—spent grains, hops, and sludges of various composition—all add to pollution-abatement headaches for brewers.

## Brewing process

The first step in the brewing process is malting. Whole barley is steeped in cool water and aerated for about 48 hr. The softened grain is then placed in germination bins at constant temperature and humidity for about a week. The germination process converts crude starch in the grain to soluble sugars and starches by enzymatic action. The barley is mechanically turned throughout the germination period to assure even sprouting.

With germination complete, the barley is transferred to drying kilns which gently toast the grain to caramelize it and stop further growth. The dry roasted grain is now called malt and needs only to be milled before it is used to give color and aroma to the finished beer.

Beer begins by mixing malt, brewer's rice, and water in large mash kettles and cooking through various temperature cycles to finish the conversion of starches into malt sugar. The cooked mash is filtered and the clear, amber malt extract is boiled with hops which add flavor and aroma. Spent hops are removed and the hot extract—called wort—is pumped to a coolship where undesirable proteins coagulate and settle out.

Cooled wort is mixed with brewer's yeast in primary fermentation tanks where malt sugars are converted to alcohol and $CO_2$. Secondary fermentation and aging complete the process and the clarified, finished beer is ready for bottling.

## Waste treatment

Waste is generated at every step of the brewing process. Although virtually all the contaminants are biodegradable, the fact that they're present in such large quantities makes treatment essential. One brewer visited by ES&T—the Adolph Coors Co.—has been able to cut pollution considerably by reusing water and reclaiming brewing by-products, rather than discharging them directly to their treatment plant.

The Coors brewery—located near Denver in Golden, Colo.—is large by industry standards. It makes about nine million barrels of beer per year for distribution in 11 western states. Until 1953, according to Coors's general manager for environmental control, Howard V. Lewis, the brewery, industries associated with it, including a porcelain manufacturing plant, and the city of Golden all were discharging untreated wastes directly into Clear Creek, a tributary of the South Platte River. In 1953, Coors built a primary treatment plant to take care of the effluent from the brewery as well as the city's sewage.

The plant didn't work very well, according to Lewis, because of heavy initial loadings and fluctuating pH.

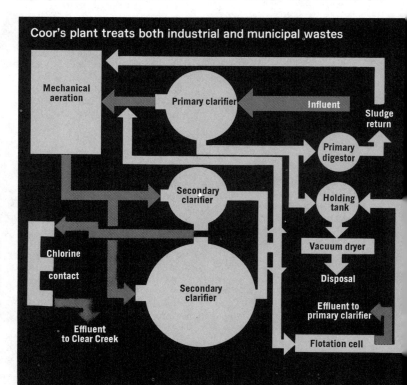

Coor's plant treats both industrial and municipal wastes

*By cutting initial loadings and
modifying sludge process, Adolph Coors
makes plant effluent sparkle*

Bits of solid waste—spent hops, grains, paper labels—choked the digesters. By adding screens and mechanical separators, Coors was able to upgrade the performance of the plant. Trapped solid wastes were pumped to anaerobic digesters modified with gas and steam recirculation systems to improve mixing.

Other problems remained, according to Lewis. Plant loadings were still extremely high, and the digesters had difficulty coping with shock loadings from batch brewing and malting operations.

Coors initiated a recovery program to cut loading and capture potentially valuable by-products. Recovering waste beer—overflow from the filling lines—rather than allowing it to flow into the treatment plant was particularly helpful, Lewis says.

Beer has a BOD of more than 125,000 mg/l., he notes, and by cooking waste beer in evaporative condensers, Coors not only reduced loading considerably,

but produced a syrup which could be used as a binder to form livestock pellets from spent grains and hops. Waste yeast was also dried, Lewis says, and made into a high-protein poultry feed supplement.

With such recovery procedures, Coors was able to reduce plant loadings from about 4000 mg/l. BOD to roughly 1600 mg/l. However, this was still not good enough since primary treatment continued to yield effluent with a BOD of about 1200 mg/l. and suspended solids content of about 125 mg/l. "Secondary treatment was clearly necessary," Lewis says.

Pilot studies indicated that activated sludge processes would need to be modified slightly because of the high carbohydrate content of the effluent. Carbohydrates promote rapid growth of *Spherotilus natans*—a filamentous plant which quickly clogs sedimentation basins. The solution to that problem, according to Lewis, was to inject small amounts of highly digested sludge di-

rectly into the activated sludge return.

Another problem was nutrient availability. The BOD:N:P ratio necessary for the secondary treatment unit to function properly, Lewis says, should be on the order of 100:5:1. But brewery effluent is both phosphate and nitrogen poor.

Wastes from the city of Golden currently supply ample phosphate, but Lewis says he must add anhydrous ammonia to make up for the nitrogen deficiency. Golden will soon join a metro municipal treatment plant, he adds, and Coors will have to supply phosphate as well.

### Sludge handling

Waste activated sludge is thickened in a flotation cell with a cationic polyelectrolyte to an average solids concentration of about 4%. The activated sludge and primary settled sludge are pumped to a holding tank which is agitated to keep the solids evenly dispersed. Prior to 1963, Coors took this liquid sludge directly to landfill in tank trucks. But, says Lewis, that proved to be a costly, inefficient method of disposal which had to be operated around the clock. The sludge mixture is now fed to a vacuum filter. Dewatered sludge is trucked to company land, where it is mechanically turned with tracked machinery and farm equipment to minimize odors and work it into the soil. Experiments are under way to test the feasibility of drying sludge further. The sludge is 44% protein, Lewis points out, making it an ideal candidate for use in animal feeds.

Final clarified effluent from the secondary clarifiers is chlorinated in a contact cell before being discharged to Clear Creek. The final effluent shows a reduction of better than 95% in BOD and better than 97% reduction of suspended solids—well within the limits for discharge set by the State of Colorado.                    HMM

**Mash kettles.** *Copper-capped vats, lined with stainless steel, are the heart of the brewery*

# Closing the loop on waste waters

*Reprinted from* ENVIRON. SCI. TECHNOL., **6**, 602 (August 1972)

The control of waste water effluents from fertilizer-producing plants is coming under more strict regulatory control, largely because of the release of the many pollutants into the nation's waterways. Various technologies have been looked at to control these effluents, but ion exchange is moving to the forefront as one of the technologies that is available to handle the wastes from actual producing plants.

The effluent from the Farmers Chemical Association, Inc. (FCAI) fertilizer plant in Harrison, Tenn., came under close surveillance by state regulatory officials in 1965. At that time, research had been done on the treatment of wastes having low nutrient levels (nutrient levels of nitrogen in the range of 10–50 mg/l.), but none of this was directly applicable to the fertilizer industry. The industry was confronted with the problem of reducing extremely large concentrations of nitrogen (1500 ppm) to levels that could be satisfactorily discharged to receiving waters.

FCAI was faced specifically with the problems of removing ammonium nitrate, a basic product of its fertilizer operation, from its waste waters and of finding ways to prevent it from entering streams and nearby waterways. Later, in 1968, FCAI was accused by the Tennessee Stream Pollution Control Board of releasing nitrogen materials to waterways, thereby causing excessive chlorine demands for purification of raw water used by the City Water Co. in Chattanooga, a short distance downstream on the Tennessee River.

## Early start

The technical man behind the scenes who played an important role in bringing the new technology to the fertilizer field is Edward (Ed) C. Bingham, director of environmental and public affairs for FCAI.

Bingham says that FCAI worked with Peter Krenkel of Vanderbilt University and the Nashville-based consulting engineering firm, AWARE (Associated Water and Air Resources Engineers, Inc.). On April 18, 1969, FCAI received an R&D grant from the federal EPA to develop a feasible treatment technique (Project 12020

EGM). With AWARE as consultant, FCAI performed laboratory tests and evaluation studies on six different processes before ion exchange technology eventually won out; each of the other five, however, was limited by certain restraints:

- microbial nitrification—slow and inefficient
- biological denitrification—inefficient and expensive
- air stripping of ammonia—promising, but results in an air pollution trade-off
- precipitation of ammonia as magnesium ammonium phosphate—problem of removing all the phosphate
- reverse osmosis—FCAI tried one vendor's equipment, but membranes were not available to give good nitrate ion rejections.

In about May 1970, FCAI discovered that ion exchange systems were in operation where the resins were regenerated by the use of nitric acid and ammonia, and about then, "We put all our hopes on ion exchange," Bingham says. "The system that came to our attention was that of a Canadian fertilizer operation—Simplot Chemical Co., Ltd., in Brandon, Man. That operation, a system designed by Chemical Separations Corp. (Oak Ridge, Tenn.), was using resins that could be regenerated by ammonia and nitric acid in the purification of well water for their boiler feed water.

## Products

FCAI produces five different fertilizer products—four at the Harrison, Tenn., plant, and an additional one at the Tunis, N.C., plant. FCAI produces ammonium nitrate (AN) for its four owner cooperatives and has a production capacity of 1000 tons a day at the Harrison plant; it will have double this capacity at its newer Tunis plant near Ahoskie, N.C., on the Chowan River. FCAI's big product item is a 30% nitrogen solution which contains AN, urea, and water. This solution is used for direct application on soils.

## Effluent characteristics

Effluent volume from the Harrison plant is typically 0.9 million gal. of waste water each day, with a concentration of 1500 ppm AN as the essential pollutant to be removed by the ion exchange technology. Chemical Separations Corp. (Oak Ridge, Tenn.), a relatively small company, was awarded the contract to design a system for

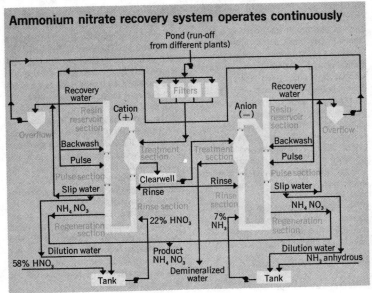

**Ammonium nitrate recovery system operates continuously**

*Fertilizer plants in Tennessee and North Carolina completely recycle waste waters from ammonium nitrate production*

**FCAI's Bingham**
*Bringing ion exchange to the forefront*

FCAI. [Chem-Seps and the engineering construction firm of Foster Wheeler Corp. (Livingston, N.J.) have agreed in principle to a merger, subject to Securities and Exchange Commission approval.]

Bingham says that Chem-Seps actually designed a system capable of handling 1.5 million gal. throughput at a lower salt loading. In this way, rainwater could be treated along with the plant waste waters. Since the drainage system in the plant consists of open ditches, rainwater would collect in the collecting ponds. The contract with Chem-Seps was let in January 1971.

During 1971, the ion exchange recovery plant was designed and installed. It began operating in December 1971 and has operated intermittently ever since. However, the plant suffered a setback on April 16, this year, resulting in an estimated damage of $30,000. Bingham says that the problems were "mechanical in nature and have been corrected." None of the problems diminished the appeal or efficiency of ion exchange technology.

**How it works**

The essential steps in the ion exchange process are:

• collecting of plant waste waters, including runoff, in large settling ponds
• filtering through anthracite coal filters to remove particulate matter
• contacting waste water with strong

acid cation exchange resin that removes the ammonium cation contaminant to less than 3 ppm
• contacting the "decationated" water with a weak base resin in the hydroxide form that removes the anion contaminant to 7–11 ppm nitrate
• regenerating cation resin with 22% nitric acid
• regenerating anion resin with 7% ammonium hydroxide
• combining backwash from each separate unit and neutralizing excess acid with ammonia.

The system operates by continuously cycling between the process mode and the pulse mode (see flow chart). During the process mode, the resin stays still and the ion exchanging and regenerating processes occur simultaneously in the separate ion exchange sections of the loop; the waste water and regenerants move in and out of their respective sections. During the pulse mode, the resin is moved from section to section by a pulse of water, and the waste water is used to pulse the resin.

Each loop contains about 450 ft³ of resin; the resin life is guaranteed by Chem-Seps for five years. Dow resins are used, but resins from other manufacturers would work just as well, according to FCAI. Bingham says that there is, of course, some breakage of resin beads in movement around the loop, but all valves in the system are of the butterfly type which sweep away the resin before closing. In this way the valves do not press on the beads to break them up.

Before signing the contract with Chem-Seps, FCAI looked at other ion exchange techniques and other companies. For example, FCAI considered batch operations using both fixed and mixed resin beds, but soon became aware that in such operations there is a tremendous resin inventory problem due to loss of resin. A critical consideration in the selection of a system was a recovered waste stream of maximum concentration to preclude additional evaporation costs.

**Recovery**

The recovered waste stream from the Chem-Seps system turns out to be

a 20% total solids solution containing at least 18% of the AN product. What happens overall is that about 6 tons of AN (100% basis) in this solution are recovered each day by the Chem-Seps system. The acid and ammonia used in regenerating the resins are combined to produce an additional 6 tons of AN in the 20% solution each day. The total amount of the AN product that is recovered, then, by the ion exchange system, is more than 12 tons each day as a 20% solution.

There is no market for the 20% solution, as such, Bingham says. But in the production of AN, nitric acid (58–60%) is combined with anhydrous ammonia in a vessel known as a neutralizer. This combination of chemicals results in the production of an 83% AN solution. This 83% solution is blended with the 20% solution from the recovery unit to make up the FCAI solution product, the 30% solution.

FCAI calculates that with the ion exchange closed loop system it would produce more than 6000 tons of AN annually and additionally. (Current price of AN is $40–45 per ton.) Of this, half is recovered from the waste stream and half is formed from the combination of the acid and base regenerants used for resin regeneration.

**How applicable**

Certainly, FCAI is the first in the industry to have licked a pollution problem that all AN producers share. Just how applicable the ion exchange recovery operation is to other AN producers is nevertheless subject to certain interpretations. There are 58 AN producers in the U.S., according to "Fertilizer Trends 1971," the biennial publication of the National Fertilizer Development Center (Muscle Shoals, Ala.). All companies that produce AN have this same waste problem. If the company makes a "solution" product (and it is estimated that about 31 of them do), then they might blend the recovered AN into that product. But other producers who do not make "solution" products are still faced with the problem of what to do with the 20% solution if they go the ion exchange recovery route. SSM

*Coupled with last year's cost estimate for organics, the majority of the chemical industry's waste water treatment cost is assessed*

*Reprinted from* ENVIRON. SCI. TECHNOL., **4**, 469 (June 1970)

## Inorganic chemicals industry:

# Clean water cost estimate

"Economics of Clean Water," the third annual report to Congress on the cost of clean water for the U.S., contains waste water treatment profiles for municipalities, animal feedlot wastes, and the inorganic chemicals industry. Required by the Federal Water Pollution Control Act, this report details the in-depth cost estimates for achieving certain levels of waste water treatment within these particularly troublesome areas.

The municipal profile, Volume I, forms the basis for the administration's request for funds to provide $10 billion for construction of municipal treatment facilities between now and 1974. The animal feedlot wastes profile, Volume II, is of special concern and interest to the Council on Environmental Quality. Volume III comments on the inorganic chemicals industry. These three volumes plus a summary report complete this year's updating.

Together with last year's profile on the organic chemicals industry, the third volume now affords the cost estimate for treating the waste waters from 80–90% of the chemical industry in the U.S.

During the five-year period, 1970–1974, the inorganic chemicals industry will have to increase its outlays nearly eightfold to reach the 100% level of treatment. On the other hand, just to keep up with its present level of treatment, the industry must increase present outlays for capital and operating expenses more than 50%, from $414.5 million in 1970 to $630.2 million in 1974. Earlier, last year's estimate found that the organic chemicals industry would have to spend $234–331 million for an earlier five-year period, 1969–73.

The inorganic chemicals profile was developed under contract by four consulting engineering firms—Cyrus Wm. Rice, a division of NUS Corp. (Pittsburgh, Pa.), which served as coordina-

tor for three other firms—Resource Engineering Associates, Inc. (Stamford, Conn.); Datagraphics, Inc. (Allison Park, Pa.); and Gurnham, Bramer, and Associates, Inc. (McMurray, Pa.). It is based on data representative of 59 inorganic chemicals plants from more than 2700 plants in the U.S.

Fred Stein, the federal project manager for Volume III, notes that this profile, as well as others, serves a number of real purposes: They

• Give the R&D people the state-of-the-art for handling specific wastes within certain trouble areas.

• Present the economic alternatives and cost curves for treatment of wastes within specific industrial segments covered by the profiles.

• Detail the manpower requirements for waste water treatment operators in this particular segment of U.S. industry.

### Treatment levels

Only costs for two levels of waste water treatment were considered, namely 27% and 100%. The 27% removal figure represents the current overall level of efficiency within the entire inorganics industry. Of course, the 100% figure represents the universal application of advanced waste treatment practices for complete pollutant removal.

Due to the fact that inorganic wastes are primarily inorganic solids that respond only to physical treat-

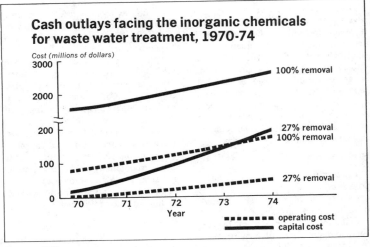

**Cash outlays facing the inorganic chemicals for waste water treatment, 1970-74**

Source: NUS.

ment methods, intermediate levels of removal efficiency are not distinguishable because there is no intervening technology. Unlike the case for organic waste waters (ES&T, April 1969, page 311), there is no series of technological plateaus through which the whole population may progress.

There is nothing magic about the 27% figure. Some existing plants have complete treatment facilities in operation today. Most have facilities only for neutralization and suspended solids removal. On the average, 27% of the industry's total pollution load is removed before discharge. So, the projected five-year estimate gives the cost for equipping the industry to remove the remaining 73% from existing plants in addition to that from plant expansions, new plants, and anticipated technological changes.

### Input-output

In its operation, this industry produces eight major groups of industrial materials. Although its products are used for variety of industrial purposes, the identity of products are far removed from consumer identity.

Including eight separate Standard Industrial Classification (SIC) numbers, the inorganic chemicals industry, for the purposes of the profiles, nevertheless includes:
• Alkalies and chlorine.
• Industrial gases (except organic).
• Inorganic pigments.
• Paints and allied products.
• Fertilizers (excluding ammonia and urea).
• Inorganic insecticides and herbicides.
• Explosives.
• Other major industrial inorganic chemicals.

### Wastes characteristics

Solids are the big problem in the treatment of this industry's wastes, both dissolved and suspended solids. In general, biological methods of treatment are not applicable. So, the physical treatment methods of equalization, neutralization, flocculation, sedimentation, and sludge dewatering are employed.

The industry's waste waters are either contaminated waters from process operations or relatively clean waters from cooling uses, general washings, and the like. Practically all plants practice the segregation of contaminated and clean wastes.

An estimated 40–80% of the industry's use of water goes for cooling purposes. In fact, many process facilities generate large amounts of thermal energy which must be removed either by the circulation of cooling waters or air, the profile points out. These contaminated wastes usually come from the following four areas:
• Electrolysis and crystallization brines.
• Washings from filter cakes.
• Spent alkalies and acids.
• Washings from raw materials.

Plants with small discharges such as paints and herbicides, for example, employ equalization and neutralization with total discharge to municipal systems for joint treatment. Approximately 4.2% of the discharge from the total inorganic chemicals industry is sent to municipal treatment facilities. This practice is not expected to change during the five-year time span.

In general, however, the industry has found that separate treatment has economic advantages in those cases where significant quantities of waste waters are concerned.

On the other hand, other segments do not make significant use of municipal systems due to the fact that their discharges are highly contaminated with chlorides, acidity, and specific inorganic species. This is true, for example, of the following segments: alkalies and chlorine, inorganic chemicals, fertilizers, and explosives.

Nevertheless, there is wide variation within the eight SIC categories. For example, industrial gases and the paints and allied products segments, SIC groupings 2813 and 2851, respectively, use municipal sewer systems rather extensively, 40% and 60%, respectively. But the waste waters for these segments are essentially clean.

### Cost criteria

The cost estimates were developed for four levels of flow rates, three levels of acidity, two levels of suspended solids, and three levels of total dissolved solids. However, the data from the 59 plants were not sufficiently complete in regard to effluent quality to construct any statistically significant relationship between the numbers of plants and effluent quality parameters such as acidity, suspended solids, and dissolved solids.

The total costs given in the profile are for the construction and operation of waste treatment facilities only. As

such, they cannot be applied to individual plants. Nor do the estimates include costs for process changeover, restriction of plant operations, or sewer segregation. Only detailed engineering studies can provide data on particular plants.

### Manpower

The manpower—operators, maintenance, and support personnel—required by the inorganic chemicals industry to achieve the two levels of treatment ranges from a low of 1826, the industry's currently employed manpower for its waste water treatment operations, through an intermediate of 2922 personnel, the manpower required by the large plants to reach a 100% level of treatment through 1974, to a high of 9365 personnel, needed in small plants by 1974 to provide the current 27% treatment level.

In general, the number of operator and maintenance personnel range from one to five. The small and large plants which normally provide 27% level of treatment would employ one operator per plant. On the other hand, the large plant that would provide 100% removal of solids normally would employ four or five operators per plant, according to the industry profile.

### Trend

In 1969, total production in the industry was 329 billion pounds of products, the profile notes. The estimated value of shipments that year valued at more than $13.1 billion. The projected 1973 output is 456 billion pounds, valued at $16.9 billion.

Expenditures for pollution control will be of greater relative significance in the inorganics segment of the chemical industry than, for example, the organics or remaining segments of the chemical industry for a number of cogent reasons. For example, the growth rate for inorganics historically has been 1.5–2.0% of the GNP. However, in recent years, the overall price index of its products has fallen 2–5%.

Nevertheless, certain segments within inorganics will grow. For example, its industrial gases segment is expected to increase 18% during the five-year time frame. As a whole, the industry is tending to concentrate in the midwest and southwest, according to the profile, reflecting the continuing trend in this industry to locate production facilities near raw materials and markets.

# Effluent control at a large oil refinery

**Robert T. Denbo and Fred W. Gowdy**

*Humble Oil & Refining Co.*
*Baton Rouge, La. 70821*

More efficient use of water
highlights the results of a
five-year abatement program
at Humble's huge Baton
Rouge refinery

*Reprinted from* ENVIRON. SCI. TECHNOL.,
**5,** 1098 (November 1971)

Humble's 450,000-barrel(bbl)/ day capacity Baton Rouge refinery is the largest in the U.S. and is considered one of the most complex in the world. The refinery manufactures a complete line of petroleum products including motor gasolines, aviation fuels, diesel and other distillate fuels, and lubricants and greases. A number of petrochemical units are located within the refinery proper.

Since the refinery came on-stream in 1909, waste water quality considerations have continually changed as more industry was installed and the population south of Baton Rouge increased. In this area, groundwater has high salinity and is not suitable for domestic use. Consequently, approximately 1.5 million people in South Louisiana depend on the Mississippi River as their source of drinking water. Louisiana has set stream standards to protect drinking water

quality, and at the request of the state, the Environmental Protection Agency Water Programs Office has conducted taste and odor control studies of Mississippi River water since late 1968. Planning to improve waste water from the refinery has been carried out in conjunction with both state and federal agencies.

The complexity of sewer and waste water systems of older refineries adds significantly to the effort and costs associated with high-quality waste water. During the five-year period ending in mid 1971, approximately 72 man-years of professional planning effort were expended to develop the current waste water improvement program at the Baton Rouge refinery (below). Projects completed or under construction amount to over $22 million in investment cost—and the program is still not completed. There is no question that the percent of to-

## Waste Water Improvement Program at Humble's Baton Rouge Refinery

| | Phase | Purpose | Planning effort (man years) | Status as of Aug. 1, 1971, % complete | Completion | Project cost |
|---|---|---|---|---|---|---|
| I | Waste water characterization | Problem definition | 6 | 100 | 1968 | Not applicable |
| II | River water replacement | To reduce waste water volume by 90% and oil discharge by 80% | 30 | 85 | 1st quarter 1972 | $19,000,000 |
| III | Taste and odor reduction | To reduce taste and odor | 6 | 80 | Lagoon to be expanded | 1,000,000 |
| IV | In-plant waste load reduction | To abate pollution at the source | 13 | 20 | 4th quarter 1972 | 1,500,000 |
| V | Monitoring and surveillance improvements | To provide faster follow-up on effluent problems | 3 | 20 | Mid 1972 | 200,000 |
| VI | Removal of suspended oil and solids | To provide pretreatment to prevent upsets in additional treatment facilities | 2 | 20 | Probable 1974 | Not available |
| VII | Additional treatment facilities | To remove dissolved organics for meeting long-range effluent quality goals | 6 | 40 | Schedule not developed | Not available |
| VIII | Rainfall detention | To detain and treat rainfall prior to discharge | 6 | 20 | Schedule not developed | Not available |
| | Total | | 72 | | | $22,000,000 plus |

### Waste water characterization

The characteristics and volumes of approximately 325 specific waste water streams were determined. This information was necessary to develop methods to reduce pollutants at the source and treatment techniques for pollutants not removed. Details were presented at the 43rd Water Pollution Control Federation meeting in Boston, Mass.

### River water replacement

By the mid 1960's, significant progress had been made toward the refinery's goal of eliminating oil losses to the Mississippi River. However, 80% of the remaining oil was associated with use of large quantities of once-through river water for process cooling. The incoming river water contained approximately 325 $yd^3$/day of silt. Fine silt particles formed oil–silt–water emulsions of about the same density as water and were not removed satisfactorily by gravity separation. These emulsions were responsible for most of the remaining oil discharged in the waste water from the refinery. It was decided to replace once-through river water with recirculated cooling water systems.

Consequently, a $19 million project was developed to replace once-through river water by installing nine

tal investment devoted to waste water control to achieve a given level of waste water quality is significantly higher for a large existing refinery than it would be for a new refinery.

Similar approaches are being employed at the Baton Rouge refinery to control air emissions and solid waste disposal. Overall, approximately 90 man-years have been spent in environmental quality planning at the refinery during the past five years.

In recent years, substantial improvements have been made in the quality of waste water from the refinery. During the 10-year period ending 1968–69, abatement efforts were directed primarily at reducing the amount of oil and phenol in waste water discharges. Marked reductions in those parameters amount to a 75% reduction in oil and 85% in phenol.

Since 1968–69, efforts have been directed at reducing total organics in the effluent with emphasis on taste and odor reduction. Intensified planning efforts in water pollution abatement were started in 1965. Overall, the approach in the Baton Rouge refinery program has stressed increasing reuse and eliminating pollution at the source. The most important step in this program has been a $19 million project to eliminate once-through use of river water for cooling by installing a recirculating system employing cooling towers. This project not only will reduce oil discharges to the river by 80% below the 1968–69 level, but will cut waste water volume by 90%.

As of the summer of 1971, this project was 60% complete. Nevertheless, substantial improvements in waste water quality were achieved from this and other projects. Improvements were greater than predicted (below) resulting from excellent cooperation of personnel at the various refinery units.

### Early start

In 1965, facilities for treating refinery waste water consisted of in-plant facilities such as: • sour water stripping • a Phenex unit for removing phenol by solvent extraction • a ballast water treating tank and • effluent treatment including a silt deoiling unit and seven pairs of modernized oil-water gravity separators, followed by a holding basin providing $2\frac{1}{2}$-hr detention time for final oil removal. A segregated system for ponding effluent from the phenol treating plant for detention and biological oxidation was in operation also.

Approximately 25% of process cooling river water is utilized on a once-through basis. The remainder of water cooling employed recirculating systems with cooling towers. Since this once-through water amounted to approximately 170 million gal/day (gpd) (90% of the waste water flow), treatment beyond primary was impractical.

In 1965, a comprehensive program for waste water improvement was undertaken at the Baton Rouge refinery. Eight phases were delineated (see previous page).

## Refinery Waste Water Improvement Since 1968–69

% reductions in absolute quantities from the 1968–69 levels

| | | Actual | |
|---|---|---|---|
| | Predicted | July | Best week |
| Oil | 77 | 62 | 81 |
| Phenol | 42 | 68 | 68 |
| BOD$_5$ | 30 | 72 | 80 |
| COD | 20 | 59 | 80 |
| TOC | 20 | 65 | 80 |
| Odor contribution | 50–70 | 60 | 70 |

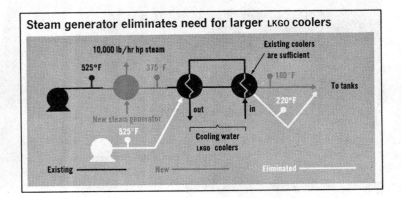

**Steam generator eliminates need for larger LKGO coolers**

cooling towers, varying in size from 4500 to 46,000 gal/min (gpm). Reducing volume from 135,000 gpm to 14,000 would also permit additional treatment to remove dissolved organics. Although a number of heat efficiency projects were found to be economically attractive, the total project has no economic return.

Water circulation in the system amounted to 135,000 gpm in 18 units throughout the refinery. Initial process design efforts were complicated by lack of adequate information on the 18-unit river water piping systems, and up-to-date drawings had to be prepared. Process design information on all cooling equipment in the river water service was also obtained, and heat removal requirements and water rates for all coolers and condensers using river water had to be completely redefined.

A highlight of this phase included selecting a makeup water source which is required to replace evaporation losses. In addition, water must

be continually withdrawn from the system, "blown down" to avoid excessive buildup of dissolved materials. Three sources of cooling tower makeup water were considered: • additional groundwater • batture wells (water present in shallow sands near the river bank) and • clarified river water. It was necessary to select the highest cost source of the three considered—clarified river water.

Investment and operating costs for clarifying river water, makeup water distribution system, and cooling towers were compared for operation at 1.5, 2, and 4 cycles of concentration. Cycles of concentration refers to the buildup in concentration of dissolved solids via evaporation where the initial dissolved solids content is 1. Therefore, 2 cycles of concentration refers to doubling the initial concentration of dissolved solids. Operation of cooling towers at 4 cycles of concentration sets the clarification plant size, giving minimum investment and minimum operating costs.

Sufficient capacity is included in the clarification plant to supply anticipated increases in water requirements for new projects through 1975. The clarification plant includes two completely independent 10,000-gpm trains for removing silt and partially softening raw river water. Each train is capable of supplying 100% of the water needs for the project. If necessary, a third train can be added to supply additional clarified water for refinery use. In the early stages of process design, several outside clarification plants were visited. As a result, care was taken to avoid a number of potential problems. Storing and feeding chemicals in liquid form eliminate maintenance problems associated with dry chemical feeders. Also, sludge pits and sludge pumps were eliminated by discharging sludge by gravity.

Two 40,000-bbl clarified river-water tanks provide 4-hr holdup at design pumping rates. A diesel engine driver, selected for the spare clarified river-water pump, also provides emergency fire water in the event of a total electrical and steam failure at the refinery.

The heat removal requirements for all coolers and condensers in the system were redefined, and a number of projects were identified to reduce cooling-water requirements. Additionally, many problem areas involving cooling water were spotlighted. A few examples that improved efficiency of existing operations include:

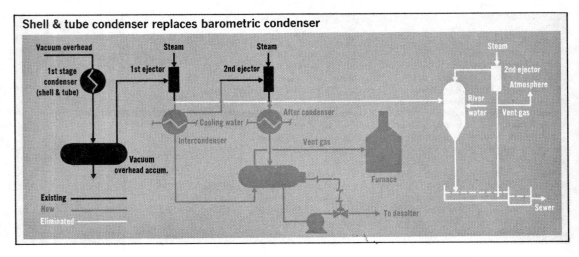

**Shell & tube condenser replaces barometric condenser**

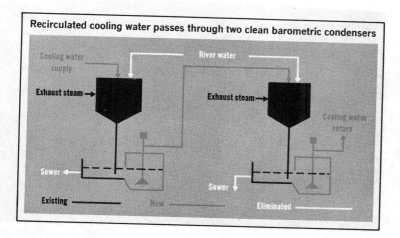

**Recirculated cooling water passes through two clean barometric condensers**

Cooling water supply · River water · Exhaust steam · Exhaust steam · Cooling water return · Sewer · Sewer · Existing · New · Eliminated

• The light coker gas oil (LKGO) coolers were inadequate (upper left), but installation of a 10,000-lb/hr high-pressure steam generator yielded an acceptable return on investment and allowed continued use of the existing coolers at a lower water rate.

• Before this project, all pipe still-vacuum tower barometric condensers used once-through river water (below left). Cooling in the second-stage barometric condenser was by direct contact with the river water. Condensed steam and hydrocarbons formed an emulsion with silty river water which interfered with the gravity separator operation to remove oil. The second-stage ejector was discharged to the atmosphere which presented an air pollution problem. Both of these problems were solved. The direct-contact barometric condenser was eliminated by installing a shell and tube intercondenser utilizing recirculated cooling water. Thus, the hydrocarbon–water–silt emulsion was eliminated. The second-stage ejector discharge, formerly vented to the atmosphere, is now condensed in a new shell and tube aftercondenser. A small amount of gas, still uncondensed at this stage, is burned in a special waste gas burner installed at an existing furnace. This eliminates a potential odor pollution problem associated with venting steam ejector gases containing a trace of sulfur compounds to the atmosphere. The extremely small amount of sulfur dioxide formed by combustion caused no atmospheric pollution problem. The steam condensate produced in the shell and tube equipment is reused as desalter water, thereby saving well water.

• In conjunction with a consulting firm, a computer program was developed to solve cooling water distribution problems in complex piping networks. Results included significant investment savings in piping and pumping equipment when compared to designing by manual calculation.

• Installing an air-fin cooler on the atmospheric sidestream at one of the large pipe stills released sufficient cooling tower capacity for the heat load of vacuum tower surface condensers.

• "Clean" barometric condensers use once-through river water for condensing steam only. This once-through river water is being eliminated by connecting these systems to new recirculated cooling-water systems. On one unit, recirculating cooling water was passed through a second barometric condenser before being returned to the cooling tower (above).

• Replacing barometric condensers with surface condensers was impractical on certain units in the lube operations area where entrainment and

carry-over of wax or asphalt would plug shell and tube equipment at cooling temperatures. In this situation, once-through water was eliminated by installing a specially designed "oily-water" cooling tower (below) which provided recirculated water to be used at the barometric condensers.

The new "oily-water" cooling tower is constructed entirely of ceramics instead of flammable materials. The ceramic tower consists of two separate sections, each designed to handle the total cooling load. Piping is provided for circulating basin water through a direct-contact steam heater to the return water distribution headers. When the accumulation of wax in the tower becomes a problem, one section can be shut down at a time for a high-temperature wash cycle. A skimming pump is provided for transporting skimmings to the emulsion treating tanks.

**Taste and odor reduction**

In 1969, a study was carried out to determine the major contributors to the taste and odor of refinery waste water. Essentially all of the specific waste water streams were reevaluated by the Threshold Odor Number (TON) test.

Waste water from the barometric condensers at the crude vacuum distillation units contributed approximately half of the total odor to refinery waste water. Fortunately, these condensers were being replaced by surface condensers at all crude distillation units as part of the river water replacement project. Odor from these sources was reduced by 95% by using the condensate from the new surface

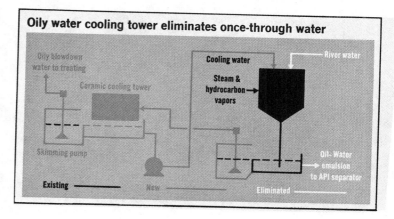

**Oily water cooling tower eliminates once-through water**

Oily blowdown water to treating · Ceramic cooling tower · Cooling water · River water · Steam & hydrocarbon vapors · Skimming pump · Oil-Water emulsion to API separator · Existing · New · Eliminated

condensers as makeup water for crude desalting operations (odor-causing agents are extracted from the water by the crude).

The volume of the existing phenol-holding pond was increased and 7-75-hp surface aerators were added to accelerate biological oxidization of any phenol that may be accidentally spilled at the lube phenol treating plant. Aeration capacity was included for treating other streams selected for odor reduction. The addition of these streams to the lagoon also provided a biological seed, so that, in the event of a spill, the time required for phenol degradation in the lagoon is reduced. Three aerators were installed in August 1970, four more were added in April 1971, and the installation of three more aerators of 100 hp each is planned by the end of 1971.

### In-plant waste load reduction

Construction of the river water replacement project was well under way in the summer of 1970. Pilot plant operations to determine treatment plant configuration and design parameters were scheduled (early 1972) to begin after 90% reduction in waste water flow following completion of river water replacement. The in-plant waste load reduction program was developed to coincide with the completion of the river water replacement project.

A set of economic guidelines was developed to compare the cost of installing facilities to correct pollution at the source with the cost of treatment facilities. The cost of treatment was based on an activated sludge plant feeding approximately 20 million gpd of waste water complete with sludge disposal. The total savings are the combined effects of overall savings in investment and overall operating costs over equipment life, discounted to present worth. On this basis, saving 1 gpm of waste water reduces cost by $550 and eliminating organic loading reduces cost by $35/lb of $BOD_5$ per day. Actually, these reductions in cost are understated since elimination at the source is totally effective for the organics removed, while treatment is only partially effective.

The waste load reduction survey included a detailed inspection of each unit. Total manpower requirement for the inspection phase of this project was eight man-years, and engineering efforts for these projects will amount to another five man-years.

One of the largest sources of excess waste water usage throughout the refinery was found to be pump jacket cooling water discharged to the sewer. Water used on the bearing jacket, gland jacket, and pedestal jacket is clean and can be returned to the cooling towers. Also, cooling water could be routed in series through the bearing jackets, gland jackets, and pedestal jackets.

Only the water used on the gland or mechanical seal may come in contact with hydrocarbon. A small needle valve to restrict flow will reduce this from up to 15 gpm/pump to an average of $1/4$ gpm/pump.

Cooling tower water used on pump jackets and for other purposes on the unit which is discharged to the sewer affects cooling tower operations by increasing makeup requirements. Most of the existing towers on well water makeup were operating at 1-1.5 cycles of concentration due to the large quantity of cooling tower water being blown down to the sewer through pump jackets and other equipment. By installing new piping systems to return pump jacket water to the cooling tower, cooling tower blowdown to the sewer from this source will be reduced by 95%. Using this approach, the cycles of concentration on the existing cooling towers, which use well water makeup, will be increased to 2 cycles of concentration or higher (below). Some of the more significant water reduction items include:

• The 4500 gpm of pump jacket water which is currently discharged to the sewer at units throughout the refinery will be reduced to approximately 200 gpm by installing necessary piping to return this water to the cooling towers.

• Approximately 600 gpm of steam condensate can be economically recovered and returned to the boiler feed water system. Besides reducing waste water treatment plant investment, it will reduce significantly the size of a future boiler feed water treatment facility.

• The volume of water used for washing alkylation reactor product can be reduced 150 gpm by installing a system to recycle the wash water and to maintain a controlled blowdown.

• Plans have been made on a trial basis to reuse cooling tower blowdown (950 gpm) in the fire and utility water system to determine if the hydrocarbon content could present a safety problem. If the operation proves to be

successful, cooling tower blowdown water from other cooling towers will be injected into the system and water saving will be increased. If this system proves workable, analyzers will be evaluated to detect hydrocarbons in the cooling tower blowdown (which would allow switching cooling tower blowdown to the sewer if the water became contaminated).

• A typical existing desalter system (upper right) on a crude distillation unit employing a quench for hot desalter brine uses approximately a total of 500 gpm of once-through water to cool hot desalter brine by direct contact. The combined brine and cooling water are discharged to the sewer through a condensing blowdown drum. The direct-contact cooling operation will be eliminated by installing shell and tube equipment. The coolant in this equipment will be makeup water for the desalting operation. This, in effect, will provide warmer water contacting crude with a resultant reduction in heat loss in the crude, will save water and reduce emulsion problems with the brine-cooling water discharge, and will provide an attractive economic return based on heat economy.

• The refinerywide taste and odor study, Phase III, spotlighted the pipe still barometric condenser water as being the largest taste and odor contributor (about half of the total refinery contribution). Installing surface condensers to replace barometric condensers decreased the volume of odorous water from 4000 to 200 gpm. However, the 200 gpm required some type of treatment for odor improvement. Reusing this odorous steam condensate as desalter water makeup not only reduced water usage but also improved the quality of the waste water discharged. Crude extracts organics from this stream resulting in 95% odor reduction, 90% oil reduction, and 40% COD reduction. No harmful ef-

### Pipe Still Cooling Tower Operation

|  | Present | 1st Step | 2nd Step |
|---|---|---|---|
| Cycles | 1.4 | 2.0 | 4.0 |
| Makeup, gpm | 700 | 440 | 293 |
| Evaporation, gpm | 200 | 220 | 220 |
| Wind loss, gpm | 13 | 14 | 14 |
| Blowdown, gpm to sewer | 487 | 206 | 59 |

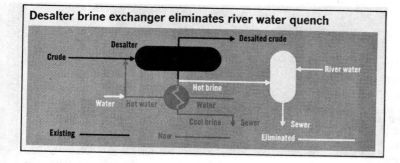

**Desalter brine exchanger eliminates river water quench**

fects have been observed in product streams as a result of this extraction.

### In-plant effects

The refinery waste water flow after river water replacement is predicted to be 14,000 gpm. The in-plant project is expected to reduce the waste water flow by an additional 50% and make significant reductions in organic pollutants. The incentive for this reduction is significant. The estimated investment for in-plant modification is $1.5 million which will reduce costs in the waste water and water treatment plants about $4.5 million (average savings to investment ratio of 3 to 1). About 70% of these projects have even higher savings-to-investment ratios. The present schedule calls for 80% completion of the projects by second quarter, 1972; completion on the remaining 20% is fourth quarter, 1972.

### Monitoring and surveillance

The current program is based mainly on 8-hr and 24-hr composite samples which are taken at key points in the system for oil, total organic carbon, and phenol analyses in the laboratory. Gas chromatographic analyzers are being used to identify soluble organic chemicals for quick tracing of problems to the source.

An effluent monitoring project, based on the application of on-stream analyzers for the semicontinuous monitoring of critical points in the waste water system, is under development. The on-stream analyzer system provides immediate indication of an upset condition, thus allowing faster followup on problems. Data from the analyzer will be fed to a computer. Printouts will be available for routine reports on the performance in the various waste water systems and for quick recall during pollution episodes. The development of this entire effluent monitoring project is expected to re-

quire at least three man-years.

### Suspended oil and solids

Efficient removal of suspended oil and solids by gravity separators was inadequate; excessive quantities remained in process water. Consequently, additional facilities for removing suspended oil and solids are expected to be required, particularly if activated carbon is selected for additional treatment. Laboratory studies are just under way to compare dissolved air flotation and various sand filtration techniques.

### Additional treatment

Facilities will be designed to remove dissolved organics from waste water after maximum volume reduction and elimination of pollutants at the source has been accomplished. Laboratory studies are currently under way comparing biological treating (employing activated sludge) with activated carbon treating. So far, activated carbon treatment appears feasible on strong refinery waste water (after primary treatment only). Several disadvantages to biological treatment may be overcome by activated carbon treating. These include sensitivity of the biological system to fluctuation in hydraulic and organic loadings, toxic shock, and organic sludge disposal problems. Moreover, biological treatment requires substantially more land than facilities employing activated carbon.

By the end of 1971, Humble expects to have sufficient data to complete a technical comparison of these two methods on Baton Rouge Refinery waste water so that a process can be selected for development.

### Rainfall detention

At present, the oily water sewer system handles any rainfall admixed with oily process water. The facilities for additional removal of sus-

pended oil and solids will require a segregated system for handling the process water requiring treatment. Such a system will leave the existing oily water sewer and separator for handling only rainfall runoff.

More study is required to define the volume of detention that should be provided and the manner of treating prior to discharge. A major problem in this study is providing space for rainfall detention because of limited land availability in the area where waste water is discharged from the refinery.

### Acknowledgment

The authors express appreciation to personnel who actively participated in Humble's pollution abatement programs.

**Robert T. Denbo,** *senior staff engineer at Humble's Baton Rouge Refinery, received his BS degree in chemistry from Louisiana State University. For the past seven years, Mr. Denbo has been involved with environmental quality control at the Baton Rouge Refinery and is presently responsible for environmental planning and development work there. Address inquiries to Mr. Denbo.*

**Fred W. Gowdy** *is staff engineer at Humble's Baton Rouge Refinery. He received his BS and MS in mechanical engineering from Louisiana State University. For the past four years, Mr. Gowdy has been project manager of the river water replacement and in-plant waste load reduction projects.*

*New biochemical plant features identification,*
*segregation, and separate treatment of wastes*
*as integral parts of day-to-day operation*
Reprinted from ENVIRON. SCI. TECHNOL., 4, 898 (November 1970)

# Waste control highlights plant design

Under construction among the endless corn fields of western Indiana is an industrial plant that is putting into practice the much talked-about principles of recycling, reuse, and conservation of natural resources. Although not yet fully on stream, this plant promises to be a shining example of enlightened industrial waste control, and it is likely to be closely watched in future months by those who want to see if what looks good on paper works out as well in practice.

The plant is a new $50 million fermentation and biochemical facility owned by Eli Lilly & Co. (Indianapolis, Ind.) and situated near Clinton, Ind., on the west bank of the Wabash River. Company officials estimate that at least $8 million of the plant's capital cost can be directly or indirectly attributed to the environmental control measures incorporated into its design. When in full operation later this year, the plant will reuse over 90% of its total water requirements. Even so, the plant will discharge 4000 g.p.m. of water to the Wabash, but that effluent will be of a quality well within the limits set by Indiana authorities, says Lilly. Indeed, the water would be of good enough quality to supply a drinking water treatment plant.

## Plant design

Eli Lilly & Co. is a large manufacturer of chemical and biochemical products for medicinal and agricultural use. When the time came for the company to enlarge its fermenting and biochemical manufacturing capacity (already on-going in several U.S. and foreign locations), it chose Clinton as a site because of its proximity to corporate headquarters (Indianapolis) and to its large manufacturing facilities at Tippecanoe Laboratories (Lafayette, Ind.). While the plant location decision was a logical one, and the sort of decision regularly made by large manufacturing concerns, another decision Lilly made was highly unusual: The Clinton plant would be designed, from the very beginning, to provide maximum environmental control. Not content with such broad, even if unprecedented, generalities, corporate management in early 1969 made some specific stipulations regarding waste control at Clinton:

• No wastes *of any description* would leave the site.

• No organic wastes would be buried either on or off the site (on-site burial of inert inorganic wastes after incineration was permitted, and will be practiced).

• Solvents and process chemicals would be recovered, even if the recovery proved uneconomical.

• Recycling of water and use of cooling towers would be maximized.

• Effluents discharged to the Wabash would not increase river temperature "by any measurable amount" and would otherwise conform to, or be better than, the most stringent requirements of Indiana regulations.

• Deep-well disposal of any wastes would not be used.

• If possible, conventional biological treatment would be avoided.

Of all these stipulations, none flies in the face of traditional plant design more than the last one. As many ES&T readers are aware, biological treatment is usually first choice for industrial waste waters as well as for domestic sewage, and the idea of any alternative is regarded as almost revolutionary by diehard sanitation engineers.

Faced with such a formidable set of guidelines, the Lilly engineering staff prospered where lesser men might have given up the ghost. "It was a real challenge," recalls Robert H. Ells II, manager of plant engineering at Clinton, "but it has been 1¾ years of adventure." Although Ells is confident that the plant will work well, there in-

## Waste disposal flow sheet

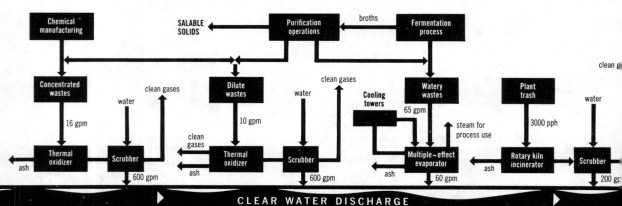

CLEAR WATER DISCHARGE

evitably are skeptics. In view of the innovative nature of the plant Ells and his colleagues have come up with, the skepticism is understandable but hopefully ill-founded.

There are several innovative keys that characterize the plant's environmental control system, according to Ells:

• Design of processes to eliminate wastes. For example, complete re-

**Engineering manager Ells**
*A real challenge*

cycling of the water contained in a high-BOD stream associated with fermentation broth from which the "activity" (the desired product, for instance an antibiotic) has been removed.

• Concentration of wastes where possible.

• Segregation of wastes at the source. Different waste streams are not mixed; each is treated in the most appropriate way.

• Reduction of hydraulic loads on the waste plants through rigorous water conservation—no flow drains are present in several areas in the plant, dilute waste waters are reused, and faucets are self-closing.

• A robot monitor records the quality of all water discharged to the river.

The plant's 75 acres are divided into five distinct areas: central services area, with offices, utilities, laboratories, and cafeteria; fermentation area; purification area; chemicals manufacturing area; and waste treatment area. Each uses water in some way.

### Waste streams

Water supply to the plant comes from three wells located on the plant site. Water withdrawn from a closed-top reservoir supplied from the wells is used for several purposes: for drinking and other domestic purposes; as boiler feedwater makeup; as process cooling water; as makeup to the purification and fermentation area process cooling water systems to supplement recycled water; for use (when deionized) for process purposes in the purification and chemical manufacturing areas; and for sprinklers. "Wastes" which need treatment before discharge or which can be recycled for use are:

• **Spent fermentation broth.** The desired product (antibiotic) is separated from the broth in a special purification process about which Lilly is not saying much, except that it is the subject of a pending patent. What the company does say is that the process can remove from the broth, in addition to the antibiotic, 35 tons per day of mycelia (solid material), a product that contains less than 1% water and that can be sold as a high-protein animal feed

supplement. Other equally important units in the purification area (also covered by a veil of secrecy) are capable of extracting impurities from the water so that it can be recycled for cooling. What is, in other fermentation plants, a difficult-to-dispose-of, high-BOD stream, is in Clinton handled without need for a biological treatment plant.

• **Plant trash and rubbish.** Although Clinton is in a sparsely populated rural region, and ample land is available for landfilling, trash at the plant site will be incinerated in a Bartlett-Snow (Cleveland, Ohio) incinerator. Ashes from the incinerator (about 600 lb./day) will be buried on the site in accordance with good sanitary landfill practice.

• **Concentrated chemical wastes.** These are produced in the chemicals and purification areas. Plant engineering manager Ells classifies them as primary or secondary—primary wastes are autogenous (that is, they are capable of supporting combustion), whereas secondary wastes are not. Both types of wastes will be burned together in a thermal oxidizer (incinerator) designed by John Zink Co. (Tulsa, Okla.). There will be two such oxidizers; both are equipped with adjustable venturi scrubbers to trap particulate matter before stack discharge. Also to be burned in a thermal oxidizer are:

• **Dilute chemical wastes.** These are predominantly the bottom product of solvent stripping columns in the chemicals area and certain unrecoverable streams, such as acetic acid and hydrogen peroxide waste.

• **Watery process waste.** "Watery" wastes are those that contain no com-

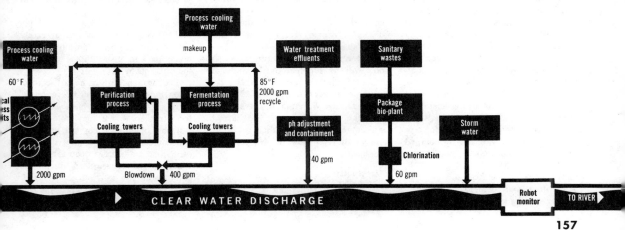

**Cooling towers.** *Fermentation area uses 85°F. water recycled at 20,000 g.p.m.*

ponents more volatile than water. At the Clinton plant they arise primarily from the fermentation area—e.g., equipment washings, innoculation media used to test sterility, foam-overs. Although these wastes could, in theory, be treated by straightforward evaporation, in practice this is made difficult by the 4% solids (dissolved and suspended organics and inorganics) the wastes contain. Scaling of heat transfer surfaces presented some difficulties which have subsequently been resolved by a multiple-effect evaporator designed and constructed under the direction of Carver-Greenfield Corp. (East Hanover, N.J.).

• **Sanitary waste.** When the Clinton plant is in full operation, over 350 people will be working there, and there will be a cafeteria in addition to the usual amenities. Lilly felt that sanitary wastes so generated would best be treated by a package unit supplied by Smith & Loveless (Lenexa, Kan.). The unit is of the activated-sludge type.

• **Process cooling waters.** Three cooling water systems are used at Clinton. The water warmed after heat transfer in cooling equipment in the chemicals area is discharged directly to the "clear water" effluent stream. This water represents about half of that discharged by the plant (6 million gallons per day) and is heated perhaps 25°F. above its initial temperature (60°F. year-round well water). The fermenter cooling system uses water that is recycled through cooling towers. The multiple-effect evaporator system also uses cooling towers. Periodically, the towers are blown down to purge any accumulated impurities.

• **Other aqueous streams.** Air is continuously supplied to the fermenting tanks to supply the aerobic reactions, and the exit air is water scrubbed to remove odors. This scrubbed water is recycled and eventually purged to the watery waste system. The water used to recharge ion-exchange resins, after pH adjustment, and the water used to scrub incinerator exhaust gases are discharged to the clear water system.

### Monitoring

Effluents from all treatment units will be monitored continuously at a central building. In addition, a robot monitor will check the quality of the combined effluent stream (clear water discharge stream) before it runs into the Wabash. Quality data (for pH, dissolved chloride, temperature, and so on) will be supplied to the Indiana Stream Pollution Control Board. Lilly estimates that no more than 2500 lb. BOD per day will enter the river, and that the 6 million g.p.d. discharged will represent no more than 0.1% of the mean river flow at the point of discharge.

### Air quality

No air pollution problems are anticipated at the Clinton plant. The steam-generating boilers are fired with natural gas, with fuel oil as backup in case of a shortage in gas supply. Stack plumes were invisible the day ES&T visited the plant. Motors driving water-chilling units are also natural-gas fired. Scrubbing of fermenter exit air and the use of negative pressure in tankage and equipment holds down the odor problems usually associated with biochemical manufacture. ES&T perceived odors, but they were not unpleasant (the plant was running at one-third capacity at the time). The thermal oxidizer units are fitted with scrubbers as is the trash incinerator.

### Prospects

In short, this new Eli Lilly plant seems likely to be the harbinger of many such plants, both within the biochemical industry and elsewhere. Although total recycle will not be achieved, that ideal will be much more closely approached than in most existing plants.

What is most encouraging about the Lilly design is that economic factors were not the sole criteria for its acceptability. For instance, the trash incinerator (a $200,000 item) could have been dispensed with and the trash taken away by a local contractor, surely a less expensive proposition. The Carver-Greenfield unit alone cost over $1 million, but Ells reports that Lilly management did not quibble over price. When told that the unit was needed to treat watery process wastes, Ells recalls that management simply said, "Do it!" It is clear that rarely have waste control engineers had so much backing from those who hold the purse strings.

The crunch will come, of course, when the plant is in full operation. If there are start-up troubles—and Ells feels that some are inevitable—there will undoubtedly be nervous moments and some wishing that pilot development time could have been longer before scale-up was required. But if this plant eventually pans out as expected, there is every reason to expect the rest of industry to follow suit.   DHMB

# Meat packing wastes respond to many treatment methods

*Reprinted from* Environ. Sci. Technol., **5**, 590 (July 1971)

**Hogs in feeding pens.** *More than 87 million hogs were slaughtered for pork in the U.S. last year, says USDA*

In 1970, some 137 million cows, pigs, and sheep marched into packing plants across the country and came out steaks, bacon, and lamb chops. The number that will make the same trip in 1971, according to the U.S. Department of Agriculture, could be 5 to 10% higher.

Meat packing—as distinguished from meat processing, which includes sausage and cold-cut manufacture, drying, and canning—gives rise to vast quantities of waste water characterized by a high biological oxygen demand (BOD), offensive odor and color, and high suspended solids content.

Blood, for example, has a BOD in excess of 150,000 parts per million. A gallon of blood discharged into a sewer has the same oxygen demand as the daily wastes from seven or eight people. Fortunately, in all but the smallest privately owned slaughterhouses not subject to government inspection, it is of significant economic advantage to reclaim blood, scrap, and other potential pollutants rather than to discharge them.

Pollution from packing operations can arise from a variety of sources. Fortunately, most wastes are biological in origin and, therefore, yield readily to conventional biological treatment processes with little or no modification necessary. Unlike the industry which must rely upon a single process for pollution control, a plant in the meat-packing industry can choose among several equally effective treatment packages, depending upon the economics of its particular operation.

Virtually all packing plants make use of a "catch basin"—a holding tank into which cleanup waters are diverted for recovery of scraps and grease. Most such holding basins have a retention time of only 20 min, according to University of Kansas economist Kathleen Q. Camin. Those plants using twice the typical 20-min detention time make a significant con-tribution to effective pollution control of plant effluents. Although the catch basin was originally installed for economic reasons—the recovery and resale of inedible waste products—it is nevertheless a valuable adjunct in pollution control.

## Treatment methods

Regardless of whether advanced treatment facilities are built by the packing plant or municipalities, there appears to be ample technology available for treating waste waters effectively and economically. Among those currently used are:

• Lagoon systems. Waste water ponds may be either aerobic or anaerobic or a combination of both types. Aerobic ponds rely upon algal growth to supply oxygen necessary to reduce BOD. Anaerobic ponds are particularly useful in the packing industry because waste waters are generally very warm and contain high concentrations of organic nutrients. Primary anaerobic lagoon treatment followed by one of more aerobic polishing ponds frequently makes sense, particularly where large tracts of land are available at reasonable costs. It is estimated that BOD reduction with a combined anaerobic–aerobic lagoon system may be as high as 90% under favorable weather conditions.

• Air flotation. A coagulating agent such as ferric chloride or alum is added to waste waters in a sealed treatment chamber, and air under pressure is introduced for a short period of time. With the iron salts acting as catalysts, the air apparently oxidizes organic components and aids in moving the floc to the top of the holding tank where it can be skimmed off. An average BOD reduction of 50% can be achieved with air flotation techniques.

• Activated sludge. Waste water from which suspended solids have been largely removed is mixed with conventional activated sewage sludge. Although BOD reduction can be quite high under optimum circumstances—

*Using "everything but the squeal" not only
adds significant profits to packing plant
operations, but reduces pollution as well*

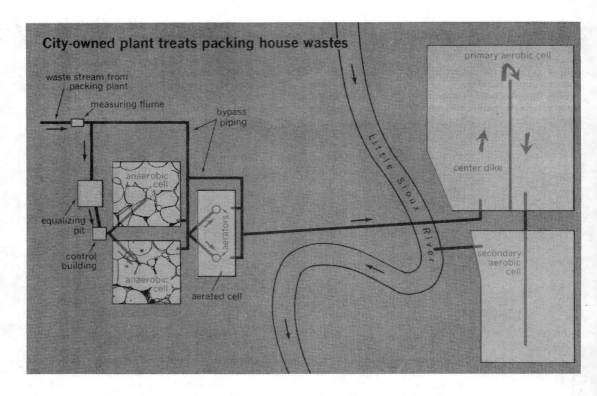

City-owned plant treats packing house wastes

reductions as high as 95% have been achieved—the necessity for using highly trained personnel and the high capital costs for the activated sludge plant make activated sludge less desirable than lagoon systems for treating meat-packing wastes. Nevertheless, there are a few activated sludge treatment plants in the packing industry, and they are operating effectively.

• Trickling filters. As in conventional installations, aerobic organisms attached to a bed of nonabsorbent filtering medium oxidize organic wastes as the effluent trickles through the filter bed. BOD removals can be as high as 90% with proper operation.

• Rotating contactors. Recently, pilot studies have been carried out with rotating disks made of high-density

Styrofoam and other inert materials which support biological growth. The disks, secured through the middle by a rotating shaft, turn in a trough carrying anaerobically treated effluent. Organisms on the disk come into contact with atmospheric oxygen as the disks turn successively through the effluent and into the air. Although some work remains to be done before the process can be considered a commercial success, a full-scale demonstration plant is nearing completion at Iowa Beef Processors, Dakota City, Neb., plant.

• Oxidation ditches. In this process, more properly known as channel aeration, waste is circulated around an oval ditch with a cage rotor. Channel aeration, originally developed in the

Netherlands, is actually an activated sludge process with additional aeration. While it has not yet gained widespread acceptance in this country, the oxidation ditch promises to be a valuable addition to more conventional methods. At least one full-scale pilot plant is in operation in the U.S., and industry spokesmen feel that a BOD reduction of at least 90% can be achieved with oxidation ditches.

• Anaerobic contact. Anaerobic processes are particularly useful in the treatment of packing wastes and the anaerobic contact process is no exception. The system consists of an equalizer tank which prevents shock loads of pollutants from disturbing the balance of the anaerobic digesters, digester tanks seeded with anaerobic sludge organisms, a degasifier to remove dissolved methane, sulfur dioxide, and other odor-causing gases and to promote settling, and a settling basin. Settled sludge from the basin is recirculated to the digesters until too much sludge has accumulated. Excess sludge is then trucked to agricultural fill. Anaerobic contact treatment is usually followed up with aerobic polishing lagoons. BOD removal is in the neighborhood of 95%.

## Construction alternatives

In addition to a variety of technologically proved treatment methods, packing-plant operators have several options available for construction of pollution control facilities.

First, they may build in-house treatment centers to purify effluents before they leave the plant. Alternatively, they may discharge waste water directly to municipal sewer systems where permitted by state law and where the municipality has the necessary treatment capacity. A third approach is to discharge wastes into a specially constructed facility built and operated by the municipality in which packing plants are located.

An interesting example of the third arrangement may be found in the recently completed Wilson Certified Foods plant in Cherokee, Iowa. The Wilson plant packs both hogs and beef on a round-the-clock basis. Although plant officials decline to give the exact number of animals killed each day, volume appears to be quite high as observed in a recent plant tour by ES&T.

Cherokee, Iowa, a city of about 7500 people, is located on the Little Sioux River—a stream highly prized by local residents for its excellent catfishing. The city's existing municipal sewage treatment plant, although adequate for the local population was built in the late 1930's and would have been insufficient to handle wastes generated by the packing plant, estimated to be the pollutional equivalent of 100,000 people.

The city agreed to build a new treatment facility and water supply for the Wilson operation. Between 1961 and 1965, construction on the plant and the treatment facility progressed concurrently. After nearly five years of continuous operation, both Cherokee and Wilson are pleased with the results.

The treatment facility is a lagoon system with the workhorse being two

anaerobic contact digesters. Waste from the packing-house operation is transported by undergound pipe to a metering flume which admits to an equalizing pit. The flume allows positive control over the effluent and minimizes shock loading of the digesters.

The effluent stream from the equalizing pit is variably split and pumped into two anaerobic cells by two 1500-gpm raw sewage pumps. Two 750-gpm sludge pumps, located in the control building, recirculate the sludge in the anaerobic cells. The cells are covered with a grease layer up to a foot in depth which effectively maintains anaerobic conditions and insulates the warm waste waters against the severe winter climate of northwestern Iowa.

From the anaerobic cells, waste water runs by gravity over baffled weirs to an aerated cell. Mechanical aerators in this cell give the anaerobic waste a boost in oxygen content before the effluent is discharged through a pipe under the Little Sioux River to the primary aerobic cell.

The primary aerobic cell has a surface area of 25 acres and water from the aerated cell is made to flow in a horseshoe pattern around a 5-foot-high center dike before discharging to the secondary aerobic polishing cell.

By the time waste water has made the rounds in the Cherokee system, according to B. D. Hester, Cherokee's director of public works, the pollution load is reduced by more than 98%.

## Wilson helps

Although the Cherokee plant seems capable of taking care of just about anything Wilson sends its way, good in-house pollution control makes things immeasurably easier for Cherokee and adds profit for Wilson.

Pollution control starts with the unloading operations and continues until the last piece of meat is trucked away. Animal wastes accumulating before cattle or hogs are taken to the killing floor are frequently hosed into special floor drains which empty into a collection pit. There, manure is screened from the water and carted away to agricultural landfill. Cleanup water—largely free of solids—is diverted to the main sewer running into the treatment plant.

Inside the plant, animals are stunned, stuck, and allowed to bleed to death while they are suspended by their hind legs from an overhead conveyor. Blood is collected in a trough

underneath the conveyor and rec[overed] for processing into animal fee[d]. Besides being a profitable operatic[n] the reduction in potential BOD loa[d]ing from blood recovery alone is pro[b]ably the most significant polluti[on] control step taken by packers.

Hogs present an additional proble[m] after bleeding. Hides are not remove[d] from hogs as they are with cattl[e]. Rather hogs are dipped into vats [of] scalding water to loosen hair. M[e]chanical paddles beat the hog and r[e]move hair which is recovered and h[y]drolyzed to yield a high-grade anim[al] feed supplement.

After hide or hair removal, the an[i]mal is eviscerated. Edible organs ar[e] segregated from inedible ones and g[o] into pet foods or other by-products[.] Of considerable importance to pollu[tion] control at this stage is th[e] paunch—the portion of a cow's multi[-]chambered stomach which whe[n] cooked becomes tripe. The paunc[h] contains large quantities of undigeste[d] or partially digested food. A 1000-l[b] cow may have as much as 50 lb o[f] "paunch manure" to be disposed of. I[n] the Wilson plant, paunch manure i[s] flushed from the stomach with water under pressure. The waste stream from paunch removal is not combined with other sewered streams from the plant, but is diverted to a separate collecting basin where paunch manure is reclaimed and sold as high-quality agricultural fill.

Other potential sources of pollution from packing plants are scraps from cutting and boning operations, inedible parts of the carcass such as ears, eyelids, or parts too bony to be of commercial value, and carcasses condemned by government inspectors. Inedible and condemned meat that does not find its way into such by-products as dog food is shredded and processed by steam cookers to render grease and sterilize the residue. Inedible greases find various uses in industrial applications, soap manufacture, and, with dried processed scraps, in high-grade livestock food. Edible and inedible rendering processes are kept separate as required by federal law and sterilized "high-grade" is stored in glass-lined silos awaiting trucking to animal feed compounders.

Cleanup water, carcass wash water, and process waste waters are diverted to a catch basin to recover grease and suspended solid scraps. Water thus treated is released to the Cherokee treatment plant for final cleanup. HMM

**161**

# Chemicals, bugs tame process wastes

*Combination of coagulation and*
*activated sludge treatment cleans*
*up plastics plant effluent*

*Reprinted from* ENVIRON. SCI. TECHNOL., **4**, 637 (August 1970)

More than 90% removal of biochemical oxygen demand (BOD) is being achieved by a $1 million waste water treatment plant newly started up by the B. F. Goodrich Chemical Co. in Pedricktown (N. J.). The facilities in the polyvinyl chloride (PVC) plant are, according to Goodrich, the most advanced of their type in the PVC industry and were financed to the tune of $364,900 by the Federal Water Quality Administration (FWQA). Partial funding by FWQA was contingent on Goodrich sharing its treatment plant design and know-how with other companies in the PVC business during the first year of plant operation.

The Goodrich PVC plant in Pedricktown is the company's sixth and represents a multimillion dollar investment for the world's major vinyl raw material producer. Emulsion, suspension, and bulk polymerization processes used at the plant are based on vinyl monomer manufactured by Goodrich in Kentucky. Water is supplied to the plant from nearby wells and deionized before use if needed for process purposes. Process effluent comprises the bulk of the waste stream that requires treatment before discharge to a tributary of the Delaware River.

## Design criteria

The waste is milky white and contains various (and varying) amounts of monomers, polymers, dispersants, organic and inorganic salts, etc., in both dissolved and suspended forms. Goodrich personnel studied raw waste from other PVC plants before construction of the Pedricktown plant, and the following parameters were found typical:

|  | (mg./l.) |
|---|---|
| 5-Day biochemical oxygen demand (BOD$_5$) | 720 |
| Chemical oxygen demand (COD) | 1285 |
| Total solids (TS) | 2000 |
| Suspended solids (SS) | 1000 |
| Volatile suspended solids (VSS) | 950 |

It was recognized, too, that these figures were highly variable due to process fluctuations.

The waste treatment plant was designed so that discharged effluent conformed to regulations stipulated by the New Jersey Department of Health and the Delaware River Basin Commission (DRBC). Chief among these requirements were:

• A minimum of secondary treatment must be provided, regardless of the quality objective for the affected stream.

• Discharges cannot contain "more than negligible" amounts of floating material, suspended matter which will settle to form sludge, toxic substances, or substances (or organisms) that produce color, taste, odor of the water, or taint fish or shellfish flesh.

• BOD reduction must be at least 85% (30-day average).

Goodrich therefore aimed at 90% reduction of ultimate BOD under winter conditions (13° C.) and based design criteria on a waste water flow of 800,000 gallons per day (g.p.d.) and the effluent characteristics listed above.

Two years were spent in research and process design for the final process now on stream at Pedricktown. Goodrich received technical assistance from Roy F. Weston and Associates (West Chester, Pa.).

It was decided early in the game that waste equalization facilities would be needed because variations in organic loading (the quality of the waste water) and in hydraulic loading (the actual volumetric flow rate) were expected. Research also showed that a preliminary step for latex solids removal would have to be included, since the presence of such solids materially reduces the efficiency of biological treatment. The waste was otherwise found to be nontoxic to a biological treatment system.

### Alternatives rejected

Activated carbon adsorption was considered as a means of reducing BOD, but adsorption capacities proved poor, a factor attributed by Goodrich engineers to the presence of long-chain organic soap molecules in the waste. Contact stabilization and trickling filter systems, too, were considered but rejected. The method chosen was a completely mixed activated sludge system, which uses a biological floc similar to that conventionally used for domestic sewage treatment. (In fact, the seed material with which the Pedricktown treatment plant started up was obtained from a municipal plant at Haddonfield, N. J.)

### Primary treatment

Although the PVC plant at Pedricktown is not yet on stream at full capacity, the waste water treatment plant is completely built and is at present treating an estimated 250,000 g.p.d.

Waste water from the polymerization process units enters first a 950,000 gal.-capacity stirred equalization tank, sufficient in size to provide one day's detention time at design throughput rate. The first treatment step (*see*

**Biological.** *Two 178,000-gal. aeration tanks treat* PVC *waste water at Goodrich* PVC *plant*

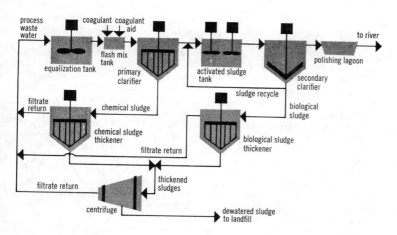

Some of this sludge is recycled to the aeration tank and the excess is thickened in a separate tank. Overflow from the secondary clarifier runs into a polishing lagoon which also receives storm runoff from the plant. Although some additional BOD and suspended solids removal may be expected in the lagoon, the secondary clarifier effluent is fully treated, and the lagoon discharges directly into the nearby tributary of the Delaware.

**Solids disposal**

Settled sludge from both chemical sludge and biological sludge thickeners is centrifuged and dewatered to about 45–50% solids. The dewatered sludge is inert and is trucked to a landfill, while the filtrate is returned to the equalization tank.

The plant is working well. Representatives of several other PVC-producing firms have studied the Goodrich treatment system and FWQA officials hope some of these firms may be able to adopt similar systems. Goodrich aims eventually at 100% reuse of its water supply but further treatment of the lagoon effluent will be necessary to meet its stringent requirements for process water. Even so, the present secondary treatment system has proved capable of removing 93–98% of the BOD from the PVC plant waste water.

*flow sheet*) is the addition of coagulant and coagulant aid in a two-compartment flash mixing tank. The ferric chloride coagulant is added to the waste, at a dosage of about 150 p.p.m., before the coagulant aid—a polyelectrolyte, either Calgon 227 or Nalco 670—is mixed in at a dosage of 1 p.p.m. About two minutes detention time is provided for mixing in each compartment. The waste then flows to the primary clarifier in which the flocculated solids settle out. This "chemical sludge" is drawn off from the bottom of the primary clarifier and is thickened (from 2.5% to 8% solids) in a separate tank.

**Secondary treatment**

Clean supernatant from the clarifier is pumped to one of two 178,000-gal. capacity biological aeration tanks in

which the oxygen-demanding components of the waste are consumed by bacteria contained in an activated sludge mass. To the supernatant are added small amounts of phosphoric acid and ammonium hydroxide which supply the sludge bacteria with nutrient phosphorus and nitrogen. (Domestic sewage normally contains enough of these essential elements to eliminate the need for nutrient addition.) Caustic soda is also added just prior to the aeration tank to keep the pH above 7.0—the $FeCl_3$ used as coagulant tends to acidify the waste, and at a pH below 7.0 filamentous organisms form at the aeration stage. Ten hours detention time is provided at this stage. Overflow from the aeration tank goes to a secondary clarifier (the plant has two) in which sludge solids—"biological sludge"—settle out.

# Raycycle cuts sulfite pulp pollution

*ITT Rayonier will change the basic pulping chemistry of its Fernandina Beach, Fla., mill and spend $38 million to avoid ocean discharge*

Reprinted from ENVIRON. SCI. TECHNOL., **6**, 596 (July 1972)

ITT Rayonier, Inc., has some good news and some bad news. First, the good news. Rayonier has developed a "major technological breakthrough" in cutting pollution from its beleaguered Fernandina Beach, Fla., sulfite chemical cellulose pulp mill. Now the bad news. It's going to cost the company a whopping $38 million to do the job.

From those two facts, there are some interesting inferences to be drawn. First, despite references to alternative pulping methods which crop up in the technical press from time to time, Rayonier obviously feels that sulfite pulping is here to stay. Such a large investment would otherwise be hard to justify. Second, the $38 million price tag gives some idea of the financial muscle necessary to clean up the pulp industry. Third, and perhaps most important, the decision points toward the industry's greater reliance on process change rather than retrofitted cleanup packages, even if the technology is still largely unproved, as is the case with Fernandina.

## Old mill

Rayonier's Fernandina mill, located near Jacksonville, Fla., is unique in several ways. It was the first mill in the world to produce chemical cellulose pulp from southern pine and it's still the only mill that extracts cellulose from southern pine by the sulfite process.

Changes have been made in the mill and its process since it was built in 1938, but the basic machinery is still the same as when it was designed. The mill, which produces 450 tpd of highly refined, bleached chemical cellulose, is a major factor in Fernandina's economy. It employs more than 450 persons, with an annual payroll of over $5 million, and an equivalent number of area residents derive income from supplying the mill with pulpwood. But along with bread and butter, the Fernandina mill contributes some 600,000 lb/day BOD to the nearby Amelia River and Atlantic Ocean. Local residents have been complaining about foam which washes back onto the beach.

By early 1970, Rayonier and the Florida Department of Air and Water Pollution Control had developed a mutually acceptable program for disposal of the Fernandina mill wastes, requiring construction of primary treatment facilities and a long ocean outfall. Rayonier estimated the cost to be in the neighborhood of $8 million. But area environmentalists objected to the outfall, and the State of Florida requested, in December 1970, that Rayonier abandon its plans for ocean disposal. Rayonier did so, and early in January 1971, the State, under section 5(b) of the Federal Water Pollution Control Act, asked the Environmental Protection Agency (EPA) to get involved in developing a more acceptable method for reducing the mill's pollution load.

A series of general and technical committee meetings followed through the spring of 1971; the meetings involved EPA, representatives of state agencies, and Rayonier. While considerable data were presented concerning conventional control techniques and costs, according to Rayonier's Jerome D. Gregoire, no definitive program emerged from the committee meetings.

After the consultations, Rayonier proposed a plan which would tack chemical recovery onto the primary treatment and ocean outfall of the original Florida-approved program. The cost of added treatment, according to Gregoire, would have been at least $23 million. Such a plan, Gregoire says, was "substantially the same" as one approved a few months earlier by EPA in the state of Washington for a similar Rayonier mill on Puget Sound. But the committee turned Rayonier down at Fernandina.

In July, EPA told Rayonier that it should seriously consider "secondary treatment" of wastes remaining after recovery. But the data on costs and effectiveness of such treatment offered to the committee, Gregoire says, were "theoretical and tentative." Rayonier retained consultants to study the feasibility of putting biological treatment facilities on the only land available at the site, a marsh adjacent to the mill.

During the studies, Gregoire says. Rayonier was monitoring "technical attitudes within the EPA and legislative developments in Washington" and the company "became greatly concerned" that the plan it was considering (primary treatment followed by recovery and aeration) would be only a temporary solution—even though it would have been a maximum effort using conventional technology and had not yet been required of any sulfite mill in the U.S. Rayonier feared that it would be faced with even more rigid control requirements in the future. In addition, the investment required would have topped $28 million. While the system would have met the BOD levels Rayonier thought EPA would require, Gregoire says, little would have been done for color, dissolved solids, and total organic carbon. That, plus the possibility of tighter BOD standards in the future, made it increasingly difficult to justify so large an investment at Fernandina when there was no guarantee that the equipment to be installed would not have to be replaced or supplemented with additional large investments, Gregoire says.

## Enter Raycycle

In December of 1971, however, the "significant breakthrough" was achieved, Gregoire says, at Rayonier's R&D facility in Whippany, N.J., where the company had been investigating the effects of process changes on the pulping procedure. Rayonier asked Florida for a month to more fully evaluate the process prior to presentation of a new proposal for Fernandina. The new process was dubbed Raycycle and laboratory testing, which included extensive testing to assure continued product quality, proved successful. Patent applications were filed and Rayonier now proposes to add Raycycle to the Fernandina Mill.

The technology of sulfite pulping includes two stages which are responsible for the major portion of wastes discharged—the digestor or cooking stage, and the hot caustic bleaching stage. Thus, a process change which removes a high proportion of liquid waste from these two stages will be most effective in controlling pollution.

Current concepts of pollution control from pulping plants call for destruction of digestor wastes by burning (with recovery) after concentration. But wastes from the hot caustic stage—which account for about one sixth of the total BOD—can't be treated that way. For those wastes, biological treatment has been the method of choice. Raycycle's advantage is that it allows concentration and burning of both digestor and bleaching waste simultaneously (see diagram). To be able to do that, Fernandina's process was switched from an ammonia-based cooking cycle to a soda-based cycle. (In sulfite pulping, either sodium or ammonium salts are heated with wood chips to dissolve lignin and carbohydrates so that wood fibers can be separated more easily.) Using sodium in the digester is then compatible with sodium in the bleach. Soda-based pulp from the digestor is drained of some of its waste liquor—but not washed—and passed directly to the hot caustic stage. After hot caustic extraction, the pulp is washed, and the wastes removed in washing are added to the digestor waste liquor. The combined wastes can then be concentrated and burned. Chemicals in the ash can be recovered.

Elimination of the washing step between cooking and bleaching saves considerable amounts of water and gives more concentrated waste. Rayonier expects to save 3–5 million gpd, depending on the type of pulp.

Another change in the process from conventional sulfite pulping processes is that caustic bleaching is moved ahead in the process cycle. The process used at Fernandina involves cooking, washing, chlorination, and bleaching. With the Raycycle process, chlorination will be delayed until after caustic bleaching, to avoid additional chemical species in the waste that would complicate recovery of materials from the burned digestor residue.

## Environmental benefits

Benefits derived from the use of Raycycle would be almost entirely environmental. It becomes economical only in comparison to the prohibitive costs of high-level secondary treatment. The investment required for Raycycle is enormous, Gregoire points out, and there is a substantial economic penalty despite the fact that chemicals and heat are recovered.

The ecological gains, however, are "striking," Gregoire says. Using Raycycle, he adds, "it becomes practical for the first time to reduce BOD to very low levels for the most refined grade of pulp." Among the environmental advantages of Raycycle are:

• reduction of BOD from 1100 lb/ton of air-dried product to 120 lb/ton of air-dried product (polishing by conventional secondary treatment could cut the figure in half again)

• reduction of total organic carbon and chemical oxygen demand by approximately 95%—that's 67% better than the maximum that could be expected with the best secondary treatment using ammonia-based cooking

• reduction of color by an estimated 90%—which beats the conventional system by 80%

• elimination of foaming from mill effluent, since foam-causing materials will be burned in the recovery furnace

• conservation of several million gallons of water each day

• reduction of air pollution since $SO_2$ and $NO_x$ emissions with Raycycle are substantially lower and easier to control

• preservation of some 200 acres of ecologically valuable marshland that would otherwise have been used for conventional aeration facilities and saving 4200 kW of electricity per day.

## There are snags

With all the apparent advantages, however, there are still some snags. Various permits are needed which depend in part on favorable public hearings, but Rayonier thinks its chances are good. And, of course, Raycycle is new and untried. The process has been successfully tested under laboratory conditions, but the practicality of the project depends on a carefully balanced economy of chemical recovery, heat, and power.

Rayonier estimates that if Raycycle gets prompt approval, the system could be in full swing by the end of 1975. That's no longer than it would take to install a conventional treatment system of similar scope, Rayonier argues. In the meantime, the company plans to build a large holding pond at the mill. The lagoon, which could be completed this year, will be large enough to store ocean-bound wastes during periods of adverse on-shore currents or pipeline breaks, minimizing the possibility of foam accumulation until Raycycle can be put into operation.  HMM

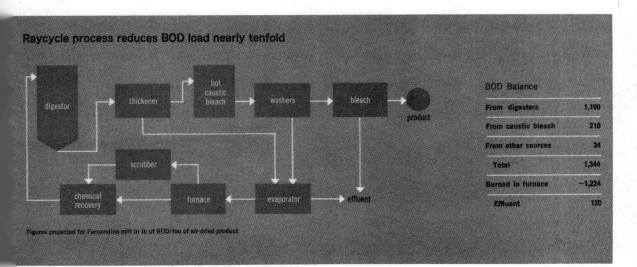

**Raycycle process reduces BOD load nearly tenfold**

digestor → thickener → hot caustic bleach → washers → bleach → product

scrubber

chemical recovery — furnace — evaporator → effluent

BOD Balance

| | |
|---|---|
| From digestors | 1,100 |
| From caustic bleach | 210 |
| From other sources | 34 |
| Total | 1,344 |
| Burned in furnace | −1,224 |
| Effluent | 120 |

Figures projected for Fernandina mill in lb of BOD/ton of air-dried product

*By tackling head-on the most pollution-prone steps in the leather tanning process, a midwestern firm solves not only its own problems, but those of its customers*

Reprinted from ENVIRON. SCI. TECHNOL., **6**, 594 (July 1972)

# Plant cuts water use, eases pollution

Today's tanning industry in the U.S. is concentrated in about 250 plants, amounts to a $5 billion enterprise, and employs about 27,000 workers, according to a federal profile of the leather tanning and finishing industry. The profile further points out that the industry's two most burdensome processes from a pollution-laden point of view have traditionally been the unhairing operation in the beamhouse and the tanning operation in the tanhouse (ES&T, April 1968, p 261).

To be sure, the tanning industry is an archaic one with many old plants. Nevertheless, the newest plant in the industry is pollution-free and is equipped with the best available pollution control technology. The new plant of the Blueside Co. at St. Joseph, Mo., handles both of the burdensome steps (unhairing and chrome tanning). It will accommodate about 40,000 hides per week, or about 2 million hides annually, when the plant is operating at full capacity next summer. The total U.S. industry's tanning capacity is 35 million hides annually.

The new plant produces a "semi-finished" product, which in the trade indicates tanning to the wet blue chrome stage, and from whence Blueside Co.

**Blueside's Lyon**
*And a more uniform product*

derives its name. The leather product is blue in color from the chrome salts used in the tanning operation.

Traditionally, many leather tanning companies are family-owned operations. Lee Lyon, president of the Blueside Co., is a fourth-generation Lyon whose great grandfather founded M. Lyon & Co., a Kansas City–based tanning operation in 1870. A Harvard graduate and sailing enthusiast of the C-Scow class boat, Lyon is the man with fore-

sight whose company has incorporated the best available technology in the Blueside operation. Blueside is a totally new company and not a subsidiary of M. Lyon & Co.

Lyon is well-known in the tanning trade; he is credited with the development of Lycoil, the process by which fleshings and trimmings are converted into profitable oils and protein feeds. He is also credited with the labor-saving, precut, preknotted and color-coded Redi-(Hide)-Rope and is a past president of the National Hide Association.

The Blueside plant is the only "semi-finisher" in existence today that operates in a big way. It is the first volume producer of wet blue hides for general sale in the U.S. "Our semifinished product represents a $3.50 value added to the hide," Lyon says. (The current price of cattle hides is $17–18 each.)

### New technology

The new plant was built in May 1971 and started operation in August of the same year. Earlier, a design study of the tanning operation waste stream was performed for Blueside by the Boston-based consulting engineering firm of Camp Dresser & McKee. The engineering firm recommended:

---

## Tanning is just one step along the way from farm animal to final product

Hides from slaughtered animals are dehaired and chrome tanned . . .

semi-finished "blue stock" is wrung prior to shipping . . .

• blending all waste water effluents together and allowing settling. At the St. Joseph plant the holding basin is 60 ft in diam and 15–20 ft deep

• removing solids from both the top and bottom of the holding tank

• acidulating the effluent and lowering the pH to 5.5 (from a pH of 9.0 or more resulting from the caustic used in the unhairing process)

• aerating the effluent to remove sulfides

• • treating the sulfide-laden airstream with caustic to recover the sulfide as sodium sulfide for reuse in unhairing.

On this last step, Blueside currently has an active grant from the EPA to demonstrate the feasibility of such removal; the EPA project officer is W. E. (Pete) Banks, of the Region VII office.

When complete in 1973, the new plant will represent a $3 million investment. In addition, all waste water streams are available for individual recycling, which is not the case in some old plants. In fact, there is the strong suspicion that no one even knows where all the waste outfalls are in some old plants.

All processes in the new Blueside plant are continuous, fully automated, and monitored. In this way, Lyon says, Blueside provides a "uniform" product, one on which "final" tanners can rely in their processing (i.e., they don't have to change their processes to accommodate the differences from batch to batch).

### New vs. old

Without question, the one main aspect of the new trend at Blueside is a significant reduction in the volume of water that is used in the tanning operation. Lyon says, "We use 1.0–1.5 gallons of water per pound of hide throughout our entire processes." In contrast, some tanneries may use 40–50 gallons of water per pound of hide to effect the same degree of tanning.

This reduced water usage is achieved by replacing archaic vats-and-paddle mixing operations with better mixing facilities. For example, hides are unhaired in a concrete truck–like mixer. In this way, hides can be brought into intimate contact with the processing chemicals using very low floats compared to those used in paddle vats.

Basically, Blueside provides a real service to other tanners who finish its "semifinished" product. The "final" tanners are also faced with pollution control updating requirements, but the pollution burden of their operations is considerably lessened once the unhairing and chrome tanning has been done.

Lyon figures that the location of his plant is another advantage to his operation. He argues that Blueside is in a good geographic location; it is intermediate between the cattle-producing areas of the U.S.—Texas, Iowa, Nebraska, Colorado, and Kansas—and final leather tanners in the Great Lakes area and on the East Coast. Thus, Blueside provides the "semifinished" processing on the way to eastern markets and "final" tanners.

Blueside is innovative in yet another way. The plant operates round the clock, seven days a week, but operators and maintenance crews are in the plant six days; they perform a three-day work week—three 12-hr shifts with four days off! SSM

. . . to final finishers who make the product the consumer buys

# POLLUTION CONTROL in the

*Reprinted from* ENVIRON. SCI. TECHNOL., **5,** 1004 (October 1971)

The majority of the iron and steel industry in the United States is centered in integrated facilities of large corporate enterprises, which are numbered among the largest industrial corporations in the country. A comparatively small plant in this industry represents a very large industrial complex. Generalizations about steelmaking operations are difficult because exceptions can be found in every mill. Steel mills range from new, modern facilities built within the past several years to older, marginal facilities built early in the century. Production units are quite large and require great quantities of water for both process and cooling purposes. Water use in the industry is the highest of any manufacturing industry, amounting to an average of 40,000 gal/ton of finished steel for all purposes.

## Manufacturing processes

Iron and steel manufacturing operations may be grouped as coke production, pig iron manufacture, steelmaking processes, rolling mill operations, and finishing operations. A single mill generally does not incorporate all combinations and variations of these operations that are possible. All operations produce waste water effluents, atmospheric emissions, and solid wastes, but the quantities and characteristics from each source vary greatly. Average water uses in the various departments of an integrated steel mill are shown on the next page. A medium-sized mill may discharge 100 million gallons of water per day. Atmospheric emissions and solid waste

generation are of similar magnitude; for example, a blast furnace may use air at the rate of more than 100,-000 ft³/min and produce slag in excess of 1,000 tons/day.

## Coke production

Most large steel mills operate by-product coke plants which produce the metallurgical coke used in pig iron production. Coke oven gas is a by-product and is used as a fuel within the mill. Crude tar, light oil, and ammonia are the other by-products of the coking operation, and these are further processed or sold depending on plant design and the marketability of specific products. Competition from the petrochemical industry has greatly reduced the profit and, hence, the incentive to manufacture by-products other than as necessary.

Principal air pollution problems from coke production are sulfur dioxide generation from combustion of coke oven gas, emissions from ovens during charging and pushing and from door and lid leaks, and emissions from waste water quenching of the incandescent coke. State-of-the-art abatement measures include removing hydrogen sulfide from the gas, oven lid and door maintenance, baffling quench towers, using clean water for quenching, and regulating coking times. Other methods under development here or already in use abroad include negative oven pressures during charging, larry car scrubbers, hood arrangements with collection ducts and gas cleaning equipment, and stack gas cleaning of sulfur dioxide.

Principal water pollution potentials in the coke plant operation are in ammonia still wastes and light oil decanter wastes which average about 44 gal/ton of coal carbonized and contain phenols, ammonia, cyanides, chlorides, and sulfur compounds. Using contaminated waste waters for coke quenching effectively prevents water pollution as long as there is no overflow from the quench system. This practice, however, results in air pollution and severe corrosion on nearby steel structures. Abatement measures include biological treatment either on site or by cotreatment with municipal sewage, chemical oxidation, and carbon adsorption. Coupled with water reuse reducing effluent volumes to a minimum, these measures can eliminate potential water pollution.

## Blast furnace operations

Blast furnaces are operated primarily to produce iron as hot metal for subsequent steelmaking processes but also produce pig iron, silvery pig iron, and ferroalloys. Sinter plants are typically operated as parts of blast furnace departments to agglomerate fine ores, blast furnace flue dust, and mill scale as part of the blast furnace burden. Slag quenching operations produce granulated or expanded slag depending on the methods used.

These various operations can produce particulate emissions in blast furnace gas, from handling blast furnace burden materials including furnace charging, from opening blast furnace pressure release valves due to slips, and from materials handling and agglomeration operations at the sinter

*Steelmaking operations produce pollution at almost
every step, but it can be controlled by proper treatment
or by conserving and reusing materials*

# Steel Industry

**Henry C. Bramer**

*Datagraphics, Inc.
Pittsburgh, Pennsylvania 15232*

plant. Hydrogen sulfide and some sulfur dioxide are generated in slag quenching. Venturi scrubbers or electrostatic precipitators clean blast furnace flue gas to less than 0.01 gr/ft³ so that major emissions are due to venting or to materials handling in the blast furnace operation. Proper hooding, venturi scrubbers, and baghouses can effectively control particulate emissions from sinter plants. Control of slips can greatly reduce blast furnace venting, and combinations of hooding arrangements and chemical additives are the most likely solutions to controlling particulate emissions from materials handling and sulfur compound emissions from slag quenching. So-called dry-quenching or using minimal quantities of water reduces the emission potential from slag operations.

Water pollution problems in the blast furnace department result primarily from gas cleaning with wet washers. Blast furnace gas-washer water contains suspended solids, cyanides, phenols, and ammonia. If the gas-washer water is completely recycled and any blowdown disposed of in dry slag quenching or on ore piles, this pollution source can be eliminated. Cyanides, ammonia, and phenols in gas-washer water result primarily from waste water quenching of coke, but some cyanide is synthesized in the blast furnace in any case.

The major solid waste problem in these operations is producing more slag than can be sold for road building. Huge slag dumps continue to build up near most steel mills. Solids recovered from blast furnace gas,

either wet or dry, or from the sinter plant are reusable as blast furnace burden; such wastes are only a disposal problem when a mill has no sintering facilities.

### Steelmaking processes

The major steelmaking processes are the basic oxygen process (BOP), electric furnaces, and open hearth furnaces which are now generally oxygen lanced. The air pollution potential of all of these processes is in the fume which is generated from the furnaces themselves and during molten metal transfer operations. Control facilities

on basic oxygen and open hearth furnaces are generally either venturi scrubbers or electrostatic precipitators while many new electric furnace installations use baghouses. Emissions of carbonaceous material known as kish, generated during molten metal transfer operations, can be controlled by evacuation hoods and gas-cleaning equipment such as baghouses. Hood designs are critically important in furnace installations; proper hood design can greatly reduce air volumes, permitting greater efficiencies and/or smaller equipment.

Water pollution problems in steel-

Water use in an integrated steel mill

1000 gal./ton

Total 40,000 gal/ton

12

25%   25%   20%   12.5%   12.5%   5%

6

Blast furnace   Open hearths   Coke plants   Hot mills   Finishing mills   Sanitary, boiler, other uses

making result from wet gas-cleaning methods and consist primarily of suspended solids. Recirculating such waste waters with cooling towers is the most generally satisfactory solution. Treatment of waste waters from oxygen steelmaking processes requires chemical coagulants or similar methods and can seldom remove all particulate matter. Magnetic agglomeration has proved to be particularly effective in treating BOP waste waters.

Recovered materials from basic oxygen process emissions represent one of the industry's most troublesome solid waste problems. High concentrations of zinc, originating in the scrap metal charge, are generally regarded as making recovered material unusable

for reuse. Leaching with acid may be a method of removing unwanted zinc, but some recent experiments in BOP operation indicate that unsuitability of the material may be more a matter of conjecture than fact.

### Rolling mill operations

Rolling mill operations start either with ingots molded from steelmaking processes and subsequently reduced to blooms, slabs, or billets or with slab production by continuous casting. An ever-increasing quantity of steel is processed by the latter, newer process. The semifinished steel is prepared for finishing in a number of ways—chipping, grinding, and scarfing—all of which remove surface defects. Scarf-

ing refers to using oxygen torches to remove surface defects and is accomplished by hand on cold steel or mechanically on hot steel as part of the rolling operation.

Finishing mills include a wide variety of mills intended to produce specified steel products—plate mills, mills for rolling rails and structural shapes, bar mills, wire mills, tube mills, and continuous strip mills. Most hot-rolled strip is pickled in acid baths, usually in continuous units, to remove surface oxides. Many other products, such as wire, bars, and pipe, are often pickled in batch-type operations.

The primary air pollution source in these operations is scarfing, hot scarf-

# Major areas of pollution in steelmaking processes

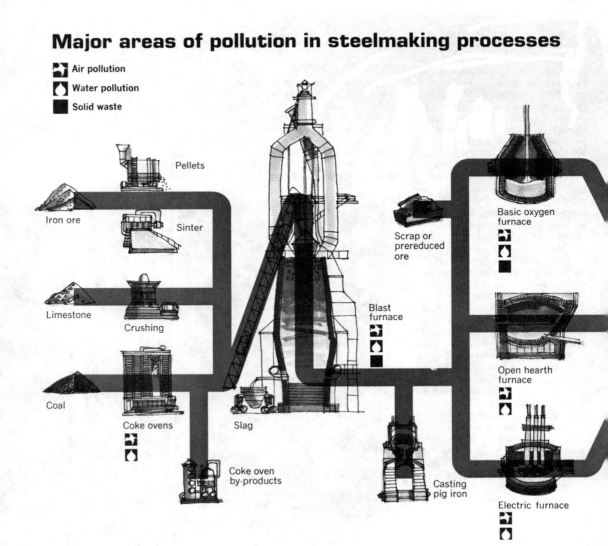

ing in particular. Airborne particles are extremely fine and difficult to remove, and when wet washers are used, a water pollution potential, of course, results. The principal solid waste problem is in sludge resulting from lime neutralization of spent pickle liquor. Lagoons full of this material, which never dries, is a major problem at many mills. Although crop ends and scale amount to large tonnages of solid material, it is all reused within the mill as internal scrap or is sintered for recycle.

Major water pollution potentials from rolling mills are suspended particles of waterborne scale, lubricating oils, spent pickle liquor, and pickling rinse water. As previously indicated,

strong pickle liquors are often neutralized with lime, and rinse waters may be treated in the same manner. Other alternatives include contract hauling, deep well disposal, and pickle liquor regeneration, which is more economically attractive and technically proved when hydrochloric acid is used instead of sulfuric. (Using hydrochloric acid is becoming the general practice.) Contract hauling only removes the problem outside the mill's boundaries, and deep well disposal is coming into disfavor in many states. Pickling rinse water treatment is analogous to acid mine drainage treatment.

Most scale and oil produced in rolling mills is recovered in crude settling chambers called scale pits. (These

chambers have always been used to prevent sewer clogging and are similar to the grit chambers in municipal sewage treatment plants.) Recirculation can be practiced to a high degree in rolling mills, and effective effluent treatments include chemical coagulation, magnetic agglomeration, and deep bed filtration. Primary oil-removal equipment, such as belt skimmers, is necessary to effectively treat rolling mill waste waters.

### Finishing operations

Finishing operations primarily consist of cold reduction, tin plating, galvanizing, chrome plating, coating, tempering, and polishing. Air pollution problems associated with these

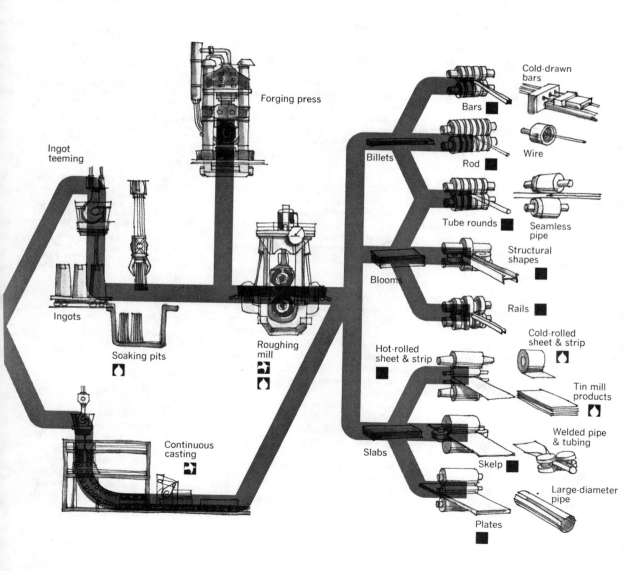

**BOP.** *These furnaces may emit air and water pollutants as well as solid wastes*

operations are negligible except for the practice of open-burning recovered oils. Solid waste problems are associated with internally generated scrap and with sludges which may result from tank cleanouts or precipitates from waste water treatments. Scrap is essentially all reused, but can result in additional waste water problems, such as reclaiming tin-plated or galvanized steel.

One of the most difficult waste water treatment problems is treating emulsified oil from cold rolling. Proprietary mineral oils have all but replaced palm oil in cold rolling, and because they are formulated to yield stable emulsions, their treatment is difficult. Since treatment is usually less than satisfactory, the highest possible degree of recirculation and reuse is necessary with these waste waters. Chemical treatment, such as iron salts and lime followed by air flotation, is a common treatment method. Various emulsion-breaking agents are used, and magnetic separators are effective in cleaning emulsions for reuse. On-site reclamation of these oils by outside contractors is a common practice.

Tin plating, chrome plating, and galvanizing waste waters present generally the same problems as do metal finishing wastes in other industries, except that the volumes are usually greater than commonly encountered in other industries. Other finishing operations do not produce significant waste water problems.

### Conservation and reuse

Conservation and reuse practices are quite productive waste control methods in the steel industry because of the sheer magnitude of the quantities involved. Insofar as coke oven gas, blast furnace gas, flue dust, slag, soluble oils, mill scale, some coke oven chemicals, and scrap are concerned, the industry has mostly practiced conservation and reuse because of obvious economic incentives. In most steel mills, however, water has neither been conserved nor reused to a significant degree, nor has reclamation been practiced for such cases as pickling acids, plating solutions, basic oxygen process dust, lubricating oils, and many coke oven chemicals such as sulfur. Air pollution control equipment is frequently required to handle excessive volumes of gas or air, particularly when operations are hooded to prevent emissions. Solid wastes, such as neutralized pickle liquor sludge or BOP dust, accumulate despite the availability of alternatives or due to the lack of simple technology development.

When the need arises, the steel industry has adopted methods which will result in effective waste control. Where water is plentiful, most mills use water on a once-through basis and in excessive amounts for most purposes. Where water is scarce, or when legal requirements have been imposed, large steel mills have reduced waste water effluents to as little as 1,000 gal/ingot ton, a fraction of the industry average. Pickle liquor recovery processes have been developed and demonstrated, as have methods for recovering all coke plant chemicals.

The steel industry has experienced great changes over the past decade as production costs have increased and vigorous competition has appeared from oversea steel mills and alternative domestic products. Many of the new production facilities, such as the basic oxygen process, larger electric furnaces, and high-speed strip mills, help reduce unit costs by increasing production, but the newer facilities result in increased pollution potentials.

Waste control problems in the steel industry are characterized by large-sized equipment, high flow rates, and large quantities of materials. Conservation and reuse are particularly effective measures for waste control throughout the industry, resulting in lessened potential pollution loads and in more efficient treatment facility operation. The industry has shown the capability to develop and apply new technology; waste control can be effectively accomplished by vigorous application of this already proved capability.

### Additional reading

Bramer, H. C., "Iron and Steel," in "Industrial Waste Water Control," Academic Press, New York, N.Y., 1965.
Bramer, H. C., "New Aspects of Water Pollution Control in the Steel Industry," paper presented at the Ohio Water Pollution Control Conference, Columbus, Ohio, 1966.
Bramer, H. C., Gadd, W. A., "Magnetic Flocculation of Steel Mill Wastes," paper presented at the Purdue Industrial Waste Conference, Purdue, Ill., May 1970.
Hoak, R. D., "Water Resources and the Steel Industry," *Iron and Steel Eng.*, May 1964.
Department of Health, Education and Welfare, Public Health Service, Division of Air Pollution, "Air Pollution Aspects of the Iron and Steel Industry," 1961.
Federal Water Pollution Control Administration, "The Cost of Clean Water," Vol III, Industrial Waste Profile No. 1, "Blast Furnace and Steel Mills," 1967.

**Henry C. Bramer** *is president of Datagraphics, Inc. While attending the University of Pittsburgh, he received his BS and MS in chemical engineering and his PhD in economics. Dr. Bramer has specialized in industrial pollution control, especially in the steel and chemical industries.*

# Water uses and wastes in the textile industry

*Reprinted from* ENVIRON. SCI. TECHNOL., **6**, 36 (January 1972)

*Milling processes use a tremendous amount of water and
may add a multitude of chemicals to waste streams*

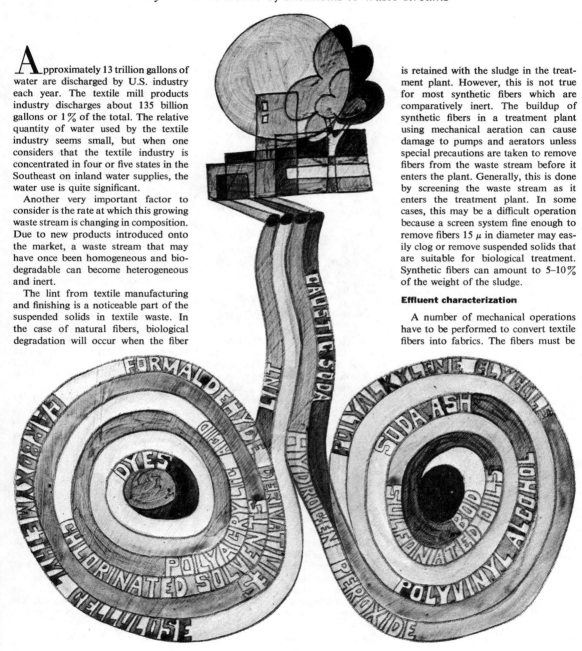

Approximately 13 trillion gallons of water are discharged by U.S. industry each year. The textile mill products industry discharges about 135 billion gallons or 1% of the total. The relative quantity of water used by the textile industry seems small, but when one considers that the textile industry is concentrated in four or five states in the Southeast on inland water supplies, the water use is quite significant.

Another very important factor to consider is the rate at which this growing waste stream is changing in composition. Due to new products introduced onto the market, a waste stream that may have once been homogeneous and biodegradable can become heterogeneous and inert.

The lint from textile manufacturing and finishing is a noticeable part of the suspended solids in textile waste. In the case of natural fibers, biological degradation will occur when the fiber is retained with the sludge in the treatment plant. However, this is not true for most synthetic fibers which are comparatively inert. The buildup of synthetic fibers in a treatment plant using mechanical aeration can cause damage to pumps and aerators unless special precautions are taken to remove fibers from the waste stream before it enters the plant. Generally, this is done by screening the waste stream as it enters the treatment plant. In some cases, this may be a difficult operation because a screen system fine enough to remove fibers 15 $\mu$ in diameter may easily clog or remove suspended solids that are suitable for biological treatment. Synthetic fibers can amount to 5–10% of the weight of the sludge.

### Effluent characterization

A number of mechanical operations have to be performed to convert textile fibers into fabrics. The fibers must be

John J. Porter, Donald W. Lyons,
and William F. Nolan

*Clemson University*
*Clemson, S.C. 29631*

combined into yarns and then the yarns into fabrics. After fabrics are manufactured, they are subjected to several wet processes collectively known as finishing, and it is in these finishing operations that the major waste effluents are produced.

**Cotton.** The consumption of cotton fibers by textile mills in the United States exceeds that of any other single fiber (Figure 1). Slashing is the first process in which liquid treatment is involved. In this process, warp yarns are coated with "sizing" to give them abrasive resistance to withstand the pressures exerted on them during the weaving operation. The principal slashing polymer used before 1960 was starch which was easily degraded biologically and should present no problem to the conventional waste treatment plant other than BOD loading.

Development of many synthetic fibers in the 1950's and their use in blended fabrics created the need for new sizes which were more compatible with the hydrophobic fibers. Some of those which are still in use are polyvinyl alcohol (PVA), carboxymethyl cellulose (CMC), and polyacrylic acid.

If an average size concentration of 10% is assumed to be present on woven fabrics, which constitute 70% to 80% of the fabrics produced, approximately 400 million lb of size per year are currently entering textile finishing waste streams. Since PVA and CMC are resistant to biological degradation, conventional treatment methods would not be expected to alter their chemical structure. While the polymer may be partially removed from the waste water by adsorption on the sludge, it is questionable whether this is an effective method of treatment.

The operation of desizing removes the substance applied to the yarns in the slashing operation, by hydrolyzing the size into a soluble form.

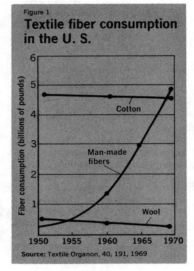

Figure 1
**Textile fiber consumption in the U. S.**

Source: Textile Organon, 40, 191, 1969

There are two methods of desizing—acid desizing and enzyme desizing. In acid desizing, the fabric is soaked in a solution of sulfuric acid, and in enzyme desizing, complex organic compounds produced from natural products or malt extracts are used to solubilize the size. Due to the unstable nature of these organic compounds, the whole bath must be discarded after each batch. Desizing contributes the largest BOD of all cotton finishing processes—about 45%.

Scouring follows desizing. In this process, cotton wax and other noncellulosic components of the cotton are removed by hot alkaline detergents or soap solutions. In most modern plants, scouring is done in conjunction with desizing rather than as a separate operation. Caustic soda and soda ash along with soaps and synthetic detergents and inorganic reagents are used to remove the noncellulosic impurities. The waste liquor will have a 0.3% alkaline concentration.

A few of the major chemical manufacturers are now offering solvent processes to the textile industry for scouring where little water is used. In these cases, nonflammable chlorinated solvents are used, and the projected solvent recovery is between 90% and 97%. However, nearly 1 ton of solvent per day per range will reach the atmosphere or waste stream.

Bleaching, the next process, removes the natural yellowish coloring of the cotton fiber and renders it white. The three bleaches most commonly used for cotton are sodium hypochlorite, hydrogen peroxide, and sodium chlorite. The bleaching process contributes the lowest BOD for cotton finishing.

Mercerization gives increased luster to cotton fabrics, but more importantly, imparts increased dye affinity and tensile strength to the fabric. The process uses sodium hydroxide, water, and an acid wash. The effluent from the overall process has a high pH and also a high alkalinity if the caustic material is not recovered. After mercerizing, the goods are sent to the dye house or color shop. The dyeing process is carried out in an aqueous bath with pH variations of 4 to 12.

In the color shop, the goods are printed with colored designs or patterns. The color is imparted to the fabric from rolling machines which contain the printing paste. This paste contains dye, thickener, hygroscopic substances, dyeing assistants, water, and other chemicals. The pollution load from the color shop comes mainly from the wash-down rinses (used to clean the equipment in the shop) and the cloth rinsings and is rather low in both volume and BOD. When a mill does both printing and dyeing, the BOD contribution of the combined processes is 17%, and the total BOD load comes from the process chemicals used.

Dyes have to be more and more

## Table I. Pollution effects of cotton processing wastes

| Process | pH | BOD | Total solids |
|---|---|---|---|
| Slashing, sizing yarn | 7.0–9.5 | 620–2,500 | 8,500–22,600 |
| Desizing | | 1,700–5,200 | 16,000–32,000 |
| Kiering | 10–13 | 680–2,900 | 7,600–17,400 |
| Scouring | | 50–110 | |
| Bleaching (range) | 8.5–9.6 | 90–1,700 | 2,300–14,400 |
| Mercerizing | 5.5–9.5 | 45–65 | 600–1,900 |
| Dyeing: | | | |
|   Aniline Black | | 40–55 | 600–1,200 |
|   Basic | 6.0–7.5 | 100–200 | 500–800 |
|   Developed Colors | 5–10 | 75–200 | 2,900–8,200 |
|   Direct | 6.5–7.6 | 220–600 | 2,200–14,000 |
|   Naphthol | 5–10 | 15–675 | 4,500–10,700 |
|   Sulfur | 8–10 | 11–1,800 | 4,200–14,100 |
|   Vats | 5–10 | 125–1,500 | 1,700–7,400 |

resistant to ozone, nitric oxides, light, hydrolysis, and other degradative environments to capture a valuable portion of the commercial market. It is not surprising, therefore, that studies on the biological degradation of dyestuffs yield negative results when dyes are designed to resist this type of treatment. The range of pollution loads of the various cotton textile wet-processing operations are listed in Table I.

Federal Water Pollution Control Administration estimates for BOD, suspended solids, total dissolved solids, and volume of waste water for 1970–82 are shown in Figure 2. The gradual decrease of the gross pollution load in coming years is based on these assumptions: new machinery, which tends to produce less pollution per unit of cloth due to water reuse and countercurrent flow designs; trends in process modification, new chemical manufacture, and better housekeeping will continue; a larger percentage of the wastes will be treated due to increased efficiency of treatment facilities; and increased state, local, and federal pressure.

**Wool.** Wool fiber consumption is the smallest of the three groups, and the trend seems to be toward less demand in the future on a percentage basis. Scouring is the first wet process that wool fibers receive. This process removes all the natural and acquired impurities from the woolen fibers. For every pound of scoured woolen fiber, 1 1/2 lb of waste impurities are produced; in other words, wool scouring produces one of the strongest industrial wastes in terms of BOD by contributing 55–75% of the total BOD load in wool finishing.

Depending on whether the fabric is classified as woolen or worsted, the remaining wet processes will vary. Burr picking and carbonizing are steps to remove any vegetable matter remaining in the wool after scouring and before dyeing.

The volume of waste water generated by dyeing, either stock or piece goods, is large and highly colored, and many of the chemicals used are toxic. The BOD load is contributed by the process chemicals used, and represents 1–5% of the mill's total BOD load. Although the following mixing and oiling step does not contribute directly to the waste water volume, the oil finds its way into the waste stream through washing. The percentage contribution to total BOD load of this process varies with the type of oil used.

Fulling or felting is another operation that does not directly contribute to the waste stream, until the process chemicals are washed out of the fabric. It is estimated that 10–25% of the fulled cloth's weight is composed of process chemicals that will be washed out in this process and discarded.

Wool washing after fulling is the

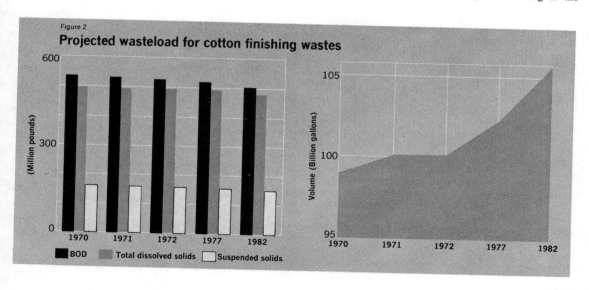

Figure 2

## Projected wasteload for cotton finishing wastes

BOD • Total dissolved solids • Suspended solids

second largest source of BOD, contributing 20–35% of the total. This process consumes 40,000–100,000 gal of water for each 1000 lb of wool fabric, and analyses show that wool, once thoroughly washed, will produce little or no BOD on being rewashed. Carbonizing the fabric or stock of fibers (with strong acid to remove cellulose impurities followed by a soda ash wash) contributes less than 1% of the total BOD load.

Wool is bleached if white fabric or very light shades of colored cloth are required; however, the amount of wool fabric bleached is rather small. With hydrogen peroxide and sulfur dioxide, bleaching the BOD contribution is usually less than 0.5%, and optical brighteners, which use organic compounds, contribute about 1% of the total BOD. In processing woolen fibers, five sources of pollution load exist—scouring, dyeing, washing after fulling, neutralizing after carbonizing, and bleaching with optical brighteners. The average values of the pollution load of each of these processes is shown in Table II. The waste water volume for woolen finishing wastes is shown in Figure 3.

**Synthetics.** This category of textile fibers has two broad classifications: cellulosic and noncellulosic fibers. The two major cellulosic fibers are rayon and cellulose acetate; the major noncellulosic fibers are nylon, polyester, acrylics, and modacrylics. Different processes to produce synthetic fibers result in varying pollution loads (Table III).

The first process in which synthetic fibers are subjected to an aqueous treatment is stock dyeing (unless the fabric is to be piece dyed). When stock dyeing is used, the liquid waste discharge will vary from about 8 to 15 times the weight of the fibers dyed. Due to the low-moisture regain of the synthetic fiber, static electricity is a problem during processing. To minimize this problem, antistatic oils (polyvinyl alcohol, styrene-based resins, polyalkylene glycols, gelatin, polyacrylic acid, and polyvinyl acetate) are applied to the yarns and become a source of water

## Table II. Pollution loads of wool wet processes

| Process | pH | BOD, ppm | Total solids, ppm |
|---|---|---|---|
| Scouring | 9.0–10.4 | 30,000–40,000 | 1,129–64,448 |
| Dyeing | 4.8–8.0 | 380–2,200 | 3,855–8,315 |
| Washing | 7.3–10.3 | 4,000–11,455 | 4,830–19,267 |
| Neutralization | 1.9–9.0 | 28 | 1,241–4,830 |
| Bleaching | 6.0 | 390 | 908 |

## Table III. Pollution load of synthetic wet fiber processes

| Process | Fiber | pH | BOD, ppm | Total solids, ppm |
|---|---|---|---|---|
| Scour | Nylon | 10.4 | 1360 | 1882 |
| | Acrylic/modacrylic | 9.7 | 2190 | 1874 |
| | Polyester | | 500–800 | |
| Scour & dye | Rayon | 8.5 | 2832 | 3334 |
| | Acetate | 9.3 | 2000 | 1778 |
| Dye | Nylon | 8.4 | 368 | 641 |
| | Acrylic/modacrylic | 1.5–3.7 | 175–2000 | 833–1968 |
| | Polyester | | 480–27,000 | |
| Salt bath | Rayon | 6.8 | 58 | 4890 |
| Final scour | Acrylic/modacrylic | 7.1 | 668 | 1191 |
| | Polyester | | 650 | |

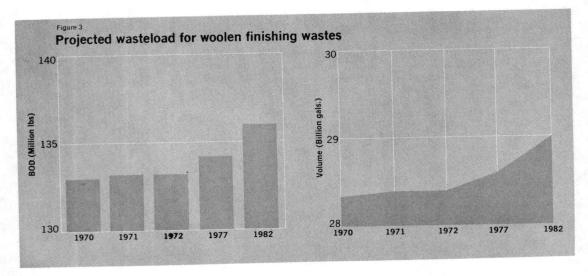

Figure 3
## Projected wasteload for woolen finishing wastes

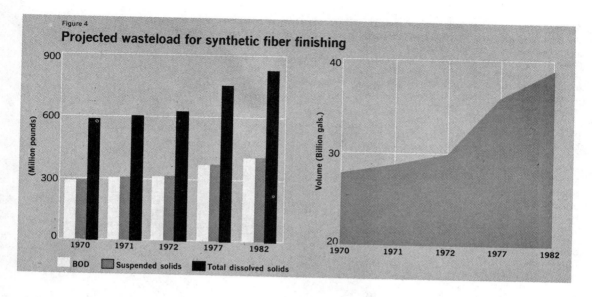

Figure 4
**Projected wasteload for synthetic fiber finishing**

BOD ■ Suspended solids ■ Total dissolved solids

pollution when they are removed from the fabrics during scouring.

Since the manufacture of synthetic fibers can be well controlled, chemical impurities are relatively absent in these fibers; and if synthetics are bleached, the process is not normally a source of organic or suspended solids pollution.

In finishing rayon, one of the synthetic fibers, scouring and dyeing are usually done concurrently in a single bath. If scouring and dyeing are the only finishing processes given rayon fabrics, an equalized effluent of 1445 ppm BOD and 2000–6000 ppm salt contained in approximately 5000 gal of water for each 1000 lb of fabric processed will be produced.

For acetate fibers, the wastes from scour and dye baths average 2000 ppm and 50 lb of BOD for each 1000 lb of acetate fabric. Typically, the bath contains antistatic lubricant desizing wastes, which contributes 40–50% of the BOD load; the sulfonated oil swelling agent, which accounts for 30–40% of the BOD load; the aliphatic ester swelling agent, which amounts to 10–20% of the BOD discharged; and the softener which has negligible BOD content.

These processes produce a composite waste of 666 ppm of BOD for each 1000 lb processed; the volume of water required to treat this amount of cloth averages 9000 gal. If bleaching is substituted for dyeing, the BOD of the discharge of the scouring and bleaching bath is approximately 750 ppm.

Nylon differs from other synthetics in that approximately 1% of the fiber dissolves when scoured. Soap and soda

ash are used in the scouring process which averages 1360 ppm and 34 lb of BOD for each 1000 lb of cloth processed. The substances present in the bath contribute the following percentages to the total BOD of the bath; antistatic-sizing compound (40–50%), soap (40–50%), and fatty esters (10–20%).

When nylon is dyed, sulfonated oils are used as dye dispersants. These dye dispersants contribute practically all of the process's BOD, which amounts to an average of 600 ppm and 15 lb for each 1000 lb of cloth dyed. However, the BOD contribution of scouring is roughly 80%, the remaining BOD being contributed by the dyeing process.

Another group of synthetic fibers are the polyesters whose scour wastes average 500–800 ppm of BOD. Processing 1000 lb of polyester fabric will produce 15.5 lb of BOD of which 90% is contributed by antistatic compounds used for lubrication sizing. Because of the high concentrations at which they are used and the inherent high rate of BOD, the emulsifying and dissolving agents used in polyester dyeing will produce high BOD loads. The rinses in polyester finishing are usually low in BOD. But, the processing of polyester uses an average of 15,000 gal of water per 1000 lb of fiber. Projected gross wasteload for synthetic fiber finishing is shown in Figure 4.

### Acrylics and modacrylics

Although these two fiber types have different physical and chemical properties, they are both subject to the same

finishing techniques. The waste from the first scour averages 2190 ppm and 660 lb of BOD per 1000 lb of processed fiber. The chemical components of the bath are the antistatic compound, which accounts for 30–50% of the BOD, and the soaps used to accomplish this process. When using acid dyes, the dye baths average 175 ppm and 5.3 lb of BOD per 1000 lb of fabric, the total BOD load coming from the dye carriers.

The final scour averages 668 ppm and 20 lb of BOD for 1000 lb of cloth. This final scour is accomplished with synthetic detergents and pine oil, which together contribute practically all the BOD. The equalized discharges will have a BOD of 575 ppm and 120.9 lb in a volume of 25,000 gal of waste water for each 1000 lb of acrylic and modacrylic fabric processed.

### Finishing

A treatment of a fabric that modifies its physical or chemical properties may be classified as finishing. Examples include permanent press finishes, oil repellents, soil release agents, low-crock polymers, abrasion-resistant polymers, fire retardants, lamination polymers, germicide and fungicide chemicals, to mention a few. A small number of these materials are biodegradable; however, most are not.

The polymers used for textile finishing are generally supplied to the finishing plant as emulsions which are sensitive to pH, salt, or agitation and may coagulate when they enter waste streams. Sewer lines may then become clogged with inert materials which have to be

**177**

removed by hand. Although the bulk of the polymer emulsion can be coagulated and removed in a treatment plant, some of it remains emulsified and is not removed by biological treatment. For complete removal of the polymer emulsion, chemical treatment is sometimes necessary. However, this is an additional step which in itself could replace much of the need for biological treatment.

Most of the finishes used for wash and wear and permanent press fabrics are manufactured from urea, formaldehyde, melamine, and gloxal compounds. Some of these products are readily degradable by microbial action; others are not. The formaldehyde derivatives can react with themselves or other chemicals in the waste stream to form insoluble products that may be removed by sedimentation.

A class of finishing chemical that has come into prominence in recent years is fire retardants. Most of the commercial fire-retardant finishes are phosphorus- and nitrogen-containing compounds. One such compound, triaziridyl phosphine oxide (APO), could present a serious problem if it got into a natural stream. The chemical reactivity of APO would facilitate its hydrolysis in a waste stream and prevent the parent compound from reaching the discharge water of a treatment plant. Whether or not these initial hydrolysis products are toxic or harmful is not known. This points to the increasing need for the characterization of industrial waste.

In the future, waste streams from different processing operations will have to be isolated and treated by either chemical or biological methods. The choice of treatment will naturally depend on the composition of the stream. By using this approach, industry will have more latitude in choosing chemicals and processes for their inherent production advantages and not their effect on pollution.

### Additional reading

"New Product Parade," *Text. Chem. Color.*, 1 (27), 55 (1969).
Federal Water Pollution Control Administration, "Industrial Waste Profile No. 4, Textile Mill Products," "The Cost of Clean Water," Vol. III, U.S. Government Printing Office, Washington, D.C. (1967).
Souther, R. H. "Textiles," Industrial Waste Water Control, C. P. Gurham, Ed., Academic Press, New York, N.Y., (1965).
Evers, D., "Effluent Treatment—What Is Involved," *Int. Dyer*, 143 (1), 56 (1970).
Fedor, W. S. "Textiles in the Seventies," *Chem. Eng. News*, 48 (17), 64, April 20, 1970.

**Donald W. Lyons,** *associate professor of textiles and mechanical engineering, is also cooperating faculty member and research advisor for the Textile Research Institute. He received his BS and PhD in mechanical engineering (Georgia Tech). His interests include industrial noise abatement, mechanics of yarn during spinning, and applying microwave heating to resin curing, among other areas.*

**John J. Porter** *received his BS degree in chemical engineering and PhD in chemistry from Georgia Tech. He is presently associate professor of textile chemistry at Clemson University. Dr. Porter's interests lie in the areas of waste water treatment, thermodynamics of dye adsorption, the reaction of cellulose finishing agents, and the morphology of fiber-forming polymers. Address inquiries to Dr. Porter.*

**William F. Nolan** *received his BS in textile management and marketing (Philadelphia College of Textiles and Science) and his MS in textile science (Clemson University). His recently completed thesis research covered the state-of-the-art of water treatment in textiles. Mr. Nolan's major areas of interest include product management, research and development, and quality control.*

# Treating lead and fluoride wastes

*Westinghouse has developed a one-step process
which treats wastes from TV picture tube manufacturing
and leaves an effluent of drinking water quality*

Reprinted from Environ. Sci. Technol., **6,** 321 (April 1972)

All industries are faced with some sort of waste disposal problem, and the electronic tube industry is no exception. Included in the electronic tube category are color television picture tubes which produce wastes specific to their manufacture. To understand how these wastes are generated and the steps taken to prevent water pollution, color television picture tube production needs to be explained.

## TV tube production

In the first place, there is quite a difference between black and white tubes and color picture tubes. A black and white tube is one glass entity, but the color television picture tube (color cathode-ray tube) has two glass sections—a face plate and a funnel. A viewing phosphor screen of color (triads of red, green, and blue dots) is laid down on the face plate. However, this operation must take place on a clean, virgin glass surface to prevent poor registration of the color dots, imperfect dots, or contamination (when these circumstances occur, the tube is rejected).

The commonly used method to clean the face plate is light etching of the surface with hydrofluoric acid washes. After several water rinses to remove the hydrofluoric acid, phosphor dots are applied on the face plate surface by a photosensitive process with exposure through a metal aperature mask. Hydrofluoric acid is used throughout the electronic tube industry to clean, etch, and prepare all-glass envelopes prior to applying the phosphor screen.

After etching, the face plate is fused to the funnel section (and will eventually become a vacuum-sealed tube) by glass frit—a lead solder glass (70–80 wt % lead). A machine applies a fillet (bead) of frit to the edge of the funnel section. The face plate is placed on the funnel, and the combined pieces are baked in a lehr (high-temperature oven). In the lehr, solvents are driven out of the frit glass which solidifies into a solid mass and binds the two sections together.

However, if the glass does have a flaw in it or the phosphor dots are not registered properly, the picture tube assembly is rejected, salvaged, and reprocessed. To reclaim the metal mask and glass sections, the sealed tube must be taken apart.

Since the weakest point in the structure is the frit, the entire glass tube is immersed in nitric acid to weaken and dissolve the frit bond. The two glass sections are mechanically pulled apart or heat shocked (with alternate hot and cold water) to finish breaking the seal. The glass parts are then cleaned with hydrofluoric acid, and the whole process begins once more.

What happens to these fluoride and lead wastes that accumulate from manufacturing and reclaiming cathode-ray tubes? Westinghouse Electric Corp. (Elmira, NY) was faced with this problem. Westinghouse produces several hundred thousand tubes yearly with the resulting liquid wastes. Of the total effluent of three million gpd, 90,000 gal are lead and fluoride wastes—10,000–15,000 gal of lead wastes and the remainder is fluoride.

This indeed is a formidable amount of lead and fluoride to handle in face of the strict water quality regulations in New York State. Since the Elmira plant does not have an adjoining body of water, NY officials have classified the effluent as a discharge to groundwater. Furthermore, receiving bodies of water in the state are classified according to best usage; in this case, usage is a source of potable water supply. Discharged effluents must consequently meet state health department drinking water standards which limit lead discharges to $1/500$ of 1 ppm and fluoride to 1.5 ppm at point of discharge.

## Treatment program

Here's how Westinghouse takes care of its lead and fluoride wastes. During the cathode-ray tube manufacturing process, the concentrated lead and fluoride wastes are individually separated from other rinse waters containing relatively small amounts of these same wastes.

Rinse waters are discharged as generated to individual collection sumps for lead and fluoride. Simultaneously, con-

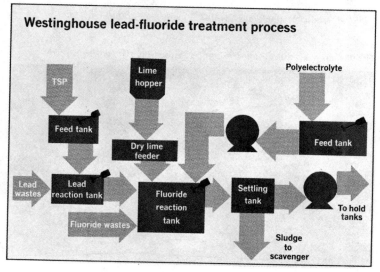

**Westinghouse lead-fluoride treatment process**

TSP · Lime hopper · Polyelectrolyte · Feed tank · Dry lime feeder · Feed tank · Lead wastes · Lead reaction tank · Fluoride reaction tank · Settling tank · To hold tanks · Fluoride wastes · Sludge to scavenger

centrated lead and fluoride wastes are continuously discharged to their collective sumps. The combined dilute and concentrated wastes then are piped to the treatment plant. Uniform concentrations of each waste are maintained by carefully metering the concentrated wastes going into the total stream.

At the treatment plant, lead wastes are fed into a tank and treated with trisodium phosphate (TSP). For best results, pH 3.6 is maintained. TSP reacts with the lead waste to precipitate lead phosphate. This treated waste, with insoluble lead in suspension, then flows into a fluoride reaction tank where the fluoride wastes are pumped in. Lime (to maintain a pH of 12) and a polyelectrolyte coagulant are added. The fluorides then precipitate as calcium fluoride.

This mixture then flows through six cascading tanks. Most of the solids are removed by settling in the first tank. The clear liquid in the last tank contains 0.2 ppm lead and 10–30 ppm fluoride which, when mixed with the remainder of Westinghouse's effluent, is within drinking water standards. Periodically, the tanks are cleaned, and the sludge (consisting of excess lime, calcium fluoride, and lead phosphate) is hauled to a landfill.

### History

Fluoride and, in some instances, lead wastes are common to several industries: electronic tube, glass, nuclear fuel, aluminum, and fertilizer. The basic technology of the treatment process used by Westinghouse is not new. Fluoride wastes have been treated with lime, and lead wastes have been treated with phosphates and other precipitants for years. In fact, other industries routinely treat the wastes

**Disposal.**  *Over 90,000 gallons of lead and fluoride wastes are treated each day*

separately or sometimes do not treat lead wastes at all, depending on the water quality standards of that particular region. Industries also vary fluoride waste treatment by the amounts of lime fed, pH, fluoride concentration, types of lime, retention time of the precipitate, and the mode of separation. The integrated treatment of both wastes used at Elmira is the new twist on the known technology.

During initial plant startup, Westinghouse neutralized its fluoride wastes in 55-gal drums. In the late 1950's, fluoride wastes were treated and separated by a rotary vacuum filter. The plant expanded when color television came into its own, and the vacuum filtration unit could not handle the increased volume and variety of wastes. Therefore, a continuous treatment substitute had to be found, and the present facility was completed in 1967.

All problems are not yet solved, however. More research is needed in several areas. For instance, the sludge that accumulates at the end of the treatment process eventually may have to be disposed of in ways other than landfilling (increasing numbers of landfill operations are closing as regulations tighten). In anticipation of this, Westinghouse researchers are looking into possible markets for fluoride and lead sludges.

Several patent applications have been made for converting fluoride sludges to cryolite and calcium fluoride. In fact, some aluminum firms build cryolite processing facilities adjacent to their aluminum processing plants.

Fluoride technology also has a few shortcomings. Treating fluoride wastes includes adding large amounts of lime and, in all probability, reclamation plants will not want excess lime. "We are working in the lab to develop other methods of treatment, particularly in cutting down on the lime used," says Westinghouse's Ken Rohrer, fellow engineer. Also, if a market develops for fluoride wastes, Westinghouse officials say that separate treatment may be reinstated to keep lead out of fluoride sludges.

Westinghouse officials agree that the treatment facilities are successful and efficient. The most difficult problem encountered is getting all noxious material to the treatment plant. Besides instituting employee education programs, Westinghouse has ensured that all open drains in manufacturing areas using large quantities of acids have either been closed or connected directly to the treatment plant.          CKL

**Production.**  *Frit-sealed cathode-ray tubes are conveyed to packing stations*

# Reusing waste water by desalination

### Reverse osmosis and ultrafiltration offer many prospects
### for municipal and industrial waste water treatment

*Reprinted from* ENVIRON. SCI. TECHNOL., **7**, 314 (April 1973)

**Fred E. Witmer**

*Office of Saline Water*
*Washington, D.C. 20240*

Progressively stricter effluent discharge standards (best available technology by 1983 and "zero" discharge by 1985) lend added impetus to water recycling. In certain areas of the U.S., notably southern California, the pressure of increasing urban growth has seriously taxed the supply of fresh water. As a consequence, adoption of "open-leg" municipal reuse techniques is imminent. In this approach, applications involving aquifer recharge, surface reservoirs, and/or irrigation are interposed between the point of discharge and the user.

A number of dissolved organic and salt contaminants are introduced during many of the industrial and municipal water-use cycles. Control of dissolved material is a necessary step in most reuse schemes; thus, desalting technology has an important and major role to play in water reuse and pollution abatement.

The Office of Saline Water (OSW), U.S. Department of the Interior, has led Federal Government efforts in desalting, having spent approximately $250 million sponsoring industrial, university, and state efforts in desalting. OSW develops promising desalting processes to the point where the technology appears commercially viable. Until recently, OSW's effort has been confined to desalination of seawater and brackish water as a source of supply for municipal and industrial usage. A varied inventory of desalting equipment exists as a result of significant efforts on the part of OSW and private industry. The fact that this equipment, with suitable modification, can be applied to water reuse applications is largely fortuitous. The maturation of desalting technology coincides with current ecological concern.

While desalting is generally a secondary objective of the reuse cycle, it is, in many instances, an essential one. Reuse water usually has low levels of salinity (500–1500 ppm total dissolved solids); as a consequence, the overall salt rejection efficiencies may be compromised below those values normally associated with the conversion of brackish water. This is not to say that high selectivities for specific contaminants, for example, trace heavy metals, toxic elements, and pesticides, should not be preserved, and indeed encouraged. In some cases, especially where potable water is to be produced, supplemental posttreatments are required to remove or render inactive certain noxious contaminants and viruses which may skirt the desalination process.

Much of the commercially available water reuse–desalting hardware represents first-generation equipment. This equipment, particularly in the reverse osmosis (RO) –ultrafiltration (UF) field, has evolved from the application of a new unit operation to a number of specialized, small-scale (less than 1 mgd) applications. In most instances the economy of scale-up, for example, the deployment of large-capacity items of equipment, has not been fully realized (Figure 1).

Industrial water reuse applications covered in this article have initially evolved, with few exceptions, from meeting waste water discharge standards rather than economic considerations. For relatively small volumes of waste which contain a recoverable by-product that may underwrite a portion of treatment costs, overall costs of 40–60¢/1000 gal are tolerable. This is not the situation for major pollution abatement applications, involving petroleum complexes, chemical plants, and municipal water supplies where the cost for reuse-desalting must compare favorably with the conventional process train of acquisition, conditioning, and waste treatment. As waste treatment processes become more elaborate to meet higher quality discharge standards, treatment costs will escalate to the point that state-of-the-art reuse-desalting becomes a tenable alternative. Obviously, the state-of-the-art reuse-desalting processes are adopted out of the necessity of timely compliance to pollution control legislation.

Reuse-desalting costs can be expected to improve as the benefits associated with scale-up, large-capacity

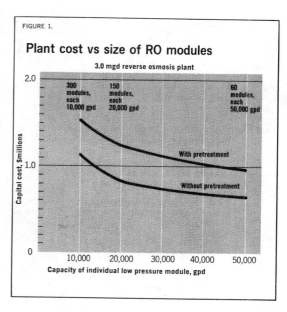

FIGURE 1.

**Plant cost vs size of RO modules**

3.0 mgd reverse osmosis plant

300 modules, each 10,000 gpd
150 modules, each 20,000 gpd
60 modules, each 50,000 gpd

With pretreatment

Without pretreatment

Capital cost, $millions

Capacity of individual low pressure module, gpd

TABLE I

## Summary of reuse-desalting commercial endeavors

| Company | Equipment description/configuration | Commercially viable at present | | | Future | | | | | |
|---|---|---|---|---|---|---|---|---|---|---|
| | | a | b | c | d | e | f | g | h | i |
| **Abcor** <br> Cambridge, Mass. | UF/RO tubular | X | X | | | | | | X | X |
| **Westinghouse** <br> Philadelphia, Pa. | UF/RO tubular | X | X | | | | | | X | X |
| **Philco Ford** <br> Newport Beach, Calif. | UF/RO tubular | X | X | | | | | | X | X |
| **Dorr-Oliver** <br> Stamford, Conn. | UF flat-plate thin film | X | X | | | | | | X | X |
| **Romicon** <br> Woburn, Mass. | UF annular thin channel | X | X | | | | | | X | X |
| **GESCO** <br> San Diego, Calif. | RO sprial wound | | | | X | X | X | X | | X |
| **Rex Chainbelt** <br> Milwaukee, Wis. | RO Du Pont hollow fine fiber | | | | X | X | X | X | | X |
| **Permutit** <br> Paramus, N.J. | RO Du Pont hollow fine fiber | | | | X | X | X | X | | |
| **Envirogenics** <br> El Monte, Calif. | RO tubular and spiral wound | X | X | | X | X | X | X | X | X |
| **Dow**[a] <br> Midland, Mich. | RO hollow fine fiber <br> UF hollow fiber dialyzer (specialty) | | | | X | X | X | X | | |
| **Monsanto**[a] <br> Durham, N.C. | RO hollow fiber | | | | X | X | X | X | | |
| **Ionics** <br> Watertown, Mass. | Modified ED unit | | | X | | | | | | |
| **Aquachem** <br> Waukesha, Wis. | Multistage evaporation | | | | X | | X | X | | |
| **Resource Construction** <br> Seattle, Wash. | Vapor recompression distillation | | | | X | | X | X | | |
| **Colt Industries** <br> Beloit, Wis. | Vacuum freezing–vapor compression | | | | X | | X | X | | |
| **Avco**[a] <br> Wilmington, Mass. | Secondary refrigerant freezing | | | | X | | X | X | | |

[a] Hardware currently under development, not commercial.

modules, and improved equipment and systems make their mark on the cost effectiveness of the reuse-recycle process leg. Much of OSW's effort in the development of large-scale brackish water conversion systems (Figures 2 and 3) is directly applicable to effecting similar improvements in reuse-desalting hardware. Thus, it is anticipated that rapid strides can be made in improving the status of reuse-desalting into a competitively viable unit operation. With the total assimilation of advanced desalting technology into the reuse-desalting process, treatment costs of 30–40¢/1000 gal (desalting is 15–20¢) appear to be a realistic target.

### Developing reuse-desalting applications

Three "reuse-desalting" applications have developed on their commercial merits, namely, ● the reuse of electropainting rinses (a); ● the fractionation and ultrafiltration of cheese whey (b); and ● the point-of-source manufacture of sodium hypochlorite as a waste treatment biocide (c). A number of feasible, specialty applications, which have resulted from pollution control considerations, are under development. These include but are not limited to ● cooling tower and boiler blowdowns (d); ● petroleum stripping waters (e); ● plating rinses (f); ● metal finishing rinses (g); ● pulp and paper spent liquors (h); and ● municipal sewage effluents (i).

In primary reuse-desalting processes, a concentrated brine stream which requires further processing is produced. Environmental considerations dictate that this stream be concentrated until dry, containable, and, it is hoped, until usable solids are produced. Depending on the content and size of the brine stream and location of the plant, a number of desalting processes are used to achieve final concentration: electrodialysis (ED), vapor recompression evaporation, multieffect evaporation, freezing, and ponding, etc. The contribution of these and similar processes to the total reuse-desalting system should not be overlooked. A summary of a selected group of manufacturers who have capabilities in the area of reuse-desalting is shown in Table I. The reuse-desalting applications are presented in order of anticipated development; paradoxically, the more difficult, large-volume applications in terms of augmenting water supply are ranked last.

### Electropainting rinses

Electropainting is widely used by manufacturing industries (automobile, truck, for instance), which require a uniform primer coat on an intricate assembly. The paint is applied by dipping the electrically charged part into a dip tank containing an emulsion of resin and pigment particles. The electrostatic charge causes the paint colloid to migrate to the surface of the part and to coalesce into a uniform film. When the object is lifted from the

**Figure 2**
*This hollow fiber reverse osmosis module is capable of desalting 20,000 gpd brackish water*

painting bath, excess paint (dragout) is removed by rinsing, and the rinse water is returned to the dip tank. The return of the rinse, coupled with paint usage, lowers the concentration of paint in the dip tank. Since a constant, optimum concentration of paint is required to maintain proper quality control, water must be continuously removed from the dip tank. In addition, certain ionic substances—chromates, phosphates, chlorates, and sulfates—are picked up during surface pretreatment operations and progressively contaminate the dip tank. Make-up paint, with surfactants (amines or hydroxides) to stabilize the paint emulsion, is added to the dip tank to maintain the paint inventory.

Removing excess water and contaminants from the dip tank may be accomplished through ultrafiltration of a side stream from the dip tank (Figure 4). The ultrafiltration unit generally consists of a tubular membrane assembly in which the paint emulsion is circulated inside the membrane tubes at 15–50 psi pressure. Rapid, highly turbulent cross flow (10–20 ft/sec) is maintained past the membrane surface to retain the paint particles in the turbulent core and prevent blockage due to the accumulation of paint on the membrane surface. The paint solids are returned to the dip tank for reuse while the excess water containing trace salts and stabilizer is used as a rinse and/or is discarded.

Broad industrial acceptance of this process has resulted from cost savings effected by a drastic reduction of paint dragout losses. Payout time for ultrafiltration equipment to realize these savings is generally four to six months.

### Fractionation of cheese whey

Cheese whey is normally discarded to sewage systems and waterways. Due to its mainly lactose sugar content (5 wt % for cottage cheese whey), it possesses a high biological oxygen demand. Whey also contains appreciable amounts of protein as casein (0.8 wt.% for cottage cheese whey). By employing a two-step ultrafiltration/reverse osmosis process (ES&T, May 1971, p 396) for fractionating cottage cheese whey into valuable protein and lactose concentrates for food products, cleanup costs of the whey wastes may be recouped by the sale of the by-products.

The ultrafiltration/reverse osmosis (UF/RO) plant must conform to normal dairy processing practice. Tubu-

**Figure 3**
*The facility above manufactures 5000–10,000 gpd spiral wound cartridges for RO modules with 50,000 to 100,000 gpd capacity*

Feed flow
Permeate flow (after passage through membrane)
Permeate out
Permeate side backing material with membrane on each side and glued around edges and to center tube

lar membrane modules are housed in cabinets where "cleaning in place" sanitizing solutions are sprayed to the outside of the tubes while also being pumped through the interior of the tubes.

A representative UF/RO system consists of a series of UF modules for treating the raw whey from a temperature-controlled feed tank. The UF system separates the larger protein molecules from the smaller sugar molecules which, along with salts, pass through the UF membrane. The protein is retained in the concentrate recirculated to the feed tank to achieve concentrates of 10–12 wt % solids in the final concentrate. The solids contain approximately a 6 to 4 split of casein to lactose. The UF concentrate is then evaporated (scraped surface or falling film evaporation) to provide the 40–50 wt % solids required for the final spray-drying operation. The lactose containing permeate from the UF section is cooled and collected in an interstage tank. It is then pressurized and pumped in a once-through mode through a tapered RO section, thereby effecting a fourfold increase in lactose concentration (6–25 wt %). At higher concentration factors, lactose begins to precipitate from solution and may scale the membrane. The lactose concentrate is fed to a conventional crystallization train consisting of such items as a scraped surface evaporator, crystallizers, centrifuges, dryers, and baggers.

Preliminary conclusions indicate that the total by-product protein and crude lactose produced will significantly offset processing and pollution abatement costs.

### On-site generation of hypochlorite

The use of oxidizing biocides, such as sodium hypochlorite, to control the biological activity of waste water is increasing. On-site generation of sodium hypochlorite is an attractive alternate to tank trucking and maintaining an inventory of reagents. Sodium hypochlorite is generated by an electrolytic membrane cell producing chlorine gas and sodium hydroxide solution which are subsequently reacted outside the cell to produce 5–10% sodium hypochlorite solution. Production and use of hypochlorite are less hazardous than direct injection of chlorine gas which is highly toxic and corrosive. Improvements in safety, reliability, and economy result from using a hydraulically impermeable cation exchange membrane.

The hypochlorite generator is composed of a series of back-to-back membrane cells mounted in a "plate and frame" type of device. The membrane cells consist of an anode (a titanium screen coated with a special metal oxide coating), an anode compartment spacer, a perfluorosulfonic acid cation exchange membrane, a cathode compartment spacer, and a cathode (expanded mild steel). Filtered, fresh water is purged through the cathode compartments, while a sodium chlorine brine solution is passed through the anode compartments. The chlorine gas and spent brine anolyte is sent to a trapped disengagement section. The chlorine gas flows overhead to a reactor where it reacts with the caustic formed in the cathode compartments to produce the hypochlorite. The hydrogen gas from the cathode compartments is diluted with air well below explosive limits and vented off to the atmosphere. Costs for this system are approximately two thirds of those associated with the purchase and delivery of sodium hypochloride to the treatment site.

### Blowdowns

Environmental discharge standards require control and reduction of dissolved solids such as sulfates and radioisotopes, such as from nuclear plants, contained in process blowdown streams. Blowdown waters originating from cooling towers and both conventional and nuclear boilers are amenable to established brackish water desalting processes. The low salinity levels generally involved, 500 to several thousand TDS (parts total dissolved solids), encourage membrane-based processes, low-pressure reverse osmosis, and electrodialysis for the initial concentration step in desalting.

However, chromate-base corrosion inhibitors are poorly rejected by some RO membranes when used with cooling tower waters. Thus the desalted permeate, if recycled to the cooling tower system, would significantly reduce the net consumption of this inhibitor.

With conventional boiler blowdown, in addition to the normal dissolved ion content, the membrane system

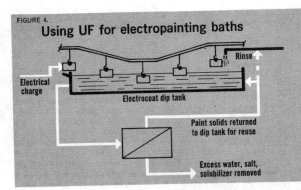

FIGURE 4.

## Using UF for electropainting baths

Electrical charge

Rinse

Electrocoat dip tank

Paint solids returned to dip tank for reuse

Excess water, salt, solubilizer removed

must remove phosphates and amines (chemical treatment agents for the maintenance of pH to suppress corrosion). In nuclear applications, borate leakage and/or carry-over from neutron absorption systems pose a potential problem since borate rejection is normally poor. However, under alkaline conditions (pH of 10.0–11.0), polyamide base membranes may adequately reject the borates.

Approximately 150 billion gal per year of cooling tower wastes are generated by industry and the utilities (3–1 ratio between manufacturing and power generation). Cooling towers, more than likely, will be used more and more within the power industry to meet thermal discharge standards. About 10 billion gal of boiler blowdown are produced annually.

### Petroleum stripping waters

About 10% of the waste water attributed to refinery operations represents waters which contain extracted ammonia, hydrogen sulfide, and phenols along with trace organics from various petroleum fractions. The presence of these compounds preclude direct discharge into a waterway. Ammonia and hydrogen sulfide are conventionally removed by steam stripping. The bulk of the phenols and organics are biologically oxidized, with considerable difficulty, prior to discharge. However, reverse osmosis coupled with appropriate chemical pretreatment and posttreatment oxidation can supplant the sometimes unreliable physicochemical-biological step. Development of improved physicochemical waste treatment processes for petroleum stripping waters appears to be a certainty.

### Plating rinses

In the plating process only a small fraction of the metal inventory—nickel, chromium, for instance—present in the plating bath is actually deposited. A large majority of these often expensive materials are chemically precipitated, discarded, and eventually lost to the environment. Membrane processes can purify low-level plating rinse wastes to concentrate the plating ions to levels where recycle to the plating bath is economically attractive.

### Metal-finishing rinses

In metal-finishing operations large quantities of water are used for cooling and lubrication. The water usually picks up trace quantities of heavy metals and organics during use. Generally this water is discharged to a receptor upon cooling, but strict pollution discharge standards relating to the toxicity of heavy metals are stopping this practice.

This situation is similar to the case of plating rinses—physicochemical processes must be utilized to obtain an acceptable effluent. Two alternatives exist: to use a desalting process, for example RO or ED which would produce a water suitable for reuse and reduce water consumption; or to use a selective exchange process, for example, ion exchange (to displace heavy metals with an acceptable ion such as sodium without a reduction in water requirements).

## Pulp and paper

The pulp and paper industry is a major water user. A wide variety of effluents results from the basic pulping processes and their numerous process streams. Strong cooking liquors (10–15% solids) are generally concentrated by multieffect evaporation and recycled within the mill or marketed. In some instances, such as for acid sulfite pulping, the volatile acids ($SO_2$, acetic, and formic acids) are combined with the condensate and comprise a major BOD source in the waste streams. Bleach plant effluents, originating from alkaline sulfate pulping, pose a difficult problem since they contain biological oxidation-resistant dissolved organics and salts (NaCl, $Na_2SO_4$). Complete control of such waste discharges by evaporation is expensive ($1.50–2.00/1000 gal). UF/RO used as a first step in the concentrator train is an attractive alternate.

Considerable research has been performed in this area. However, the research for promising results over a reasonable service life (1 year) has been hampered by: • high back-osmotic pressures associated with many of the waste liquors which require processing pressures of 600–800 psi; • high fouling tendencies possessed by these liquors which may result from the complexing of lignin sulfonic acids with polyvalent cations at the membrane wall; and • development of reliable tubular RO hardware which must endure severe operating conditions.

Progress is being made—dynamically formed lignosulfonate membranes formed on UF supports, in lieu of integral cellulose acetate membranes, have the potential to stop the leakage and failures associated with the harsh membrane cleaning cycles normally used with high fouling feeds.

## Municipal sewage effluents

Wide-scale practice of municipal water reuse depends on need, costs, and public acceptance. A number of communities across the U.S. could improve the quality of their water supply by adopting reuse-desalting techniques. In several instances, a reuse-desalting system is superior to expanding the water supply via "straight" desalination of a brackish water source.

Southern California is an area where municipal water supply limitations appear to be limiting growth and economic development. The critical situation in the area has encouraged a number of water agencies to consider seriously development of alternate water supplies, including reuse. Ground water supplies in this region are being severely taxed and becoming subject to seawater intrusion. Sewage plant effluents, in conjunction with desalting techniques, are scheduled for reuse in charging and maintaining ground water aquifers. One advantage that reuse-desalting via aquifer recharge has for municipal applications is that while it "closes the loop," it does so in a manner psychologically acceptable to the general public. Several sewage reuse-desalting schemes are under consideration, including direct recycle:

| Location | Process |
|---|---|
| Orange County, Calif. | Sewage effluent aquifer recharge diluted with desalted seawater |
| Escondito, Calif. | Desalt sewage effluent prior to aquifer recharge |
| Oceanside, Calif. | Sewage effluent recharge, desalt on withdrawal from aquifer |
| Ventura, Calif. | Desalt sewage effluent, sorb and oxidize organics, and recycle |

Conventional biological treatment is an integral part of each of these candidate reuse systems.

A possible variation for municipal recycle is to utilize UF to remove suspended organics and clarify the water for recharge to an aquifer and to use RO as a complementary step to demineralize and purify the water when taken from the aquifer. A secondary posttreatment chlorination of product water should, of course, be performed as a precautionary measure.

Many of the waste treatment physical-chemical unit operations costs (flocculation, sedimentation, clarification, or absorption) are volumetric throughput controlled. By increasing the concentration and reducing the size of the process stream early in the process, proportionately smaller capital equipment costs and operating expenses are incurred. In many instances the efficiency of the process is improved; higher clarity supernates, more effective biological digestion, and direct oxidation, to name a few, are some of the anticipated outgrowths dealing with higher concentrations of contaminants.

One potential application concerns dewatering and disposing of sewage sludge which accounts for approximately 30% of sewage treatment costs. Sewage sludge is highly voluminous and generally 94% water. Pressure cooking of sludge (Porteous Process) greatly enhances the dewaterability—the sludge is only 50–60% water, and sludge volumes are reduced nearly an order of magnitude. However, the resultant supernate is heavily loaded with dissolved organics, and if recycled to a conventional treatment plant, an appreciable increase in biological loading occurs. RO and UF provide an attractive alternate to biologically processing the organically charged supernate.

The cost effective utilization of desalting processes in reuse applications can be achieved by optimizing the total system rather than simply suboptimizing the desalting process per se. Important progress has been made in desalting feeds that possess a high fouling potential: Recently, UF-dialysis has been used to remove side product salts from high-molecular-weight polymer suspensoids. The desalting process, such as RO or UF, may be used to create concentrated waste or process streams which may be dealt with in a more intensive fashion than the highly dilute waste or process streams normally encountered. The ability to produce and subsequently treat grossly concentrated waste streams, opposed to the treatment of dilute feeds, opens new and exciting possibilities in the area of industrial and municipal waste processing.

**Fred E. Witmer** is presently a chemical engineer in the Membrane Processes Division, Office of Saline Water, U.S. Department of the Interior. His work is specialized in brackish (low salinity) water desalting processes. Previously, Dr. Witmer was with Hydronautics, Inc., as a project engineer in developing oil scavenger equipment and reverse osmosis hardware.

**185**

# Desalters eye industrial markets

*Reprinted from* ENVIRON. SCI. TECHNOL., 4, 634 (August 1970)

A potential market of over $1 billion for desalination equipment sales to industry is foreseen by an economist at the Office of Saline Water (osw). Speaking to attendees of a conference of the American Water Works Association in Washington, D.C., Eric D. Bovet, economic adviser to osw, spelled out reasons why he believes that equipment and processes originally developed for desalination of seawater and brackish waters can more than pay their way in the treatment of many industrial wastes.

Bovet's optimism is based in part on the results of a study undertaken for osw by Aqua-Chem, Inc. (Waukesha, Wis.). Aqua-Chem was asked to identify industries in which liquid wastes met two simple requirements:

• A by-product must be recoverable from the waste through one or more desalination processes.

• The commercial value of the by-product must equal or exceed the processing cost. Desalination has, in the past, been aimed almost entirely at the production of potable water, and the aim of osw is partly to expand horizons for the whole range of processes developed for that purpose, many with the assistance of osw funds.

## Prime markets

The Aqua-Chem study identified six large potential markets for desalina-tion technology in industries where wastes meet the above requirements.

In **the cheese industry,** the major barrier to reclamation of the whey produced as a by-product is its salt content (about 1%). According to the Aqua-Chem study, a combination of reverse osmosis and electrodialysis process steps could produce a protein supplement grade of whey that would command about 25 cents per pound in the marketplace. An eventual market for desalination processes in excess of $150 million is predicted for the cheese industry; perhaps 10% of this might be expected to materialize before 1975, according to the study (*see chart*).

The potential market in **the pulp and paper industry** is greater than in any other, says the osw study. $75 million worth of electrodialysis equipment alone is predicted for use in recovering organic compounds from spent sulfite liquor (see also ES&T, November 1969, page 1147, for an account of the use of electrodialysis in the pulp and paper industry). Even larger markets are foreseen for reverse osmosis in the treatment of bleach plant effluent and dilute wash water.

**The iron and steel industry** could benefit from the use of electrodialysis techniques in the treatment of spent pickle liquor, according to the study. The total industry market is estimated at $140 million. (Alternative processes for the recovery and regeneration of pickling acid and iron are described in ES&T, May 1970, page 380.)

Electrodialysis also has a large potential market in **the plating and metal finishing industry,** where spent plating solutions and rinse waters must be treated to remove metals and toxic acids. (Electrolysis techniques, too, are being pushed into this market—ES&T, March 1970, page 201, describes one such process.) The market may be as big as $50 million, says osw.

Evaporation (not exclusively a desalination process, of course) is seen as representing a potential $50 million market in the **nuclear power generating industry** and a possible $400 million market in the treatment of **acid mine drainage.**

In all, Aqua-Chem's study predicts the existence of possible markets worth almost $1.5 billion for evaporation, reverse osmosis, and electrodialysis processes, with almost 10% of the market realizable within the next five years. Whether this exceedingly large market is as tangible as equipment makers might hope remains to be seen, of course. But Bovet emphasizes that the Aqua-Chem study ferreted out only those instances where revenue from by-product sales would equal or exceed operating expenses, including amortization of the initial capital investment. In other words, osw feels that these markets can be penetrated without any need for additional laws against water pollution. Should laws tighten further—as they probably will—the potential markets will be even greater. And, Bovet continues, the great possibilities in the giant chemical and petroleum industries were not even considered in the Aqua-Chem work.

What must be considered as a counterweight to boundless optimism regarding the osw projections is the absolute certainty that, if the markets do indeed develop, desalination processes will not have the field to themselves. As ES&T readers will be aware, a whole range of alternative techniques is already being lined up to take advantage of any move by industry to recover its by-products or otherwise defray the costs of water pollution control.

## Desalting processes have market potential in several industries

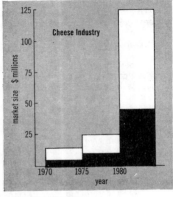

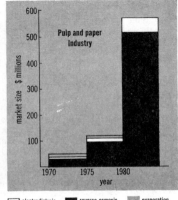

Source: Office of Saline Water

☐ electrodialysis   ■ reverse osmosis   ▨ evaporation

# Microstraining water and waste water

## Invented during WW II, microstrainers are now marketed for effluent polishing as well as preliminary treatment of raw waste water

*Reprinted from* ENVIRON. SCI. TECHNOL., **7,** 104 (February 1973)

Microstraining, a form of simple filtration, was invented in England during World War II to remove fibers from the effluent of a munitions plant. The process was simple—intercepting solids by direct screening —and was accomplished by mounting a fine wire cloth on the periphery of a revolving drum and passing the effluent through the revolving drum.

After the war, microstraining was developed commercially by Glenfield & Kennedy Ltd., London, England for treating drinking water supplies. In 1950, microstraining was applied to secondary treated sewage effluent to upgrade it to tertiary standards (effluent polishing).

Since then, Crane Co. (King of Prussia, Pa.), an international firm in the field of water and waste water treatment, has acquired Glenfield & Kennedy Ltd. and further developed microstraining technology. Today, more than 600 microstrainer installations filter two billion gallons of water each day in 32 different countries (300 microstrainers are operating in operating in the U.S.). It's used for municipal drinking water treatment as well as municipal and industrial waste water treatment.

Only six companies in the U.S. produce microstraining equipment. It's really a different type of process— not quite sewage treatment and not quite water treatment. It can supersede waste treatment processes or can complement them by effluent polishing (tertiary treatment).

### How it works

By definition, filtration is a process to separate solids from liquids by passing the liquid through a network of wires, threads, or other porous membranes, or through porous beds of granular material, such as sand. As the liquid passes through the filter material, certain sizes of solids are directly intercepted by the medium itself and indirectly arrested by solids already held or matted on the medium.

When the process is performed without the aid of chemical or biological controls, it is called simple filtration. Using chemical controls results in the equivalent of rapid sand filtration; biological controls lead to the equivalent of low sand filtration.

Microstraining clarifies liquids by filtering the greatest possible amount of suspended solids, especially microscopic solids. Crane Co. uses a specially woven 316 stainless steel wire cloth (with apertures ranging from 60 down to 23 microns) mounted on the periphery of a revolving drum arranged for continuous backwashing. Stainless steel is expensive but more durable and corrosion resistant. "We have experimented with plastic cloth," says E. W. J. Diaper, manager of the municipal water and waste water treatment department of Crane Co., "but the plastic strands tend to stretch and reduce filtration efficiency."

Enclosed in a reinforced concrete tank, the drum is submerged in flowing water to approximately three fourths of its depth. Raw water enters through the open upstream end of the drum and flows radially outward through the microfabric which intercepts suspended solids. The pressure from the difference in water level (head loss) between the inside and outside of the drum (see illustration) pushes the water through the drum. The mat, created as the solids build up on the microfabric, will trap many particles smaller than the aperture size of the mesh.

The intercepted solids are carried upward on the fabric on the inside of the drum beneath a row of wash water jets spanning the full width of the fabric. From there, these wastes are flushed into a receiving hopper on the hollow axle of the drum. The solids move by gravity out of the microstrainer for ultimate disposal.

Water for backwashing is drawn from the downstream side of the unit. Drum rotation and backwash are continuous operations and are adjustable either manually or automatically with the flow rate. Crane offers microstrainers with capacities up to 30 mgd; for larger flows, multiple microstraining units can be installed.

The only real problem in microstrainer operation is slime buildup on the drum which "blinds" the fabric in some water works and in most sewage installations. Crane researchers have overcome this drawback by placing high-intensity ultraviolet irradiation equipment over the drum which inhibits bacterial and other organic slime growth.

If the incoming raw water contains iron or manganese, a film of iron or manganese oxide may be deposited on the microfabric. However, the corrosion-resistant drum can be cleaned by an inhibited acid cleanser.

### Municipal applications

Water supplies which are virtually free of color and extraneous matter may be distributed for consumption after being microstrained and sterilized. Also, in the municipal water treatment field, microstraining can be used to clarify water prior to sand fil-

**World's largest.** This microstrainer, now operating in Chicago, treats 15 mgd

tration, explains Diaper. Microstraining reduces the load on the plant, reduces wash water requirements, maintains plant capacity under overload conditions, increases output to meet high demands, reduces the amount of chemicals used and resulting sludges, and is economical to operate. In 1960 a microstrainer was installed in Denver, Colo., to treat 100 mgd of drinking water for Denver residents.

In municipal waste water treatment, microstraining can be applied to the final filtration or polishing of sewage effluent. Remaining suspended solids and BOD can be reduced by 90%. There are now 56 microstrainer installations in the U.S. in operation or under construction which treat sewage or industrial effluents.

Chicago's North Side Sewage Treatment Plant at Skokie, Ill., also employs a microstrainer—the largest one in the world. In fact, the 10 × 30-ft drum treats 15 mgd, reducing the suspended solids from 18 to 5 ppm. To increase capacity, the periphery of the microstrainer drum was corrugated (making it equivalent to a 50-ft-long drum).

Two Pittsburgh, Pa., suburbs are operating microstrainers for tertiary treatment of municipal sewage effluent. Similar units are in operation or under construction in Michigan, Ohio, Illinois, Florida, Massachusetts, New York, Tennessee, Maryland, Kentucky, Minnesota, North Carolina, Pennsylvania, and New Jersey. Microstrainers being considered for Euclid, Ohio, will have a twofold purpose. In dry weather, sewage effluent will undergo tertiary treatment, and, during wet weather storm flows, the storm sewer overflow will be microstrained. This new method of treatment is being considered by other cities.

### Industrial applications

Microstrainers are also used to treat industrial water supplies. One of the first installations in the U.S. (1957) was used by the Anaconda Co. in Montana to protect high-pressure pumps used in the hydraulic debarking of logs. In Thurso, B.C. (Canada), a redesigned water system using microstrainers for a paper mill reduced total water costs by 2.5¢/1000 gal. with an annual savings of $45,000 per year in chemicals.

In Port Alice, B.C., two units remove microorganisms (Copepoda) and organic debris from up to 20 mgd of lake water supply for a paper mill. The shrimp-like crustacea caused discoloration in the stock prior to installation of the microstraining equipment in 1957.

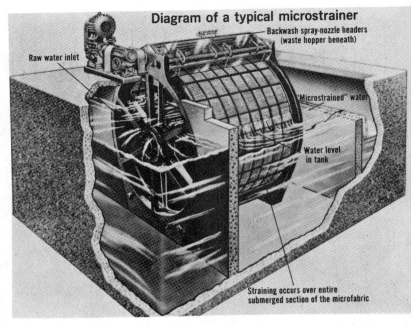

**Diagram of a typical microstrainer**

Backwash spray-nozzle headers (waste hopper beneath)

Raw water inlet

"Microstrained" water

Water level in tank

Straining occurs over entire submerged section of the microfabric

About a year ago a paper mill in Turners Falls, Mass., installed a microstrainer that removes 80% of the suspended solids and BOD from paper and board mill effluent. Since no chemical additive is needed in the treatment, the reclaimed material can be returned directly to the process. Normally, these solids are concentrated to approximately 2% of the total filter flow. When heavier loads are anticipated, the microstrainer is operated downstream of the primary clarifier.

### Expansion to other fields

Crane Co. researchers are now looking to nutrient removal applications for the microstrainer. Results thus far show that by using ferric chloride as a chemical coagulant, microstrainers can remove 60 and 90% of the phosphorus from treated effluent. A pilot plant operation at a distillery is testing algae removal from the water supply for the plant. The microstrainer removes 96% of the algae—a record which even beats the sand filter used for the same purpose.

"We're even using the microstrainer to remove oil from water," continues Crane's Diaper. "We've applied for a patent on this process." After water with an oil content of 200 ppm passes through the microstrainer, the discharged water only contains 2 ppm oil.

A growing application for the microstrainer is treating storm sewer overflows. Crane Co. was awarded a government contract several years ago to operate a small microstrainer on storm sewer overflows in Philadelphia, Pa. The 800-ppm suspended solids in the storm discharge was reduced to 40 ppm with a very high flow rate—45 gal./min/ft$^2$. "I think the microstrainer stands head and shoulders above other forms of treatment for versatility, simplicity, and economy," summarizes Diaper.

### The competition

Microstraining is in direct competition with sand filtration. Although sand filters can remove a few parts per million more of suspended solids, microstrainers have several inherent advantages.

Total installation and operating costs for microstrainers are about 1.5¢ per 1000 gallons for a 10-mgd plant which compares favorably with other tertiary treatment methods. For example, sand filtration runs about 4¢ per 1000 gallons, and carbon absorption has a price tag of 8¢ per 1000 gallons.

Also, these microstrainers require only about one third the space that sand filters need, and land is often at a premium. Sand filters often require additional equipment for proper operation: holding tanks to ensure constant flow to the filter and pumping equipment to lift the incoming water for the necessary 10-ft head loss for sand filter operation. Besides adjusting to effluent flow and requiring only 6–18 in. head loss, microstrainer characteristics include instantaneous startup and continuous filtration. CKL

# Debugging physical-chemical treatment

*Exactly where such water treatment fits in will become*

*evident as 17 plants go on-stream in the next few years*

Reprinted from Environ. Sci. Technol., **6**, 984 (November 1972)

To paraphrase a recent TV commercial, you can take the bugs out of a waste treatment plant—for example, design a physical-chemical treatment (p-ct) plant—but you can't take p-ct as a panacea for bug treatment.

Is p-ct a substitute for biological treatment? Under what conditions does it work? Just where does p-ct fit into the U.S. arsenal of water pollution control treatment methods? These questions and many more were the gist of a recent conference sponsored by the International Association of Water Pollution Research and the American Institute of Chemical Engineers and hosted by Vanderbilt University.

To be sure, the Nashville, Tenn. conference brought up many more questions than answers. More than anything else, perhaps, it urged a closer look at the consequence of actions that have been taken over the past few years and that will result in some 17 programmed p-ct plants coming into operation later in the decade.

Certainly, p-ct is no panacea for domestic waste water treatment but it will fit in some, at this time largely unproved, place in the nation's cleanup scheme.

There are often-cited advantages including claims that it:

• will remove certain toxic materials

• can handle overloaded situations

• is not subject to the vagaries of weather and other unpredictable (yet often encountered) conditions

• requires smaller land mass requirements than biological systems

• requires lower initial capital investment than a biological system, but, understandably, higher operating costs.

P-ct proponents argue that in this day of aerospace technology and precision, why should we leave our waste water treatment plants to the uncertainties of the bugs and unpredictable operational upsets. What is more, they cite that in this age of recycle, the chemicals used in p-ct plants would be regenerated, recycled, and reused.

But what may sound like a pat solution to U.S. water pollution problems is far from reality. Despite the fact that about 17 p-ct plants are either on the

**Vanderbilt's Wesley Eckenfelder**
*Interrelating bio and p-c treatments*

drawing board or in early construction phases, the fact remains that no full-scale plant is in operation today.

The proposed p-ct plants range from the large, 50-mgd plant for Cleveland (ES&T, September 1972, p 782) to the smallest, a 1-mgd plant for Rosemount, Minn., which will probably start operating early next year. Plans for a 54-mgd plant for Alexandria, Va. are in the final design phase. Sometime later, of course, construction bids for the others will be announced.

## How it works

Basically, p-ct is quite simple; there is a minimum of five steps—pretreatment, clarification, filtration, adsorption, and disinfection. Take a few physical methods, add a few chemical methods, and you in essence come up with the newest waste water treatment process known as p-ct. What first has to be done is to remove solids by the physical methods of screening, clarification, and filtration. But before clarification is performed, chemicals may be added to help with the precipitation of pollutants including phosphorus. Later, an adsorption process, usually performed in an activated carbon column, removes dissolved organic material that escapes the above steps, and is followed by disinfection.

The way that p-ct came into the waste water treatment limelight in the first place probably stems from the fact that in this day of changing (always stricter) standards and increasing requirements

for removal of pollutants and nutrients, physical-chemical treatment was needed as an add-on (often called "tertiary") process. This "advanced waste treatment process" had to be employed to remove nutrients, initially phosphorus but more recently nitrogen, from biological treated secondary effluents. Later, emphasis was placed on more efficient systems for handling domestic wastes, with p-ct being a prime contender for systems in need of low initial capital investments and minimum land requirements. The additional guarantee of removing phosphorus, and perhaps removing nitrogen, was an added attraction. Somewhere along the line, economic considerations entered the field.

Rumors have it that the Environmental Protection Agency (EPA) put pressure on the p-ct route so that the agency could realize the greatest mileage from expenditure of its construction grant monies. The initial capital investment for a p-ct plant is much less than that for a biological treatment plant. In this way EPA's financial investment—which goes entirely for capital construction costs rather than operating costs—could be made in more plants. What truth, if any, there is in this rumor is left to the environmental speculator.

Of course, a majority of a p-ct plant's operating cost goes for chemicals—in many cases lime but in others alum, ferric chloride, polyelectrolytes, or combinations of these coagulants. From the viewpoint of at least some operators, the present pressure to adopt chemical processes is an attempt to limit the amount of capital costs which may be partly reimbursed by federal or state grants by shifting overall costs on to local authorities in the form of high operational expenses. On the other side of the coin, sludge handling equipment represents a large portion of the capital investment for a p-ct plant.

Of course, both the carbon that is used for dissolved solids removal and the lime used for coagulation customarily are regenerated and reused. Thermal regeneration of carbon in a multiple hearth furnace is the process that is farthest along and the method contemplated for use in planned construction. Carbon losses, regenerated carbon characteristics, and costs are all well documented. Sometimes in smaller operations the thermal regeneration is done in a rotary kiln.

### Applications today

One U.K. spokesman at the symposium, for example, noted that there was great interest in p-ct because of the special conditions that the British must live with, including a requirement to treat sewage before disposal to an estuary or ocean. Sweden also is employing a number of p-ct methods. The reason the Swedes chose the p-ct route is that their waste water treatment plants must remove phosphates by 1975. Today, there are 110 plants with phosphate removal capability in Sweden; there will be 230 by 1975. Finland has 45 plants for phosphorus removal, Switzerland 40, and so on.

An example of recreational reuse in the U.S. is at Santee, Calif., where the treated effluent from a 7.5-mgd plant is used for swimming after chlorination. At the South African Pulp and Paper Industries plant at Springs, South Africa, secondary effluent from a 27-mgd plant is cleaned up for industrial reuse.

The only case in which waste water is converted to potable water is at Windhoek, South Africa. The treated effluent from a 4.5 Ml/day plant makes up 14% of the water supply of Windhoek, and as much as 40% during the winter months. The South African spokesman said that water supply and water demand in South Africa are on a collision course. It is an absolutely basic fact that more water sources must be provided to sustain growth in his water-short country, he said.

The health implications of waste water reuse for consumption as drinking water are simply unknown. Nevertheless, it has been estimated by EPA that about 200 million people in the world today are drinking water that can meet neither the World Health Organization standards nor the Public Health Service standards for drinking water.

### Looking ahead

What now seems to be of more immediate and urgent concern is identification of the factors that optimize biological treatment, so that the add-on advanced waste treatment process can indeed be chosen to make the best overall economical and reasonable system. Under certain operating conditions, both phosphorus and nitrogen can and have been removed by the biological process. But p-ct proponents are quick to point out that removal of either nitrogen or phosphorus, or even for that matter BOD, never occurs with any predictable deliverable efficiency. To confuse the choice of treatment methods even further, researchers are now finding biological activity on activated carbon columns.

A growing awareness of energy requirements is now for the first time being considered in the design of waste water treatment plants. There are engineers who point out that pump requirements, recycling operations, and the like are energy-using processes. Of the unit processes which are candidate replacements, and which are under the limelight, the low-energy users may well win out in the long term.

The real problem with p-ct, despite its theoretical advantage, may come in the actual operation of the plant. While it is true that no 100% physical-chemical plant operates today, operators at existing pilot plants find p-ct much more difficult to operate. The logistics of handling the chemical supply alone are staggering, so much so that the operation becomes troublesome. It is to be hoped that, before all designed plants go onstream, p-ct will not turn into an operational nightmare.

In addition to the operational difficulties and chemicals handling, there is no escaping the fact that operating costs of p-ct plants simply are not recovered. At this time, too, no one knows with sufficient assurance whether the heavy dosing of water with chemicals may not in fact be introducing trace quantities of impurities in treated water which would be in excess of water quality standards. The unknowns far outweigh the knowns. Despite the uncertainties, the needs for high levels of waste treatment are so pressing that p-ct will continue to be developed, albeit cautiously and under critical scrutiny.　SSM

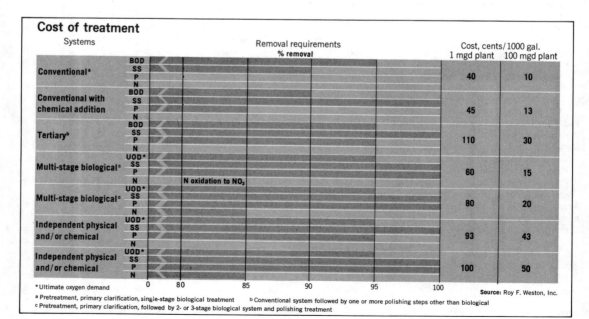

**Cost of treatment**

Source: Roy F. Weston, Inc.

* Ultimate oxygen demand
a Pretreatment, primary clarification, single-stage biological treatment
b Conventional system followed by one or more polishing steps other than biological
c Pretreatment, primary clarification, followed by 2- or 3-stage biological system and polishing treatment

Fish may be raised in warm water discharged from steam electric power plants, but success depends upon costs and profits

*Reprinted from* ENVIRON. SCI. TECHNOL., **6**, 232 (March 1972)

# Thermal aquaculture: engineering and economics

## Aquaculture included in power plant design

Water inlet

Nuclear plant site

Processing and freezing plant

Blended stream conveyed to aquaculture area

Pipeline transferring fish to processing and freezing plant

Discharge channel

Hatchery

Effluent discharge basin

Sump for collecting mature fish

Distribution basin

Channel separators

Thermal effluents from electric power plants have been a subject of increased environmental concern, and national power demands may double over the next decade. For every kilowatt of electrical energy generated, more than 1 and as high as 2 kW of low-grade thermal energy will be produced and discharged into water or air as a waste.

This waste heat could be considered a resource and examined as to how it may be utilized productively. Very little has been done to demonstrate the value of this waste heat in biological applications, particularly in agriculture and aquaculture.

Much has been published on the potential and possible limitations of utilizing thermal effluents to enhance the culture of aquatic species. Small-scale experiments have been described for shrimp and pompano in the United States, a variety of fin-fish in Great Britain, multispecies culture in Japan, and carp culture in the USSR. However, large-scale application of these effluents for aquaculture will be dependent on its commercial viability. (Figure 1 illustrates increasing demand.)

### Status of aquaculture

Aquaculture is a term that in recent times has come to imply a degree of environmental control over the culture medium such that fish yields are enhanced by orders of magnitude. Most impressive are the yields in running water culture with intensive feeding practiced by the Japanese—800,000 to 3 million lb/acre/year. By contrast, hunting wild species on U.S. coastal waters by conventional gathering methods may yield only about 20 lb/acre/year.

**191**

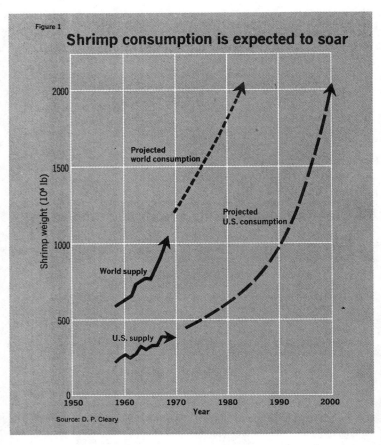

**Figure 1**

## Shrimp consumption is expected to soar

Shrimp weight ($10^6$ lb)

Projected
world consumption

Projected
U.S. consumption

World supply

U.S. supply

Year

Source: D. P. Cleary

**William C. Yee**
*Oak Ridge National Laboratory*
*Oak Ridge, Tenn. 37830*

**Aquaculture.** *Japanese shrimp, raised in warm water, may sell for $11–12/lb*

The yield figures demonstrate the potential of aquaculture, but there are also problems. Aquaculture as a technology is still in its infancy. Intensive culture of marine fish species is generally confined to the warm months of the year. The population in culture is dependent on the natural nutrient concentration in drainage from rivers and estuaries. No universally suitable artificial food has been developed yet, and yields may be subject to drastic curtailment by predator attack when the facility is not isolated from the sea.

The thermal discharge from a power plant is a potential source of warm water for maintaining optimum temperature ranges in aquatic environments for year-round fish culture and is a source of flowing water for intensive aquaculture. At large power stations of the 500–1000 MW size, flow volumes of hundreds of thousands of gallons per minute are available.

Precedence for the use of thermal effluents in aquaculture is only of recent vintage. Mollusks such as oysters are currently being cultivated year round on a commercial scale by Long Island

Oyster Farms, Inc. (division of Inmont Corp.) using the coolant water of the Long Island Lighting Co. at Northport, L.I.

Intake water, ranging from 40° to 70°F during the year, is warmed by 12° to 18°F and discharged at the rate of 150,000 gpm into a 7-acre lagoon. A continual source of warm water permits year-round culture. Baby oysters from a controlled environmental hatchery are placed on trays, and racks of these trays are immersed into a discharge canal from four to six months. A "finishing" stage follows in which oysters are transplanted to cold water areas to mature. Overall, the growing period is cut in half from about five years under natural conditions to about 2.5 years under culture conditions.

Catfish are being cultured commercially the year round at the fossil-fueled power plant of the Texas Electric Service Co. at Lake Colorado City, Tex. Cages of catfish are put in effluent water in the plant discharge canal that is about 70–75°F in the wintertime. Yields are reported to be the equivalent of 100 tons/acre/year with intensive feeding.

Experimental and/or developmental work in thermal aquaculture includes the following projects: catfish culture at the Gallatin Power Plant of the Tennessee Valley Authority by Trans-Tennessee Industries, oyster culture at Pacific Gas and Electric Plant in Humboldt Bay, and lobster culture at several institutions including a California group (San Diego Gas and Electric Co. and Mariculture Research Corp.) and the Department of Sea and Shore Fisheries, the State of Maine. Florida Power Corp. and Ralston Purina Co. recently announced a joint five-year program to evaluate thermal effluents for the cul-

**192**

ture of high-value aquatic species including shrimp.

The Japanese have pioneered in the use of thermal effluents for aquaculture. Shrimp, eel, yellowtail, seabream, ayn, and whitefish are among the aquatic species that are being evaluated. Since 1964, at least six generating stations have established demonstration programs. In one reported experiment at a power plant in Matsuyama, shrimp cultured during the summer in controlled temperature ponds had a weight gain of 1.2 times that of shrimp cultured in ponds with no temperature control; in winter, growth as measured by weight gain was seven times that in ambient temperature water.

The English commenced tests on the culture of plaice and sole on an experimental basis in 1966 using the thermal discharge from a nuclear plant in Scotland. With some temperature control and supplemental feeding, fish growth from the egg to a 1½-lb size and above was attained in less than two years, which is less than half the time required in nature.

### Design and cost factors

Fish cultivation in a dynamic system could be a way to utilize a resource that is currently discarded as a waste from steam electric power plants. A flowing stream provides a culture system with a more uniform concentration of dissolved oxygen. Biological oxygen demand is minimized because fish wastes and excess foods are flushed away, and, the system is more responsive to temperature control.

Only the Japanese have developed the technology of culturing shrimp (*Penaeus japonicus*) into a commercially viable operation. Extensive culture is practiced in numerous bays and inlet areas diked off from the sea, while intensive culture with flowing water has only recently been demonstrated.

In the Gulf of Mexico, shrimp growth under natural conditions is limited to the time period between April or May to about October or November when the water temperature ranges from 70° to 85°F. Growth virtually ceases when the water temperature drops below 70°F and does not resume again until the next spring when the water temperature warms up. However, if warm water were available on a year-round basis, two crops instead of one might be produced annually (Figure 2). An increase in the temperature of the water environment from 68° to 78°F increases the growth rate by more than

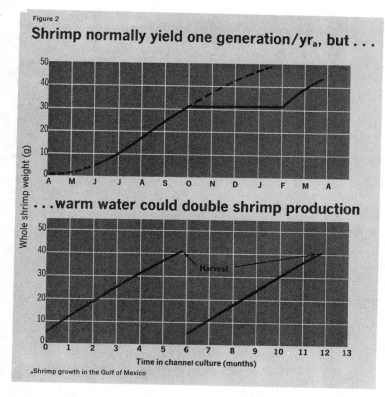

Figure 2

## Shrimp normally yield one generation/yr[a], but . . .

## . . .warm water could double shrimp production

Whole shrimp weight (g)

Time in channel culture (months)

[a] Shrimp growth in the Gulf of Mexico

80%, implying that better control of the water temperature might even produce three crops annually.

A conceptual design is based on the shrimp growth curve but is equally applicable, in principle, to culturable finfish species. As in Japanese culture, a period of up to 60 days is allotted to hatchery growth where mass cultivation from the egg to shrimp fry is done under highly controlled conditions. At this stage, the shrimp population should be developed sufficiently to tolerate the rigor of channel culture with a greater than 90% survival rate.

### Continuous culture

For intensive culture of shrimp on a continuous basis, a channel is divided into 26 pens of constant width and increasing length (Figure 3). The surface area of the first pen is proportional to the area under the first segment of the growth curve at the stage where channel culture of shrimp commences. Each succeeding pen area is proportional to the corresponding area formed by a segment of the growth curve.

Shrimp fry are introduced into the first pen of a channel and cultivated for a set period of time until they attain a

weight density (g/ft²) that is common to each succeedingly large pen. They are then advanced to a second pen, and more fry are stocked in the first pen.

Each succeeding week, shrimp are moved forward one pen until the end of the growth period in the 26th pen. They are now ready for harvest, the channel system is in equilibrium, and, ideally, it should be possible to harvest a uniform-sized product, week after week. A dependable source of supply of shrimp fry is assumed, so that, in principle, the number of new young shrimp introduced into the channel system is the same as the number of mature shrimp being harvested.

Flowing water is essential to the practice of intensive fish culture. The linear flow rate has to be high enough to sweep away fish wastes but low enough so that shrimp can maintain their position at the bottom of the channel with minimal expenditure of their food intake energy. The flow rate also has to be high enough to maintain temperature control of the water stream and to minimize atmospheric effects on heat loss.

Weight density is also an important consideration. In the continuous type of operation proposed, there is quite an economic incentive to devise culture procedures such as maintaining constant weight density throughout the culture period which will enhance yield per acre. Aeration is also included in the design so that fish population is not limited by the dissolved oxygen content of the water.

Feed from an external source is assumed so that culture population is not dependent on the nutrient concentration in the water stream. In Japanese culture, specific feeds have been developed for the larval and the post-larval stages of shrimp development. Beyond these stages, low-value fishes constitute the main diet of the shrimp in channel culture. This is suitable for culture economics in Japan where, during the off-season, live species sell on the retail market for $8/lb or more. In the United States, however, premium quality shrimp might be priced at one third of this value. Low-value fish as feed would not be economic in a large-scale continuous culture operation except under unique circumstances, such as a fish processing plant located in the vicinity of the aquaculture facility.

## Cost analysis

A hypothetical integrated aquaculture facility that includes a hatchery operation, culture in channels to a marketable size, and a processing plant that converts cultured shrimp to a raw headless frozen product is considered in a cost analysis. A conceptual design is proposed for a plant that is sized to distribute 1000 million gpd of water at 80°F through 17 culture channels in parallel flow. Electricity is furnished from a nearby power plant to drive pumps to blend ambient temperature water with thermal discharge water and to aerate channel water to maintain optimum culture conditions the year round. Power is also furnished for processing and freezing the cultured product. Yields are projected to be 10 million lbs annually when shrimp weight density can be maintained at 110 g/ft² over a total water surface of 400 acres.

Capital costs are divided up into land development costs which are variable according to the site chosen, and equipment costs which are fixed. The site-sensitive costs include a water conveyance system to the installation, culture channel construction, and a discharge system. Major capital items for the integrated facility include mechanical equipment to divide each channel into pens, utility and aeration equipment, instrumentation, hatchery, food pelletizing facility, dewatering and harvesting equipment, and shrimp processing and freezing equipment. The total capital cost for such a facility is estimated to be more than $47,000/acre (of which $27,000/acre is for site development).

Annual operating cost is estimated to be $0.80/lb of headless frozen shrimp. Shrimp feed accounts for 60% of the cost, assuming that the food cost is $0.10/lb and the food conversion ratio is 3 lbs of dry feed fed/lb of wet meat produced. The remainder of the cost is associated with labor for cultivating and processing shrimp, utilities, processing materials, plant overhead, and contingencies.

With this capital and operating cost schedule as a basis, a sensitivity analysis can be prepared to show the effects of site-sensitive costs, food conversion ratio, wage rate and labor productivity, price of product, and shrimp yield on annual production costs. This is a function of return on investment for an idealized capacity of 10 million lb/year. A standard case for this analysis is: site-sensitive cost of $27,000/acre, a food cost of $0.10/lb and a food conversion ratio of 3:1, a labor wage rate of $2.00/man-hr, and a labor productivity of 100,000 lb of fish handled/man-year. Annual investment costs include only recovery and return on investment and interest on working capital for 90 days. Depreciation is taken over a 15-year period. (Taxes and insurance have been omitted since these items vary from site to site.)

This approach can be used to indicate a range of conditions under which thermal aquaculture might be commercially viable as well as some conditions where such a venture might result in a deficit for the business operation.

One cost factor that was not included in this analysis is the potential expense of effluent waste treatment at the facility. Primary and perhaps secondary treatment may be necessary before discharging the effluent to receiving waters.

### Marketing and economics

The potential for thermal aquaculture can be examined both from an economic and from a marketing point of view. With a source of warm water to culture continuously, the unit cost of production can be reduced significantly below that of a seasonally cultured product. As an example, producing catfish seasonally in open ponds has been estimated to cost about $0.30/lb while culture in flowing water on a continuous basis has been projected to be $0.20/lb.

Land availability, large inventory of thermal effluent, and easy access to electric power are attractive synergisms between a power plant site and intensive aquaculture. Nuclear power stations are

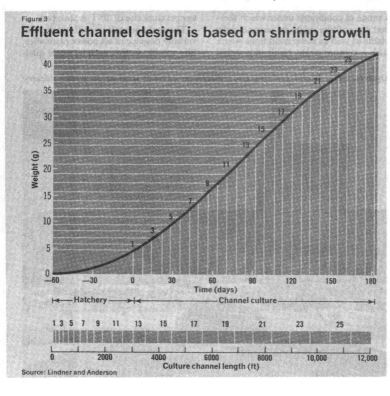

Figure 3

## Effluent channel design is based on shrimp growth

Source: Lindner and Anderson

required by law to have an exclusion area around the plant site. Since the operating utility has complete control over this area, which may be hundreds of acres, there should be ample land to build an integrated facility. Unit size of a new nuclear station would be at least 500 MW, and the coolant water requirement might be 615 million gpd if a water temperature rise of 20°F is specified.

Thermal aquaculture will not be universally possible at all power plant sites. Site specific factors such as water quality, single-unit or multiple-unit power stations, and plant shutdown and startup patterns would be factors in determining the feasibility of such an operation. For example, coolant water sources should not contain toxic quantities of trace metals. Or economic ways of removing objectional impurities must be available. Biocide residuals in the thermal effluent must be below levels that would be toxic to aquatic species in culture. Provisions may have to be made to cope with unscheduled shutdowns, when sudden changes in coolant water temperature might adversely affect the entire crop of fish in culture. The cost of sewage treatment of fish wastes in the culture facility effluent may be an important factor in the economics of the degree of intensiveness of fish culture. Nuclear plant thermal effluents used for aquaculture must be taken before the stream is used as a diluent for any radioactivity discharged from the installation. These are some of the major factors that would determine the commercial viability of thermal

**Fish.** *Heated discharge water can easily be diverted into tanks for fish farming*

aquaculture for a particular power plant site.

Thermal aquaculture has a potential that remains to be demonstrated. Since thermal effluents are waste products of the utility companies, a regulated-industry, technical feasibility may be only one of several considerations to be reckoned with to prove its commercial viability. Legal and regulatory hurdles on a federal, state, and local level and marketing considerations may also be equally important factors. Ultimately, site-oriented demonstration projects will have to show the revenue-producing potential of thermal aquaculture and the extent to which it can assist in helping to pay for the cost of dissipating waste heat from an electric power plant. The re-

source value of this waste might then be estimated, and a liability of the present might be turned into an asset of the future.

### Additional reading

"Chemurgy—For Better Environment and Profits," Proceedings of 32nd Annual Conference, Chemurgic Council, 350 Fifth Ave., New York, N.Y. 10001, 1971.
Proceedings of the Conference on Beneficial Uses of Thermal Discharges, Albany, N.Y., September 16–18, 1970, p 51. New York State Department of Environmental Conservation, Albany, N.Y., 1971.
"Joint Venture Aims at Thermal Effluent Use," *Amer. Fish Farmer*, **2** (8), 19 (1971).
"The Catfish Industry—1971," *Amer. Fish Farmer*, **2** (4), 12 (March 1971).

**Oysters.** *Long Island Lighting Co.'s discharge canal is used as an oyster nursery*

**William C. Yee** *is task force leader in the U.S. Atomic Energy Commission's Environmental Impact Reports Project at Oak Ridge National Laboratory. Dr. Yee has been engaged in waste utilization and waste treatment work since 1959 and has obtained four patents in these fields. Prior to his present position, he spent three years investigating productive uses for waste heat from power generating stations, not only in the technical sense, but also from the legal, regulatory, and product marketing points of view. The research in this article was sponsored by the U.S. Atomic Energy Commission under contract with Union Carbide Corp.*

Detection of mercury in water courses and in fish life above the standards established by the U.S. Public Health Service and growing concern with lead as an environmental contaminant stemming from its wide usage as a gasoline additive have sparked general interest in heavy metals as potential hazards in environmental control. The metals of most immediate concern are: chromium, manganese, iron, cobalt, nickel, copper, zinc, cadmium, mercury, and lead.

These metals are widely distributed in materials which make up the earth's surface. Igneous rocks, for example, typically average about 5% iron and contain other heavy metals at various levels ranging down to about 20 ppm lead, which is a relatively scarce element. These rocks are constantly weathered and leached by rainwater, and yet the natural runoff in rivers, even in extreme cases such as the Colorado River, is remarkably free of dissolved heavy metals.

Several of the so-called trace metals fill essential roles in life processes—e.g., manganese, iron, cobalt, copper, zinc, and molybdenum, while others such as mercury and lead are regarded with more suspicion as potential cumulative poisons. A remarkable compatibility exists, however, between the chemistry of heavy metals as encountered in nature and living organisms.

Heavy metals occur largely in natural mineral form as sulfides, oxides, carbonates, and silicates. These natural compounds are usually insoluble in water and only very slowly broken down by weathering and exposure to rainfall and groundwaters. For example, rainwater containing dissolved carbon dioxide attacks basic rocks such as peridotite which may contain 50% magnesium oxide and selectively dissolves magnesium in association with the bicarbonate ion, while iron, which may be similarly dissolved, oxidizes to the ferric form and precipitates as highly insoluble ferric hydrate even

# Removing
# heavy metals
## from waste water

*Reprinted from* ENVIRON. SCI. TECHNOL., **6**, 518 (June 1972)

**John G. Dean and
Frank L. Bosqui**

*Dean Associates
North Scituate, RI 02857*

**Kenneth H. Lanouette**

*Industrial Pollution Control, Inc.
Westport, CT 06880*

*Pending water pollution legislation requires heavy metal
removal not only before industrial wastes are discharged into
navigable waters, but also prior to ocean or land disposal*

when pH is as low as 2. Other heavy metals tend similarly to follow the behavior of iron and precipitate in the oxide residue while magnesia is carried off in groundwater as bicarbonate hardness. The order of precipitation from dilute solutions as the pH is raised is as follows:

| Ion | pH | Ion | pH |
|-----|-----|-----|-----|
| $Fe^{3+}$ | 2.0 | $Na^{2+}$ | 6.7 |
| $Al^{3+}$ | 4.1 | $Cd^{2+}$ | 6.7 |
| $Cr^{3+}$ | 5.3 | $Zn^{2+}$ | 7.0 |
| $Cu^{2+}$ | 5.3 | $Hg^{2+}$ | 7.3 |
| $Fe^{2+}$ | 5.5 | $Mn^{2+}$ | 8.5 |
| $Pb^{2+}$ | 6.0 | $Co^{2+}$ | 6.9 |

Very small heavy metal ion concentrations can be expected in neutral solutions. Ferric ion concentration in neutral water solution may even be far below the parts per billion (ppb) range subject to practical measurement. Iron is an extreme case but even lead shows extreme insolubility; for example, lead ion concentration at equilibrium with water containing 5 ppm or more of carbonate ion is less than 1 μg/l. or 1 ppb.

The heavy metals are, for the most part, responsive to practical treatment methods which have already been developed and utilized for water purification and metal recovery operations. Treatment methods which should be considered include chemical precipitation, cementation, electrodeposition, solvent extraction, reverse osmosis, and ion exchange.

## Chemical precipitation

The most generally applied treatment method, particularly where complex chemical compounds are not involved and economic recovery is not a con-

sideration, is the typical lime treatment plant (Figure 1), because of its relative simplicity and low cost of precipitant. Removing such metals as copper, zinc, iron, manganese, nickel, and cobalt requires almost complete precipitation as the hydroxide with no special modifications.

For cadmium, lead, and mercury, precipitation may be incomplete, however, and a modified flowsheet employing soda ash (for lead) or sodium sulfide (for cadmium and mercury) may be required. Where chromium is present, reducing the solution with sulfur dioxide, ferrous sulfate, or metallic iron before lime treatment is necessary. Chlorination may be needed to break down complex organic metallic compounds before chemical precipitation.

Where strong acidic wastes exist, part of the neutralization with limestone may be somewhat less expensive than lime. However, limestone must be evaluated carefully for each acid waste since it may not be effective as theoretically indicated owing to particle coating, need for fine grinding, and pH limitation of calcium carbonate. For example, nickel sulfate–sulfuric acid solutions treated first with crushed limestone, then with hydrated lime, show:

|  | Soln A | Soln B |
|-----|-----|-----|
| $NiSO_4$, meq/l. | 4 | 100 |
| $H_2SO_4$, meq/l. | 4 | 100 |
| Nickel, ppm | 117 | 2,935 |
| pH | 1.8 | 0.8 |

Agitation in a rotating reactor for 30 min with −10 +40-mesh crushed limestone at a rate of 100 g/l., filtered:

| pH | 4.9 | 4.9 |
|-----|-----|-----|
| Nickel | unchanged | unchanged |

Agitation of filtrate with two times equivalent of the nickel with calcium hydroxide for 30 min, filtered:

| pH | 10.0 | 11.1 |
|-----|-----|-----|
| Ni, ppm | <1 | <1 |

In the first step, coarsely crushed limestone readily reacts with the free acid but does not precipitate nickel. In the second step, hydrated lime precipitates the nickel below the detection point with sulfide ions. The nickel hydroxide precipitate settles quite readily and filters without difficulty.

The economic attraction of limestone is readily apparent when cost comparisons are made on an equivalent basis of contained calcium oxide (CaO). However, it is rarely possible to utilize the cheapest material—lump limestone—to anything approaching 100% effectiveness of its contained CaO. This is not true for pulverized limestone where neutralizing some mineral acids can be accomplished with high efficiency and pH below 5. To produce a neutralized effluent with a pH of 7 or above, either hydrated lime or quicklime (pebble lime) must be used.

## Electrodeposition

Some metals found in waste solutions can be recovered by electrodeposition techniques, using insoluble anodes (Figure 2). Acid pickling of copper as done in wire drawing provides an opportunity for applying this method. Typical practice involves dipping the wire coils at intermediate stages of drawing in hot solutions of acid (10% $H_2SO_4$, for simple cleaning or 10% $H_2SO_4$–5% $HNO_3$, $Na_2Cr_2O_7$ for other oxidant for pickling). In cleaning, oxides are removed from the surface, but when oxidants are used, the metal is etched. From 0.5 to 3% of the total copper fed to such operations may be dissolved. The copper content of solutions increases as the acid is consumed, and the dissolution rate declines. Finally, solutions must either be processed or replaced.

Spent solutions resulting from sulfuric acid cleaning of copper may be saturated with copper sulfate in the presence of residual acid. These are ideal for electrowinning where high-quality cathode copper can be electrolytically deposited while free sulfuric acid is regenerated. Anolyte from electrodeposition cells, after minor adjustments, can be recycled to the cleaning operation while cathode copper can be exchanged for credit to the basic supplier of copper bars. This type of recovery operation can be profitable while at the same time solving the bulk of the effluent disposal problem.

After the acid dip, copper coils must be thoroughly washed. Dragout (or carry forward) of acid copper sulfate solution can be minimized by use of a countercurrent system of first dipping in make-up water, but the final washing, to be effective, inevitably produces a large volume of dilute acid–copper sulfate solution. This same solution may also carry finely divided cuprous oxide which is not readily soluble in dilute sulfuric acid solutions. Typical composition of the wash effluent may average 0.5 milliequivalents (meq) of copper per liter (0.5 meq Cu/l. = 15 ppm Cu) and 0.5 meq/l. $H_2SO_4$, along with other elements present in plant water supply.

Ion exchange is a possibility for recovering copper from weak solutions as a copper sulfate solution of sufficient concentration to warrant charging to the electrolytic recovery system. The presence of free acid as well as other metallic ions can reduce the efficiency of this operation and also not recover cuprous oxide which may approach colloidal dimensions. Thus far, straightforward lime or lime-soda precipitation seems to provide the greatest practical promise. The general procedure is to dose the water effluent continuously with hydrated lime which may be supplemented with calcium carbonate or soda ash (much as employed in hot lime-soda treatment of bicarbonate hard water).

Liquid-solids separation may be accomplished after appropriate reaction time by thickening and filtration or centrifuging. Solids are mainly a mixture of copper hydroxide and basic carbonate plus calcium sulfate and carbonate. In some instances, these solids have been sold for the copper content, but more frequently must be disposed of as landfill. The filtrate carries some sodium sulfate but is suited for repetitive reuse in many plant operations or direct discharge to a diluting stream or sanitary sewer.

## Cementation

Another method to remove metals from waste streams, particularly where metal recovery is desirable, is cementation. Contracting a metal-bearing solution with the correct metal powder or scrap will precipitate certain selected metals as metallic "sponge" (Figure 3). In practice a considerable spread in the electromotive force between metals is necessary to ensure adequate cementation capability.

The most commonly used cementation metal is iron—often in the form of shredded detinned cans—and in the case of copper-bearing waste streams,

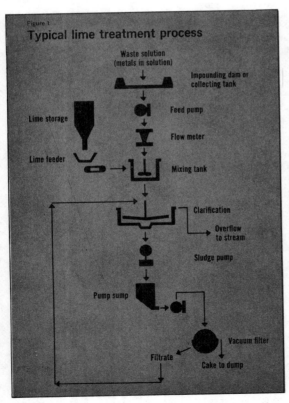

**Figure 1**

**Typical lime treatment process**

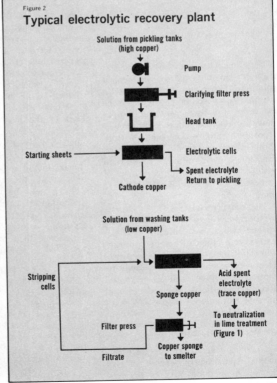

**Figure 2**

**Typical electrolytic recovery plant**

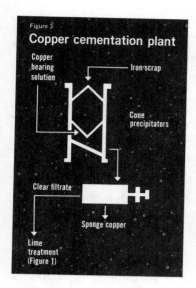

**Figure 3**

## Copper cementation plant

Copper bearing solution → Iron scrap

Cone precipitators

Clear filtrate

Sponge copper

Lime treatment (Figure 1)

settlers" are used, and countercurrent flow, where feed and solvent enter at opposite ends of the system, is preferred.

### Ultrafiltration

Reverse osmosis or ultrafiltration consists of semipermeable membranes which act basically as "molecular sieves" permitting soluble compounds having various molecular size ranges to pass through their pores. The membranes are synthetic organic materials and are frequently laminated. When set into pressurized ducts, these elements permit continuous flow, with the filtrate passing into parallel chambers. Pressures range from 50 to 600 psi, and 50 gal./ft$^2$/day, is the capacity. Power required for pumping amounts to about $^1/_4$ to $^1/_2$ hp/1000 gal. treated.

This technique has promise for removing and recovering metal ions from waste solutions, for, on a test scale, it successfully treated tertiary stage biological sewage and produced high-purity effluents.

### Ion exchange

Developed to a high degree of efficiency for the recovery of uranium from its ores, ion exchange resins are well-known. In addition to systems that handled clarified solution, the so-called "resin-in-pulp" method, in which the resin is introduced directly into the ore pulp, works successfully. After separating the loaded resin, the metal is recovered in a more concentrated form by elution with suitable reagents.

Previously, industrial waste treatment has not responded well to this form of processing. This has been attributed to destructive effects of certain impurities on the resins, interfering ions, limited

offers the possibility of recovering most of the copper as a salable by-product called cement copper. The operation is carried out in an acid solution and results in dissolution of iron which is later removed by lime precipitation along with any residual copper.

A process employing cementation has been developed by the Connecticut Research Council and is installed in at least one brass mill for recovering copper from waste pickling liquor. Using zinc dust as a precipitant for gold and silver from cyanide solutions is a widely used cementation process that has interesting potential for recovering certain small amounts of metals, such as cadmium, mercury, and lead from industrial wastes.

### Solvent extraction

One treatment scheme, liquid ion exchange, used in chemical and metallurgical industries involves extracting a particular metal from solution by contacting the solution with an organic reagent. This reagent reacts preferentially with the heavy metal ion of interest and converts it to a form soluble in appropriate organic solvents. A water-immiscible solvent for the organic reagent, such as kerosene, is intimately mixed with water, and then the two phases are separated. Acid-treating the organic fraction then releases the metal in a concentrated, water-soluble form which can be handled by conventional recovery methods.

This process, which has been adopted successfully by the uranium and copper industries, has possibilities for recovering metals from industrial wastes. A number of different types of "mixer-

## Heavy metals found in major industries

| | Al | Ag | As | Cd | Cr | Cu | F | Fe | Hg | Mn | Pb | Ni | Sb | Sn | Zn |
|---|---|---|---|---|---|---|---|---|---|---|---|---|---|---|---|
| Pulp, paper mills, paperboard, building paper, board mills | | | | ● | ● | | | | ● | | ● | ● | | | ● |
| Organic chemicals, petrochemicals | ● | | ● | ● | ● | ● | | ● | | | | ● | | | ● |
| Alkalis, chlorine, inorganic chemicals | ● | | ● | ● | ● | ● | | ● | ● | | ● | ● | | | ● |
| Fertilizers | ● | | ● | | ● | ● | | ● | | ● | ● | | | | ● |
| Petroleum refining | ● | | ● | | ● | ● | | ● | | | ● | ● | | | ● |
| Basic steel works, foundries | ● | | | ● | ● | ● | | ● | ● | | ● | ● | | ● | ● |
| Basic non-ferrous metals–works, foundries | ● | | ● | ● | ● | ● | | ● | ● | | ● | | ● | ● | ● |
| Motor vehicles, aircraft–plating, finishing | | ● | | ● | ● | ● | | ● | | ● | | ● | | | ● |
| Flat glass, cement, asbestos products, etc. | | | | | ● | | | | | | | | | | |
| Textile mill products | | | | | ● | | | | | | | | | | |
| Leather tanning, finishing | | | | | ● | | | | | | | | | | |
| Steam generation power plants | | | | | | | ● | | | | | | | | ● |

**Note:** plastic materials, synthetics; meat products; dairy products; fruits and vegetables; grain milling; beet sugar; beverages; and livestock feedlot industries have no heavy metal discharges.

loading capacity, high cost of operation, etc. If a selected purified dilute stream from pretreated waste were contacted with a suitable resin, however, a successful recovery step might result.

### Activated carbon adsorption

Similar to the resin-in-pulp technique, activated carbon adsorption, however, employs activated carbon instead of synthetic resins. Used successfully in commercial operation for extracting gold from cyanide solution, activated carbon adsorption has a number of advantages over ion exchange.

Relatively coarse (10 mesh) activated carbon is contained in stainless steel cylindrical screens which are submerged in the pulp, and the flow of suspended carbon from one screen to the next is countercurrent to pulp flow. The loaded carbon is then eluted with hot cyanide solution or sodium sulfide which dissolves the gold for later precipitation with zinc dust, and the carbon is recycled. The carbon is periodically reactivated by heating it in a small rotary kiln. Using activated carbon for metal removal has considerable promise in industrial waste treatment for removing the last trace of metal (in the range perhaps of 1 to 2 ppm) following electrodeposition or cementation, such as in the processing of electroplating wastes.

Each waste treatment problem must be regarded as a special case demanding a thorough study of the chemistry and economics involved. One of the critical factors, for instance, is the metal concentration in the solution to be treated. The various treatment schemes discussed are practical only within certain ranges of metal concentration in the feed solution. For example, processes that might be employed for various feed concentrations of copper are:

| Feed soln, g/l. Cu | Treatment scheme |
| --- | --- |
| 100 to 10 | electrowinning |
| 20 to 1± | cementation or electrostripping |
| 1 to 0.01 | lime precipitation, solvent extraction |
| 0.2 to 0 | ion exchange, activated carbon |

As discussed earlier, copper can be concentrated in very weak solutions by ion exchange or solvent extraction to a level where electrowinning or electrostripping might be applied. In the case of "stripping cells," copper is recovered as a fine "sponge" which falls off the cathodes and can be collected as a commercial product by vacuum filtration.

### Additional reading

R. H. Evans (Infilco), "Precipitation of High Density Metallic Hydroxides for Recovery or Disposal."

P. D. Kemmer et al. (Nalco), Chemical Treatment of Wastewater from Mining and Mineral Processing," *Eng. Mining J.,* April 1971.

P. D. Kostendader et al., "High-Density Sludge Process for Treating Acid Mine Drainage," Bethlehem Steel Corp., 1970.

W. A. Parsons, "Chemical Treatment of Sewage and Industrial Wastes," National Lime Association, 1965.

W. Teworte, "Economic Aspects of Recovery of Minerals from Effluents," *Chem. Ind.,* May 1969, pp 565–74.

**John G. Dean** *is manager and general technical director of Dean Associates. Dr. Dean has specialized in the chemistry of metals, including their extractive chemical metallurgy, refining, electrochemistry, and conversion to derivatives for applications in many fields.*

**Frank L. Bosqui,** *senior associate and project supervisor, has experience in South African gold and copper fields. His specialties include ore preparation, beneficiation, solids-liquid separation, furnacing, leaching, metal refining, precious metal recovery, and extractive metallurgy.*

**Kenneth H. Lanouette** *is vice-president of Industrial Pollution Control, Inc. He has extensive experience in water and waste water treatment plant engineering and sales and has presented papers on waste treatment and biofiltration. Address inquiries to Mr. Lanouette.*

# Tube settlers up clarifier throughput

*Ranks of inclined plastic tubes could save the day for overloaded water treatment plants if they perform as well in long-term trials as they have in pilot studies*

Reprinted from ENVIRON. SCI. TECHNOL., **6**, 312 (April 1972)

Here's the answer to a waste treatment plant operator's dream—how to double the capacity of his clarifier at a mere fraction of the cost of installing a second one.

New technology? Actually, it's more like refined technology. The prototype unit was built around the turn of the century. There was only one problem—it didn't work very well.

The flat, parallel plate settler, first built in 1904 has given way to the inclined tube settler—which works very well indeed. It works so well, in fact, that in pilot plant operations, and the initial industrial field installations, it has been possible to double flow rates through a clarifier without impairing effluent quality. And in most clarifiers, it's possible to install an inclined tube settling module in existing equipment without adding to space requirements.

## How it works

To understand how the inclined tube settler works, imagine a cylindrical tube, 1-in. in diameter and 10 in. long, filled with liquid in which settleable solids are suspended. If the tube were shaken and then held vertically, solids would have to travel almost 10 in. before coming to rest on the bottom. If, on the other hand the tube were held horizontally, the solids would have to travel only 1 in. to settle out. Settling in the horizontal tube would be approximately 10 times faster than in the vertical tube.

The problem with the horizontal tube—and with the earlier parallel plate settler as well—is that the tube quickly becomes plugged with solids. To avoid that, one could tip the tube—incline it—and the solids would slide out the bottom.

To make a continuous system, one would need only to cement several tubes together, incline the entire bank, and introduce at the bottom particle-laden water which would flow upward through the tubes. Solids would settle out, slide down the tubes and out at the bottom, and clarified effluent would be discharged over the top. Add a few minor engineering niceties and the result is a module which upgrades the efficiency of a precipitator system considerably.

A fairly typical configuration of the tube settler unit installed in a precipitator is shown at left. Waste water is introduced into the coagulation chamber along with suitable conditioners and flocculating agents and mixed by a rotor on the bottom of the tank. As coagulation proceeds, a sludge blanket is formed below the tube settler module. Flocculation is completed in the sludge blanket. Smaller particles become attached to larger particles rather than being carried out with the effluent. Sludge solids tend to collect in the bottom of the unit where they may be concentrated and removed.

As the effluent rises through the tube settler module, settling occurs in about the first one third to one half of the tube length, solids drop back into the clarifier, and clarified effluent spills out

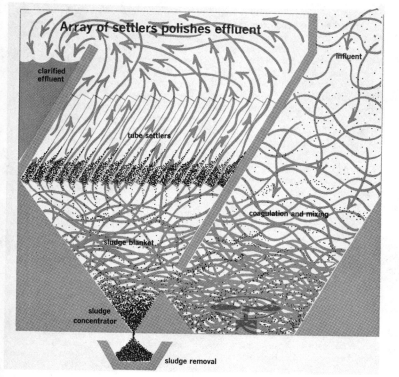

Array of settlers polishes effluent

clarified effluent

influent

tube settlers

coagulation and mixing

sludge blanket

sludge concentrator

sludge removal

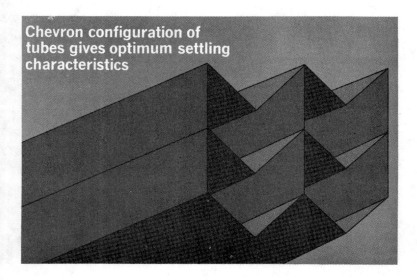

the top of the tubes and into the discharge stream.

### Tube design important

Although the settler concept is basically a simple one, performance depends heavily upon tube design. Flow characteristics are important because, for settling to occur, liquid flow needs to be laminar, and turbulence must be minimized. In the bottom section of the tube, flow is largely turbulent. As drag exerted on the liquid by the walls of the settling tube increases with tube length, turbulent flow changes to laminar flow.

Flow rates through the tube can be increased as long as the laminar flow characteristics are not upset. Generally, the higher the flow rate, the smaller the tube diameter (and therefore the higher the surface area on the liquid in proportion to its velocity) must be.

The flow characteristics of a tube are optimized by relating them to the Reynolds number, a dimensionless number which describes the general flow characteristics of a liquid through a tube. (With a Reynolds number less than 2100, flow is usually, but not necessarily, laminar. With a Reynolds number greater than 2100, flow is necessarily turbulent.)

The Reynolds number is related to fluid characteristics by the equation:

$$\text{Re} = \frac{D\rho V}{\mu}$$

where $D$ is tube diameter, $V$ is velocity of the fluid, $\rho$ is the density of the fluid, and $\mu$ is the viscosity of the fluid. The above equation holds true for circular tubes. For noncircular tubes, the equivalent diameter is substituted for the tube diameter, according to the equation:

$$\text{Equiv diam} = \frac{4 \times \text{area}}{\text{perimeter}}$$

Since the highest velocities are obtained by decreasing the equivalent diameter, the optimum shape of the tube is one with a cross-sectional area of maximum perimeter.

Tube shape is also an important consideration in determining settling characteristics. The tube height should be as short as possible to minimize settling distance. Uniform settling distance is desirable so most particles have the same settling time. Circular tubes, for example, are not efficient because particles entering at the top of the tube have a greater distance to settle than those entering at the sides.

Tube shape should permit nesting so that there is no wasted space between tubes in the unit. Again, circular tubes are not particularly efficient because of the large amount of dead space between tubes in the settler array.

One design which maximizes settling characteristics is the Chevron design, according to William A. Beach, research associate for The Permutit Co. Permutit's Chevron Tube Settler module is an array of nested 24-in.-long extruded polystyrene tubes with a cross-sectional chevron shape (see diagram). The 1-in. chevron configuration, Beach says, has the highest perimeter of any common shape of the same area. Tube height is less than tube width and settling distance for particles entering anywhere along the top of the tube is the same. An added advantage, according to

John R. Anderson, Permutit's vice president for research and development, is that the v-groove promotes optimum sludge compaction and flow.

From studies in pilot installations, and about a dozen industrial field installations, Permutit is optimistic that the tube settlers will provide significant improvement over existing conventional clarifiers. With proper engineering and design, Beach says, clarifiers can operate at rates between 4 and 5 gpm/ft². Without tube settlers, these same clarifiers are only rated at about 1-1.5 gpm/ft².

### Markets

Despite their promising future, one wonders why tube settlers are not more widely used today. Besides Permutit, at least two other U.S. companies (Graver Water Conditioning Co. and Neptune Microfloc) market proprietary tube settler units. Several installations are successfully operating in Canada, according to William Wachsmuth, president of Ecodyne, Ltd.

The main reason for the relatively poor market penetration, Beach says, is that the technology has been refined only in the past three years. Although there are a handful of operations scattered around the country using tube settlers, Permutit prefers to await further field testing before launching a major marketing effort. Still unanswered, Beach says, are questions of longevity. How, for example, will the buildup of algae or precipitates on walls affect the units' performances?

Furthermore, the units have not been tested on a wide variety of industrial process effluents—although there is no reason to expect that they would not do well, judging from preliminary data. Applications in municipal areas are similarly not well tested, although it appears that tube settlers would best be adapted to polishing secondary effluent rather than primary effluent.   HMM

Magnesium carbonate flocculation with regeneration
of lime and coagulant is under full-scale investigation
in Montgomery, Ala., after successful pilot runs

# Recyclable coagulants look promising for drinking water treatment

*Reprinted from* ENVIRON. SCI. TECHNOL., **7**, 304 (April 1973)

For the last half century or so, drinking water has been purified by roughly the same methods centering around alum or iron coagulation. The gelatinous nature of the hydrolysis products of aluminum or iron salts has served very well to trap floating solids and make possible their removal from water.

But the same gelatinous properties have made it difficult to dewater the sludges left over from water treatment. A newly developed water treatment process, developed by A. P. Black, professor emeritus of chemistry and environmental engineering at the University of Florida (Gainesville), and Cliff Thompson, a partner in the Montgomery, Ala.–based environmental consulting firm of Thompson and Tuggle, could change all that. Demonstrated under contract to the Environmental Protection Agency, the process uses magnesium carbonate as the coagulant. What makes the magnesium salt so highly attractive is that it's recyclable. Magnesium treatment also promises better quality water and easier sludge handling, although under current practices, magnesium carbonate coagulation costs more than conventional lime-alum or iron systems.

## Developmental work

The process is the outgrowth of work done earlier by Black with softening plant carbonate sludge at Dayton, Ohio. Thompson, who recently received his Ph.D. under Black's direction, is now in charge of a full-scale test of the process at the 20-million gpd Clarence T. Perry water purification plant, just outside Montgomery, Ala.

The team carried out tests on natural and synthetic waters—water samples prepared by adding known amounts of contaminants to distilled water—in small jar tests and large 55-gal drum tests to work out the chemistry of precipitation before putting the process on stream at a 50-gpm pilot plant at Montgomery.

They finished the developmental work and began operation of the Perry plant in June of 1971. Within a couple of months, the evaluation phase of the work will be complete. EPA officials held a preliminary technology transfer meeting at the Perry plant in February.

The basic chemistry of the purification process is relatively simple. A lime slurry is added to raw water which contains either naturally occurring magnesium bicarbonate or magnesium carbonate which has been added. The addition precipitates magnesium hydroxide and calcium carbonate. The magnesium hydroxide acts in a manner similar to the hydrolysis products of iron or aluminum salts and forms a floc which settles impurities out of the water.

In essence, the treatment system is a combination softening and purification process. Softening takes place by raising the pH of water by adding lime to convert all carbon dioxide and bicarbonate alkalinity to the carbonate form according to the reactions:

$$CO_2 + Ca(OH)_2 \rightarrow CaCO_3 + H_2O$$
$$Ca(HCO_3)_2 + Ca(OH)_2 \rightarrow 2CaCO_3 + 2H_2O$$

In the case of magnesium salts, the carbonate is converted to the hydroxide form by adding more lime:

$$Mg(HCO_3)_2 + Ca(OH)_2 \rightarrow MgCO_3 + CaCO_3 + 2H_2O$$
$$MgCO_3 + Ca(OH)_2 \rightarrow Mg(OH)_2 + CaCO_3$$

When magnesium is present as noncarbonate hardness, calcium is exchanged for magnesium and there is no change in the total hardness:

$$Mg^{2+} + SO_4^{2-} + Ca(OH)_2 \rightarrow Mg(OH_2) + Ca^{2+} + SO_4^{2-}$$

## Reclaiming magnesium

Once the settled sludge is removed from the process, it may be dewatered and landfilled—a practice followed with conventional sludges. But with magnesium as the coagulant, it's possible to do more. The sludge can be carbonated by bubbling carbon dioxide through it, which solubilizes the magnesium selectively as magnesium bicarbonate. The magnesium bicarbonate can then be reclaimed from the filtrate obtained by vacuum dewatering and returned to the process. The filter cake, consisting chiefly of clay and $CaCO_3$, may then be landfilled or treated one step further. The lime values can be reclaimed from the sludge, and the carbon dioxide given off from the calcining of $CaCO_3$ to produce CaO for lime makeup can be used in the recarbonation process. At the same time, reclaiming the lime values further reduces the amount of sludge that must be landfilled. The sludge that's finally left is almost entirely clay—easily handled and environmentally stable.

The magnesium salt favored in the water treatment process is the carbonate trihydrate ($MgCO_3\cdot3H_2O$) form. As the trihydrate, magnesium carbonate does not add to the total dissolved solids content of the water.

Unfortunately, the salt is not commercially available since there are presently no large industrial uses for it. But the trihydrate could very well be available in sufficient quantities

## Flow scheme of Perry Plant comparing MgCO₃ and alurn processes

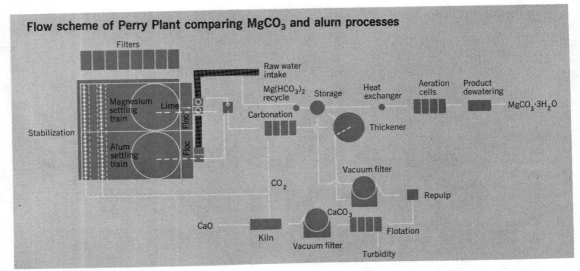

from cities which operate large-scale softening plants.

Cities which must treat water with relatively high magnesium content—including major metropolitan areas such as Chicago, Detroit, Cleveland, Washington, Indianapolis, Philadelphia, and Pittsburgh—would be able to recover large quantities of magnesium carbonate trihydrate which they could sell to offset the cost of water treatment. Thompson and Black estimate that about 150,000 tons per year of magnesium carbonate trihydrate would be produced by 20 cities with high magnesium water. By switching to the recycled magnesium carbonate process, these cities would not only be able to reduce their treatment costs, but would solve their present sludge disposal problems as well.

There are many advantages to magnesium carbonate treatment, Thompson points out. The magnesium carbonate system is at least as effective as iron or aluminum in removing turbidity and organic color from surface waters. Neither the base exchange capacity of turbidity causing clays nor the level of turbidity present significantly affects optimum coagulant dosage.

Unlike the gelatinous, ephemeral flocs formed with aluminum or iron salts, the magnesium hydroxide flocs are better formed and heavier since they are loaded with calcium carbonate. There is, therefore, more rapid and complete settling with less carryover of the floc in succeeding stages of water treatment.

The process carries a natural disinfection bonus with it. With alum and ferric sulfate treatment, viruses are removed to a certain extent by coagulation but can be recovered live from the floc. Because the pH of

the magnesium carbonate system is so high, however (11.0–11.5), the viruses are not only removed by the flocculation step, but are destroyed by the alkalinity. Bacteria are also removed by the process, eliminating the need for rechlorination in many applications.

The high pH also promotes removal of virtually all the iron or manganese that may be present in some surface waters.

### In operation

The magnesium carbonate process is in full-scale operation today at the Perry plant and residents of Montgomery have been drinking water purified by the process for about 9 months. The Perry plant is the ideal proving ground for the process since the plant consists of two parallel 10-million gpd purification trains, one of which uses conventional lime-alum treatment while the other uses the magnesium carbonate process. Much of Montgomery's water comes from wells and the plant is seldom run at peak capacity, but magnesium carbonate has been used on a large enough scale here to prove that it works reliably in day-to-day operation.

The Perry plant does not necessarily plan to convert to the MgCO₃ process, however, even when all the bugs have been worked out, Thompson says. The final utility of the process may depend more on federal regulations than anything else. For all its advantages. the magnesium carbonate process is more expensive than conventional treatment, in the Montgomery application—unless the cost of sludge disposal is fully accounted for. And much depends upon regulations for sludge treatment which the EPA may adopt.

While the process allows recovery and reuse of both lime and magnesium carbonate, Black and Thompson estimate that there are more than 4000 water plants in the United States that will not find lime recovery possible because of the small amounts of lime required at those plants. The carbon dioxide produced by recalcining figures high in total costs, however, since bottled CO₂ can come high. For plants that could not afford to make their own CO₂ for the recarbonation step, Thompson suggests that an alternative source of CO₂ might be found. Carbon dioxide could come from a diesel fuel or natural gas–burning engine, he suggests, or even from power plant stack gases or other industrial source.

The cost differences between the magnesium carbonate and lime-alum systems may be more apparent than real, Thompson says. For example, at the Perry plant, the cost of lime–alum treatment, without provision for sludge handling, is $8–10/million gal. The cost for magnesium carbonate treatment, with purchased CO₂, high calcium lime, and purchased MgCO₃, without recycling of lime is about $19/million gal. If dolomitic lime could be used and could serve as a magnesium source, the costs would come down to about $12/million gal. If a CO₂ source— say from a diesel engine or power generating plant could be incorporated, the cost would tumble even further. On-site power generation, with use of CO₂ in stack gases could put the operating costs at about the level as conventional lime–alum systems. "Of course, if we add the costs of sludge handling, treatment costs are already in the same range," Thompson says. HMM

204

*Reprinted from* ENVIRON. SCI. TECHNOL., **7**, 20 (January 1973)

# Canals cool hot water for reuse

Hailed nationally as an example of what an electric utility can do to make itself compatible with the surrounding environment, the Florida Power & Light Co. (FPL) has come up with a unique concept for power generation and cooling water requirement at its Turkey Point plant located 25 miles south of Miami. The "perfect marriage" of industry and environment is attributable to McGregor Smith, a noted conservationist and FPL's chairman of the board who recently passed away. It was his idea to use the vast acreage required for a power plant site for as many beneficial uses as possible—wildlife reserves, nature trails, public beaches, picnic areas, scout camps, and sea survival training.

The Turkey Point plant is not FPL's newest, but it certainly is the one that caused the greatest stir and gained the most attention. Basically, there are two conventional oil-fired plants and two nuclear plants at Turkey Point. Construction of the two conventional plants (432 MW each) started in September 1964; the first unit went on line in April 1967, and the second a year later. FPL announced construction of the two nuclear units (760 MW each) in November 1965, and construction of the first of them began in April 1967. This unit received an operating permit from the U.S. Atomic Energy Commission last July and only last month turned out power for the first time. The second nuclear unit is scheduled for initial startup in about a year.

## Legal injunction

In effect, the first two plants were operating and the others were under construction before any concern about thermal pollution had been registered. As everyone knows by now, the debate on thermal effects was mounted early in 1968. Turkey

Point became a focal point in the national debate which resulted in a suit by the Justice Department against FPL and a great deal of legal proceedings.

The Justice Department suit sought a legal injunction on the operation of the plant—both temporary and permanent stoppage of electricity generation. Although operation of the plant was in fact never stopped, the federal judge who heard the arguments for the temporary injunction personally visited the Turkey Point plant and ruled that there was no irretrievable damage. If there

*Florida Power and Light Co. finds that there is more than one way to cool water used in making electric power, but every answer depends on the locale*

were damage, it was minimal and retrievable, he ruled, and FPL plans that were then under way on construction of a six-mile canal with discharge into Card Sound, the underbelly of Biscayne Bay, would create a "benign situation."

The consent decree into which FPL entered with the Justice Department on September 10, 1971, applies to the entire cooling water system for the plant—both the conventional and nuclear units at Turkey Point. It was reached after some 26 meetings with government officials over a 15-month period and was heard in the U.S. District Court for the Southern District of Florida.

Even after the thermal effects issue emerged nationally, the only law or numerical guideline that FPL has ever had to go on was a Dade County requirement of 95°F as a limitation.

The decree, a very elaborate and very carefully detailed document, spells out every phase of the cooling requirements. For example, there are detailed specifications on the operation of the cooling system with regard to velocity, volume, and temperature of the water, as well as requirements for monitoring. The decree also spells out terms for the operation of an "interim system." In effect, FPL is operating under terms of the "interim system" now.

## Closed system

The end of 1974 is the goal for construction of the entire cooling capacity at Turkey Point (see map). When Dade County passed its ordinance requiring the temperature to be no higher than 95°, the company started construction of the six-mile canal which was designed to meet the 95°F limitation. The precise cost of this construction has not been broken down by FPL. An alternate system—a closed circuit—will be completed by the end of 1974.

In the interim system, the cooling water comes into the plant, flows through and cools the condensers, and flows out the six-mile canal.

In the closed system, the makeup water comes into the plant in the same fashion, but the water will circulate through this cooling system like a giant radiator. By the time the water has worked its way through the network, which has an accumulated length of 168 miles, it will be cool enough to be recycled—i.e. used again. Occasionally, makeup water will have to be brought in from Biscayne Bay and periodically, there must be some purging of the system

**205**

—for example, if and when there is a salt buildup.

Of course, this additional cooling capacity does not come scot-free. The total project cost on the cooling canal system adds an estimated expense of $30 million to the Turkey Point plant. All told, some 38 canals are involved in the construction; eight have been completed, the other 30 are expected to be completed by the end of 1974.

Dade County is in the throes of issuing a $36 million revenue bond issue which will cover the Turkey Point plant. It's the first use of these bonds in Dade County. The county commission of Dade County has voted to move ahead on the necessary steps for the bond issue but they were not complete at press time. The bonds are financially backed by FPL; they are tax-free municipal bonds and in that sense the federal and state governments have come forth with a mechanism for making financing available.

Acquiring the land mass for the construction of the canal system is a saga in itself. FPL had to purchase an enormous chunk of land; Turkey Point is now more than 12,000 acres. Initially, there was some disagreement on "bulkhead line" (boundary location) because of the marsh land on Card Sound. It was necessary to take steps to make the land useful other than for cooling purposes. In working with the state, FPL deeded to Florida certain wilderness areas which will be preserved in their natural condition. In return, the state settled the question on the location of bulkhead line which then permitted construction to proceed on the cooling system.

### Other cooling solutions

Before the canal system won out as the cooling solution at the Turkey Point plant, a number of other techniques were considered but discarded for one or more reasons:

● cooling towers were ruled out because construction on marsh land was not feasible, nor was there any U.S. experience in the use of salt water in utility cooling towers

● deep-well injection was deemed not practical in the marsh area.

FPL has 10 major electric utility plants in the state, and has newer ones under construction. The utility is taking preventive measures against thermal pollution from these plants. But it is important to note that cooling requirements are dictated by the environmental requirements of the area where they operate. There is no uniform solution to every situation.

This summer, a new unit went into the Sanford plant; FPL had older units at Sanford, but one new unit

came into service this summer and another is under construction and will go into service next year. A man-made cooling lake, about 1100 acres in area, will take care of cooling the water at Sanford. FPL also has units under construction in Manatee County on the west coast. There is also the Hutchinson Island nuclear plant, which has become known as the St. Lucie plant. Since the plant is located on the ocean, an ocean-to-ocean once-through cooling system is used.

In terms of both the environment and technology, the Turkey Point case is a classic example of society's shifting priorities. For years, the

call for electric power was for more power at the lowest cost. Suddenly— almost overnight—priorities shifted to beautification and environmental protection, which are in many ways in conflict with low-cost power.

FPL has never had a rate increase in its 47-year history but in December 1971 it filed for one. A decision is still pending before the Federal Power Commission. Inflation and the impact of environmental expenditures are very much a part of the case behind the need for higher rates. Simply put, projects like the cooling canals that are designed to protect the environment cost money.    SSM

FPL power plants

St. Augustine
Palatka
Daytona Beach
Sanford
Cape Kennedy
Cocoa
Hutchinson Island
Rivieria
West Palm Beach
Ft. Myers
Boca Raton
Port Everglades
Hollywood
Lauderdale
Miami Beach
Miami
Cutler
Turkey Point

"Interim system" cooling at Turkey Point plant

Turkey Point plant
Water leaves plant 10–15°F warmer
Water from plant is cooled by mixing with cooler bay water
Cooling water enters plant intake channel at same time water for dilution enters canal
Biscayne Bay
Cooled water travels six miles in canal cooling further enroute; travel time is approximately 4.5 hr
Water cannot leave canal in excess of 95°F under Metro Ordinance
Card Sound

"Closed system" cooling at Turkey Point plant

Turkey Point plant
Inlet
8 canal finished
Status of construction today
30 canal to go
5.5 miles
6 miles
Outlet
Target date—late 1974

# Membrane processing upgrades food wastes

*Cheese whey has been causing pollution problems, but selective membranes may stop pollution as well as provide potential marketable products*

Reprinted from ENVIRON. SCI. TECHNOL., 5, 396 (May 1971)

Little Miss Muffet had a taste for curds and whey, but dairymen and cheese processors do not find whey so appealing. Too much whey is the problem. Faced with increasing amounts of waste whey that can no longer be overlooked or dumped into streams, cheese processors are searching for adequate disposal or reclaiming methods.

Whey, a greenish-yellow waste fluid from cheese manufacturing, can be classified as "acid" or "sweet." Whole milk, used in natural or processed cheeses such as cheddar, results in a sweet whey (pH 5–7). Acid whey (pH 4–5) is a by-product from cottage cheese that is usually produced from skim milk. Both wheys have a high biological oxygen demand (BOD content—30,000 to 45,000 mg/liter) which makes disposal to streams undesirable and, in some cases, unlawful.

Also, many municipal sewage treatment systems have difficulty handling this waste adequately. The BOD in 100 pounds of whey is equivalent to the waste produced by 21 people every 24 hours; in other words, a cheese plant producing 100,000 pounds of whey per day requires waste treatment equivalent to that of a city of 22,000. Many of the (usually rural) communities in which cheese plants are located have only primary treatment facilities that cannot reduce BOD sufficiently. Since it is difficult to handle whey in a single unit, many municipalities refuse to accept it into their disposal systems.

## Waste whey

At present, 5–10 pounds of whey is left from each pound of cheese processed. In total, 22 billion pounds of whey is produced annually in the U.S. According to the U.S. Department of Agriculture, about one third of this whey is now used in human food or as a supplement to animal feed. That leaves about 14¾ billion

pounds for the cheese manufacturer to contend with.

But this "waste whey" is not really a complete waste. Although whey itself is 93–94% water, the solids portion consists of approximately 74% lactose, 13% protein, 8% ash, 3% lactic acid, and 1% fat. Actually, discarding whey as a waste is a loss to the world's food supply. For example, in 1970, over 400,000 pounds of lactose was imported each month, verifies R. A. Anderson, executive secretary of the National Cheese Institute. Furthermore, protein from whey can be equal to or of a higher quality than that in nonfat dry milk, says D. A. Vaughn, USDA.

If the disposal methods at hand are undesirable or inadequate and recovery procedures unestablished, what can be done with 22 billion pounds of fluid whey each year? In answer to this question, the Industrial Pollution Control Program in the Water Quality Office (formerly Federal Water Quality Administration) is encouraging and partially supporting R&D projects to control pollution from acid wheys, sweet and acid whey rinse water, and wheys from small cheese plants in rural areas. Costs of ongoing WQO-supported projects now total over $7 million (WQO participation amounts to about $2 million). One such project for controlling pollution from acid whey was undertaken by the Crowley Milk Co., Binghamton, N.Y.

In previous years, whey has been utilized in various manners: • used in its original form for animal feeding; • condensed or dried for whey solids or lactose recovery and used in human or animal food; and • deposited on land where nuisance or water pollution would not occur. Each of these methods has its disadvantages and cannot adequately dispose of large amounts of whey continuously. Even the present product recovery methods (evaporation and drying, precipitation of protein and crystallization of lac-

tose, electrodialysis, gel filtration, or fermentation) have their drawbacks and limits when faced with such huge quantities. Another major difficulty has been breaking the product into fractions for more diverse use. As is, whey's uses are limited.

The most common whey disposal methods have been runoff into the local sewage system (as previously mentioned) and high-temperature drying to produce whey powder. Although sweet whey powder is a salable product, it is usually just a break-even proposition and still adds more than pure water to local sewage systems. Acid whey drying, however, is a technology mastered, at this time, by few.

### The Crowley project

The Crowley Milk Co. project, headed by Robert R. Zall, experimented with membrane processing cottage cheese whey for pollution abatement. "The use of membranes is not new," explains Zall, "but the way that we are using them is rather unusual." The membrane processing is a two-stage program—ultrafiltration and reverse osmosis—simultaneously fractionating and concentrating the whey to produce protein, lactose, and a low BOD effluent.

The WQO grant calls for dividing the project into two phases—design, construction, and operation of a pilot plant; and, depending upon the outcome of the pilot phase, construction and operation of a full-scale plant.

The ultrafiltration section, a low-pressure system operating at 60 psi and at 120° F removes the protein from the whey, and when operated in a batch method, can concentrate it 20-fold or more. The cellulose acetate membranes—supplied by Abcor, Inc. (Cambridge, Mass.)—in this ultrafiltration system have a 20-Å pore size (one angstrom unit is one ten billionth of a meter). They operate below 50 psi and selectively retain 97–98% of the real proteins (α-lactalbumin, lactoglobulin, and higher molecular weight proteins in whey). The protein concentrate can be recovered from the membrane unit by displacing it with water or by draining.

The deproteinized solution from the ultrafiltration unit, containing water, lactose, amino acids, lactic acid, salts, and vitamins, then passes to a 90° F reverse osmosis membrane system. The high-pressure reverse osmosis system has two purposes: concentrating a lactose stream and reducing effluent BOD. Another use is concentrating the whole whey, again reducing BOD. This reverse osmosis membrane (American Standard, Hightstown, N.J.), also cellulose acetate, has a 3-Å pore size and operates at 800 to 1000 psi. As water is removed, the lactose concentration is increased, and the effluent is mostly water with some minor contaminants, such as salt and lactic acid. This can be handled in most waste treatment systems. "We have reduced the BOD content from the incoming whey in excess of 97%—from 35,000 mg/liter to less than 1000 mg/liter," continues Zall.

What about the lifetime of the membranes? They have to stand continuous pressures without fatigue or leaks. "The membranes used for the pilot plant show no signs of degradation after operating for a year," comments Zall. The membranes could conceivably last years. A five-year life is being projected for depreciation and economic replacement purposes, Zall adds.

As for leaks, these can be detected easily enough, however. The whey entering the system is a greenish color (caused by riboflavins) but leaves the reverse osmosis unit as a clear, colorless liquid. If the effluent is tinted (the riboflavins should have been removed), there is a leak in the system. Another simple method used to check for leaks is conductivity. A number of salts are in the deproteinized whey influent; if there is a leak in the system, the salt level increases markedly in the effluent and shows up by increasing conductivity in the permeate.

## Success

The Crowley pilot plant operation processed 1000 lb/hour or 20,000 lb/day of acid whey. Since this unit was successful, phase two of the project, a potential 300,000-lb/day plant, will be built; in fact, negotiations for construction are now under way.

There are other advantages for the dairyman in this scheme. The machinery and equipment used for whey processing can be operated much like conventional dairy equipment and can be cleaned to satisfy present health standards. Effective processing of acid whey "has been moved from a laboratory phenomenon to a real-world situation," summarizes Zall.

Successful completion of phase one in December 1970 resulted in another step toward the complete waste recycling that is becoming a necessity. Zall concludes from phase one that:

• Membrane processes can be economically attractive for treating cottage cheese whey by deriving protein concentrates through ultrafiltration and lactose through reverse osmosis.

• The BOD of raw cottage cheese and whey can be reduced from 35,000 mg/liter to less than 1000 mg/liter.

• Protein concentrates up to 20% protein (80% protein on a dry solids basis) can be generated in a single step (ultrafiltration) but will probably be generated in a batch method.

• Twenty-percent lactose concentrates (75% lactose on dry solids basis) are produced by reverse osmosis of the ultrafiltration permeate.

• Whole cottage cheese whey can be concentrated to approximately 25% solids by reverse osmosis without prior deproteinization (ultrafiltration) but will require special cleaning techniques to prevent membrane fouling.

• Protein and lactose products have no coliform count and low plate counts for other microorganisms when the equipment is properly cleaned and operated.

• Plant capital cost for a 250,000-lb/day installation is projected to be $610,000 with annual operating costs of $196,000.

• Installation and operation of the potential 300,000-lb/day plant could yield a high return on investment before taxes, corresponding to rapid plant payout. (This depends, of course, on sales of the various products which are recovered. Although individual products at this time tend to be high priced, there is still difficulty in identifying markets and penetrating them.)

After recovering or reprocessing the fractions from whey, these ingredients have to be marketed or the technology involved is wasted. Still the largest whey users are manufacturers of animal and poultry feeds. In the pharmaceutical line, lactose is used in antibiotics and as a carrier or extender in pills.

Future uses will determine whether whey and its products can compete economically. Several new markets seem to be appearing, however. Whole whey itself is being considered as a partial substitute for nonfat dry milk. Lactose greatly intensifies the flavor of foods, much like monosodium glutamate. Protein can be utilized as a human food supplement.

Technical developments for cheese whey reclamation and marketing may not cause dairymen to exactly savor Miss Muffet's favorite dish, but the outlooks for pollution control, waste reclamation, and profit seem much brighter.                        CEK

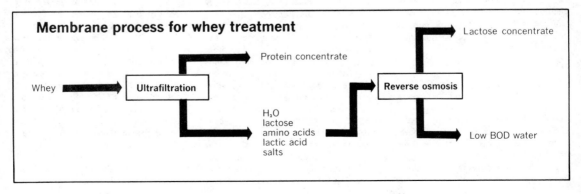

## Membrane process for whey treatment

Whey → Ultrafiltration → Protein concentrate

H₂O lactose amino acids lactic acid salts → Reverse osmosis → Lactose concentrate / Low BOD water

**Jacob I. Bregman**

*Water Pollution Research and Applications, Inc.*
*Washington, D.C. 20036*

# Membrane processes gain favor for water reuse

*Recent studies hold promise*
*of cost reductions and improved*
*efficiency in many applications*

Reprinted from ENVIRON. SCI. TECHNOL., 4, 296 (April 1970)

A phenomenon is occurring in the U.S. that would have been unbelievable a few years ago. We are running out of clean water. This is disconcerting because our country is blessed with many large streams and rivers. The U.S. abounds in lakes varying in size from farm ponds to the mighty Great Lakes. We have one of the most highly developed dam and reservoir systems on earth. Our average rainfall is sufficient to water most of the nation plentifully. Nevertheless, our population explosion and industrial growth have made the need to clean up and reuse our waters a new and painful fact of life.

The U.S. Water Resources Council predicts that municipalities which now use 23.7 billion gallons of water a day will need more than 50 billion gallons a day by 2000. In other words, our cities will need more than twice as much water within the short span of 30 years. Water demands by industry make these figures seem almost insignificant. Already using twice as much water as cities, U.S. industry, by the year 2000, will be demanding almost three times its present consumption.

To varying degrees, depending upon the state of development and population, this is the situation facing most the countries of the world—how to make the same amount of water stretch farther. This requirement calls for the development of new treatment processes, since conventional water and waste treatment processes are only partially effective against an expanding list of compounds resulting from our advanced technology. Dilution water is becoming less and less available, and stream self-purification treatment does

## Reverse osmosis for industrial wastes

| Capital costs[a] | 10 m.g.d. | 50 m.g.d. |
|---|---|---|
| Desalting equipment | $2,037,500 | $ 9,323,710 |
| Pumps, piping, valves | 1,194,850 | 4,653,000 |
| Pretreatment and feed intake system | 252,000 | 1,072,000 |
| Membrane casting, energy recovery, chemical injection systems | 164,000 | 510,000 |
| Instrumentation and electrical equipment | 420,000 | 1,770,000 |
| Buildings, land, and storage facilities | 332,850 | 761,000 |
| Miscellaneous and financing costs | 397,950 | 1,957,220 |
| Total capital costs | $4,799,150 | $20,036,930 |

| Operating costs (cents per 1000 gal.) | | |
|---|---|---|
| Annual fixed costs | 12.0 | 10.0 |
| Supplies, labor, maintenance | 23.1 | 15.1 |
| Total product water costs | 35.1 | 25.1 |

[a] Basis: Costs in 1968 dollars, 800 p.s.i.g. operating pressure, 70% product water recovery, and membrane life of one year

## Municipal tertiary treatment

| Capital costs[a] (thousands of dollars) | Conventional plant | Conventional plant plus electrodialysis | Conventional plant plus reverse osmosis |
|---|---|---|---|
| Primary and secondary treatment | $ 6,626 | $ 6,626 | $ 6,626 |
| Surge pond | — | 405 | 405 |
| Carbon filters | — | 2,492 | 2,492 |
| Electrodialysis unit | — | 6,025 | — |
| Reverse osmosis unit | — | — | 4,020 |
| Product water storage | — | 332 | 332 |
| Buildings, land, and services | 2,119 | 2,964 | 2,964 |
| Engineering and miscellaneous costs | 1,560 | 3,120 | 2,786 |
| Total capital costs | $10,305 | $21,964 | $19,625 |

| Operating costs (cents per 1000 gal.) | | | |
|---|---|---|---|
| Fuel and electricity | 1.6 | 3.5 | 4.8 |
| Chemicals | 2.6 | 5.2 | 5.5 |
| Supplies | 1.0 | 5.4 | 7.1 |
| Brine and solids disposal, labor, and miscellaneous costs | 11.5 | 23.3 | 21.7 |
| Total product water costs | 16.7 | 37.4 | 39.1 |
| Net costs of demineralization | — | 20.7 | 22.4 |

[a] Basis: 30 m.g.d. plant at 1968 prices with 7.1% capital charges

**209**

not work for many compounds that resisted waste treatment. The result is that tertiary treatment must be utilized if polluted water is to be made usable again. Nine tertiary treatment research and demonstration projects were operational in 1968 in various parts of the U.S., and 11 others were in the design or construction stage. Evaluation of these processes has proceeded to the point where preliminary cost estimates can be made. Some typical tertiary treatment costs appear in a table accompanying this article.

The cost of the removal of organic and inorganic nutrients by tertiary treatment range from 26-30 cents per 1000 gallons, depending on the techniques employed; this compares to 11 cents per 1000 gallons for secondary treatment. However, the waste water reclaimed by the higher degree of treatment has an economic value depending on the reuse application selected for it. This introduces an off-setting cost factor which is generally comparable to or higher than the cost of tertiary treatment.

There are economics of scale for the electrodialysis and reverse osmosis processes for removal of dissolved inorganics. Current work by the Office of Saline Water (OSW) indicates that a 30 m.g.d. plant in a Southern California location could produce renovated water from influent sewage at 21 cents per 1000 gallons for the electrodialysis process, and at 22 cents for the reverse osmosis process, at 1968 prices.

**Reverse osmosis**

Recent research indicates waste water reclamation by reverse osmosis offers great promise for substantial reductions in cost as well as marked improvements in efficiency. The Federal Water Pollution Control Administration (FWPCA) is sponsoring a large number of reverse osmosis waste water reclamation projects. Results have been obtained on the use of the process for a variety of waste effluents.

Reverse osmosis orginally was developed for the conversion of brackish water to fresh water, and is now in practical field use for that purpose. To evaluate the results that we will ex-

amine for waste waters, values obtained by OSW in a typical brackish water field test can serve as a baseline.

At an operating pressure of 600 p.s.i.g., and with a product to waste ratio of 3:1, the total dissolved solids (TDS) in this study drops from 5170 p.p.m. to 320 p.p.m., a reduction of almost 94%, and the hardness, iron, manganese, phosphate, and sulfate are removed almost quantitatively.

An example of converting polluted water to a usable state by reverse osmosis is a study on acid mine drainage water (Norton, W.Va.). This cooperative effort between FWPCA and OSW shows that the low pH, high iron sulfate water is upgraded substantially with practically all of the iron and a good deal of the acid removed.

Another cooperative FWPCA-OSW reverse osmosis project is a field test (Fresno, Calif.) on high salinity irrigation return flow. Results to date show reduction in TDS from 4890 p.p.m. to 340 p.p.m. with elimination of almost all of the hardness and sulfate. A similar study on agricultural runoff (Firebaugh, Calif.) is being carried out by FWPCA, the Bureau of Reclamation, and the California Department of Water Resources. In this test, the DuPont Permasep permeator system is converting the 6000 TDS agricultural drain water to a soft potable water meeting U.S. Public Health Service (USPHS) standards. Conversion of the feed water from slightly alkaline to slightly acidic values increases the unit's ability to reduce nitrates from 30% to 70-80%.

**Paper mill effluents**

One of the most severe water pollution problems in the U.S. is that caused by effluents from paper mills. A number of states restrict the location where new mills may be established and insist on far better cleanup of the used water than hitherto has been the case. For that reason, the Pulp Manufacturers Research League has been examining reverse osmosis for application to these problems. Excellent cleanup has been obtained on small-scale units, in that the major con-

taminants of one of the main effluents, spent sulfite liquor, can be controlled. The ability to achieve 95% or greater reduction in TDS, chemical oxygen demand, and color is particularly encouraging, since these are the main sources of concern in paper mill pollution control requirements.

Other applications for which reverse osmosis is being tested include treatment of photographic process wastes, maple syrup and fruit juice concentration waste, chemical and biological warfare agent removal, virus and bacteria control, laundry waste, waste water recovery in spacecraft, and sewage plant effluent renovation.

FWPCA began evaluating reverse osmosis for renovating municipal waste water in 1963, just two years after the development of an asymmetric membrane at the University of California. Early laboratory efforts by Aerojet-General Corp. showed the necessity of pH control and prior removal of dissolved organic materials to prevent fouling of the membrane surface and the resulting rapid decline in efficiency.

The first study also indicated that the cellulose acetate membranes may be biologically degraded by waste water. The product quality, however, was excellent, with rejection of more than 90% of the contaminants from a secondary effluent. Since this study, FWPCA, cooperating with the Los Angeles Sanitation District, has investigated reverse osmosis in equipment with capacities up to 10,000 gallons per day at the Pomona advanced waste treatment pilot plant. The project has included studies on tubular membrane units produced by Havens Industries and Universal Water Corp., and on spiral-wound modules from Gulf General Atomic. A flat-plate unit from Aerojet-General also is being evaluated at FWPCA's pilot plant (Lebanon, Ohio).

These studies have verified the necessity of pH control, but have shown that, under proper operating conditions, the membranes are relatively stable. FWPCA has not been able to overcome the flux decline caused by the organic materials but, in recent months, has developed membrane

# Federally funded reverse osmosis projects

| Contractor | Project title | Contract | Description |
|---|---|---|---|
| Douglas Aircraft<br>Newport Beach, Calif. | Use of improved membranes in tertiary treatment | $ 84,800 | Evaluation of chemically modified cellulose acetates membranes, developed by contractor, for improved flux and stability |
| Gulf General Atomic<br>San Diego, Calif. | Membrane materials for waste water treatment | 65,138 | Preparation and testing of three new types of membranes: cellulose acetate with varying acetyl contents, polyvinyl pyrrolidone-polyisocyanate copolymers, and membranes formed from polymer dispersions or latices |
| U.S. Atomic Energy Commission<br>Oak Ridge, Tenn. | Application of hyperfiltration with dynamically formed membranes to treatment of municipal sewage effluents | 85,000 | Study of membrane support materials of different pore sizes and the use of filter aids and film-forming additives |
| Syracuse University<br>Syracuse, N.Y. | Mass transfer analysis in reverse osmosis | 19,738 | To obtain analytical expressions for rate of production, effect of natural convection, and optimum geometrical configuration |
| Eastern Municipal Water District<br>Hemet, Calif. | Removal of dissolved solids from reclaimed water in ground water recharge project | 225,000 | Use of reverse osmosis to maintain satisfactory concentration of salts and refractories in recycled groundwater |
| Gulf General Atomic<br>San Diego, Calif. | Waste water reclaimation | 112,410 | Provide two spiral wound reverse osmosis modules for operation at the Pomona, Calif., test facility of the Los Angeles County Sanitation District |
| Aerojet-General<br>El Monte, Calif. | Renovation of municipal waste water | 203,299 | Lab scale studies with flat membrane cells on several grades of municipal waste water representing a broad spectrum of effluent quality |

## Paper mill waste treatment

| Constituent | Before treatment | % Reduction by reverse osmosis |
|---|---|---|
| Solids | 1,461 gm./l. | 95 |
| Color platinum—cobalt method | 74 | 99 |
| BOD | 5,100 p.p.m. | 92 |
| COD | 17,400 p.p.m. | 95 |
| Inorganics | 650 p.p.m. | 99 |

## Brackish water treatment by reverse osmosis process
(p.p.m.)

| Constituent | Feed | Product | Waste |
|---|---|---|---|
| pH | 6.7 | 5.3 | 6.8 |
| Total dissolved solids | 5,170 | 320 | 20,600 |
| Total hardness as CaCO$_3$ | 1,880 | 20 | 7,350 |
| Ca$^{+2}$ | 360 | 7.6 | 1,400 |
| Mg$^{+2}$ | 240 | 0.5 | 940 |
| Na$^+$ | 900 | 110 | 3,400 |
| HCO$_3^-$ | 340 | 12 | 1,150 |
| SO$_4^{-2}$ | 630 | 0 | 2,580 |
| Cl$^-$ | 2,020 | 170 | 7,850 |
| Total alkalinity | 280 | 10 | 940 |

Operating pressure—600 p.s.i.g.
Product to waste ratio—3:1

## Demineralization of irrigation return
(p.p.m.)

| Constituent | Feed | Product | Waste |
|---|---|---|---|
| pH | 7.9 | 6.0 | 7.4 |
| Total dissolved solids | 4,890 | 340 | 7,420 |
| Total hardness | 1,290 | 15 | 2,060 |
| Ca$^{+2}$ | 274 | 6 | 505 |
| Mg$^{+2}$ | 147 | 0–1 | 194 |
| Na$^+$ | 1,020 | 102 | 1,620 |
| SO$_4^{-2}$ | 2,690 | 43 | 4,370 |
| Cl | 371 | 107 | 532 |
| NO$_3^-$ | 43 | 32 | 54 |
| SiO$_2$ | 44 | 13 | 62 |

Operating pressure—600 p.s.i.g.
Product to waste ratio—1:1

## Acid mine drainage
(p.p.m.)

| Constituent | Feed | Product | Waste |
|---|---|---|---|
| Total dissolved solids | 1,235 | 4 | 4,650 |
| pH | 2.9 | 4.5 | 2.4 |
| Total iron | 101 | 0.1 | 352 |
| Aluminum | 24 | 0.1 | 92 |
| Sulfate | 820 | 0.5 | 3,000 |
| Magnesium | 104 | 0.1 | 370 |
| Calcium | 218 | 0.2 | 810 |

Operating pressure—600 p.s.i.g.
Product to waste ratio—3:1–4:1

## Agricultural waste water reclamation
(p.p.m.)

| Ion | Feed | Product | Reject |
|---|---|---|---|
| Ca$^{+2}$ | 331 | 4.8 | 483 |
| Mg$^{+2}$ | 172 | 1.9 | 245 |
| Na$^+$ | 1,308 | 111 | 1,755 |
| SO$_4^{-2}$ | 3,596 | 88.0 | 4,885 |
| Cl$^-$ | 450 | 117 | 578 |
| NO$_3^-$ (as N) | 11.4 | 3.7 | 20.6 |
| SiO$_2$ | 46.3 | 30.7 | 52.3 |
| Total hardness | 1,760 | 26.6 | 2,487 |
| Total dissolved solids | 6,046 | 400 | 8,280 |
| pH | 5.1 | 4.9 | 5.3 |

**Inspection.** *Cellulose acetate is a common material for reverse osmosis membranes*

## Tertiary treatment costs

| Process | Cost (cents/1000 gallons for 10 m.g.d. plant) |
|---|---|
| Conventional | |
| Primary treatment | 7.5 |
| Activated sludge | 11 |
| Filtration | |
| Microscreening | 1.5 |
| Coarse media | 2.5 |
| Fine media | 3.5 |
| Phosphate removal | |
| Mineral addition to aerator | 3 |
| Coagulation, sedimentation | 3.5 |
| Coagulation, sedimentation, filtration | 7 |
| Ammonia stripping | 1.5 |
| Granular carbon adsorption | 3.5–8 |
| Dissolved inorganic removal | |
| Electrodialysis | 14 |
| Reverse osmosis | 25 |
| Ion exchange | 25 |

cleaning procedures that are both effective and economically attractive. These procedures employ commercial laundry presoak formulations containing enzymes.

Studies are continuing at Pomona and Lebanon and at the laboratories of selected contractors. In addition, an extensive effort incorporating all the physical configurations presently on the market is in the planning stage. This will result in the installation of five reverse osmosis units on the grounds of the conventional waste treatment plant (Hemet, Calif.). A wide variety of pretreatment processes also will be available so that their effect on the reverse osmosis process can be determined.

Aerojet-General Corp. has conducted a study of the decline in product water flux from various municipal waste streams. The use of certain proprietary additives has been shown to inhibit solids deposition in some cases, and, therefore, minimize flux decline.

FWPCA previously had tested the application of electrodialysis to sewage treatment (Lebanon, Ohio). The results with both reverse osmosis and electrodialysis have shown that it is technically feasible to convert secondary sewage effluents to potable quality. Based on these results, preliminary cost estimates have been prepared for three types of 30 m.g.d. plants:

• A combined sewage treatment reverse osmosis plant.

• An integrated sewage treatment/electrodialysis plant.

This comparison indicates that the net cost of producing potable water from secondary sewage effluent is 22.4 cents per 1000 gallon for reverse osmosis and 20.7 cents per 1000 gallon for electrodialysis.

A 1400 gallon per day spiral-wound module was evaluated (Bergen County, N.J.) on secondary sewage effluent. Rejection of all species except nitrogen compounds was 90-95% or better. The rejection of TDS was about 96% while ABS, odor, and turbidity were removed completely. The biological oxygen demand (BOD) was reduced from a range of 8-26 p.p.m. to 1 p.p.m. Rejections of nitrate and ammonia were about 70% and 90%, respectively. Water flux was about 8 gallons per day per square foot with only a negligible decline during the test.

The same equipment showed interesting results on bacteria during tests on contaminated Potomac River water. The high *E. coli* count was reduced to negligible values. Tests at the Point Loma sewage treatment facility (Calif.) showed that high quality water was obtained even from primary sewage effluent, but that the problems of membrane fouling and module clogging were much more serious than in secondary effluent. Another test on dilute radioactive wastes showed that the beta and gamma activity in a waste stream contaminated primarily with uranium and thorium and their daughter products was reduced by a factor greater than 10,000.

Dorr-Oliver installed a 3000 gallon per day sewage treatment plant (Sandy Hook, Conn., August 1967) which had a reverse osmosis unit following a grinder, aerated holding tank, and an activated sludge tank. Another plant was installed at Norwalk (February 1968), where the effluent from a biofiltration plant treating 20,000 gallons per day of domestic sewage went through the reverse osmosis unit. Performance of both systems was similar. Typical data indicated that 90% of the time, the BOD was less that 14 mg./l., and the suspended solids were completely removed. Fifty percent of the time, the coliform density was less than 100 organisms per 100 ml. and 85% of the time, it was less than 1000 per 100 ml. Less than five units of color were obtained consistently. Suspended solids tended to settle in the membrane forcing a shutdown for cleaning three times in seven months. The membranes showed a continuing decrease in the flux restored by each cleaning cycle, until after seven months, only 35% of the initial flux could be attained.

A rather unique but critical problem that lends itself to solution by use of reverse osmosis is the conversion of hospital wastes to potable water. A study of this problem by AiResearch on a 46,000 gallon per day system indicated that kitchen and other hospital wastes, excluding human waste, can be treated to recover more than 90% of the waste as potable water by a sequence of flotation, filtration, and reverse osmosis. The overall rejection of dissolved salts across the membrane exceeded 99%. The product water was low in TDS but has some free $CO_2$ and a pH in the range of 3.5-4.0.

Activated carbon was used to remove traces of taste and odor forming compounds and surfactants. After pH adjustments and dosing with calcium hypochlorite solution, potable water containing 5 p.p.m. residual-free chlorine was obtained.

Studies on supplying industrial water by reverse osmosis treatment of either secondary sewage effluent or industrial wastes show that costs of 20-35 cents per 1000 gallons are reasonable to expect. This could have substantial appeal since industry today pays from 12.5-40 cents per 100 gallons for water with up to 1000 p.p.m. dissolved solids. Sometimes, this cost must be supplemented with additional treatment costs to meet specifications.

## Problems to be solved

There are still a number of problems that need solution so that costs of sewage and industrial water reclamation by reverse osmosis techniques may be made even more attractive. Active research on these problems is underway, and many interesting answers should be available soon. This research is being carried out by industry and government, since both have a large stake in the outcome.

One obvious need is for tailor-made membranes for the retention or passage of specific materials. The experiments I have described were carried out with commercial units with membranes designed primarily for the conversion of brackish water to fresh water. What is needed is a whole new family of membranes tailored to the chemical and bacteriological content of sewage waste. For instance, the reduction of ammonia, phenols, detergents, and carbon chloroform extract with conventional membranes does not seem to be good. Without doubt, cellulose membranes are relatively permeable to ammonia. In some cases, phenols have negative rejections, and actual permeate enrichments as high as 20% have been observed. The question of rejection of the new linear alkyl benzene sulfonates needs to be resolved, since the USPHS standard for potable water is only 0.5 p.p.m. Carbon chloroform extract is a combination of many types of organic matter, and cellulose acetate membranes are relatively permeable to a number of the low molecular weight organic compounds. It is likely that waters reclaimed by reverse osmosis from many sources will not pass the USPHS standard of 0.2 p.p.m. carbon chloroform exhaust.

Reverse osmosis membranes in use today tend to be susceptible to fouling by organics or other trace contaminants. This problem still must be overcome, either by new membranes or by devising satisfactory *in situ* cleaning techniques.

Membrane fouling, coupled with membrane compaction reduces membrane flux and the effective service life. The membrane flux reduction pattern appears to occur in three stages. First, there is a small rapid initial compression when the membrane is put under pressure, resulting in a flux decrease to 85-90% of the original value. The second phase of flux reduction appears related to the buildup of solids and inclusion of waste molecules in the membrane openings. The third phase is probably caused by the slow compaction of the film or cake. Therefore, work is needed on the mechanism of membrane fouling to reduce the effect of cake buildup on the flux, and to increase the washing or self-regeneration characteristics of the system.

A variety of approaches have been taken either to prevent flux decline or to restore the flux. Methods used with some effectiveness include aeration of the primary effluent, diatomaceous earth filtration of both effluents, and partial redesign of the membrane-containing part of the system to prevent the solids buildup. OSW has found it possible to maintain the average flux of the present membranes at a fairly high level by periodic cleaning of membrane modules with chemical solutions. As an example, a membrane module whose flux has degraded from 13 gallons per day per square foot to 6 gallons was restored to its initial value by recirculating 5% sodium hydrosulfite for four hours. Acetic acid and enzymes also have shown promise as membrane cleaning agents. These procedures have doubled the effective service live of the present membrane to two years.

**Jacob I. Bregman** *is President, Water Pollution Research and Applications, a position he has held since 1969. Previously (1967-9), he was deputy assistant secretary, Water Quality and Research, U.S. Department of the Interior, and commissioner, Ohio River Valley Water Sanitation Commission. Bregman received his B.S. from Providence College (1943), his M.S. (1948) and Ph.D. (1951) from Polytechnic Institute of Brooklyn. The author of "Corrosion Inhibitors," Bregman is a member of* ACS, *American Water Resources Association, Water Resources Research Council, N.Y. Academy of Sciences, Sigma Xi, Phi Lambda Upsilon, and American Inst. of Chemists.*

# Ion exchangers sweeten acid water

*Spanking new treatment plant*

*in coal region removes mine drainage contaminants*

*from a community's water supply*

*Reprinted from* ENVIRON. SCI. TECHNOL., **5**, 24 (January 1971)

A piece of countryside scarred by strip mines is not a pretty sight: a visit to many parts of Appalachia will testify to that. But quite apart from the esthetic destruction that years of grubbing for coal has wrought on the landscape, a more insidious evil has appeared as an unwanted by-product of mining activity—acid mine drainage.

How acids form inside coal mines is a matter of some scientific debate—in the research pages of this publication and in other forums—but it is broadly accurate to say that the sulfur in coal is capable of being oxidized to sulfuric acid in the presence of moisture. In any event, conditions in abandoned mines and within the stripped ground cover left after strip mining machinery has moved on to greener pastures seem to be ideal for acid formation—with unfortunate consequences for any nearby body of water.

The unwanted side products of coal mining have reached such staggering proportions that several states in the eastern U.S. coal belt have been forced to undertake almost Herculean efforts to keep them in check. (See "States make headway on mine drainage," ES&T, December 1969, page 1237.) In Pennsylvania, for example, Operation Scarlift has been launched to spearhead a multipronged attack on mining's legacy of pollution. Pennsylvania voters in 1967 authorized a $500 million bond issue for solution of problems associated with coal mines. The state's Department of Mines and Mineral Industries has $200 million with which to attack problems of mine drainage.

Some of this $200 million has found its way to the small community of Smith Township, some 20 miles west of Pittsburgh, in a part of the state where signs of years of strip mining are all around (the city park and the high school in Burgettstown are built on reclaimed strip mining land). Smith Township is spending $700,000 to solve a problem it shares with many small communities in Appalachia—you can't drink the water because it is contaminated by acid mine drainage.

## Contaminated

Not that Smith Township's water is acid—its pH is in fact around 8, distinctly alkaline. But mine drainage is indeed the cause of the community's undrinkable water. Acid leaching out of mine formations is neutralized by the dolomite, which occurs naturally in this rocky landscape. The result: the township's Dinsmore reservoir is full of water containing a staggering 1000 ppm of sulfates, largely calcium sulfate, and several hundred ppm of carbonates. The water's alkalinity is accounted for by the presence of the smaller quantities of calcium and magnesium carbonates.

Joe Abate, secretary of the Smith Township Municipal Authority, recalls that the community has for years sought ways to reduce the solids content of its water. Studies in the mid-60's conducted in conjunction with the Dow Chemical Co.'s ion-exchange group (then part of Nalco Chemical Co.) initially led the municipal authorities to consider ion exchange as a potential way to clean up their water supply. With the availability of state money in 1968, the township could afford to build a plant embodying an efficient process.

## Ion-exchange process

The process that was finally arrived at for the 500,000-gpd Smith Township water treatment plant, now undergoing final shakedown tests, is based on Sul-biSul technology of the Dow Chemical Co. and on a resin handling system developed by Chemical Separations Inc. (Oak Ridge, Tenn.). It employs two ion-exchange steps and an intricate method for regenerating and transporting ion-exchange resins.

In outline, the process operates like this: the incoming raw water is fed to a cation exchanger, where the cations (positively charged ions—$Ca^{2+}$, $Mg^{2+}$, etc.) are exchanged for the hydrogen ion in a mixed bed of Dowex HCR-W strongly acidic and Dowex CCR-2 weakly acidic cation-exchange resins. The liquid leaving the cation exchanger contains sulfuric acid and carbonic acid. $CO_2$ gas is allowed to escape from solution (carbonic acid quickly dissociates) and the liquid, now predominantly composed of sulfuric acid, is passed through the anion exchanger. This exchanger contains Dowex SBR strongly basic anion-exchange resin. The resin is initially in "the sulfate form," and passage of the acid-containing water results in removal of bisulfate ions from the water onto sites on the resin, which then attains "the bisulfate form" (hence, Sul-biSul). The effluent water, when filtered and chlorinated, contains less than 250 ppm of dissolved solids, and doubtless will taste like vintage champagne to those in the township.

## Regeneration

While such an overall description illustrates how the ion exchangers work in broad terms, the secret of the process, in practice, almost certainly relies as much on the way the resin beds are regenerated as on the specific ion-exchange properties of the resins themselves. Since, with passage of raw water (through the cation exchanger) and acids (through the anion exchanger), the resins become loaded and lose their effectiveness for further ion exchange, they must at intervals be regenerated.

This means that the hydrogen ion

**Purification·** *Effluent from cation exchanger (not shown) contacts anion-exchange resin in column (far left). Spent resin is regenerated, rinsed, and recycled*

must be put back into the cation-exchange resin, and that the anion-exchange resin must be reconverted from the bisulfate to the sulfate form. Obviously, to stop the whole process while regeneration is carried out would severely limit the overall throughput rate of the whole plant.

What Irwin Higgins, technical director of Chemical Separations, has done is to design a system that regenerates resin in a short cycle during which the resin bed is not being loaded. After regeneration, a brief hydraulic pulse drives the regenerated resin into the loading zone in the ion exchanger (such a pulse is used in both anion and cation exchangers, though not necessarily at the same time, of course).

To achieve complete regeneration and to avoid contaminating product with regenerant, an automatic control and monitoring system is designed into the plant. Work during the plant shakedown period has been aimed at finding exactly the right timing for the regeneration cycle.

Ion exchange would be a very cheap process in many instances were it not for the necessity to buy regenerating materials. At the Smith Township plant, lime is used to regenerate the anion-exchange resin and sulfuric acid is used to return the hydrogen ion to the cation resins. Both regenerants are fairly cheap—perhaps $20/ton for CaO and $30/ton for $H_2SO_4$—but, as Higgins explains, some unique wrinkles make the most of the cheap materials. For instance, the Dowex CCR-2 weakly acidic cation-exchange resin is used primarily because of its strong affinity for the hydrogen ion—i.e., its easy regenerability—rather than because it is useful in removing calcium and magnesium cations from the raw water. According to Higgins, presence of this resin in the cation exchanger makes it possible to use sulfuric acid in near stoichiometric amounts for regenerating the mixed-resin bed. Normally, regenerant acid must be used in considerable excess, obviously leading to increased regeneration costs.

To regenerate the anion resin, lime is used, together with a saturated calcium sulfate solution produced as a result of regeneration of both anion and cation exchangers. The calcium sulfate alone would be sufficient to reconvert the anion-exchange resin to the sulfate form, according to Higgins, but lime is used to neutralize the sulfuric acid produced by regeneration. From the economic standpoint, Higgins points out that only enough lime to neutralize this acid needs to be used. If the acid produced by regeneration could be discarded somehow (back into the reservoir, for instance!), no lime at all would be needed. But, of course, the acid cannot be discharged. Even so, chemical costs will be low: Lou Wirth, Dow's manager of ion exchange, estimates these costs at less than 0.25 cents per 1000 gallons.

The only fly in the ointment that Joe Abate and the Smith Township authorities are concerned with is that their plant, in addition to the potable water, will produce large amounts of calcium sulfate. Although present plans call for the settled calcium sulfate to be trucked away and landfilled, Abate is imbued by the true spirit of ecology: "We'll find a use for that stuff, somehow," he vows.

### Economies

The ion-exchange process at Smith Township represents a considerable economy for the community—water treatment costs will be brought down to about 20 cents per 1000 gallons from approximately 50 cents with their old lime and soda ash process, which produced much less drinkable water, as well as being more costly. But when one considers that it costs $700,000 to provide potable water for what amounts to a population of about 10,000 people, one wonders how much money will have to be spent just to clean up acid-contaminated waters in mining areas. One wonders, too, whether numbers such as the $500 million in the Pennsylvania bond issue may not just be drops in an almost bottomless bucket.　　　DHMB

# Ozone bids for tertiary treatment

*Its effectiveness against waste*
*effluents makes many companies see renewed*
*vigor in a 70-year-old technology*

Reprinted from Environ. Sci. Technol., 4, 893 (November 1970)

In the quest for water quality, technologists are taking second looks at treatment processes that have lain dormant for some time, in the hope that state-of-the-art advances or a changing economic or social climate may open significant new applications. Ozonation, which underwent its peak development soon after its commercial introduction some 70 years ago, is one such process. It has achieved stature in industrial applications; its most common use today is in the production of long-chained esters, and it is also used in the manufacture of carbon black, as a sterilizing agent in food and beverage manufacture, and in industrial odor control. Research laboratory use is another well-established application.

Traditional uses aside, several recent commercial developments point to increased interest in tapping the oxidizing potential of ozone for waste water treatment:

• About a year ago, Air Reduction Co. received an FWQA grant for pilot-plant studies on the use of ozone for treating sewage plant effluents. At the time the grant was announced, Air Reduction stated that the costs of ozone generation had been reduced to where it is competitive in waste treatment.

• In October of last year, Crane Corp. received a license from La Companie des Eaux et de l'Ozone (CEO) (Paris, France) for U.S. distribution of Otto process ozonation units. Crane now offers the CEO units in combination with microstraining for tertiary water treatment.

• Chromalloy American Corp. (Hawthorne, Calif.) has developed the Purogen activated oxygen system—basically an ozone process—which the company markets through Water Treatment Corp., a subsidiary. Chromalloy hopes the units will find application in a wide range of water treatment applications.

• Purification Sciences, Inc. (PSI) (Geneva, N.Y.), a two-year-old manufacturer of ozone generators, has recently completed pilot-plant studies on effluents from an activated carbon treatment process at New Rochelle, N.Y. PSI president J. Robert Costello tells ES&T that the results of the test, aimed at disinfection of the process effluents, were "very encouraging;" furthermore, says Costello, the primary barrier to ozone utilization—its cost—has been overcome, and it is competitive for waste treatment applications.

• The Welsbach Corp. (Philadelphia, Pa.), the major U.S. producer of ozone generators, admits to a keen interest in water treatment. Dr. Carl Nebel, manager of ozone systems development for Welsbach, points out that, "Over five million pounds of ozone are produced annually in the U.S. for industrial applications, most of it on Welsbach generators. Because of this long experience in ozone technology, we've worked closely with consultants preparing plans for new water treatment plants, and will continue to do so."

Ozone's great attractiveness in water treatment is its high oxidative capacity, second only to that of fluorine. As such, it is effective against a wide range of chemical and biological pollutants, usually at relatively small dosages. But because of its reactivity and instability, and the low rates of conversion from air or oxygen, ozone must, unlike other chemical oxidants, be generated on site as needed.

## Generation

The only commercially feasible method for ozone generation is by electric discharge. Oxygen or air is passed through a narrow gap between two electrodes; also interposed between the electrodes is an insulating material of suitable dielectric properties. When a high-voltage alternating current is applied to the electrodes, a portion of the oxygen stream is converted to the allotropic form of oxygen, ozone. All commercial ozonators use this basic principle, but usually show some differences, mostly proprietary, in electrode configuration and circuitry.

Air or oxygen can be fed to the generators, although oxygen is preferred in large installations because the output is doubled when pure oxygen is used. Even so, ozone concentrations in the stream from the units rarely exceed 2%. At this conversion ratio, typical costs for ozone generation range between 6 and 8 cents per pound generated.

Ozone's potential in water treatment falls into two categories: potable water supplies and waste water effluents. Treatment of municipal water systems was one of the first major uses of ozone, particularly in Europe, where much of the early ozone technology was developed. Paris, for example, ozonates about a third of its public water supply in lieu of chlorination, and the practice gets smaller, though significant, in other European countries. In the Americas, both Canada and Mexico use the process to some extent, but usage in the U.S. has been slight. The city of Whiting, Ind., has been ozonating its municipal water supply since 1939; and, in the mid-fifties, Philadelphia installed what, at the time, was the largest ozone unit in the world to treat its water supply. The Philadelphia unit was discontinued in 1961, however.

## Advantages

The dominant method for sterilization of water supplies is, of course, chlorination, in spite of the fact that ozone offers several advantages. For one thing, on-site generation eliminates the hazard associated with transporting and handling large amounts of chlorine. Tests have shown that ozone destroys both bacteria and viruses; chlorine is ineffective against viruses. Furthermore, as Welsbach's Nebel points out, ozone is more effective against the major taste- and odor-causing compounds in raw water—phenols and amines. Chlorination merely converts these to compounds that are more refractive to oxidation, he says, and the trend to recycling and reuse of water will compound this problem. Small amounts of chlorine would still be necessary for residual protection against contamination since ozone has a half-life in water of only 20 minutes. But in Nebel's view, "There is no reason for any water supply system not to use ozone for complete sterilization, and adding only enough residual protection as is necessary."

Proponents of ozonation for waste water treatment can point to a few successful applications to industrial effluents to bolster their cause. The waste treatment plant of the Boeing Co. at Wichita, Kan., uses ozone for the oxidation of waste cyanides and phenols. Initial consumption of ozone, when the plant was first built over 10 years ago, was 120 lb. per day, and the plant capacity has since tripled. And the British Petroleum Co. refinery at Bronte, Ontario, uses 190 lb. of ozone per day to remove phenols from the effluent of a biological oxidation plant.

**Tertiary treatment**

Although the oxidative capacity of ozone is considerable, it is not likely to supplant biological oxidation as used in municipal sewage treatment. Units big enough to supply the required amounts of ozone are just not practical. But, as Nebel points out, ozone has considerable potential in tertiary treatment concepts. "Ozone is capable of higher reductions of residual biochemical oxygen demand (BOD) and total organic

carbon (TOC) than carbon adsorption polishing, and is fully cost competitive," he says, and offers other advantages to boot. The process of ozonating effluents introduces considerable amounts of air or oxygen into the waste, thus increasing the dissolved oxygen content of the receiving stream. Chlorination of the effluents, which is sometimes practiced, is unnecessary since ozone effectively sterilizes the effluents as it destroys dissolved contaminants.

Whether ozone makes any headway as a water treatment practice will depend on numerous factors that transcend the usual problems of bringing new technology to bear on a field dominated by well-established practices. For one thing, over the years, ozone has developed a bad "image." Many engineers have the impression, established in the early days of the industry, that ozonators are bulky and cumbersome, in spite of the considerable advances in their design and construction. Also, ozone is generally thought of as a dangerous compound—which it is, in quantity—but on-site generation of the small amounts needed for continuous water treatment makes it less hazardous than other materials or processes.

**Ironic**

A much more subtle factor is the popular conception of ozone as an air

pollutant, even though its role in photochemical smog formation is in no way comparable to its use under controlled conditions for water treatment. This last point is especially ironic in view of the fact that one of the major uses for ozonation today is in air pollution control—for the removal of odors in sewage treatment plants and chemical processing.

In spite of its advantages, considerable market development would seem to be necessary for wholesale advances of ozonation into the water treatment field. But it's a moot point whether the required promotional effort can be mounted in the next few years. Ozonators are manufactured by a number of small companies which, although aware of the call for new technology to solve environmental problems, do not have the necessary resources to shoulder the development effort by themselves. Major electrical companies conceivably could fill this role, but interestingly enough, none of them has actively entered the ozonator market, although they undoubtedly have the expertise. Thus, with a handful of companies competing for what is at present a limited market, intercompany relationships are more likely to be concerned with proprietary rights and price competition, rather than with industry-wide promotion of ozone processes.　　　　　PJP

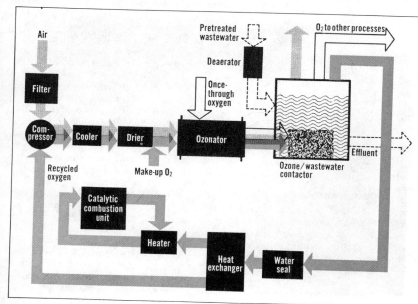

Ozone for waste effluent treatment can be generated from air or oxygen, and the hypothetical scheme (left) illustrates the flow chart and associated equipment for each case. For once-through oxygen use, which is usually not economic, the ozonator feed does not need pretreatment, and the spent oxygen can often be used for other processes. With air or recycled oxygen, the feed stream must be chilled and dried for maximum conversion. When recycled oxygen is used, the clarified waste is vacuum de-aerated to remove dissolved nitrogen, which would otherwise find its way into the recycle stream and reduce the conversion rate. The efficiency of ozone treatment can sometimes be improved with multistage contactors

# FWQA steps up tertiary treatment study

*Pilot plant will evaluate*
*four separate approaches*
*to nutrient removal*

Reprinted from ENVIRON. SCI. TECHNOL., 4, 550 (July 1970)

**Adsorption.** *Carbon columns remove last traces of dissolved organic compounds*

Advanced waste treatment is a fairly general term for a broad technological area and in its various usages is often applied to anything from modification of existing treatment practices to such comparatively exotic techniques as gamma irradiation or reverse osmosis. At the Federal Water Quality Administration (FWQA) and the District of Columbia's advanced waste treatment pilot plant in Washington, D.C., however, the common working definition of advanced treatment is an integrated system that can reduce residual pollutants lower than conventional primary and secondary treatment. Within this context, the FWQA-D.C. pilot plant, completed about three years ago, has been studying a wide range of individual treatment steps, including lime precipitation, mineral addition, activated carbon adsorption, oxygenation processes, ion exchange, and sludge handling procedures. Just recently, the pilot plant has undertaken its most ambitious program to date—a side-by-side evaluation of four treatment combinations which show the greatest potential for high level treatment, including better than 90% nutrient removal capability. The study has only recently gotten underway, but Fred Bishop, supervisory research engineer at the pilot plant, confidently says, "We expect to have some fairly definitive results ready for publication in the next ten months."

## Four processes

All of the treatment sequences under investigation include processes that have, as yet, found little or no use in municipal waste water treatment. Three of the sequences include conventional primary treatment and secondary biological processes, followed by tertiary nutrient removal steps. Another, an independent physical–chemical system, relies primarily on nonbiological means for both organic and nutrient removal.

The main features of the treatment sequences—which qualify them as advanced waste treatment—are the capabilities for phosphate and nitrogen nutrient removal. Lime precipitation for phosphate removal, the step for which the pilot plant has accumulated the most data in its two-and-a-half years of operation, is common to all three systems, in either of two basic modes of operation. In the two-stage process, consisting of two clarifiers and a recarbonation tank, the calcium carbonate produced in the recarbonation step is settled in the second sedimentation basin. In single-stage operation, need for the second clarifier is eliminated by precipitation of the excess calcium ions in the first clarifier with sodium carbonate. In one sequence, mineral addition is used for phosphorus removal, wherein aluminum or iron salts are added in the secondary aeration step. Two alternate approaches, biological or physical–chemical, are available for nitrogen removal.

## Conventional-tertiary

In the first conventional-tertiary process, after primary sedimentation of the raw waste water, the effluent is biooxidized, either by conventional aeration or with pure oxygen. After settling the secondary solids, the effluent is treated with lime in a two-stage clarification step. Following pH adjustment, the effluent from the second clarifier is treated in a stripping column for nitrogen removal, as ammonia. The final step prior to chlorination and discharge is activated carbon adsorption in a polishing column, to remove the last of the dissolved organic matter. In this process, as well as in the others using lime, recovery of the lime used for phosphate removal is possible. Calcium carbonate from the clarifiers is calcined to regenerate the lime for reuse and produce carbon dioxide for use in the carbonation steps.

The second of the conventional-tertiary processes differs from the first in the nitrogen removal step. Following solids removal from the biooxidation process, the effluent is then reoxygenated to nitrify it, converting the nitrogen from the ammonia form to nitrates. This is followed by single-stage lime precipitation and then by biological denitrification in an anaerobic bacterial process, (as opposed to aerobic bacterial treatment in the secondary aeration). Methanol is used as food for the dentrifying bacteria, and the denitrification may be carried out in a tank or in activated carbon columns. A third system being studied uses biological nitrification–denitrification, but alum is added in the first biooxidation stage to remove phosphorus, eliminating the lime precipitation step from the system.

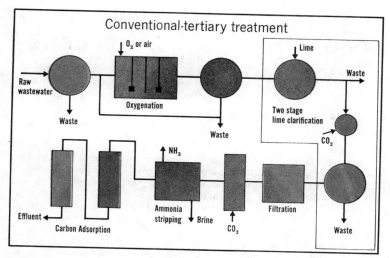

## Conventional-tertiary treatment

O₂ or air · Oxygenation · Waste · Raw wastewater · Waste · Lime · Two stage lime clarification · Waste · CO₂ · Filtration · NH₃ · Ammonia stripping · Brine · CO₂ · Carbon Adsorption · Effluent · Waste

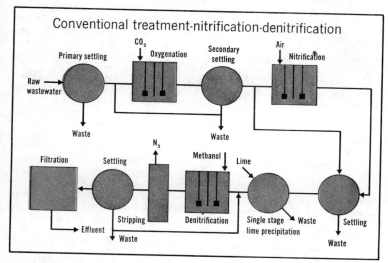

## Conventional treatment-nitrification-denitrification

Primary settling · CO₂ · Oxygenation · Secondary settling · Air · Nitrification · Raw wastewater · Waste · Waste · N₂ · Methanol · Lime · Filtration · Settling · Stripping · Denitrification · Single stage lime precipitation · Waste · Settling · Effluent · Waste · Waste

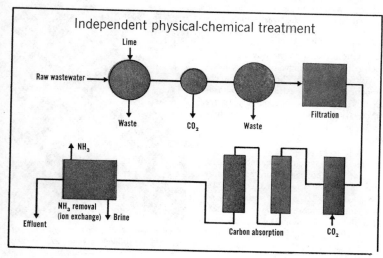

## Independent physical-chemical treatment

Lime · Raw wastewater · Waste · CO₂ · Waste · Filtration · NH₃ · NH₃ removal (ion exchange) · Brine · Effluent · Carbon absorption · CO₂

### Nonbiological treatment

The other process being evaluated, the independent physical–chemical treatment, bypasses the conventional primary and secondary treatment steps. Instead, the incoming raw waste water is treated with lime in a two-stage clarification process. The effluent from the phosphate removal step is treated in carbon adsorption columns and then undergoes physical–chemical nitrogen removal. The pilot plant staff is studying the possibility of using ion exchange for removing the nitrogen as ammonium ions in conjunction with Battelle Memorial Institute, which has set up an ion exchange unit at the pilot plant, under an FWQA grant. However, ammonia stripping still remains a possible alternative for nitrogen removal in this process but must be employed immediately after lime precipitation.

The independent physical–chemical process, on the surface, appears to be the most attractive, in view of its simplicity, low capital cost, and small land area requirements. Indeed, this process has already been the subject of considerable enthusiasm both within FWQA and elsewhere. But according to pilot plant supervisor Bishop, such speculation is as yet premature. "We can't make any substantive comment on the superiority of any of these systems on the basis of our limited results so far," he says. As to projected costs, he wryly notes that "they'll be more expensive than conventional treatment," and then adds that although the systems are relatively new and untried, they comprise most methods that have reasonable potential for municipal waste water treatment.

Bishop and his staff hope to have some preliminary results on their work ready for presentation this fall and more detailed results in a series of papers to be presented at a meeting of the American Institute of Chemical Engineers in Houston next March. "We hope to have some reliable cost estimates by then," says Bishop.

**219**

# Electrolysis speeds up waste treatment

Reprinted from ENVIRON. SCI. TECHNOL., **4**, 201 (March 1970)

*New electrical process hastens destruction*
*of contaminants in liquid plating wastes*

A new wrinkle to the chemical industry's ancient art of electrolysis may provide those who have been seeking a cheap, simple means for treating dilute liquid wastes with just the technique they've been looking for. At least Stauffer Chemical Co.'s specialty chemicals division thinks so. The division has just reached an agreement which will give it exclusive rights to market a new electrolytic process developed by Resource Control, Inc. (West Haven, Conn.), to the metal finishing industry.

Stauffer, as a supplier of chemical products for the plating industry, had been looking for some time for an effective, low-cost way to treat dilute plating wastes, so that its customers in the industry would have a practical means for dealing with a growing water pollution problem. At the same time, Resource Control, Inc. (RCI), a relatively small company without a large marketing force, was seeking outlets for its new process which previously proved effective in treating plating wastes at several U.S. sites.

## Waste problem

The metal processing and finishing industries produce large volumes of dilute wastes. These arise in the rinsing operations carried out to remove chemicals from metal parts. Since there may be as many as 15 separate steps in a finishing operation, a typical shop has numerous rinse water streams, each contaminated by different chemicals. Commonly, the chemical contaminants which must be removed before the water is discharged are cyanide, hexavalent chromium and heavy metals (silver, cadmium, zinc, copper, etc.).

The usual way of removing the contaminants is by chemical treatment: Cyanides are oxidized by sodium hydroxide and chlorine, the chromium cation is reduced by sulfur dioxide, and metal bearing wastes are chemically neutralized to precipitate the metal as an insoluble salt.

## New twist

Although electrolysis has been used in the past to treat metal finishing wastes, it has been found to be a slow and costly method. Generally performed batchwise, the oxidation and reduction reactions necessarily have been slow because the electrical resistance of the waste is high, particularly when the concentration of contaminants becomes small and the liquid is ionized only slightly.

RCI has managed to overcome the problem of high resistance by using a cell in which the space between anode and cathode is packed with carbonaceous material. (The exact nature of the material is not being revealed, but RCI says that it is cheap enough to be discarded after several months of use, consists of small particles, and takes up only 50% of the available cell volume.)

Avery Smith, president of RCI, refers to the bed as semiconductive. For some reason, the presence of the particles does not short-circuit the anode and cathode, as might perhaps be expected. What does happen, or so RCI believes, is that each individual particle gains, at different parts of its surface, a positive and a negative charge. Oxidation and reduction of contaminants in the waste then take place at the anodic and cathodic sites on each particle. The result is that the amount of current flowing is essentially independent of the conductivity of the waste, and the chemical reactions which destroy the contaminants take place more quickly (more than 1000 times as fast, RCI claims) than in conventional electrolysis using the same applied voltage (typically, 12 volts d.c.). Power costs are cut, too.

The new system is capable of reducing the concentration of cyanide ion in a waste stream from 20 p.p.m. to less than 0.5 p.p.m. with a corresponding reduction in copper content (in the case of copper cyanide plating) from 40 p.p.m. to less than 1 p.p.m. Similar reductions in hexa-

valent chromium are claimed by the company, which also has calculated that the system costs 95% less to operate than conventional chemical treatment.

RCI presently is manufacturing mobile, easily installed units rated at either 5 or 10 gallons per minute. These sizes are deemed adequate for the average metal finishing shop, but larger units can be custom built, according to RCI. Capital outlay called for is less than $20,000 for the typical shop.

## Applications

RCI's Smith is enthusiastic about the possible uses of his company's electrolytic technique for the treatment of a wide variety of liquid wastes—in fact, any waste where the contaminants are capable of being oxidized or reduced to a harmless or easily disposable form. Whether the method will find wide application, of course, remains to be seen, but, if its attractive economics in the metal finishing field can be taken as a guide, it certainly will be a competitor to better established waste treatment processes.

**Neutralized.** *Stripchart records pH of metal plating wastes after electrolysis*

# Adsorption process eases acid recovery

*New continuous technique has been used with success on spent aluminum cleaning solutions*

Reprinted from ENVIRON. SCI. TECHNOL., **6**, 687 (August 1972)

Pennsylvanians seem to be famous for encouraging frugality. "Waste not, want not," and "A penny saved is a penny earned," Benjamin Franklin said.

Another Pennsylvanian agrees. Dr. Leslie E. Lancy, founder and president of Lancy Laboratories, Zelienople, Pa., says, "To protect the environment, waste nothing. Discharge only that which absolutely cannot be reused." That philosophy, coupled with his belief that cleanup should begin near the beginning of a process and not at the end, is reflected in Lancy's new phosphoric acid recovery unit. The unit was recently unwrapped—but only partially unwrapped—for newsmen on the occasion of the dedication of the company's newly expanded research and development facilities near Pittsburgh.

Lancy Laboratories, a part of the chemical group of Dart Industries, specializes in waste treatment design and engineering for metal-finishing systems. Lancy's phosphoric acid unit is designed to clean up acid-bearing wastes from aluminum brightening and finishing operations.

In the past, acid drag-over from brightening baths contaminated with aluminum phosphate and diluted by rinse waters had to be chemically neutralized before discharge. Recovery processes have been uneconomical to date, despite the fact that commercial-grade phosphoric acid costs about $140/ton.

Lancy's technology recovers better than 75% of the acid used. The recovered dilute acid (25–30%) needs only to be concentrated for reuse. The aluminum content of the concentrated acid is negligible (about 1–2 g/l.) Lancy says, so the recovered phosphoric can be returned directly to the manufacturing process. Proper neutralization of the acidic aluminum phosphate fraction which cannot be returned to the process stream yields a sludge that is suitable for fertilizer. Alternatively, aluminum phosphate could be recovered for sale to the pigment industry, Lancy believes.

## Sorption process

The proprietary package plant developed by the Lancy group is based on sorption. The heart of the unit is a U-shaped 23-ft-tall column packed with ion exchange resin. Although the packing is an ion exchange medium, the separation process is strictly adsorption and desorption. No ion exchange takes place, Lancy says, and the resin serves only as a contacting medium. Lancy is currently using Rohm and Haas's 410 resin, but he says that other anionic resins can be used as well.

The continuous loop process functions like this: A pulse of spent acid enters the

**Pilot.** *Small unit features 2-in. diameter columns, 70-gpd throughput*

loop at a given point on one of the legs of the U-shaped tubes at preset intervals. Phosphoric acid separates out in one direction and phosphate separates out in the other. When separation is complete, the products are eluted from the resin and the procedure is repeated. With the elution, the resin bed is moved countercurrently to the liquid flow direction. The 2-in. diameter columns used in the pilot operation permit a flow rate of about 70 gpd at 30–50 psi and ambient temperatures. The largest plant contemplated, Lancy says, would be on the order of 200 gpd.

Lancy is chary with details of the process because of possible loss of international patent rights due to premature disclosure, but a key advantage of the technology would seem to be that recovery is not complicated by introducing chemicals into the column to desorb the acid. Once the capacity of the resin to adsorb phosphoric acid is reached, according to Lancy, the resin is simply backflushed with water to elute the acid fraction and the aluminum phosphate fraction.

Costs of the Lancy process are largely for capital expenditure. It would cost $70–100,000 to build a full-scale phosphoric acid recovery unit, Lancy says, although the firm has not yet decided whether it will build the plants and sell them outright or merely lease them. Operating costs are confined to water and electricity, and resin is replaced a bit at a time as the backflushing wears it down. Turnover time for the resin would be about five years. "An operator would be nice," Lancy says, but no special training or qualification would be necessary to run the unit properly.

Although he declined to put a price tag on savings effected by recycling the acid, Lancy anticipates that they will be "considerable." He sees applications for the phosphoric acid process in the automotive and appliance and construction fields, and expects to extend the basic technology to other acid-bearing wastes as well. Next in line is a nitric-hydrofluoric acid recovery system for cleaning up stainless steel pickling wastes although, he admits, there are still some problems to be solved.

Fundamental to all Lancy's systems is his belief that "good waste treatment can't be done at the end of the line." It's easier to clean up concentrated wastes of a special type than it is to create large quantities of heterogeneous sludges that often present disposal problems of their own. And, as he adds, it's more profitable. HMM

# Spinoff processes aid sewage treatment

*Technology from the chemical and aerospace industries is showing the way to achieve much cleaner discharges—but the price is high*

*Reprinted from* ENVIRON. SCI. TECHNOL., 5, 756 (September 1971)

Sewage treatment technology is finally on the move. That's the conclusion one draws when considering some recent developments that have brought technology from "space-age" and chemical process industries into what was once the exclusive domain of the traditional sanitary engineer.

Several factors have influenced the long overdue move toward technical sophistication in the treatment of domestic sewage:

• Firms with technical expertise but shrinking defense-related markets are casting around for new fields to conquer.

• Tightening effluent standards around the country are making the conventional biological secondary treatment process, with its limited capabilities for reducing BOD and nutrient levels, look like something that has outlived its usefulness.

## Propellants to sewage

Consider the case of Thiokol Chemical Corp. (Bristol, Pa.), the industry leader in production of solid propellant rocket motors for such defense-related programs as the Minuteman and Poseidon missiles. The company has been affected by aerospace cuts and, although it is still very much committed to the aerospace field, has made several diversification moves.

One such move involves Thiokol's Wasatch Division (Brigham City, Utah), which, one year ago, began looking at the sewage treatment field. An important factor in Thiokol's search for a suitable entry into the field, says Paul Nance, manager of advanced pollution control systems for the company, was an electrolytic cell used in the aerospace industry.

The cell in question is used for the manufacture of ammonium perchlorate, an important chemical oxidizer for solid propellants. Dubbed the PEPCON cell, it is patented by Pacific Engineering and Production Co. (Henderson, Nev.), with which Thiokol has an exclusive licensing agreement.

In its sewage treatment systems,

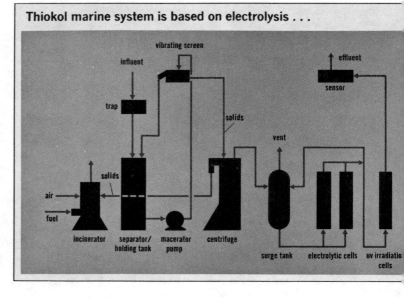

**Thiokol marine system is based on electrolysis . . .**

influent · vibrating screen · effluent · trap · sensor · solids · vent · solids · air · fuel · incinerator · separator/holding tank · macerator pump · centrifuge · surge tank · electrolytic cells · uv irradiatio cells

Thiokol doesn't use the cell to make ammonium perchlorate, but rather to generate sodium hypochlorite—which decomposes to yield atomic oxygen for oxidation of dissolved organic matter. To hasten the decomposition reaction, and so to speed up the entire treatment process, Thiokol uses either ultraviolet light (in a cell process developed and licensed by Midwest Research Institute, Kansas City, Mo.) or a catalyst bed containing a proprietary bed material developed by Thiokol. Which method is actually used depends on the application.

The system Thiokol foresees for on-board, marine sewage treatment illustrates the type of new thinking that firms entering the sewage treatment field are bringing to it:

Sewage, carried in brine in the marine application, is screened and conveyed to a centrifuge of conventional design—see flow sheet. Clear liquid from the centrifuge flows through a number of PEPCON cells, either in series or in parallel, and then through cells irradiated by uv light. Conditions can be arranged so that the final

effluent meets standards determined by the Environmental Protection Agency (EPA) under authority of the 1970 Water Quality Improvement Act (see ES&T, May 1970, page 379). These standards haven't yet been finalized, but are likely to be around 150 mg/l. for suspended solids and 100 mg/l. for BOD.

Centrifuged solids are burned with auxiliary fuel—diesel or fuel oil—in a fluid-bed incinerator of Thiokol design. Although there is no specific provision for offgas scrubbing in the marine system, Thiokol says that this can be arranged for land-based units.

The biggest advantage of the Thiokol marine system is its compactness. A 6000-gpm system—big enough to handle a 200-man sewage load—could be accommodated in a space 6 × 6 × 6 ft and would weigh less than 2 tons. One possible disadvantage is its power demand; the PEPCON cells draw perhaps 500 A at 6–7 V dc, and the 200-man system would require up to 30 kW. Thiokol has a $584,000, two-year contract with the U.S. Navy to develop and demonstrate the system (which so far has been tested in a 50-man pilot plant).

Thiokol recognizes that the marine system above is inherently too expensive for use on pleasure boats (the system described above would cost around $40,000). For these, Nance suggests a system that would include a "filter-incinerator" and a catalytic re-

The importance to AWT and to the future of sewage treatment technology is equally real. For AWT is 80% owned by Hercules, Inc. (Wilmington, Del.), another company in the solid propellant business, and a firm that has made deliberate strides to increase its penetration into the pollution control business (see ES&T, May

be of drinking water quality, although in the Freehold project it will apparently be discharged. The company predicts 95% removal of BOD and phosphate.

Solids from the primary screen, clarifier, and "magnetic filter" will be partly dewatered in a sand filter and then conveyed with sand into a fluidized bed incinerator (trademarked Fluidhearth by AWT) where the solids are burned off at 1400–1500°F. The incinerator will be fitted with cyclone and water scrubber to check particulate emissions. In the 50,000 gpd system at Freehold, there will be just 10 lb of inert ash per day to dispose of, says Procedyne.

An interesting wrinkle to the AWT system is that the fluid-bed incinerator will be used, two days out of the month, to regenerate activated carbon for reuse as an adsorptive medium.

With a capital cost of $450,000, the system is certainly expensive, but AWT Managing Director John Floyd points out that subsequent systems could be built for less, especially since 50,000 gpd is "about the minimum economic size."

ITT Levitt, for its part, sees the Freehold installation as being essentially only temporary, and expects that when the housing development is sewered, about five years from now, the treatment plant will be disassembled and reerected at another housing site where it is needed.

## ... while AWT land system stresses physical-chemical treatment

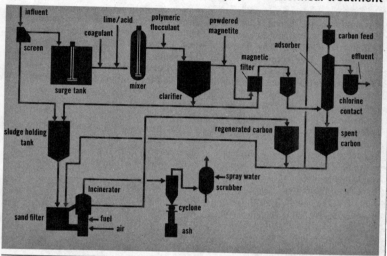

**Self-contained community**

New sewage treatment technology, too, is at the heart of a housing development planned by ITT Levitt & Sons, Inc. (Lake Success, N.Y.), the giant home builders. Under a $750,-000 project that is equally funded by Levitt, EPA, and AWT Systems, Inc. (Wilmington, Del.), Levitt will build a 125-home community in Freehold Twp., N.J., and equip it with its own sewage treatment plant, based on AWT technology.

The importance to Levitt of such a development is very real: It enables houses to be built in an area that is not presently served by a trunk sewer (and incidentally, would help the company to build where there are sewers, but where the treatment plant is already overloaded, a situation that led to a building moratorium in several parts of the U.S. last year).

1970, page 390). The other 20% of AWT is owned by Procedyne Corp. (New Brunswick, N.J.), an outfit experienced in fluidized solids technology—a field that is relatively well-known in the chemical industry but which is only just beginning to make its mark in pollution control.

The system that AWT has put together for the ITT Levitt Freehold project may be a sign of things to come in sewage treatment; like the Thiokol systems described above, the AWT system shuns biological treatment in favor of physical-chemical methods. In essence, what the Freehold system will involve is this:

Screened sewage will have added to it an inorganic coagulant and a polymeric flocculant (and, possibly, lime or acid for pH control)—see flow sheet. After clarification, powdered magnetite will be added and the residual solids removed in an electromagnetic field. The remaining impurities—dissolved organics—are removed by adsorption onto activated carbon. The chlorinated effluent from the carbon column will, according to AWT,

**Pressure needed**

Whether newer and more technically sophisticated sewage treatment methods ever become anything more than of purely technical interest seems to depend primarily on the federal government. It's not just a matter of the EPA's providing research and development funds—though that, to be sure, is probably a necessary stimulus. But the newer systems are more expensive than older, less efficient, biological methods, and are likely always to remain so. Unless there is federal pressure to increase the stringency of effluent standards, there will be no real need to apply the newer methods.

It seems certain, however, that this pressure is intensifying, and with the congressional penchant for writing a requirement for the latest and best pollution control technology into federal law, the prospect of a major revolution in the disposal of domestic sewage is more than just pie in the sky.　　　DHMB

*Reprinted from* ENVIRON. SCI. TECHNOL., **7,** 209 (March 1973)

# Curtailing pollution from metal finishing

Not only can the quantity of these
waste waters be reduced, but waste recovery
decreases overall treatment costs

**James F. Zievers**
**Charles J. Novotny**
*Industrial Filter & Pump Mfg. Co.*
*Cicero, Ill. 60650*

Seven types of pollution are commonly associated with the normal metal-finishing operation. These wastes include cyanides, hexavalent chromes, pH fluctuations, oil, heavy metals, phenols, and phosphates.

Cyanides and hexavalent chromes must be treated because they are toxic, and this is usually done by alkaline chlorination and low pH sulfonation, respectively. Wide fluctuations in pH need to be corrected since such fluctuations can be considered corrosive and hence destructive. Alkaline or acid pH correction agents usually handle this problem.

Oil is inflammable and dangerous; and by covering the surface of water bodies, it prevents life-giving oxygen from entering the water. Depending on the quantity of oil involved, it may be skimmed or otherwise taken from the water surface, or it may be "broken"—normally at low pH at relatively higher temperatures. The oil fraction is usually recovered, and the aqueous fraction is treated further to required water quality standards.

Heavy metals, usually metallic hydrates, are considered toxic, depending on the quantity, by water pollution control authorities. Sedimentation, clarification, and/or filtration can be used to remove heavy metals from metal-finishing wastes.

Depending on the type of metal-finishing operation, phenols may be present in the effluent. These odoriferous compounds may be treated by adsorption onto activated carbon. If the quantity of phenol in the effluent is great enough, it may be recovered by scrubbing and distillation.

The last major pollutants from metal finishing are phosphates, which the majority of the scientific community believes contributes to eutrophication of rivers and lakes. Phosphates can be removed by a number of methods. When they are in relatively small concentration and consequently appear most difficult to handle, adsorption on activated carbon is an effective method of removal; they can also be precipitated with lime.

## Waste plant design

Six procedures in metal-finishing operations can be followed to result in the smallest treatment plant possible handling the least possible quantity of wastes. The first item, "housekeeping," involves personnel discipline and morale. Maintenance of a clean plant will normally guarantee against injurious and obnoxious spills, mixes, and process losses that would otherwise be expensive.

The second routine is segregation which includes identification of all sources of waste that must be treated and segregation of those sources into channels permitting sensible treatment of the various components of the overall waste flow in the most efficient manner. In other words, cyanide-bearing wastes should be separated from chrome-bearing wastes. It further entails handling concentrated dumps (periodic discharges of very strong wastes) so that those dumps can be bled slowly into running rinses for treatment; this removes shock loads, overdesign of dosage equipment, etc.

The next method of preparation is scheduling. If water is to be recovered, all concentrated dumps can be scheduled for treatment for a limited period of time per week, or per unit period of operation, and running rinses can be treated the rest of the time. Since the amount of dissolved salts in running rinses is relatively small, the cost of recovering water will be relatively small. During the period when the dumps are used, the waste can be neutralized and still safely discharged to the receiving water body.

The fourth consideration in waste plant design is reducing effluent volume, which can be commonly accomplished by employing countercurrent rinsing and increasing the number of countercurrent rinses or by using conductivity control on rinse water. Next, reduction of intensity is accomplished by switching the type of plating bath used. For example, shifting from a high cyanide bath to a low cyanide bath reduces the intensity of cyanide in process to $\frac{1}{10}$ of its previous concentration. Mist suppressants are commonly used in conjunction with chrome baths for similar purposes. In some cases, the substitution, for example, of cyanide baths, can lead to difficulty in breaking down chelating agents utilized in the substitute. The sixth technique is termed change of chemistry. For example, the use of copper sulfate baths as a substitute for copper cyanide can eliminate the need for cyanide treatment in some cases.

The importance of waste plant design and, of course, reducing the amount of water entering the plant and the amount of water leaving the plant via the waste system is shown by the average changes in raw water cost and in sewer charges in the U.S. Water is expensive, and plants should be operated to use as little water as possible.

There are 12 commonly used methods for treating metal-finishing wastes. Dilution was more commonly used in past years where, for example, an effluent flow containing 25 ppm of cyanide could be diluted with sufficient water to drop the total cyanide in the effluent to whatever the tolerable limits were. However, with rising water costs and tighter control on allowable limits by regulatory agencies, this technique will probably be used less and less.

The second technique, containment, is lagooning some

materials for eventual disposal by evaporation. In other cases, drumming of materials such as solvents or radioactive materials may be used. Drumming will probably remain as a widely used containment technique for many years; lagooning, however, may fall into disuse as tighter restrictions on materials entering the subsoil are enforced and as prices of land available for lagooning continue to rise.

Chemical conversion includes cyanide destruction by alkaline chlorination, chrome reduction by low pH sulfonation, and pH adjustment for conversion of soluble heavy metals to heavy metal hydrates for subsequent disposal by other techniques. These are widely accepted as general treatment techniques and can be made economically feasible by careful selection of the chemical equation to be followed and by use of the most ubiquitous and lowest priced chemicals.

Using combustion as a disposal technique may involve a metal finisher attempting to solve a liquid pollution problem and producing a subsequent air pollution problem. Combustion has been used in the past with oil, solvents, and some contaminated carbons. Evaporation, and/or reverse osmosis are both used for the recovery of fairly concentrated metal baths for reuse.

Ion exchange can be used to concentrate wastes to reduce treatment plant size but is more commonly used for metal recovery. Often ion exchange and evaporation will be used in conjunction with each other to recover metal baths for reuse. Ion exchange is also widely used for preparing treated effluents for reuse in the treatment plant itself.

In some cases, extremely difficult acids and some ammonia compounds (weak ammonia liquors from steel plants) are more easily disposed of by pumping them into a deep well drilled into suitable strata in the subsoil. In the U.S., such wells are usually drilled to a depth of 2500–4000 ft. Permission must be obtained from the cognizant government authority prior to deep-well injection, and the effluent is normally filtered to a high degree of

## TABLE I
## Metal finishing wastes treatment techniques

| TREATMENT | COMMENT |
|---|---|
| **Dilution** | With rising water costs and stringent limits, probably will be used less and less |
| **Containment** | Lagoons for some materials; drums for solvents |
| **Chemical conversion** | Cyanide destruction; chrome reduction |
| **Combustion** | Oil–solvents–carbon |
| **Evaporation/reverse osmosis** | Recovery of fairly concentrated metal baths for reuse |
| **Ion exchange** | Capture of metals for recovery; purification of water for reuse |
| **Deep well disposal** | Difficult acids, some ammonia compounds |
| **Sedimentation** | Natural "fall-out" of hydrates, etc. |
| **Clarification** | Promoted "fall-out" of hydrates, etc. |
| **Aeration** | Reduction of BOD |
| **Filtration** | Removal of suspended solids—practically a "must" for recovery or reuse |
| **Combination** | Viz use of fly ash to aid removal of heavy metal hydrates |

**Waste treatment.** *Chromic acid can be profitably recovered from metal finishing wastes by an atmospheric evaporator (above), and high quality rinse water is produced by a two-bed ion exchange unit (below). A packaged two-stage cyanide destruction system (right) prepares wastes, by chemical conversion, for disposal by other techniques*

clarity prior to disposal to prevent plugging the well. Although well disposal is fairly widely practiced, six states still do not allow it.

The eighth treatment technique is sedimentation or the natural precipitation and settling of hydrates and other suspended matter in a relatively quiescent pond—the effluent from which is then fed to the receiving water body. From time to time, the ponds must be dredged and the sludge disposed of by dumping or piling.

Because of the rising cost of land, clarification, rather than sedimentation, is used. Precipitation of hydrates and other suspended matter is promoted in a hydromechanical device to which polymeric flocculants, alum, etc., are added to induce the sedimentation of materials. Underflow from a clarifier will normally contain 1½–3 wt % solids. Overflow will normally contain a liquid with a clarity equivalent to perhaps 15–30 ppm of suspended solids, which is normally satisfactory for discharge to a receiving water body.

Aeration is normally used for reduction of BOD. Filtration is used to remove suspended solids and can be accomplished with pressure- or vacuum-type equipment. Filtration is practically a must when recovery or reuse of treated waters is being considered. The recovered waters must be suitable for rinse purposes; suspended solids cannot be tolerated. When vacuum equipment is used, accumulated solids on the filter are normally discharged at about 80% moisture. Solids from pressure filters are usually about 50% moisture. Effluent clarity will average 0–3 ppm suspended solids.

The last treatment technique, combination, will use two obnoxious materials together to accomplish one mutually compatible end. For example, fly ash from a stack scrubbing operation can be used as a filter aid and precipitation promoter for the filtration of heavy metal hydrates. Each of the two items, the fly ash and the metal hydrates, constitute a problem, and yet when combined each helps the other to solve the problem.

### Waste treatment and profit

Can waste treatment be profitable? Probably not overall, but some methods certainly can be less expensive than others, and recovery credits can go a long way toward alleviating treatment cost debits.

First, a look at costs is in order. Figure 1 illustrates up-to-date initial capital costs (including installation) of various metal-finishing waste treatment plant components. The graph illustrates the economic importance of reducing individual waste flows to the lowest possible volume. Those same initial capital costs have been amortized over 10 years and included with chemical and labor costs to produce the operating cost curves shown in Figure 2. Of special interest in Figure 2 is the cost-reducing effort of change in plating bath chemistry, such as the use of low-cyanide baths. Further reductions in operating costs can be made, for example, by using $SO_2$-bearing flue gases in treatment of chrome-bearing wastes.

Recovery of acids and/or metals may result in waste treatment producing financial credits. Such recovery usually involves evaporation and may or may not involve ion

226

exchange equipment or reverse osmosis equipment. If we assume evaporation system is properly sized and operated, a metal finisher may expect to recover 95–96% of the valuable product present in his rinse stream. Recovery can comprise an average of 40% of total usage.

Recovery by evaporation is practiced, for example, in conjunction with phosphoric acid used in pickling steel. In recent years, cation purification and evaporation have been widely used profitably to recover chromic acid.

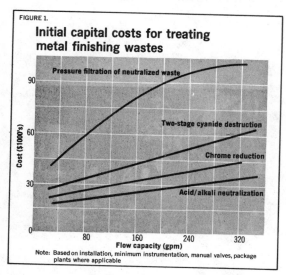

FIGURE 1.

## Initial capital costs for treating metal finishing wastes

Note: Based on installation, minimum instrumentation, manual valves, package plants where applicable

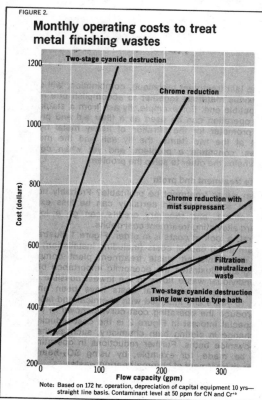

FIGURE 2.

## Monthly operating costs to treat metal finishing wastes

Note: Based on 172 hr. operation, depreciation of capital equipment 10 yrs—straight line basis. Contaminant level at 50 ppm for CN and Cr$^{+6}$

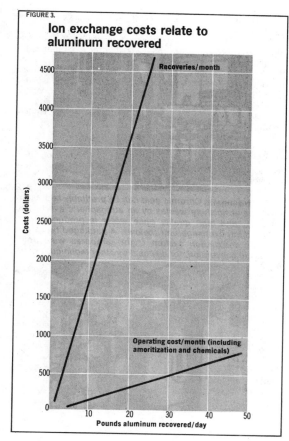

FIGURE 3.

## Ion exchange costs relate to aluminum recovered

Countercurrent rinsing can reduce volume of rinse flow which in turn reduces evaporation costs.

Ion exchange can also be used to create a less expensive waste problem. For example, in the new hard-coat anodizing bath (in metal finishing), dissolved aluminum ties up some of the anodizing acid. Part of the bath can be discarded to the waste treatment plant and replaced with new solution to maintain the aluminum concentration at an acceptable level.

An alternate method is to remove dissolved aluminum with a cation exchange system. The waste regenerant goes to the waste plant just as discarded bath would have gone. However, ion exchange allows reuse of the anodizing bath, which is valued at $1.00/gal.

The monthly cost of such an ion exchange system (including amortization and chemicals) is a function of the amount of aluminum which must be removed each day. The value of the recoveries—the value of the bath which would have been discarded without ion exchange—indicates the profits of the system or the reduction in the cost of waste treatment (Figure 3).

Recently a large automobile manufacturer installed two ion exchange systems to produce high-quality water for rinsing purposes and to recover and concentrate nickel sulfate solution which in turn is reused in the nickel plating baths. By utilization of a multiple countercurrent regeneration system, 2.4 lb of nickel/ft$^3$ of cation resin installed was recovered. In the larger (120 gpm) system, the pay-out time is 2.1 years, and for the smaller (80 gpm) system, pay-out time is 2.4 years which includes cost of capital equipment, installation, and operation.

Where smaller flows are anticipated, reverse osmosis can be used to achieve similar economies.

After all of the alternatives for recovery have been explored and those that are most feasible for each individual finisher have been applied, there will still be a treated effluent flow. If we assume dumps are scheduled for a fixed period each week, the average concentration of dissolved solids in the effluent will be approximately 500 ppm. Water for rinsing purposes (except critical rinses) is satisfactory if the level of dissolved solids approximates 300 ppm. Depending on combined raw water and sewer charges, and further depending on the analysis of the raw water, it will often be feasible to recover effluent waters by ion exchange. Raw water and sewer charges are rising very rapidly and apparently will continue to rise, making water recovery a thing that must be reassessed continually.

A survey of 21 U.S. and European cities made in 1972 revealed that as of January 1972 the average of combined raw water and sewer charges was $0.460 per 1000 gal. Water recovery by ion exchange is feasible if the effluent flow approaches 300 gpm. Other factors can make such recovery feasible sooner. For example, if raw water analysis is such that treatment would be required, recovery could very well be the most attractive alternative.

Some researchers have advocated using ion exchange for water recovery by applying ion exchange directly to mixed running rinses, and have cited operating experience that indicates, for example, that cyanides (not captured by weak base resins) do not reach dangerous levels in the recycled water. Should such a procedure be acceptable to local safety regulations, it would make water recovery economically feasible at flow rates as low as 40 gpm.

Continuous ion exchange equipment is no more costly initially than conventional equipment and is approximately 30% less expensive to operate. With continuous ion exchange equipment, water recovery just may approach profitability.

### Additional reading

J. P. Homrok, *Prod. Finish.*, p 78, September 1967.

J. F. Zievers, R. W. Carin, F. G. Barclay, *Plating*, p 1171, November 1968.

Robert F. Weiner, *ibid.*, p 1354, December 1967.

R. D. Ross, "Industrial Waste Disposal," Reinhold, New York, N.Y., 1968.

W. J. Lacy, A. Cywin, *Plating*, December 1968.

**James F. Zievers** *is vice-president of Industrial Filter & Pump Mfg. Co. He has been with the firm for 20 years and has served in both sales and executive capacities. Mr. Zievers is a member of five professional societies and is a registered professional engineer in all states. Address inquiries to Mr. Zievers.*

**Charles J. Novotny** *is director of technical sales, Industrial Filter & Pump Mfg. Co. He has worked in sales and ion exchange for several years. Mr. Novotny has authored many technical papers, holds several patents, and is a member of five professional societies.*

# Reclaiming zinc from an industrial waste stream

*Reprinted from* ENVIRON. SCI. TECHNOL., **6,** 880 (October 1972)

*NVF Co. has built a reclamation plant which not only cleans up its discharge stream, but also provides a basic chemical for the firm's vulcanized fiber operation*

Red Clay Creek, a small meandering stream in the state of Delaware, slowly winds through forests and meadows. What sets it off from most other picturesque streams is the fact that Red Clay Creek has fish living in it for the first time in 50 years.

Few industries are located on Red Clay Creek; one of these is NVF Co. which has its origin in a flour mill and saw mill built in 1763. NVF manufactures 45% of the world's vulcanized fiber at

**Cleaned up.** *Five rolls of paper (right) pass through a zinc chloride bath and are bonded together to form vulcanized fiber. Until recently, wash containing zinc was discharged to Red Clay Creek (above) making it toxic to fish*

Yorklyn, Del. Vulcanized fiber is a laminated plastic made from cotton cellulose paper layers bonded by chemical treatment and converted into a homogeneous material possessing unusual physical, electrical, and mechanical properties. It can be sawed, punched, drilled, milled, turned, shaved, or formed; it's available in sheets, rolls, coils, rods, tubes, or fabricated parts; and it's one of the strongest materials per unit weight known.

### Vulcanized fiber

When vulcanized fiber is manufactured, zinc compounds are used in and discharged from the operation. Here's how vulcanized fiber is manufactured: Originally, rags, scraps from textile manufacturing, and old clothing—all 100% cotton—were used in the process. These cotton materials were cooked and beaten (in several steps) to remove any color from the fabric and to shred the cloth into individual fibers.

In most cases today, purified wood pulp is used instead of rags (only one NVF plant processes rags) since the availability of cotton is variable and the colored cooking wastes are serious pollutants. Wood pulp is treated in much the same manner as rags to produce a slurry containing cellulose fibers.

After dewatering, the fibers are com-

## NVF zinc reclamation flow diagram

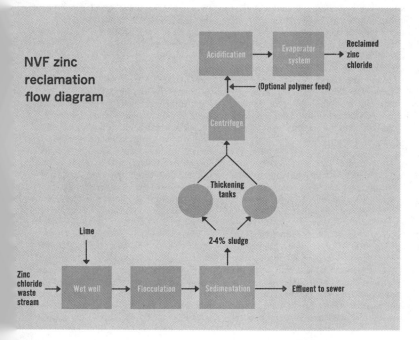

pressed into sheets of paper and wound onto large rollers. Three or more rolls of this paper are then pulled through a gelatinous zinc chloride bath which bonds the layers of cellulose paper together into a homogeneous material of desired thickness. The paper—now one layer—passes through a countercurrent washing tank (paper is pulled against the water flow) where the zinc is leached out of the vulcanized fiber.

The fiber is then dried, pressed, and ready for shipping or further processing. This cellulose vulcanized fiber is chemically pure, containing no resins or bonding agents.

The environmental problem occurs in discharging the wash water containing dissolved zinc. Vulcanized fiber has been produced at the Yorklyn, Del., location for many years; in the early years, the waste waters were discharged untreated into Red Clay Creek. Zinc is a potentially toxic substance, and fish are especially sensitive to it. For the past 50 years, the Creek was essentially sterile below the plant.

Eventually an evaporator system was built to reclaim zinc salts. Water from the washing process went through the evaporator, and zinc chloride was concentrated for reuse. However, some wash water and condensates discharged still contained zinc.

In the early 50's, Delaware officials met with NVF representatives to discuss cleaning up Red Clay Creek, but mainly in terms of wastes from NVF-owned paper mills separate from the fiber plant. NVF built an upflow clarifier that removed 80% of the BOD, reducing visible organic pollution. Effluent conformed to the state standards, but the stream was still essentially void of life.

In the late 1960's the treated and untreated discharges from the NVF fiber plant failed to meet revised state requirements for maximum acceptable zinc concentrations. The Delaware Department of Natural Resources and Environmental Control limited zinc discharges to 0.3 ppm/day if discharged directly to the stream and 1.0 ppm/day if discharged to the county sewer system.

## Zinc reclamation

To meet state standards, NVF Co. began constructing a $1.2 million zinc reclamation plant, sewer, and pumping stations capable of handling one-half million gallons of waste water per day. In March of this year, the zinc reclamation plant was in full operation, handling 350,000–400,000 gpd. Waters containing 1.0 ppm zinc are pumped through an NVF-built sewer that connects with the county sewer system. The NVF sewer from Yorklyn was designed to serve simultaneously as the sanitary sewer system for the area. Last October,

when the zinc reclamation plant was formally opened, NVF transmitted the title of the sewer line ownership to New Castle County. The pumping stations also will be eventually owned and operated by the county.

The zinc removal problem was a tough one, according to company officials. Little has been done in the area of zinc reclamation—only 14% of the zinc used in the U.S. today is recycled. The NVF plant was the first in the vulcanized fiber industry to undertake and succeed in removing zinc from its waste waters, says Bill Philhower, NVF division technical manager. The only other industry with similar problems is the rayon industry.

The zinc reclamation process is simple and concise: NVF chose alkaline precipitation to do the job. The basic process itself is well known, but not when applied to zinc removal.

Waste water containing 100–300 ppm dissolved zinc is brought from all points of the plant into a wet well where lime is added to control pH (8.5–9.5) and maximize precipitation. Next the precipitated zinc slurry is pumped to the zinc reclamation plant and into a flocculating chamber.

About an hour later, the slurry moves to long rectangular settling basins where the precipitated zinc salt settles as a sludge. The 2–4% sludge passes through thickening tanks.

The sludge passes from the thickening tanks through a centrifuge where polymers are added to enhance precipitation. At this point the sludge slurry is 15–25% total solids. As the sludge leaves the centrifuge, it is treated by hydrochloric acid (to acidify the alkaline solution and form zinc chloride) and is returned to the evaporator system for further concentration before it is reused in manufacturing vulcanized fiber. Liquid drawn off during the settling period is discharged to the county sewer system in concentrations well within limits set by the state.

What about the economics of recycling? The plant now recycles 50,000 pounds of zinc chloride per month for roughly the same price tag as the purchase price of fresh zinc chloride. Investment costs won't be recovered monetarily, but there are fish in the stream.                              CKL

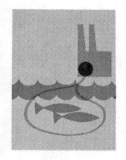

# Thermal discharges: ecological effects

Degradation to aquatic ecosystems from cooling water discharges hasn't occurred at some power plants; however, future expansion will make waste management more difficult

*Reprinted from* ENVIRON. SCI. TECHNOL., **6**, 224 (March 1972)

During the past several years, public interest in environmental quality as it relates to central-station power generation has intensified. The continued dominant role of thermal power plants to meet expanding electrical demands has focused attention on the effects of power plant–heated effluents on aquatic life.

Thus, one of the most important questions being asked today is, "What are the environmental effects resulting from waste heat additions to rivers, lakes, estuaries, and oceans?" Possible thermal effects are of concern to sports and commercial fishermen who want game and commercial species of fish available for their enjoyment and livelihood; conservationists who want the ecosystem preserved in its "natural" state; government regulatory agencies that set water temperature criteria and standards; and various users of water for cooling purposes who must discharge heated water within certain criteria and standards.

### Water use

Estimated projections indicate that future electric power requirements in the U.S. are expected to double approximately every 10 years. Even though hydroelectric power generation is expected to increase, steam-electric power (including both fossil- and nuclear-fueled plants) is expected to supply over 90% of the requirements in 2020 (Figure 1). By the year 2000, nuclear power will supply over 50% of the energy produced.

Of utmost importance to the steam-electric power industry is available water for condenser cooling. Estimated

water use and projected requirements, by purpose, for the U.S. was forecasted in the 1968 report of the Water Resources Council (Table I). In 1965, the steam-electric power industry used approximately 33% of the total water withdrawals. In 1980, the electric power industry will use about 44% of the total water withdrawals, and the forecast for

water withdrawal for the year 2020 will be 67% of the total. Projected consumptive use (nonreusable) of the total water withdrawal is about 23%, while projected consumptive use for water withdrawal for steam-electric power is only slightly greater than 1%.

Waste heat rejected to cooling water will be a function of the thermal ef-

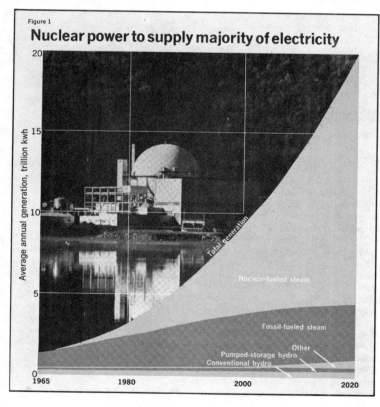

Figure 1

## Nuclear power to supply majority of electricity

Average annual generation, trillion kwh

Total generation

Nuclear-fueled steam

Fossil-fueled steam

Other
Pumped-storage hydro
Conventional hydro

1965    1980    2000    2020

**Arthur A. Levin, Thomas J. Birch, Robert E. Hillman, and Gilbert E. Raines**

*Battelle Memorial Institute*
*Columbus Laboratories*
*Columbus, Ohio 43201*

By utilizing projections of both fossil- and nuclear-fueled electrical generation capacity, data on thermal efficiencies of steam-electric plants, and water withdrawal forecasts, the quantity of waste heat that will be dissipated into the condenser cooling waters of steam-electric plants can be determined. The total quantity of waste heat discharged to condenser cooling waters by the electric utility industry will more than double from the year 1967 to the year 1980. The contribution of heated effluents from nuclear-fueled power plants in this time period increases from 1% to 45%, while contribution of heated effluents from fossil-fueled power plants decreases from 99% to 55%.

These waste heat values should be placed in proper perspective. For example, the total quantity of water used for steam-electric power for 1980 (assuming once-through cooling water) is estimated to be 193 million gallons per day, while the estimated annual heat rejection for steam-cycle systems for the same year is 11,700 trillion Btu's. This quantity of heat, assuming a once-through cooling cycle, will raise the temperature of the cooling water approximately 20°F. Temperature increase in the condenser cooling water for condensers installed in the past ranges between 10° and 30°F. Thus, the estimated 20°F rise in once-through condenser cooling water seems to be a reasonable estimate although this will vary according to each specific site location.

### Site Studies

Most studies directly concerning the effects of heated effluents on aquatic

## TABLE I. Estimated water withdrawals in the U.S.

### (Million gallons daily)

| Type of use | Used 1965 | Projected requirements | | |
|---|---|---|---|---|
| | | 1980 | 2000 | 2020 |
| **Rural domestic** | 2,351 | 2,474 | 2,852 | 3,334 |
| **Municipal** (public-supplied) | 23,745 | 33,596 | 50,724 | 74,256 |
| **Industrial** (self-supplied) | 46,405 | 75,026 | 127,365 | 210,767 |
| **Steam-electric power:** | | | | |
| Fresh | 62,738 | 133,963 | 259,208 | 410,553 |
| Saline | 21,800 | 59,340 | 211,240 | 503,540 |
| **Agriculture:** | | | | |
| Irrigation | 110,852 | 135,852 | 149,824 | 160,978 |
| Livestock | 1,726 | 2,375 | 3,397 | 4,660 |
| **Total** | **269,617** | **442,626** | **804,610** | **1,368,088** |

Source: U.S. Water Resources Council

ficiency of the particular steam-electric plant. With the steam temperatures currently in use in large fossil-fueled plants, the maximum theoretical thermal efficiency is slightly above 60%. The thermal efficiency of the best operating fossil-fueled plants is presently about 40%.

Because of a lower thermal efficiency for nuclear plants (about 33%), cooling water requirements are presently greater than for fossil-fueled plants of the same electrical generation capacity. Approximately 10% of the gross waste heat is dissipated directly to the atmosphere through the stack in the fossil-fueled plant, while none is dissipated in this manner for the nuclear-fueled plant. Thus, about 50% more waste heat is rejected to the condenser cooling water from the nuclear plant.

Any method of reducing waste heat discharged into aquatic ecosystems would be useful where a temperature rise in receiving waters is unacceptable. Several options can reduce waste heat discharged from steam-electric plants into the aquatic ecosystem. Although thermal efficiencies from fossil-fueled steam plants have reached a plateau, molten salt breeder reactors and high-temperature gas reactors should increase thermal efficiencies for nuclear plants almost to 45%. However, these improvements will probably not be available for at least a decade. Since a dramatic increase in thermal efficiency for steam-electric plants is not forecast for the immediate future, recycling or retaining condenser cooling water may be necessary to reduce waste heat effects on aquatic ecosystems.

biota at the site of electrical power generating stations are relatively recent, and few results have been published to date. Most field investigations are presently in progress.

Continuing studies of the ecological effects of thermal discharges have been conducted at the Hanford Nuclear Complex on the Columbia River (Wash.). These studies conducted over the last 25 years were mainly oriented toward the salmonid fishes because of their high value to the Columbia River commercial and sports fisheries. Although the temperature of the undiluted reactor effluent would be lethal to the fish, waste heat discharged by the Hanford reactors to the Columbia adds only a relatively small heat increment to the widely variable seasonal river temperature (less than 40°F to greater than 65°F). Also, because of the hydraulic characteristics at the outfall and the swimming behavior of the fish, many seaward migrant salmonids may be swept to cooler waters and not actually experience the direct effluent plume.

Laboratory and field studies concerning biological effects of Hanford waste heat on salmonids shows no demonstrable evidence of damage to the salmonid resources. There simply has not been any evidence to indicate kills or unreasonable risks despite a long history of heated discharges from the Hanford reactors. However, direct extrapolation of Hanford's results to another site, even in the Columbia River system, must be made only with due consideration for the uniqueness of each ecosystem as the snow-fed Columbia River is a large, cool river and not typical of many U.S. river systems.

The Chalk Point fossil-fueled steam generating plant on the Patuxent River (Md.) has been studied since 1963. Two 335-MW units use estuary water for condenser cooling with a once-through cooling system. The condenser cooling water temperature increase is designed to be 23°F under winter operating conditions and 11.5°F during summer conditions. While no major detrimental effects of thermal additions have been noted, changes have occurred in various populations which may be attributed to heated cooling water discharges. Epifaunal populations in the intake and effluent canals of the Chalk Point plant provide a number of interesting results. Among them was: • a higher rate of production was found in the effluent canal than in the intake canal during all months studied • average production in the effluent canal was nearly three times as great as production in the intake • an increase in the maximum size of the barnacle, *Balanus*, was noted in the intake and effluent canals over those in the Patuxent River itself. During July and August, the warmest months, there was a decline in the number of species in the effluent canal and the anemone, *Sagartia*, and, the tunicate, *Molgula*, were not noted in the effluent canal, although both were in abundance just outside the effluent canal.

The power plant has not added enough heat to the Patuxent River to exceed the thermal tolerance of the zooplankton species studied. On the other hand, phytoplankton destruction and productivity suppression have been reported in the cooling water supply of the Chalk Point plant, although chlorination may be partly responsible for the mortality. Also, oysters in the Patuxent River have high copper levels. The rate of copper uptake in the oysters could have been enhanced by the water temperature increase, or copper concentrations in the water may have increased due to operation of the Chalk Point plant. However, no major effects on growth, condition, or gonad development were shown by oysters on natural bars near the plant.

At the Contra Costa Power Plant (1298 MW) on the San Joaquin River, (Calif.), studies showed that passing young salmon and striped bass through cooling condensers was far less hazardous than screening them at the intake. At the same plant, young salmon could tolerate an instantaneous temperature increase to 25°F for 10 min with no mortality.

At the Morro Bay Power Plant (1030 MW) (Calif.) on the Pacific Ocean, healthy populations of the pismo clam, *Tivela stultorum*, have been maintained over the full 13 years that the plant has been in operation.

The Humboldt Bay Nuclear Plant (172 MW) in California is the first nuclear plant in the U.S. utilizing estuarine waters for cooling and is located on the Pacific Ocean about five miles from an important shellfish area. Studies at Humboldt Bay showed that the elevated temperature regime of the discharge canal was favorable for the natural setting of native oysters (*Ostrea lurida*), cockles (*Cardium corbis*), littleneck clams (*Protothaca staminae*), butter clams (*Saxidomus giganteus*), gaper clams (*Tresus nuttalli*), and a half dozen other bivalves (even though some passed through the plant's condenser system).

The effects of heated discharges from the Connecticut Yankee Nuclear Plant into the Connecticut River (Conn.) are examples of a well-documented study started in 1965, about 2 1/2 years before the plant began operation. The plant was designed to produce 562 MW with a temperature rise of 20°F in the condenser cooling water. The major thermal study areas were fish studies; benthic organisms studies; bacteriology, microbiology, and algae studies; hydrology studies; and temperature distribution predictions and measurements.

The Connecticut Yankee Plant has now been in operation for about four years. No drastic changes have been observed to date in the overall ecology of the Connecticut River as a direct result of the addition of thermal effluents.

However, a statement in the summary of all the environmental studies that were done at Connecticut Yankee, emphasizes that as yet no information is available on the possible sublethal effects of the thermal discharge. Although no fish kills have occurred since the plant operation began, the white and brown bullhead catfishes undergo a marked weight loss (average of 20%) in the warm water of the effluent canal despite a constant availability of food in the canal.

Studies are being conducted at Turkey Point in Biscayne Bay, Fla., where two fossil-fueled units of 432 MW each are in operation, and two nuclear plants of 721 MW each are scheduled to begin operation. Heated effluents from the plant have reduced the diversity and abundance of algae and animals in small areas adjacent to the mouth of the effluent canal. Many plants and animals in a 125-acre area where temperatures have risen 4°C (7.2°F) above ambient have been killed or greatly reduced in number. In a second zone of about 170 acres, corresponding to the +3°C (5.4°F) isotherm, algae have been damaged, and species diversity and abundance have been reduced. In the latter area, mollusks and crustaceans increased somewhat, but the number of fishes decreased.

*DISCHARGES. Power plants, such as this one (right), use tremendous amounts of water for process cooling and usually discharge the then heated water into the body from which it came. To assess effects of heated effluents on aquatic biota, . . .*

*. . . Battelle scientists sampled organisms in the receiving waters (second from top), simulated river ecosystems mathematically (above), and compared oyster growth before and after power plant startup (left) prior to outlining the conclusions discussed*

Studies at the Martins Creek Plant on the Delaware River (Pa.) showed that the heated waters appeared to have attracted fish and enabled them to actively feed throughout the colder months of the year to a greater extent than they normally would, although there was no conclusive evidence that heated waters actually increased fish production or growth rates.

Studies at the Petersburg, Ind. Plant (220 MW) on the White River (Ind.), report that there is no evidence that any adverse effects on fishes, such as death, impaired growth, insufficient reproduction, increased disease, and movement or lack of movement are being observed at Petersburg or in the entire White River with the exception of fish movement away from water above 93 °F.

The White River has a sandy bottom and is quite turbid. The principal pollutants are floodwater and suspended material in the water. The major aquatic species at Petersburg are the spotfin shiner, bullhead minnow, spotted bass, longear sunfish, gizzard shad, carp, and white crappie. Since sand and silt are deposited when floodwaters recede, researchers who studied the White River believe that money for thermal pollution abatement could be better "applied to the certain and very real need for flood and bank control."

### Recommendations

The result of several ecological studies around actual operating power plants is that, with a few exceptions, there has not been any major damage to the aquatic environment from the heated effluents of existing power plants. However, in the future years, as larger power plants become operational, accompanied by multiple units at a single site, environmental management of heated effluents at these sites will become more difficult.

Standards for limiting the thermal loads imposed on aquatic systems have evolved with the expansion of the electrical generating industry. However, without feasible alternate methods to produce electrical power without waste heat, there are only a limited number of alternatives. At one extreme is employing methods which recycle cooling water and add no waste heat to natural waters. This extreme is not required to ensure well-balanced aquatic communities. The other extreme is to permit unlimited thermal loading on aquatic systems which would, no doubt, be disastrous (based on the projected use of marine and freshwater resources for industrial cooling purposes). The only option remaining is discharging waste heat to waters in amounts approaching the assimilative capacity of the waters in question. Heat generated beyond those amounts will have to be dissipated by methods which recycle cooling water. Based on the knowledge available at the present time, the last option seems to be the only reasonable approach.

Pursuing this course requires total commitment to determine the assimilative capacities of freshwater and marine resources. Management and surveillance programs will be essential as will cooperation between industry and regulatory agencies. Many factors contribute to receiving capacities, and requirements for producers of waste heat will be highly variable depending on their location. Power plant sites should be chosen with the advice of competent ecologists, and base line ecological surveys should begin as soon as a suitable site is selected.

While lethal effects of heated water discharges on fish and other aquatic organisms should present little problem, assuming proper discharge procedures, the sublethal effects of these heated water discharges may produce significant changes in populations. These sublethal effects could produce physiological changes that would decrease growth rate and prevent reproduction. Future studies should be designed to obtain a better understanding of sublethal effects.

The entire food chain is of extreme importance in the balanced aquatic ecosystem. Particular aquatic organisms or plants that fish eat can be affected by waste heat from power plants. Eliminating a single component of this ecosystem would affect the feeding and growth of organisms on all higher trophic levels.

Data are not yet sufficient to permit a proper understanding of the dynamics of this ecosystem. Many laboratory studies have led to understanding many of the physical-chemical functions of aquatic organisms as well as dispersion in water systems. Consequently, regulations based on these studies will be designed to minimize all possible risks of catastrophic kills of desirable organisms. Field studies are necessary to determine the "real-life" mechanisms occurring in the aquatic ecosystems. While laboratory studies are a necessary part of understanding, extrapolating laboratory measurements to field conditions must be done cautiously.

Answers to considerations which could alter regulations will have to be provided from nongovernmental sources such as the electric utilities. As the assimilation capacity of the environment is reached, it is increasingly important to consider long-term effects. Modest investment programs looking at the ecosystem to develop and verify predictive capabilities could themselves pay handsome dividends.

To utilize more fully the assimilative capacities of natural waters to dissipate waste heat, greater ecological management will be required, and operators of steam-electric stations will have to play an important role. In addition to considering effects of heat rejection during normal plant operation, attention must be focused on the effects of temperature changes, even though the actual temperatures may be below the lethal limit.

An effort should be made to establish the assimilative capacity of all natural waters to be utilized for cooling purposes. Based on predictions from the biological, chemical, and physical studies, limiting conditions should be established to accommodate the idiosyncrasies of each site. There is no substitute for on-site experimentation utilizing the resident populations and the local water. After a new unit comes on-line, a less intense program of surveillance should become a matter of routine at all plant sites.

As more of the larger power plants become operative and as more sites are required, the ability to predict response of the aquatic ecosystem to the heated water discharges must be improved. The systems approach to study ecosystem dynamics offers a valuable tool to individuals who make decisions concerning siting and design criteria for power plants.

Criteria and regulations can only be altered with confidence when accurate predictions can be made. The pre- and post-construction studies by the utilities, if expanded to consider predictive aspects, offer an opportunity to obtain needed data on the system and to verify the predictions.

The satisfactory performance of existing steam-electric plants supports the belief that controlled amounts of heated water can be added to aquatic systems without producing adverse biological consequences. Therefore, in the absence of evidence of damage to the ecosystem involved, it would be difficult to justify requiring steam-electric stations, which have been operating for some time, to install cooling devices because they are not meeting newly adopted state or federal regulations. A careful investigation of the issue at each specific plant site should be done prior to any action being taken.

In order to understand the dynamic behavior of the aquatic ecosystem, some long-term studies are required. Of course, there are many and varied types of aquatic ecosystems so that typical rivers, lakes, estuaries, and ocean systems should be studied in a variety of climates. Industry, and, in particular, the steam-electric industry, should participate in these studies since the power plants will be the major waste heat contributor to the aquatic ecosystem. Waste heat from the power plants will become a more significant discharge to the aquatic ecosystem in the future. It may be that the effects of waste heat could be beneficial when other pollutants, such as sewage and industrial waste, are limited or removed (as reported for the Thames River in England).

Although there has been no apparent major damage to the aquatic ecosystems by cooling water discharge, there have been ecological changes. The complex interrelationships of species, populations, and communities in an ecosystem is the result of years of evolutionary trial and error. Therefore, although no major mortalities are noted, shifts in species diversity or abundance might upset delicate balances which exist, and results might not be known for years.

There are some bodies of water presently capable of accommodating more thermal loading without incurring adverse effects on the aquatic biota, while the assimilative capacities of some others have already been exceeded. Thus, it is imperative to evaluate dynamic changes which are presently taking place in aquatic ecosystems, and to be able to predict what is likely to occur as the electrical generating capacity of the nation increases.

### Additional reading

Coutant, C. C., "Thermal Pollution—Biological Effects," *J. Water Pollut. Contr. Fed.,* **43**, 1292–1334, 1970.

Levin, A. A., Birch, T. J., Hillman, R. E., Raines, G. E., "A Comprehensive Appraisal of the Effects of Cooling Water Discharge on Aquatic Ecosystems," Battelle Memorial Institute, Columbus Laboratories, Columbus, Ohio, 45 pp, 1970.

Krenkel, P. A., Parker, F. A., Eds., "Biological Aspects of Thermal Pollution," Vanderbilt University Press, 407 pp, 1969.

Parker, F. A., Krenkel, P. A., Eds., "Engineering Aspects of Thermal Pollution," Vanderbilt University Press, 340 pp, 1969.

**Arthur A. Levin,** *Senior Environmental Advisor, has been with Battelle Columbus Laboratories* (BCL) *since 1967. He is presently responsible for coordinating* BCL'*s environmental and health programs. Address inquiries to Dr. Levin,* BCL'*s Wash., D.C. Operations, 1755 Mass. Ave., N.W., Wash. D.C. 20036.*

**Thomas J. Birch** *is a research limnologist with* BCL. *His work on thermal effects includes laboratory experimentation and field investigations at electric generating stations.*

**Robert E. Hillman** *is presently chief of Battelle's W. F. Clapp Laboratories in Duxbury, Mass. He is responsible for ecological studies in marine environments adjacent to power plants.*

**Gilbert E. Raines,** *chief of* BCL'*s Ecology and Environmental Systems Division, is specifically interested in mathematically describing the response of aquatic biota to thermal stimuli.*

*After many years of apathy and indifference,*
*private citizens, industry, and government are now*
*advocating national policies for . . .*

# Groundwater pollution and conservation

*Reprinted from* ENVIRON. SCI. TECHNOL., **6,** 213 (March 1972)

Remember the days when a man hired a dowser or water witch before digging a new well? The dowser would point a forked witch hazel twig toward the ground and be mysteriously drawn to the spot where the well should be dug. His chances of predicting where water could be found were quite good, for "groundwater" can be found almost anywhere beneath the earth's surface at varying depths (see map). In fact, over 97% of fresh water in the U.S. is underground.

Most people are unaware of or oblivious to groundwater, since you can't water ski on it, fish or swim in it, or gaze across it to watch a sunset. Actually, surface water at one point may be groundwater at another, and then emerge again at a third point as surface water.

To understand groundwater and its role in the environment, its characteristics must be explained. Groundwater is water beneath the earth's surface in a geological formation that supplies wells and springs. Groundwater moves through and is stored in an aquifer—the porous geological formation containing water. The best aquifers are layers of gravel, sand, sandstone, limestone, or even nonlayered rock that has sizeable and interconnected openings such as some lava rocks. Clay, shale, and crystalline rocks are usually poor water carriers but may yield at least enough water for domestic and stock uses in areas where there are no better aquifers.

Aquifers may be a few feet to hundreds of feet thick and may underlie a few acres or many square miles. Most aquifers are local in extent; however, the Dakota Sandstone, for instance, (in the West) carries water across several states.

The water table is the top level of groundwater or the zone of saturation—the area in which all pore spaces in the rocks are saturated with water. Between the land surface and the water table is an area which hydrologists call the zone of aeration where water moves downward from the land surface to the water table. This zone may contain a little water, but it cannot be obtained through wells and is held to soil particles and rock by capillary action.

Groundwater moves through permeable rock and around or in between impermeable ones. Just like surface water, it takes the path of least resistance down as far as it can go. This downward flow can be stopped by solid rock, clay, or a denser fluid. If held between two such impermeable layers, groundwater may be under enough pressure to create artesian flow when opened by a well. Groundwater movement is quite slow and is measured in ft/year as contrasted with streams measured in ft/sec.

### Importance of groundwater

Groundwater discharges into springs, streams, rivers, lakes, swamp areas, geysers, and wells. Water flow can range from a few gal/hr to thousands of gal/min. Man taps these groundwater sources for a variety of urban, industrial, and rural uses.

Aquifers can be recharged naturally or artificially. Under natural conditions, aquifers are recharged with water moving from regions higher than the discharge area by the means of rain, snow, streams, etc. After the zone of aeration is saturated, the aquifer itself is recharged.

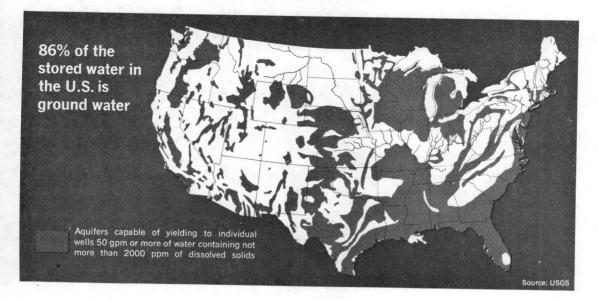

86% of the stored water in the U.S. is ground water

Aquifers capable of yielding to individual wells 50 gpm or more of water containing not more than 2000 ppm of dissolved solids

Source: USGS

**Wells.** *Groundwater supplies 95% of domestic water used in rural areas*

Artificial recharging is a valuable conservation tool and is used to counteract excessive water withdrawal from an aquifer. Sometimes wells are drilled just for recharging; in other areas, irrigation helps to recharge the groundwater aquifer.

On Long Island (NY) and in the Los Angeles area, aquifers are being recharged to protect against intrusion of salt water. The Hanover Canning Co. (Hanover, PA) and other food processors use spray irrigation to dispose of liquid wastes; this contributes to crop nourishment and groundwater recharge. Storm water runoff is used for recharging in Orlando, FL and Fresno, CA. A number of recharging projects are under way in the arid and semiarid regions of the U.S.

Why is groundwater so important and such care taken to recharge it? Groundwater supplies 20% of the fresh water used in the U.S. (61 billion gal/day). Of the nation's 100 largest cities, 20 depend entirely on groundwater for their public water supply, and 13 use both groundwater and surface water. Twelve states obtain more than 50% of their statewide public water supplies from groundwater.

Furthermore, industry uses 7.7 billion gpd, rural areas use 45.2 billion gpd, and urban areas use 8.1 billion gpd. In fact, more than 95% of the rural population in the U.S. uses groundwater for its domestic water supply. The arid and semiarid Southwest is almost entirely dependent on groundwater.

In the West and Southwest, groundwater depletion is a problem. There, as in most states, more water is being withdrawn than is being recharged into the aquifer; in hydrologic language, the water is being "mined." Because of natural recharging, groundwater is a renewable resource; but when it's being steadily mined, water is used on a deficit basis.

In areas of Texas, New Mexico, Kansas, Colorado, Arizona, and California, "it is being mined at an alarming rate," says Bob Aitken, EPA's International, Interagency, and Intermedia Standards Coordinator. Because of this, some farms and ranches have been abandoned in Texas and New Mexico. Years ago, water was found 10–30 ft below land surface and, in some cases, even flowed above land surface. But as the water table lowered, wells had to be dug deeper and deeper until water lifting costs outweighed profits derived from the land.

**Contamination**

Constant withdrawal of groundwater is not the only problem. Groundwater pollution is increasing, even though groundwater has always been considered insulated against contamination.

Groundwater, purer and cleaner than surface water, is protected naturally by an excellent filtering system—soil, clays, rock particles, etc.—that removes suspended solids, bacteria, and, to a large extent, viruses. If that filtering system is overloaded or bypassed, the aquifer itself may become polluted.

"Pollution of groundwater has always been with us, like all other water pollution problems," continues EPA's Aitken, "but the result is different. If a stream is polluted and then that contaminating source is removed, the stream can be flushed in about a week, more or less (although reversing pollution effects is not that simple). For a large freshwater lake, the retention time is about 100 years. When an underground aquifer is polluted, however, the waste is retained for 200 to 10,000 years. U.S. Geological Survey officials say that water supplies for future generations could be jeopardized by groundwater pollution.

Surface disposal of domestic and industrial waste, seepage from septic tanks, mine drainage, feedlot wastes, deep-well disposal, sanitary landfills, and agricultural chemicals—the same type of pollutants that affect surface water—have a greater impact and a more prolonged effect on groundwater. In some areas, groundwater aquifers are quite sensitive to pollutants, while in others, due to soil and geological conditions, contamination is limited.

Industrial waste disposal into surface waters can pollute aquifers fed by streams. For example, phenol wastes discharged into the Caloosahatchie River made it necessary to abandon wells near Ft. Myer, FL. On Long Island, nitrates—from fertilizers, rain picking up nitrates from the air, and septic tanks—were detected in groundwater.

**Regulation**

Nationwide, groundwater charting and investigation is presently under the auspices of the U.S Geological Survey in the Department of the Interior. However, there is only limited federal control over groundwater regulation; the USGS merely advises and supplies information on groundwater resources. By and large, control is left up to states and municipalities which, unfortunately, usually attempt regulation only when there is a shortage of water, according to USGS officials. A number of basin management programs are presently in the making.

One major groundwater management problem is the question of riparian right and local, state, or federal control. In some areas (appropriation states), if a beneficial use for groundwater is established, then no subsequent user can deprive the original user of his prior water rights. In other areas, permits are required before drilling of any kind, yet other regions have no regulation at all. These are some of the inconsistencies that EPA would like to eliminate by promulgating a national policy for groundwater protection.

The Water Pollution Control Act Amendments of 1970 apply to navigable streams and mention groundwater only once, but S. 2770 and H. 11896 (the controversial bills now in Congress) are a huge step forward from the point of protecting the subsurface environment in general and groundwater in particular.

The major flaw, environmentalists point out, has to do with deep-well injection (usually for disposal purposes). At press time, both bills give the EPA administrator responsibility and authority to administer waste inputs in the ground with the exception of oil and gas injection wells. The debate centers around the bills' language that excludes, through definition, oil, gas, or water injection associated with the word pollutant. (For more on deep-well disposal, see ES&T, February 1972, p 120.)

The groundwater issue is still undecided. Although groundwater has been considered a separate species of water, it is not isolated. It has direct communication with atmospheric water, lakes, rivers, and oceans. Many streams, at their high input stages, recharge groundwater, and when streams are at low flow, the entire water source may be groundwater. Groundwater is important. CKL

# Multipronged attack on photo wastes

*Conventional oxidation methods, new process chemistry, mechanical aids, and resource recovery help photofinishers to meet water quality standards*

Reprinted from ENVIRON. SCI. TECHNOL., 5, 1084 (November 1971)

Want to guess how many rolls of film were processed in the United States last year? Neither do Lloyd E. West or Thomas W. Bober of Eastman Kodak's photographic technology division. Although just how many square feet of film and paper or how many gallons of processing solution find their way into incinerators, dumps, or sewers each year is not known, it's safe to guess that the numbers are growing.

It's not the volume of materials used by photofinishers, however, that complicates pollution control for the industry, say West and Bober. What counts is the diversity of chemicals and processes. Solutions of thiosulfate or acetic acid, for example, have high biochemical oxygen demands. Other chemicals, like phosphates, ammonia, and nitrates, contribute to eutrophication. Still others—among them ferricyanide, dichromate, and borate—may be toxic to fish or plants under certain conditions.

While individual process wastes do differ widely, what comes out of a photofinishing plant is typically a mixture of wastes, considerably diluted by water.

Mixed processing wastes are generally low in suspended solids (20–50 mg/l.). pH is on the alkaline side—ranging from 7.5–9.0—and well within limits usually prescribed by local sewer codes. Temperatures of mixed effluent range from 70–85°F. Odor and color are low, the effluent presents no fire hazard, and only trace amounts of oils, if any, are present. Since photoprocessing is wet chemistry, wash water may make up from 50–98% of the final sewered effluent. A few processing chemicals neutralize each other, and the resulting mixed effluent—when low in volume compared with the total amount of municipal sewage carried—may be sufficiently "treated" simply by dilution.

Nevertheless, as sewer codes tighten, dilution as a pollution control technique is rapidly losing ground. Certain heavy metals—notably hexavalent chromium—are prohibited in some sewers. As water becomes more expensive, there is greater incentive to concentrate waste—by processes such as reverse osmosis—and recycle process water.

According to West and Bober, pollution from the photofinishing industry is being attacked on four fronts:
• Upgrading conventional biological and chemical waste treatment processes at the plant site and developing new disposal techniques in the laboratory
• Recycling and reuse of processing chemicals, raw materials, and water
• Developing new processes which pollute less
• Using squeegees to reduce carryover of processing solutions from tank to tank

## Biodegradability

Most of the chemicals used in photographic processing are biodegradable. Work done by Kodak at its experimental 20,000-gpd activated sludge plant indicates that $BOD_5$ reductions for 24–48-hr treatment periods are 70–95%. Package activated sludge plants or, where there is room, aerobic lagoons, can significantly reduce the BOD of mixed effluent.

Such facilities are not usually economical, however, where the film processor has the option of tying into municipal sewers. Furthermore, it may actually be advantageous for municipalities to accept processing wastes from photofinishers. While both photo wastes and municipal sewage support bacterial growth, the mixture of photographic processing wastes in amounts up to 10% of the amount of synthetic domestic sewage may in some instances result in a superadditive effect with better biological digestion.

Reclaiming material for reuse also reduces pollution. The most obvious candidate for recycling is silver because of its inherently high value. Although silver is a toxic metal in ionic form, West emphasizes that there is no free silver in photographic processing wastes. Silver is present as a thiosulfate complex, he points out; the complex is not toxic to sewage sludge organisms and is converted to the sulfide by bacterial action. Silver sulfide also is nontoxic and can be removed with the sludge from biological treatment plants.

The amount of silver that remains in photo products after processing varies widely depending upon the type of product and image density. For black and white film, anywhere between 10–60% of the silver originally present will remain after processing. Virtually all the silver used in color film will be removed by processing. The silver that is removed winds up in bleaches and fixers where it can be reclaimed by precipitation with sodium sulfide, by electrolysis, or by

**Bug breeder.** *Kodak scientist adapts bacteria to mixtures of synthetic sewage and photofinishing wastes*

**Printout.** *Sample calculation shows waste strength for hypothetical photofinishing plant*

```
                        EASTMAN KODAK COMPANY
       CALCULATED COMPOSITION OF PHOTOGRAPHIC PROCESSING EFFLUENT FOR
       LEW PHOTO, ANYWHERE, USA                              JULY 13, 1971

                                   MACHINES           HOURS/DAY
                                   PER PROCESS         OPERATED
       PROCESS                     ----------          --------
       -------
       EKTACHROME E-4                  1                  6.00
       EKTAPRINT C                     1                  8.00
       LKTAPRINT R                     1                  4.00
       KODACOLOR C-22                  2                  5.00
       b+w FILM                        1                  4.00
       b+w PAPER                       1                  6.00

       TOTAL EFFLUENT FOR COMBINED PROCESSES WHEN OPERATING SIMULTANEOUSLY
```

| CHEMICAL IDENTIFICATION | EFFLUENT MG/L | EFFLUENT LBS/DAY | COD MG/L | BOD5 MG/L |
|---|---|---|---|---|
| • ACETATE | 164. | 35. | 166. | 134. |
| ALUMINUM | 2.2 | 0.54 | 0. | 0. |
| + AMMONIUM | 62. | 13.0 | 70. | 53. |
| + BENZYL ALCOHOL | 29. | 6.8 | 0. | 0. |
| + BORON | 18.5 | 4.0 | 0. | 0. |
| BROMIDE | 13.7 | 2.8 | 0. | 8.3 |
| CARBONATE | 40. | 9.3 | 11.6 | 1.8 |
| CITRATE | 15.3 | 4.2 | 16.4 | 0.01 |
| COLOR DEV AGENT CD-3 | 18.2 | 3.9 | 1.7 | 0.16 |
| DIETHYLENE GLYCOL | 1.1 | 0.23 | 4.2 | 13.8 |
| EDTA ++ | 4.9 | 1.1 | 1.6 | 30. |
| + ETHYLENEDIAMINE | 1.4 | 0.25 | 42. | 3.2 |
| + ETHYLENE GLYCOL | 34. | 5.5 | 13.8 | 0. |
| FERROCYANIDE | 37. | 16.6 | 42. | 0. |
| • FORMALIN | 75. | 0.59 | 5.6 | 1.2 |
| HYDROQUINONE | 2.9 | 0.03 | 0. | 0. |
| IODIDE | 0.05 | 0.01 | 0. | 0.30 |
| IRON | 1.6 | 0.29 | 2.0 | 0. |
| KODAK ELON DEVELOPER | 2.0 | 0.46 | 0. | 0. |
| MAGNESIUM | 7.5 | 1.8 | 3.2 | 14.0 |
| NEUTRALIZ AGENT NA-1 | 28. | 6.3 | 0. | 0. |
| + NITRATE | 20. | 4.3 | 0. | 0. |
| + PHOSPHATE | 25. | 6.0 | 14.0 | 117. |
| SULFATE | 70. | 14.9 | 0.37 | 0. |
| • SULFITE | 0.36 | 0.08 | 146. | 0.74 |
| THIOCYANATE | 260. | 58. | 0.62 | |
| • THIOSULFATE | 2.2 | 0.36 | 1.4 | |
| + ZINC | 1.8 | | | |
| OTHER ORGANICS | | | | |

| | FLOW RATE L/MIN | VOLUME GAL/DAY | COD MG/L | BOD5 MG/L | BOD5 LBS/DAY |
|---|---|---|---|---|---|
| TOTAL EFFLUENT | 285. | 25200. | 543. | 378. | 83. |

+ MAY BE ACCEPTABLE FOR DISCHARGE INTO A MUNICIPAL SEWER, BUT NOT ACCEPTABLE FOR DISCHARGE DIRECTLY INTO A STREAM.

• CONTRIBUTES TO HIGH OXYGEN DEMAND. SECONDARY TREATMENT MAY REDUCE BOD5 TO PERMISSIBLE LEVELS.

CALCULATIONS ARE BASED EITHER ON USE OF KODAK FORMULAS AND REPLENISHER RATES OR ON DATA SUPPLIED BY THE PROCESSOR.

++ SODIUM SALT OF ETHYLENEDIAMINE TETRAACETIC ACID, EXPRESSED AS AN ION.

passing the effluent through a steel wool–filled cannister where silver is chemically exchanged for iron. For all but the smallest photoprocessors, silver recovery makes economic sense and helps conserve a scarce resource.

Though not, perhaps, as profitable as recovery of silver, regeneration and reuse of ferricyanide-containing bleaching agents are important because they reduce the amount of free cyanide that could be formed later in streams. Ferricyanide bleaches convert metallic silver in emulsions back to silver halides.

In mixed process effluents, a number of reducing agents, such as sulfite or thiosulfate, are already present to react with residual ferricyanide. Thus, ferrocyanide, but not ferricyanide, is discharged in the final effluent. Ferrocyanide passes unchanged through municipal treatment plants into waterways, where, in the presence of sunlight and oxygen, cyanide may eventually be liberated. In no cases is free cyanide present in photoprocessing effluents, but the possibility that cyanide may be formed in streams makes bleach regeneration a sound pollution control practice.

### Regeneration methods

Several processes are available for regenerating ferricyanide bleach. The most common technique involves the addition of potassium persulfate, but good quality control procedures and analytical facilities are required. In some processes, a simple packaged test is sufficient while others may require more precise analysis. Recent work indicates that ozone can be used to regenerate ferricyanide in large laboratories where the capital expenditure for equipment can be justified. The ozone method would also eliminate sulfate buildup in some bleaches, extending their useful life.

Like bleach, some developers can be regenerated to reduce the amount of chemicals to be dumped down the drain. Ion exchange methods, which remove bromides that build up in developers, have been applied in several large processing laboratories and have shown an economic advantage in some.

Development of new processes, which reduce the number of chemicals needed, also contributes to pollution abatement. A simplified color printing process—the Plus-3 system—has recently been introduced by Kodak. The new system uses three processing baths to replace the five-solution chemistry currently in use. Each unit of sensitized paper produces only about half the BOD load of the older paper. The bleach-fix can be regenerated by bubbling air through it, and silver recovery is enhanced.

Kodak has prepared a number of technical guides for photofinishers faced with pollution problems. In addition, the company has recently made available one of its computers to provide a pollutant inventory for photofinishers based on production load (see sample printout). The service is free to processors requesting it, through Kodak technical sales representatives.

As a base for its program, Kodak has examined some 100 chemicals for fish toxicity, theoretical oxygen demand, chemical oxygen demand, BOD biodegradability, human health factors (including respiratory and dermatologic profiles) and effect on sludge organisms.                    HMM

# Cleaning up oil spills isn't simple

*Reprinted from* ENVIRON. SCI. TECHNOL., **7,** 398 (May 1973)

Remember the *Torrey Canyon* disaster when 119,000 tons of Kuwaiti crude oil flowed into the sea off the coast of Britain, the well blowout at Santa Barbara, or, more recently, the largest inland oil spill in the history of the U.S.? High waters resulting from Hurricane Agnes released 6–8 million gallons of black, highly metallic waste oil and sludge from an oil reclamation plant into the Schuylkill River (near Philadelphia, Pa.). These incidents and the other 10,000 oil spills that occur annually add up to more than 10 million gallons of oil forced into the environment, according to some sources.

"Unfortunately, there is no known estimate of the total quantity of oil left in the environment as a result of these spills," emphasized Henry D. Van Cleave, the federal EPA's chief of the Oil Branch, Division of Oil and Hazardous Materials, at the recent 1973 Symposium on the Prevention and Control of Oil Spills (Washington, D.C.). Estimates based on state-of-the-art cleanup technology indicate that as much as 80% of the oil accidentally discharged into the environment remains and is not removed. For example, a 10,000-gal. spill near a river in South Carolina disappeared; not even a sheen was evident at the cleanup site on the river. Out at sea, an estimated 86,000 metric tons of "tar-ball" petroleum residues are floating in the northwest Atlantic Ocean alone.

## Spills and cleanup

Spills occurring in oceans or bays are usually blamed on ships, especially tankers, since these incidents can be major catastrophes. In fact,

60% of total world oil production is transported by sea, according to Esso Research & Engineering Co., and this percentage is likely to increase in the future. And the U.S. Coast Guard predicts that one out of every nine tankers will suffer an accident each year.

Oil finds its way into the oceans in other ways also: natural seepage; land drainage resulting from careless discharges of oily wastes from industrial operations; routine ship discharges from deballasting, bilge pumping, and tank cleaning; wharf-

**Many techniques are being developed to contain and clean up oil spills, but there is no fail-safe all-inclusive answer**

vessel transfer of cargoes; as well as tanker groundings and collisions.

Inland oil spills not only result from natural seepage or land drainage from industrial operations, but also from storage tank overflows, pipeline leaks, broken sludge pond dikes, inadequate waste disposal, and similar situations. Inland spills vary enormously in size, type of oil, location, and "there is no universal cure," says W. E. Betts, Esso Research Centre, Berkshire, England. The seriousness of the spill depends upon the geological structure of the

ground, the proximity of water supplies, and difficulty of cleanup.

Although regulatory agencies and industries alike emphasize avoiding oil spills, no spill prevention scheme is yet fail-safe. As a result, research is geared toward developing methods for direct recovery of oil from the water's surface. However, two basic methods are used when possible. First of all, floating barriers or booms contain the spill, which can then be skimmed directly using specially designed pickup heads, weirs, pump systems, and oil/water separation equipment; the other technique is removing oil with the aid of sorbent materials.

Direct oil skimming, a physical method, does not require that anything be added to the oil slick, but this method is usually unsuccessful when wave heights approach two feet and when current velocity ranges from two to three knots.

Chemical sorbents, on the other hand, are not affected to such a degree by adverse weather and sea conditions. In fact, sorbent action is enhanced by mixing, which is required for effective performance. In calm seas, mixing must be done mechanically, which is often difficult.

Sorbents, when applied early in an oil spill incident, reduce the spread of the slick, and the oil/sorbent mixture is easier to contain than the oil slick alone. Some problems arise, however, when dispersing the sorbent; wind can make uniform dispersal next to impossible. Also recovering or harvesting the oil/sorbent mixture is not simple.

In large spills, substantial amounts of sorbents are required, and recovery of sorbents on a large scale in open waters has never been attempted, explains James D. Sartor, Woodward-Envicon, Inc. (San Francisco, Calif.).

In tanker accidents, three techniques have been used, with varying success, to prevent or minimize oil discharge—transfer of cargo (depending upon proximity of aid and prevailing conditions), towing the damaged vessel to a nearby port for repairs (which can't always be done), and burning the oil at sea (limited success).

### R&D control technology

Obviously, technology for cleaning up oil spills inland or at sea is not down pat, but much research is presently under way in hopes to remedy the situation. One area of development is the use of chemical agents to disperse oil spills. This method is controversial due to the performance variation and toxicities of the chemicals involved. Proper mixing is essential in chemical dispersion, but mixing by boat propellers or fire hoses, for instance, is laborious and time consuming. Researchers are presently working with dispersants that may require little or no mixing.

In tests, chemical dispersants have increased the toxicity in bioassay tanks because dispersing the oil supply simply exposes test organisms to a higher concentration of oil. With dispersion, oil is not recovered from the environment. The ultimate fate of an oil spill after it has been chemically dispersed is still unknown.

Furthermore, surface film-forming chemicals are being tested as oil-collecting agents. These chemicals minimize the spreading of the oil slick or drive the oil back into a thicker layer after it has already spread, which decreases the area affected by the oil spill. All oil recovery devices operate more efficiently on thicker oil layers. This technique might be best applied to harbor spills; it is still in the testing stage.

Crude oil gelation, converting the liquid oil into a rigid solid, to minimize or prevent the loss of oil from a distressed tanker, is being studied by Esso Research & Engineering Co. in Florham Park, N.J. The procedure involves the chemical reaction of two organic liquid gelling agents dissolved in the oil to form a gelled compound that entraps the oil without chemically bonding it. The resulting gel would float as a coherent mass if it were extruded from a ship or if it escaped during a tanker breakup.

Gel strength is adversely affected by factors reducing the conversion of the gelation reaction such as reduced temperature, brief reaction residence time, presence of impurities, nonstoichiometric ratio of re-

agents, and reduced concentration of the gelling agents. Treatment is a bit costly also—$4.50 per barrel of crude oil.

Polyurethane foam is being tested by Shell Development Co. (Houston, Tex.) as a sorbent for spilled oil. The foam can be regenerated for use by mechanical squeezing, which separates the oil and the sorbent. Officials from Hydronautics, Inc. (Laurel, Md.), agree that the system is feasible and that 90% of oil in slicks can be recovered. For a 3000-gph system, Hydronautics projects an initial cost of $40,000–80,000 and $100–200/hr of operation. Polyurethane foam is also recommended by EPA over other sorbent materials.

Scientific Associates, Inc. (Santa Monica, Calif.), and University of California researchers claim that a free vortex (for example, the currents produced by stirring a glass of water) can be used to recover oil (and sorbents) from high seas, stream, or harbor waters. The oil slick, drawn into the center of the vortex due to water flow induced by an impeller or other rotating hardware, could be removed by pumping. The researchers claim that it is effective in oil film thicknesses ranging from less than 0.001 in. to more than 5 in.

Certain naturally occurring mixed microbial cultures break down some fractions of crude oil. Several firms are looking into commercial production of such bacteria for use in cleaning up spills.

Ocean Systems, Inc. (Reston, Va.), has developed an air-transportable, two-weir–basin recovery unit that will recover up to 2000 gpm oil and water in seas with waves of 8 ft and currents of three knots. The unit is presently undergoing full-scale testing.

Also for use in high seas is the U.S. Coast Guard's Air Delivery Antipollution Transfer System (*ES&T*, June 1971, p 512). For thicker oil slicks, a disc drum unit on a catamaran recovers oil, and a weir–basin system is used for thinner slicks. The weir thickens the concentration of the oil, which then is held in the basin. The Coast Guard is also studying containment barriers since present booms or barriers fail to contain oil in currents above two knots. The U.S. Navy has initiated a multi-million-dollar program to evaluate equipment used in oil spill control (*ES&T*, Feb. 1973, p 93).

### Cold climate and separation

Oil spill cleanup in cold climates or in winter weather can present special problems with mixtures of oil and snow or ice. Sorbents, dispersants,

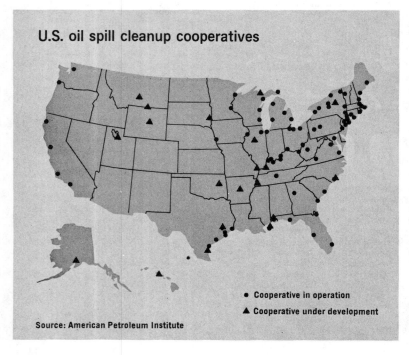

## U.S. oil spill cleanup cooperatives

● Cooperative in operation

▲ Cooperative under development

Source: American Petroleum Institute

and surfactants are useless in cold climates. The Coast Guard is developing a program to respond to Arctic spills by 1980 and have interim response capability by 1975.

After oil slicks are collected, the oil and water must be separated prior to utilization or disposal. A number of techniques are under development. Centrifuges perform well with a variety of oils. The Naval Ship and Development Center (Annapolis, Md.) is experimenting with the physical separation process coalescence as a final polishing step in an oil/water separation system. Abcor, Inc. (Cambridge, Mass.) is touting ultrafiltration to produce highly purified water from oil emulsions.

## Oil spill cooperatives

Several years ago the first oil spill cleanup cooperatives were formed to contain and clean up oil spills. The member companies pool their resources to purchase booms, oil skimmers, pumps, boats, and other equipment. They also established a system of communications and adopted a comprehensive port emergency plan.

Since 1967, cooperatives have been organized not only in the East, West, and Gulf Coasts, but on many inland waterways as well. The American Petroleum Institute reveals that 84 cooperatives are in operation, and 17 others are being developed.

Cooperatives follow several forms. Industry-wide cooperatives include the oil companies in an area; community-wide cooperatives consist of oil companies, other companies, government agencies, and public organizations. Finally, subscription cooperatives hire a local contractor to supply equipment, materials, and key manpower.

With the myriad of oil spill cleanup equipment on the market and being developed, oil spill cooperatives have made quite a few gains in the past six years:

- specialists are better prepared to determine the nature and seriousness of the spill as well as deciding what equipment will be required for containment and cleanup
- specific individuals have responsibility and authority to take action against spills
- contingency plans assure that all participants in a cooperative will be able to take quick effective action
- the reservoir of experience permits selective use of tools and approaches for each different type of spill
- substantial research points toward more effective means of preventing, containing, and cleaning up spills. CKL

243

# Oil spill technology makes strides

Reprinted from ENVIRON. SCI.
TECHNOL., 5, 674 (August 1971)

*Second joint industry-federal government conference
shows progress on all fronts—prevention, containment,
removal, treatment, and cleanup—in the last 18 months*

Oil perhaps is the most widespread of any pollutant in the world's waters—the rivers, the estuaries, and the oceans. The present national and international attention to future oil spills is none too early, at least considering the facts that:

• Each day, 15 billion barrels of oil are moved, and by 1980 the number increases to 18 billion barrels.

• Each year, an estimated 2 million tons of oil enter the oceans from tanker cargoes.

• Last year, there were 10,000 spills.

With the longest wetted coastline of any country, the U.S. understandably is not only involved but is setting the example with its oil spill national contingency plan, which now is being emulated by other nations. But other nations are concerned for different reasons. African nations, for example, are concerned because traffic off their coasts is 20 times more than that off the coast of any country. The United Kingdom, to mention another, is concerned because 75% of the known collisions has occurred in the English Channel.

In the past 18 months, the petroleum industry and the federal government, which shares in the oil spill responsibilities, have made some progress in all areas of the U.S. arsenal—contingency plans, international conventions, removal equipment, treating agents, and the like—since the first oil spill conference in December 1969 (ES&T, February 1970, page 97). At least this was the consensus at the recent second conference, the Joint Conference on Prevention and Control of Oil Spills, which was sponsored by the American Petroleum Institute (API), the Environmental Protection Agency (EPA), and the U.S. Coast Guard (CG).

## Operations

The operational rationale on oil spills has not changed in principle in the past 18 months. The various lines of defense remain much the same as they were at the first conference—namely, prevention, first; containment, second; removal, third; and treatment, fourth.

## Prevention

Everyone agrees that prevention of oil release is the route to go, but few agree with the myriad proposals that have come along. To proceed with zero release of oil on the high seas is idealistic, for example, but the real snag with this proposal is that few ports have holding and disposal facilities for receiving such wastes.

Nevertheless, prevention is the marchword of today. Unofficially, the EPA goal is to reduce oil spills 80% by 1975. In addition, the NATO-CCMS (Committee on the Challenges of Modern Society) ministerial conference last December 4 reached agreement on oil spill prevention by 1975 and no release of oil into the high seas by 1980.

The two IMCO conventions—the intervention and liability conventions (ES&T, February 1970, page 99)—which have lain dormant for the last one and a half years are now before the Senate. Ratification by the U.S. would be a first step for dealing with oil spills at the international level. The majority of witnesses who testified at recent Senate subcommittee hearings on the conventions indicated that although these conventions are not perfect, early adoption by the U.S. would set the precedent for other nations to follow. What perhaps is even more encouraging is the fact that the API spokesman Herbert A. Steyn, Jr., indicated that the attitude of European nations was favorable for adoption of the liability convention. Other spokesmen included those from the American Institute of Merchant Ships (AIMS) and the National Audubon Society.

Some senators noted that another IMCO convention has been proposed which is concerned with the formation of a convention fund. The formation of this fund will be the subject for the forthcoming IMCO meeting in December in London. While some senators favor withholding U.S. approval on the two earlier conventions until the third one has been incorporated, other senators foresee further delay in such a course of action.

Although member governments would not finance the liability insurance under the convention, this would be done by the P&I clubs (the protection and indemnity insurance underwriters for tanker liability), the member governments would, however, certify that the vessel had insurance liability.

IMCO, the Intergovernmental Maritime Consultation Organization—the specialized agency of the United Nations that provides cooperation in all shipping matters—has proved to be the appropriate international forum for oil spill problems since it was formed in 1959. More than likely it will remain the forum for oil despite a number of other forthcoming environmental meetings including the 1972 U.N. meeting on the Human Environment in Stockholm.

Meanwhile, on the home front the Coast Guard last month proposed new regulations for oil spills in the *Federal Register*. After a 60-day period for comment and a public hearing, the regulations will perhaps be finalized by the end of this year. Most likely, the regulations will:

• Prohibit the pumping of bilge and ballast waters into the navigable waters and waters of the contiguous zone.

• Prohibit the carrying of oil in bulkhead sections of tankers.

• Restrict the size of tanks in supertankers. If adopted, new tankers built after Jan. 1, 1972 would have to comply with this requirement.

## Containment

With regard to equipment, oil spill personnel realize that today's equipment is more substantial than earlier equipment. The bolstering of equipment comes essentially from recognition of the fact that the equipment must deal with forces of nature; such an appreciation was not evidenced at the first conference.

Even though the nation has better booms, a better understanding of removal techniques, and more field data with treating agents than earlier, what need becomes glaringly apparent now is the integration of the various combatants into an optimum system to handle the spill. For example, when to use one technique in preference to another? Under what operating conditions is it better to use one removal system in preference to any other such system or for that matter when to use treating agents in preference to removal?

### Removal

Design studies have been completed on a number of removal devices. But additional funds now are needed for prototype development of these devices, followed by later field tests. To be sure, a better understanding of the oil removing capabilities of some promising new pieces of removal equipment was described at the conference. Many of these will be scaled and will undergo field tests in the near future. Such devices, for example, include:

• A floating weir skimmer device—Esso Ltd.,

• Free vortex recovery system—Scientific Associates, Inc.,

• Oleophilic belt scrubber—Shell Pipe Line Corp.,

• Rotating disk oil-removal system—Atlantic Research Systems Div.,

• Rotating disk removal device—Lockheed.

The Lockheed device will be scaled to a device 8 ft in diameter and 10 ft wide. It will have a recovery capacity of 1500 gal of oil per minute, will be installed on a catamaran, and tested in West Coast field trials later this year. With an earlier model of the device, water pickup was less than 1% under quiet sea conditions and approximately 25% under No. 4 sea conditions, according to a spokesman

at the Washington, D.C. conference.

Without question, the largest device at the exhibit was a beach cleaning device of Meloy Labs (Springfield, Va.). Its mobile beach cleaner uses a froth flotation method of separating oil from the sand, a method that was adapted from the minerals-processing industry where the process has been used with great success. Field tests began last October; some 40 have been logged to date. The unit which is capable of cleaning 700 tons of contaminated sand per day, equivalent to 1 mile of beach 32 ft wide and 1 in. deep, will undergo additional tests later this year at the U.S. Navy site at Dam Neck, Va. Earth-moving equipment is used to bring the contaminated sand to the cleaner (see photo).

Nevertheless, it remains to be seen whether the federal government or industry cooperatives will be the purchasers of this $75,000 item. Perhaps, the cleaner will be made available on a lease basis. Surely, the cleaner is a replacement for straw as the No. 1 material for beach cleanup, albeit a frightfully expensive one.

### Treatment

The number of chemical combatants remains essentially at 400, the number reported at the earlier conference. When the cleanup operation resorts to treatment techniques four choices are available: sorb it, sink it, burn it, or disperse it. Although sorbents and burning agents seem to hold promise under certain conditions, there seems to be little hope for sinking agents or dispersants.

Fifty sorbent materials have been investigated for their oil-removing capacities by the Naval Ships Research Lab. The materials include synthetic organic foams, which incidentally head the list of sorbents with an oil-

sorbing capacity of 70 grams of oil per gram of material, polyurethane and urea–formaldehyde materials, and hydrocarbon polymers such as polyethylene and polypropylene.

Burning also seems to have a place in the oil spill cleanup action. The method could, for example, be considered for use under high seas conditions. In fact, the Coast Guard reported on open seas burning trials using a few promising burning and wicking agents. But their experiences to date indicate the need for further studies and experiences not only with better ways to apply the agents but a real need for better ignition methods. The use of the Cab-O-Sil agent, for example, required operators to wear protective inhalation equipment, which is awkward at least under sea conditions. Another operational technique in the burning category that remains to be checked involves containment of the slick prior to attempted ignition of the oil.

The problems with sinking agents and dispersing agents are many. Not only must the dispersants be effective but additional constraints have now been placed on such agents. The material must be biodegradable and have a low BOD value. An earlier Battelle literature study showed that all dispersants were toxic at a level of 5 ppm. However, dispersants could be used under one of three extentuating circumstances, one of which cites the protection of endangered species. Dispersants could also be used when recovery was impossible.

Now is the time for further perfection of equipment, field tests, and the gaining of operational know-how with such an arsenal so that if and when an unpredicted dramatic oil spill recurs then the U.S. could stand ready to muster an unprecedented and convincing cleanup action. SSM

**Beach cleanup.**
*Each day the Meloy mobile unit can clean a 1-mile strip of beach, 32 ft wide and 1 in. deep, but it's an expensive substitute for straw*

# Oil spills:
# An environmental threat

*Reprinted from* ENVIRON. SCI. TECHNOL., 4, 97 (February 1970)

*Petroleum industry and federal government share oil spill responsibilities*

What will the year 1970 hold in the way of a major oil spill incident? Hopefully, nothing (and, hopefully, not Alaska). But previous years have produced significant incidents—the *Torrey Canyon* grounding (1967), the *Ocean Eagle* spill in San Juan Harbor (1968), and the Santa Barbara Channel blowout (1969).

It is true that the nation is alerted to the dangers of widespread pollution by such disasters. But the U.S. still does not have adequate oil spill technology, and has not yet provided the means for bringing an adequate technology into being. Such are the conclusions of the recent report by the President's special panel on oil spills, comprised of 12 distinguished scientists and engineers and chaired by the President's science adviser, Lee A. Du-Bridge.

Had it not been for the Santa Barbara blowout (January 28, 1969), perhaps the petroleum industry would not have seen fit to hold this recent conference. Sponsored by the American Petroleum Institute (API) and the Federal Water Pollution Control Administration (FWPCA), the Joint Conference on Prevention and Control of Oil Spills (New York City) drew more than 1100 attendees to hear 30 papers on the subject and to see the wares of more than 40 exhibitors; not surprisingly, 10% of the attendees came from foreign countries.

But a massive spill such as those that receive headlines is not the only concern of the U.S. and oil companies around the world. "Recently, the U.S. Coast Guard estimated that the nation may be experiencing a spillage of polluting materials to U.S. waters approaching 10,000 incidents annually, with oil leading all other categories by a ratio of about three to one," says Kenneth E. Biglane, director of FWPCA's Office of Oil and Hazardous Materials.

The U.S. arsenal of oil combatants includes the National Contingency Plan, industry cooperatives, physical corralling and removal devices, chemicals, industry and government sponsored research and development contracts, and industry and government

**Slick.** *Oil can spread on water to 1/100th of an inch over a 25 square mile area in eight hours making removal operations extremely troublesome*

members who stand ready to assume their responsibility in cleanup.

The problem boils down to one of how the U.S. proceeds with this arsenal, and this concern causes differences of opinion. "Some would have the public and others believe that a single technique or system, applied to either the control or cleanup of the oil spill, is the ultimate goal in the problem. Those of us in the FWPCA do not happen to share this view and are recommending that we continue to concentrate our research and operational efforts along the lines of multiple systems response as opposed to single systems panaceas," Biglane elaborates.

### Prevention plans

The National Contingency Plan—completed in September 1968 and approved in November of the same year—is the first item in the U.S. arsenal. What the plan does nationwide is to provide a mobilization scheme in the event of a major oil spill incident (ES&T, July 1968, page 512). Other nations have similar lines of defense. Last March, Britain and eight other countries joined in signing the North Sea Pact. "The Board of Trade has

**Aerial view.** *Santa Barbara leak spreads oil over a wide area of channel water*

been charged with the responsibility of dealing with oil on the seas in excess of one mile from United Kingdom coasts," says H. Jagger, pollution control coordinator, Esso Petroleum Co., Ltd. (London).

On the industry scene, oil companies now are assisting one another on oil spills. Referred to as cooperatives, an agreement is made in advance by an oil company to help a neighbor who has an oil spill. "Today, the petroleum industry has 23 cooperatives fully operational and another 27 in the development stages," says Gulf Oil Corp.'s Ernest Cotton.

## Removal

Once the oil is on the water it becomes a difficult and expensive operation to remove it. The next line of defense is corralling and containment of the oil. If this is successful, the physical removal of the oil from the water by vacuum line transfer, skimming devices, or separators should be tried next. And if the oil gets on the beach, use straw to absorb it. Other than these rather elementary rules of operation, there are no remedies.

Oil spreads quite rapidly from a catastrophic release. Comdr. W. E. Lehr (U.S. Coast Guard) estimates that oil has been known to spread 1/100 of an inch over a 25 square mile area in eight hours, for example. Consequently, for the oil removal equipment to be effective, it must be modularized and capable of aerial transport and delivery within an extremely short period of time. Furthermore, the equipment must constitute a total system with all parts being compatible. Earlier operational experience

**Removal.** *Vacuum lines are used after floating booms corral the spilled oil*

revealed that, in some instances, booms from different manufacturers could not be joined together. "Additionally, the booms should be able to perform efficiently in 40 m.p.h. winds, currents of two knots, and 10 foot waves," says Lehr. "A readily transportable system will be ready in early 1970."

Chemicals also are in the oil combatants arsenal; more than 400 are on the market today for dealing with spilled oil. U.K. spokesman J. Wardley Smith, of the Ministry of Technology, notes that, despite the large quantities of the toxic solvent emulsifier mixtures which were used around the Cornish coast at the time of the *Torrey Canyon* incident, no reduction in the total weight of fish caught was observed, nor was there a reduction in the weight of crabs, shellfish, and the like landed in the two subsequent years.

However, these chemicals are not recommended for use in rivers, lakes, and estuaries. Nevertheless, a dispersant toxicity test procedure has been developed which should help in the further evaluation of these materials, according to Clarence M. Tarzwell, director of National Marine Water Quality Laboratory (Narragansett, R.I.). Hopefully, it will be issued as a standard test early this year.

Another procedure which has been proposed for evaluation was developed at FWPCA's Water Quality Laboratory (Edison, N.J.). Thomas A. Murphy notes that results from the procedure, referred to as the Simulated Environmental Tank (SET) test, correspond well with results from field tests. In Murphy's presentation, data are given for five families of dispersants (each chemically distinct) with four test oils.

## Research and development

Both the federal government and the petroleum industry sponsor a number of projects which aim at increased knowledge of how to deal with oil spills. Similarly, the British government and the U.K. oil industry support an ongoing R&D program.

Last year, API spent more than $600,000 on seven industrial R&D projects; reports are due early this year. A Battelle Memorial Institute report contains a compendium listing all available agents for oil spill cleanup, giving information on the manufacturers' recommendations, availability, cost, physical properties, etc. An Arthur D. Little, Inc., report determines where various types of agents can be used best. According to Little's

findings, the only operationally feasible method of oil pollution cleanup appears to be physical absorption by the use of straw.

"The federal R&D program in oil and hazardous substances spills is emerging from the planning stage and beginning to make progress," says Allen Cywin, director of FWPCA's division of applied science and technology.

Today, 55 federal contracts are active at a funding level of about $10 million. These studies can be divided into the following categories:
- Prevention of oil (17).
- Surveillance (11).
- Control of spills (14).
- Effects of spills (8).
- Beach and shore restoration (3).
- Miscellaneous (2).

Western Co. (Richardson, Tex.) is investigating techniques for gelling tanker cargoes to prevent oil loss in the event of an accident. American Oil Co. (Chicago, Ill.) has been awarded the largest federal oil contract, nearly $1.75 million, for demonstration of a 30 m.g.d. plant for treating refinery effluent by a combination of biological conditioning, chemical coagulation, and air flotation. Another large contract, about $1.5 million, has been awarded to National Oil Recovery Corp. (Bayonne, N.J.) for demonstration of a process for the complete conversion of crankcase waste oil into useful products without producing pollutant materials.

Other federal contracts have been awarded for the development of oil sensors, beach cleanup, and better mechanical recovery and separation devices. For beach cleanup, a sand cleaning process is being developed by Aerojet General Corp. (El Monte, Calif.). A full-scale 30 ton per hour unit will be demonstrated this summer, Cywin asserts.

## Industry steps

"Until recently, oil pollution was looked upon mostly as a problem associated with tanker operations," says L. P. Haxby, chairman of the API subcommittee on oil spills cleanup. "The API Board of Directors has, thus far, appropriated $1.2 million to fund the work of the subcommittee through 1970. Most of the money will go for R&D on new methods of cleanup."

The load-on-top method of retaining oily residues on board for disposal in proper facilities ashore is proving beneficial for eliminating pollution from tanker discharges.

In addition, the industry voluntary plan for tanker liability is well underway, according to the API spokesman. "More than half of the free world's tanker owners are now enrolled in the plan," Haxby notes. "Referred to as Tanker Owners Voluntary Agreement concerning Liability for Oil Pollution (TOVALOP), the plan provides liability at $100 per gross registered ton or $10 million, whichever is smaller. Under the plan, participating tanker owners accept responsibility to reimburse any national government for the costs it incurs in cleaning up oil spills they negligently cause."

### Cleanup costs

At the recent conference, Thomas H. Gaines (Union Oil Co. of Calif.) noted that Santa Barbara cleanup costs ran to about $4.5 million. In addition, pending legal actions against Union Oil approximate $2.5 billion.

Although the governments of France and the U.K. brought suits against the owners of the *Torrey Canyon* for $22 million, the settlement was made out of court last November for $7.2 million.

On the average, cleanup costs for massive spills are approximately $1 per gallon, according to FWPCA's Biglane. But the costs may be a bit higher for inland waters. The Battelle Northwest report notes that the cost for removing oils from harbors ranges from $1.35-3.00 per gallon. Other data are presented in the Little, Inc., report which notes that the direct costs for oil cleanup range from $1700-4100 for small (1000 gallons) harbor spills, from $64,000-115,000 for a medium (100,000 gallons) offshore spill, and from $4.5-8.5 million for a large (10 million gallons or 30,000 tons) offshore spill.

### IMCO

Two international conventions on oil spills and cleanup are the most important recent achievements of the Intergovernmental Maritime Consultative Organization (IMCO), the specialized agency of the United Nations that provides cooperation in all shipping matters, and involves 68 member governments (ES&T, July 1968, page 510). At its International Legal Conference on Marine Pollution Damage, held in Brussels last November, IMCO adopted two conventions which include authority for:
• Right of coastal states to intervene in oil spill casualties if their shoreline is threatened by the incident.

**Absorption.** *Straw proves its usefulness as Number 1 material for cleanup*

• Civil liability to both coastal states and owners of shoreline property for oil cleanup.

"The purpose of this conference was to finalize the conventions which were in the draft stage by IMCO's legal committee for the last two years," says Louis P. Georgantas, a member of the U.S. delegation from the State Department. These conventions are significant because they hold ship owners and operators to strict liability and are broader than pending U.S. legislation which does not cover third party liability, and is limited only to the ability of the U.S. government for oil cleanup.

The first convention applies only to the high seas, outside the territorial limit, so that this international convention does not interfere with the rights of a coastal state. It is a fact that this convention effectively would have taken care of the situation presented by the *Torrey Canyon,* for example, when it grounded off the English coast. Under this convention, the U.K. government would have had authority to take immediate cleanup action and prevent further damage from the oil being spilled. Precious time was lost in the *Torrey Canyon* incident over the question of legal authority to take action against the grounded tanker.

The second convention imposes strict liability on owners and operators with certain exceptions including acts of God, war, and negligence of the coastal state. Limits of liability are set at $134 per gross registered ton or $14 million, whichever is the lesser. The surprising event at the Brussels conference was the fact that the London P&I Clubs, the insurance underwriters for tanker liability, were able to secure financial backers for underwriting such policies.

Peter N. Miller, the spokesman for the insurance underwriters, was able to ensure that these limits of liability specified in the second convention were insurable.

In practice, under the second convention, compulsory financial responsibility is required, and certificates to this effect would be issued by the state of registry verifying that the vessel had financial liability to cover spills of any oil by that vessel. Also, a state could exclude a ship from its port if the ship did not have this certificate of financial liability. A state also could exclude ships from noncontracting states (not parties to the convention) unless they otherwise had obtained a certificate of insurance.

By way of an example of strict liability, Georgantas cites the case of a moored tanker containing oil that is rammed by a second cargo ship carrying dry goods, resulting in the spill of oil from the tanker. The moored vessel is held liable for the damage caused by the oil, but the tanker's owner then can take legal action against the owner of the dry goods vessel to recoup his losses. In any case, the victims are compensated.

Ratification by each IMCO member government is the next step for these conventions. "Although 49 governments participated in the Brussels conference, only 20 had authority to sign the conventions for their governments, and 16 did so, including the U.S. and leading European industrial countries —Germany, France, Italy, and the U.K. Although the Scandanavian countries—leaders in tanker tonnage—did not sign these conventions, their representatives voted for them.

In the U.S., ratification will proceed in the following manner: The State Department will submit a request to the U.S. Senate asking advice and consent for the U.S. government to become a party to the convention. Then, after hearings, the Senate will vote on the proposal. A two thirds vote by the Senate and signature by the President will make the conventions binding on the part of the U.S.

When ratified, these conventions become binding internationally. "The coastal states intervention convention becomes binding when 15 member governments ratify the convention, and the civil liability convention becomes binding after eight IMCO member states ratify the convention, but five of the eight must have more than 1 million tanker tonnage each," Georgantas says.

# New process detoxifies cyanide wastes

*There are several effective methods available for the destruction of cyanides in plating wastes; a new one claims in addition simplicity and ease of operation*

Reprinted from ENVIRON. SCI. TECHNOL., 5, 496 (June 1971)

Time was when each issue of every respectable mystery magazine featured at least one corpse whose untimely demise could be attributed to a peculiar form of halitosis—a characteristic odor of bitter almonds. As poisons went, they just didn't make them much more sinister than cyanide.

For the chemical process industry, on the other hand, there are few compounds that surpass cyanides as versatile reagents. The widespread use of cyanides—particularly sodium cyanide and hydrocyanic acid—and the extreme toxicity of the free cyanide ion, however, make adequate detoxification of cyanide-laden effluents a matter of considerable environmental importance. Fortunately, there is good technology available for cleaning up cyanide waste, although application to some industrial requirements is more successful than others.

Industrial waste streams from such varied sources as ore extracting and mining operations, photographic processing, coal-coking furnaces, and synthetics manufacturing plants contain cyanide. A major source of cyanide in process effluents—although not, by far, the largest—is the electroplating industry.

Electroplaters use cyanide baths of various formulations to hold ions such as zinc and cadmium in solution; the metal ions are then electrodeposited on ferrous metals. "Drag-over" of the plating solution—which contains cyanide ions and metal–cyanide complexes—contaminates rinse baths. Contents of the baths, together with spent plating solution (which may contain several species of metal ions), must be detoxified before effluents can be discharged. Since cyanide is toxic to bacteria in activated sludge, municipalities often place stringent restrictions upon the cyanide content of effluents to be discharged to sewers, forc-

ing dischargers to find and use an effective cyanide destruction process.

## Detoxification processes

The process most frequently used to decompose cyanides is alkaline chlorination, which may be accomplished in two ways. First, sodium hypochlorite may be added directly to the waste stream. Cyanide is then rapidly oxidized to cyanate. Alternatively, hypochlorite may be generated in solution by the addition of chlorine gas and caustic soda. One decides on the basis of economics which of the two methods to choose. Chlorine gas treatment is about half as expensive as the direct hypochlorite treatment, but handling is more dangerous and equipment costs are higher.

While the resultant cyanates are much less toxic than cyanide, they are nonetheless unacceptable to a large number of control authorities, and the trend is toward stricter regulations. Cyanates, however, can be completely converted to carbon dioxide and ammonia by acid hydrolysis. But lowering the pH of the effluent to 2 or 3 to destroy cyanates involves yet another neutralization of the acidic solution before the effluent can be discharged.

Another detoxification procedure—electrolytic oxidation—is sometimes applicable when cyanide concentrations are very high. The effluent is subjected to electrolysis at 200°F for one or two days. Initially, cyanides are completely oxidized to carbon dioxide and ammonia but, as the process proceeds, the electrolyte becomes less capable of conducting electricity and the reaction may not go to completion. Some cyanate may be formed and some chlorination may still be necessary. Recently, advances made in this type of electrolysis have been claimed to obviate some of the difficulties usually encountered (see "Electrolysis

speeds up waste treatment," ES&T, March 1970, page 201).

Other types of cyanide detoxification have been proposed as well. Several ion-exchange methods have been explored, but have not yet proved practical. Ozonation, too, shows some promise as a substitute for alkaline chlorination. Proponents of the ozonation process point out that it would be cheaper and more easily controlled than direct chlorination. Ozonation oxidizes cyanides only to the cyanate stage, however, and further oxidation of the cyanates is too slow to be industrially useful.

Another process which may one day be useful is radiation detoxification. Gamma rays can be used to bombard cyanides and destroy the triple bond between carbon and nitrogen. The irradiation process reportedly converts cyanides all the way to carbon dioxide and nontoxic nitrogenous products; it could be applicable to continuous processing as well as batch operations.

## Kastone

Recently, however, E.I. du Pont de Nemours and Co. has introduced a novel process which should appeal primarily to small plant operators using cyanide baths to plate zinc or cadmium onto ferrous metals. Called the Kastone process, it oxidizes cyanide from plating wastes to cyanates and simultaneously precipitates zinc or cadmium complexes which can be removed by simple filtration.

The key to the process is Kastone, a proprietary peroxygen formulation. Kastone contains 41% hydrogen peroxide with trace amounts of stabilizers and a small amount of a patented catalytic compound in aqueous solution. Although Du Pont declines to reveal the exact nature of the catalyst, company spokesmen say it works to break up metal–cyanide complexes "like a

**249**

chelating agent in reverse."

A major advantage of the Kastone process, according to Du Pont, is that it uses "bucket chemistry" that can be performed by technicians with a minimum of training. Here's how it works in practice:

Cyanide-laden zinc or cadmium plating wastes are collected in holding tanks, where the wastes are agitated—either mechanically or with conventional spargers—and heated to about 120°F. Cyanide content of the rinse water is checked with a simple silver nitrate titration of an alkaline sample by use of a *p*-dimethylaminobenzal-rhodanine indicator system. Based on the titration, a calculated amount of formalin (a 37% solution of formaldehyde) is added to the waste rinse tank and agitation is continued for an additional five minutes. A sufficient amount of Kastone (as determined by silver nitrate titration) is added to the holding tank and agitation is continued for one hour. Cyanide concentration is rechecked for residual amounts and, if necessary, plating rinses are treated further. The waste water can be dumped through a conventional filter system which removes reasonably

dense floc containing metallic oxides, hydroxides, and carbonates.

The reaction products from the Kastone detoxification of cyanide are many and complex. Their nature and amounts depend upon cyanide concentration, solution pH, and temperature (see below). The main products from, say, a zinc cyanide plating rinse are cyanates and some glycolic acid derivatives. Glycolonitrile may be an initial reaction product, but Du Pont chemists suspect the product—which might be expected under the given conditions from the reaction of formaldehyde and cyanide—is rapidly hydrolyzed to glycolic acid amide, then to glycolic acid and the ammonium ion. Both glycolic acid and the amide are biodegradable, Du Pont points out. In fact, the company says, glycolic acid is a normal constituent of sugarcane juice.

Du Pont isn't worried about the cyanates either. First of all, acid hydrolysis (mentioned earlier) is easily enough accomplished. But, Du Pont points out, effluent from chemical plants is usually on the acid side anyhow. If waste streams are mixed, cyanates should be of little consequence. Even in the case where Kastone-

treated and filtered effluents were to be discharged directly to sewers, Du Pont says that cyanates would be in active equilibrium with urea and are known to be biodegradable.

## Metal discharges

Du Pont points out that most states have set stringent requirements for discharge of metals—among them cadmium, zinc, chromium, and copper—into streams. The Kastone process offers an important advantage over alkaline chlorination in the removal of metals, the company claims.

Zinc hydroxide, formed as a result of processes other than Kastone, is difficult to remove by filtration, the company says. The catalytic ingredient in Kastone, on the other hand, somehow favors the formation of metal oxides over hydroxides. Du Pont claims that the oxides can be removed by ordinary filtration equipment which even the small or medium-sized electroplater already has. In the case of cadmium plating operations, the easy removal of cadmium oxide could figure importantly in resource reclamation, since cadmium electroplating compounds are sufficiently expensive to make it worthwhile to reclaim them.

Du Pont thinks the Kastone process should be competitive with any other cyanide detoxification process around today. It works well on cyanide concentrations ranging from 100–1000 parts per million (measured as NaCN). It costs about 78 cents to detoxify one pound of sodium cyanide with the Kastone process as compared with about 70 cents per pound by the sodium hypochlorite method. Alkaline chlorination with caustic and chlorine gas is only about half as expensive—about 34 cents per pound—but Du Pont figures it has an advantage in safety and equipment investment. In many cases, suitable equipment is already in the plater's shop and can be used with only slight modification. Du Pont says investment costs for Kastone can be between 40–60% of those required for alkaline chlorination with chlorine gas. Operating costs can be as much as 25% below those for alkaline chlorination with hypochlorite, Du Pont adds.    HMM

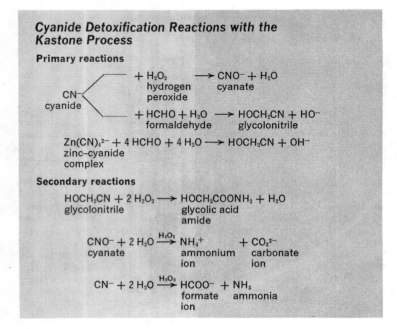

### Cyanide Detoxification Reactions with the Kastone Process

**Primary reactions**

CN⁻ (cyanide)

$$CN^- + H_2O_2 \longrightarrow CNO^- + H_2O$$
(hydrogen peroxide) (cyanate)

$$CN^- + HCHO + H_2O \longrightarrow HOCH_2CN + HO^-$$
(formaldehyde) (glycolonitrile)

$$Zn(CN)_4{}^{2-} + 4 HCHO + 4 H_2O \longrightarrow HOCH_2CN + OH^-$$
zinc–cyanide complex

**Secondary reactions**

$$HOCH_2CN + 2 H_2O_2 \longrightarrow HOCH_2COONH_2 + H_2O$$
glycolonitrile     glycolic acid amide

$$CNO^- + 2 H_2O \xrightarrow{H_2O_2} NH_4{}^+ + CO_3{}^{2-}$$
cyanate     ammonium ion    carbonate ion

$$CN^- + 2 H_2O \xrightarrow{H_2O_2} HCOO^- + NH_3$$
formate ion    ammonia

# Metals focus shifts to cadmium

*Although it's a potentially hazardous pollutant*

*and many questions remain to be answered,*

*there will probably be no "cadmium crisis"*

Reprinted from ENVIRON. SCI. TECHNOL., 5, 754 (September 1971)

The Japanese call it "itai-itai byo"—the ouch-ouch disease—because of the excruciating pain accompanying a bizarre skeletal disorder known as osteomalacia. In those afflicted with itai-itai, bones may become so fragile that mere coughing is sufficiently traumatic to fracture ribs.

For years in the aftermath of World War II, farmers in Japan's Jintsu River Basin, where itai-itai was most prevalent, had complained that waste water from a zinc mine, discharging directly into the river used for irrigating rice paddies, was spoiling crops. But it wasn't until last year that Okayama University's Jun Kobayashi established the common denominator between the frequently fatal itai-itai and failing rice crops. The culprit was cadmium—a metal intimately associated with lead and zinc deposits and long known to be harmful to man.

Although toxicologists now believe that various nutritional deficiencies play a critical role in the full-blown itai-itai syndrome, the discovery that populations could be exposed to toxic quantities of cadmium by contaminated water caused a nation still jittery over the mercury pollution of Minamata Bay to shudder anew at the prospect of yet another heavy metal disaster. And more than a little concern was expressed in the U.S. last year over findings that showed many municipal water supplies contain significantly more than the 0.1 parts per million of cadmium considered to be "safe" for drinking.

As with mercury, very little was known about the fate and distribution of cadmium in the environment. What was known was scary enough. What was not known was even scarier.

It was known, for example, that cadmium has a very long biological half-life in man and therefore tends to accumulate in the body. Estimates ranged from 10–25 years, compared with about 70 days for methyl mercury. It was also known from industrial toxicology that cadmium could cause severe liver and kidney damage, pulmonary disease, and death.

In the absence of solid toxicological data, speculative analogies between mercury and cadmium were common. Was there, for example, a process similar to the microbial methylation of mercury which would concentrate cadmium in the food chain? Fortunately, that hypothesis appears to be incorrect. Nevertheless, disturbing questions about the amount of cadmium discharged into air and water and the effects of chronic exposure to cadmium remained.

In an effort to answer some of those questions the Environmental Protection Agency (EPA) commissioned studies to determine just how serious a threat cadmium poses. Although more work remains to be done, a picture is beginning to develop that is not quite so black as might have been imagined. The consensus of opinion is

that although cadmium is a potentially dangerous pollutant, there is no reason to believe that a "cadmium crisis" is imminent. Cadmium is not "another mercury."

## Distribution

Cadmium is a relatively rare metal—the earth's crust contains an average of only 0.5 ppm. It occurs chiefly as the sulfide, greenockite, in various zinc, lead, and copper ores. Nearly all the cadmium produced in the world is a by-product of zinc smelting. Recycling is virtually nonexistent. Only 4% of the 10.6 million pounds of cadmium produced in the U.S. during 1968 was derived from reprocessing cadmium-containing scrap.

Cadmium does not degrade in the environment. As greater quantities of cadmium are refined, more and more of it becomes available to interact with man. Domestic consumption of cadmium was more than 15 million tons in 1969 (the last year for which official statistics are available). That's a whopping 13% jump over the figure for 1968. All indications point to further sharp increases in consumption for the foreseeable future. And because nearly everything produced by cadmium-using industries sooner or later winds up on the junk pile, the potential for environmental contamination is considerable.

Studies funded by the EPA provide a fairly complete inventory of sources and amounts of cadmium emissions into the atmosphere. Parallel studies for cadmium in water supplies, however, are lacking. Nearly 4.6 million pounds of cadmium was emitted into the atmosphere in the U.S. during 1968, according to a study prepared for the National Air Pollution Control Administration in 1970. Lee J. McCabe, director of Epidemiology and Biometrics for EPA's Office of Water Programs says there's "some but not much" cadmium in drinking water. Just how much is discharged into waterways is not known.

By far the largest single user of cadmium is the electroplating indus-

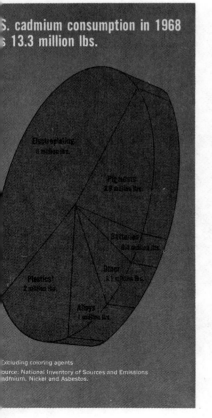

U.S. cadmium consumption in 1968 was 13.3 million lbs.

Electroplating
6 million lbs.

Pigments
2.9 million lbs.

Batteries
0.4 million lbs.

Plastics
2 million lbs.

Other
1.1 million lbs.

Alloys
1 million lbs.

Excluding coloring agents

Source: National Inventory of Sources and Emissions Cadmium, Nickel and Asbestos.

**Pounds of cadmium emitted to atmosphere in U.S. in 1968**

| | | |
|---|---:|---:|
| Mining and metallurgical processing | | 2,100,530 |
| mining | 530 | |
| Cd separation from ore | 2,100,000 | |
| Incineration and disposal | | 2,440,000 |
| plated metal | 2,000,000 | |
| radiators | 250,000 | |
| other | 190,000 | |
| Industrial reprocessing | | 33,528 |
| pigments | 21,000 | |
| plastics | 6,000 | |
| alloys | 5,000 | |
| batteries | 400 | |
| miscellaneous | 1,128 | |
| Consumptive uses | | 14,630 |
| rubber tires | 11,400 | |
| motor oil | 1,820 | |
| fertilizers | 910 | |
| fungicides | 500 | |
| | Total | 4,588,688 |

Source: National Inventory of Sources and Emissions

try. The metal imparts an attractive, corrosion-resistant finish to ferrous metals. Virtually no cadmium escapes into the atmosphere from electroplating operations; how much gets into water from plating baths is anybody's guess. Hopefully, says McCabe, such information will be forthcoming in the next couple of years as discharge permits, required by the Army Corps of Engineers, are processed (see "What's the U.S. Army doing in water pollution control?" ES&T, December 1970, page 1101).

Other important applications for cadmium include pigments, plastics stabilizers, alloys, and batteries. Together with electroplating, these five product groups account for more than 90% of the cadmium used domestically. A wide variety of minor applications—television picture tube phosphors, fungicides for golf courses, control rods and shields in nuclear reactors, and curing agents for rubber, among others—accounts for the remaining 9–10%.

## Emissions

Most cadmium emissions into the atmosphere fall into two general categories—disposal and metallurgical processing. Disposal includes incineration and recycling of ferrous scrap and accounts for more than half the cadmium put into the air each year. The biggest chunk is lost in reclamation of scrap steel and recovery of copper from automobile radiators. Incineration of plastic products—ranging from bottles to baby pants—also adds cadmium to the atmosphere.

Metallurgical processing emits nearly as much cadmium to the air as disposal does. Although actual mining operations contribute only slightly to atmospheric cadmium levels, roasting and sintering of ore release vast quantities of cadmium into the air. Smaller amounts of cadmium in the air come from a variety of manufacturing processes as well as the use of certain fertilizers, fungicides, motor oils, and rubber tires.

Against this backdrop of emissions data, EPA commissioned three Swedish veteran cadmium watchers from Stockholm's Karolinska Institute to prepare a report on the health effects of cadmium in the environment. The trio, Lars Friberg, Magnus Piscator, and Gunnar Nordberg, recently presented the results of their review to a symposium at the University of Rochester School of Medicine.

The Swedish report says that although the body contains only $1\mu g$ of cadmium at birth, a 50-year old, 70-kg (155-lb) "standard man" has a body burden 30,000 times that amount (30 mg). Excessive exposure to cadmium can damage the liver, kidneys, spleen, or thyroid, but the amount of cadmium needed to produce severe damage is not known. Ambient air contains only a small amount of cadmium, the report says, but in the vicinity of certain factories, concentrations can be much higher. Cadmium is absorbed either by inhalation or ingestion and human beings are exposed to substantial amounts in food, according to the report. A differential absorption rate,

however, makes inhalation of cadmium more dangerous than ingestion. Only about 5% of ingested cadmium is absorbed by the body; estimates for absorption of inhaled cadmium run as high as 40%, but actual amounts are probably lower. Recent work by Harold G. Petering at the University of Cincinnati indicates that people who smoke as few as 10 cigarettes per day —or the equivalent in other tobacco products—are exposed to concentrations of cadmium from 10–100 times greater than those in the ambient air. Worse yet, smoke from smoldering tobacco that is not inhaled—the sidestream—contains more cadmium than the mainstream inhaled by the smoker. Thus, everyone in the same room with a smoker is exposed to elevated cadmium levels.

Cadmium has also been implicated in cardiovascular disorders and cancer, although the Swedish team emphasizes that there is no conclusive statistical evidence linking cadmium with these diseases.

### Research needs

The Friberg-Piscator-Nordberg report lists several research priorities including:

• The "urgent and immediate" need for experiments on dose–response relationships considering absorption and excretion rates, particle size, and exposure routes;

• Studies on concentration of cadmium in various organs relative to total body burden;

• Development of better indicators of cadmium levels in the body;

• Further work on teratogenic or carcinogenic effects of cadmium, both alone and in combination with naturally occurring chelating agents such as EDTA and NTA)

• The definition of mechanisms for accumulation in food chains; and

• Metabolic studies establishing the role of essential trace metals in cadmium accumulation.

There is no doubt that cadmium has "caused trouble," according to Robert J. M. Horton, senior research advisor for EPA's Air Programs Office, Division of Effects Research. But, he adds, "I can't think of any reason why there should be a panic reaction." What steps will be taken by the EPA to control cadmium remain uncertain, but Friberg offers some advice: "All such long half-life substances should be controlled to the greatest extent possible." HMM

**Deceiving appearance.** *Innocent looking body of water could contain enough mercury to contaminate fish, fowl, and man*

# Mercury in the environment

*Reprinted from* ENVIRON. SCI. TECHNOL., **4**, 890 (November 1970)

*Where is it?*

*How did it get there?*

*What can we do about it?*

After being present on earth for 4.5 billion years, mercury is now the center of public attention and concern. This newly acquired publicity is notorious in many respects—the liquid metal is marked as the culprit in stream pollution, fish and animal contamination, and human sickness. Where is mercury? This question was asked at the recent Environmental Mercury Contamination Conference sponsored by Michigan State University and the University of Michigan. The answer was summed up in one word—everywhere.

Both man and nature share the responsibility for mercury production. In the natural environment, mercury is found in soil, air, and water. David Klein of Hope College in Holland, Mich., states that mercury levels in the northeastern surface soils have a natural upper limit of 0.04 parts per million (p.p.m.) ± 0.02 p.p.m. In air, mercury is found naturally in concentrations of only a few nanograms (ng.) ($10^{-9}$ grams) per cubic meter (m.$^3$). For example, during a breezy day in the San Francisco Bay area, the atmospheric concentration of mercury is

2 ng./m.$^3$. However, under smoggy conditions this concentration is as high as 20 ng./m.$^3$. As for mercury in water, the natural level is 0.06 parts per billion (p.p.b.) in northeastern U.S.

Monitoring and analysis throughout the country now show that mercury is present in amounts deemed excessive in all realms of the environment. A limit of 0.5 p.p.m. of mercury in fish that will be consumed by humans has been set by the Food and Drug Administration (FDA). The Canadian Food and Drug Directorate also uses the same standard. The present Canadian and U.S. bans in the Great Lakes region, enacted earlier this year, result from analyses showing concentrations up to 7 p.p.m. in fish.

### Forms of mercury

Various facts must be understood before effective action can be taken on the mercury situation. First, mercury is cumulative and, once it is in the system, it is only slowly discharged. Thomas W. Clarkson, University of Rochester, contends that it takes about a year to establish a balance between mercury intake and mercury excretion (assuming a steady mercury-containing diet). Then, it takes another year for 99% of the accumulated mercury to be removed.

Another factor to consider is the form of mercury present in water. Methyl mercury and other alkyl forms,

primarily the most harmful, are similarly toxic. Phenyl and inorganic mercuries are less toxic and less harmful than the methyl form; but when released into water, these forms can be converted into methyl mercury. "The ability to methylate mercury is not unique; various organisms are able to achieve this," explains A. Jenelöv, scientist for the Institute for Vatten ach Luftvardforskning in Sweden. Elemental mercury can be oxidized under aerobic conditions to produce divalent mercury, which is then converted to methyl mercury. Phenyl mercury reacts also to form divalent mercury, which in turn changes to methyl mercury. From this point, the methyl mercury enters the food chain through sediment ingestion by organisms.

Almost all mercury found in animal tissues is of the methyl type. The accumulation of this mercurial form results in contamination—i.e., fish and other organisms would contain a large amount of potentially toxic mercury in their systems. Of the methyl mercury present in food, 98% is absorbed by the tissue, but the absorption rate of inorganic mercury by tissue is only 1%.

These facts point to the importance of determining the form of mercury present in tissue, rather than just analyzing total mercury content. Alkyl mercury compounds differ from the phenyl and inorganic mercuries both

**253**

chemically and in their effects. The methyl (alkyl) mercury compounds are unique in that their toxic effects can cause irreversible damage to the central nervous system. Although the phenyl and inorganic mercuries are also poisonous, their damage is always reversible and can affect other tissues such as the kidneys or intestines.

### Methylation

Various factors influence the rate of methylation in a body of water. Swedish research shows that in the winter and early spring, methylation occurs rapidly, while during the summer and fall months, methylation decreases. This is owing to the increased ability of the aqueous organisms present in the winter and spring to methylate the different forms of mercury. Also, waters of low pH contain more methyl mercury than those with a high pH. Furthermore, aerobic conditions are quite important in the methylation scheme. The presence or absence of one or more of these conditions greatly influences the rate and amount of methylation.

### Hazards

Accumulation of excess mercury can result in disastrous situations.

After conversion of the various mercury forms into methyl mercury, the alkyl mercury enters the food chain, progressing from ingestion by bacteria to consumption of mercury-contaminated fish and game by man. Waterfowl, particularly fisheaters such as the great blue heron and tern, have had mercury concentrations so high as to effect decreased fertility and poor egg quality.

Because of the 0.5 p.p.m. mercury limit in fish, commercial fishing was banned in many areas early this year, particularly in the Great Lakes region, by both the U.S. and Canada. Even the tourist industry has suffered because of warnings of mercury poisoning in lakes and streams.

Effects of mercury poisoning can be reversible, irreversible, or fatal. Reversible results, meaning total recovery without residual effects, include fatigue, headache, inability to concentrate, and impairment of memory. Symptoms bordering on irreversible injuries include blurred vision, tingling and numbness of fingers, finger ataxia, and emotional irritability.

Definite irreversible effects can be manifested in the loss of motor control (hands, locomotion, speech) and sensory operations (auditory and visual).

Fatal results of mercury poisoning are preceded by involuntary mobilization, blindness, mental and emotional deterioration, and loss of consciousness.

### Sources

How does mercury initially get into the environment? The uses of mercury must be noted to determine the source of discharge. Frank M. D'Itri, Michigan State University, states that mercury can be assimilated into the environment either directly or indirectly. Direct mercury release includes the waste effluent from a manufactured product and the incorporation of mercury into a product. The use of fossil fuels, such as coal (reported to contain 0.5 p.p.m. mercury), and the burning of mercury-containing paper constitute indirect means of mercury release. Additionally, accidental misuse must also be considered. The most well-known source of mercury contamination is the chloro-alkali industry, originally discharging 0.25 to 0.5 lb. of mercury per ton of caustic soda produced into the environment.

Mercury is also used as a catalyst in the manufacture of the plastics urethane and vinyl chloride, and the chemical acetaldehyde. The electrical in-

## U.S. consumption figures consistently point to mercury losses

| Year | Production | Import | Export | Consumption | Inventories | Discrepancy |
|------|-----------|--------|--------|-------------|-------------|-------------|
| 1944 | 37,700 | 19,600 | 700 | 42,900 | 80,900 | . . . |
| 1945 | 30,800 | 68,600 | 1,000 | 62,400 | 83,900 | −33,000 |
| 1946 | 25,400 | 13,900 | 900 | 31,600 | 39,900 | −50,800 |
| 1947 | 23,200 | 13,000 | 900 | 35,600 | 19,300 | −20,300 |
| 1948 | 14,400 | 32,000 | 500 | 46,200 | 30,200 | 11,200 |
| 1949 | 9,900 | 103,000 | 600 | 39,900 | 21,000 | −81,600 |
| 1950 | 4,500 | 56,000 | 400 | 49,200 | 35,600 | 3,700 |
| 1951 | 7,300 | 47,900 | 200 | 56,800 | 30,200 | −3,600 |
| 1952 | 12,500 | 71,900 | 400 | 42,600 | 34,400 | −37,200 |
| 1953 | 14,300 | 83,400 | 500 | 52,300 | 27,000 | −52,300 |
| 1954 | 18,500 | 65,000 | 900 | 42,800 | 22,500 | −44,300 |
| 1955 | 19,000 | 20,400 | 500 | 57,200 | 10,000 | 5,800 |
| 1956 | 24,200 | 47,300 | 1,100 | 54,100 | 22,300 | −4,000 |
| 1957 | 33,400 | 42,000 | 1,900 | 52,900 | 20,600 | −22,300 |
| 1958 | 36,800 | 19,000 | 700 | 54,600 | 35,100 | 14,000 |
| **Totals** | | | | | | |
| 1945–1958 | 274,200 | 683,400 | 10,500 | 678,200 | −45,800 | −314,700 |

$+ \quad - \quad - \quad \triangle \quad =$

Source: *Industrial & Engineering Chemistry*, Vol. 51, No. 2. Figures in flasks (one flask contains 76 lb.)

dustry used over one million pounds of mercury in batteries, silent switches, high-intensity street lamps, photocopying machines, and fluorescent lights.

D'Itri also points to uses of mercury in the amalgamation industry, pharmaceuticals that produce diuretics and antiseptics, the cosmetics industry, and the paint industry, which uses mercury as an antibacterial agent and fungicide. Manufacturing magnets for electronic components releases elemental mercury. Fish kill can be caused by commercial laundries' use of a mildew inhibitor that contains mercuric acetate. Mercury discharged into Minamata Bay, Japan, by a plastics plant contaminated the Bay and its fish population. Over 100 people were severely poisoned by mercury.

Pulp and paper mills stopped using phenyl mercuric acetate as a slimicide in 1959. However, these mills still use sodium hydroxide, which contains a fraction of mercury, Clarkson adds.

Sewage treatment plants may have 50 to 60 lb. of mercury in use, says William Turney of the Michigan Water Resources Commission. Trickling filters in these plants contain mercury seals as well as seals in other instrumentation, switches, and flow meters. Seal breakage can cause mercury loss.

Reclaiming gold by using mercury vapor detectors increases the vapor level of mercury in the ambient atmosphere. In addition to the presence of mercury in coal, up to 21 p.p.m. mercury has been found in crude oil in the Santa Barbara area.

Included also in the controversy is the agricultural use of mercurial compounds in the treatment of seeds. Phenyl mercuric acetate, as contrasted with the alkyl mercury compounds, does not create a pollution hazard, says Wayne F. Gustafson, president of Gustafson Manufacturing, Inc. However, there was the Huckleby family poisoning in New Mexico, which resulted from their eating pork from hogs fed with treated grain.

In 1969, over 6 million pounds of mercury was consumed (see box). Clarkson asserts that two problems would be encountered if a strict ban was placed on usage of this magnitude. First, the mercury release into water would be reduced by only 50%. Man and nature equally share the discharge of mercury into the environment with the normal loss from rocks (geologically) averaging 50%. A second problem would be the dependence of some

industries upon mercury. Without mercury, these industries would collapse.

Mercury loss has been noted for years (see box). From 1945 to 1959, there was an inventory discrepancy of over 314,700 flasks (one flask contains 76 lb.) of mercury. Where are these 314,700 flasks of mercury?

## Analysis

The determination of the amount of mercury present can be accomplished by testing air, urine, blood, hair, water, sludge, and sediments. The various methods that can be used to ascertain the amount of mercury present in a specific material include:

• Emission spectrography.
• Mass spectroscopy.
• X-ray spectrophotometry.
• X-ray absorption.
• Atomic absorption.
• Atomic fluorescent flame spectrometry.
• Neutron activation (destructive or nondestructive).
• Polarography.
• Radiometry.

Of these sophisticated methods, two are preferred—atomic absorption and neutron activation. The other methods have one or more factors limiting their analytical capacities. In the atomic absorption method, mercury is taken in any form, vaporized, and then analyzed by ultraviolet light. The neutron activation process consists of bombarding the sample with neutrons, and converting the mercury to a radioisotope form which can then be easily measured.

Instrumentation for field and portable use are being developed for analyzing air and water. Olin Corp. and Geomet (Rockville, Md.) have both designed instruments for rapid mercury detection and analysis.

## Restoring waters

Four methods for restoring mercury-contaminated lakes have been suggested, although each has its drawback(s), points out Jernelöv.

• The mercury-rich sediment can be covered with inert clay or other absorbing materials. This process is both expensive and inefficient.
• Dredging the lakes to remove the sediment, but expense and lake size make this method impractical.
• Addition of iron sulfide or hydrogen sulfide will form mercury sulfide which does not methylate, but addition of these chemicals may kill the fish.
• Addition of nitrogen to raise the pH and change the methylation process has the disadvantage of spreading the pollutant over a larger area. The concentration is reduced, but the area of pollution is greater.

To control contamination, D'Itri sees two approaches—regulating all forms of mercury and controlling strictly the alkyl mercuries. Jernelöv agrees, suggesting that processes be changed and effluents treated as research and technology improve.

## The mercury situation

Industrial plants have reduced mercury discharge into waterways by 86% since this summer. Secretary of the Interior Walter J. Hickel says that 50 plants in the U.S. were dumping 287 lb. of mercury per day in July, but by September, this amount was reduced to 40 lb. Also, another 79 companies that use mercury had no detectable discharges.

Air pollution bill S. 4358, which passed the Senate in September, includes a clause authorizing the Secretary to establish standards for "hazardous agents" which include mercury. At present, there are no air pollution standards for mercury.

How serious is the problem? Says Klein, "Mercury is more serious than DDT and less serious than nuclear testing." Leonard J. Goldwater, Duke University, contends that there is no reason to declare a public health crisis; the situation is potentially dangerous and requires study. The FDA stated that although an increasing amount of mercury has been discovered in foods, the situation has not reached the crisis stage.
                                        CEK

---

## 1969 mercury consumption in U.S. tops 6 million pounds

(thousands of pounds)

| | |
|---|---|
| Electrolytic chlorine | 1572 |
| Electrical apparatus | 1382 |
| Paint | 739 |
| Instruments | 391 |
| Catalysts | 221 |
| Dental preparations | 209 |
| Agriculture | 204 |
| General laboratory use | 126 |
| Pharmaceuticals | 52 |
| Pulp and paper making | 42 |
| Amalgamation | 15 |
| Other | 1082 |
| **Total** | **6035** |

Note: Consumption based on 76-lb. flasks.
Source: *Chemical & Engineering News*, June 22, 1970

# Water pollutant or reusable resource?

*Reprinted from* ENVIRON. SCI. TECHNOL., **4**, 380 (May 1970)

Spent pickle liquor: it may not be the steel industry's number one pollution headache—that distinction probably belongs to particulate emissions from furnaces—but it is a prodigious problem all the same. Spent liquor is an aqueous solution containing from about 0.5–10% acid and perhaps 12% iron, either undissolved or in solution. If discharged untreated to a waterway, it is a two-way menace, first for its acid content and second for the iron it contains. Depending on the chemical and physical form of the iron, the liquor may turn the water muddy brown, deposit slime, and exert a strong oxygen demand. Even many of the conventional ways of treating this particular effluent cause further pollution problems.

The ways in which this pollutant arises, is being treated, and could be treated effectively may all serve as useful guides to those in other industries which may be faced with a similar potential water pollution hazard.

## Pickling

The reason the steel industry is faced with a spent pickle liquor disposal problem is simple: huge amounts of pickling acid—35 to 45 lb. for every ton of steel—are used in the finishing operation by which oxide scale is removed from the surface of the steel product. It is estimated that 8–15 gallons of spent liquor are produced along with each ton of steel pickled. Since 50 million tons of steel are pickled every year in the U.S., spent liquor is produced at the rate of a half billion gallons per year.

In the modern steel industry, many products such as strip and wire are generally continuously pickled by being passed through a series of tanks; the tanks are interconnected so that each succeeding tank along the path of travel contains acid of progressively increased strength (see sketch). The last tank in the line contains the fresh acid. Spent liquor—in which the acid content is so low and the iron content so high that it has practically lost its pickling effectiveness—is bled off the first tank in the line.

One of two acids, sulfuric or hydro-

chloric, is the choice for most mild steel pickling lines today, although nitric, phosphoric, and other acids are used to treat the lower tonnage products such as stainless and special steels. At one time, sulfuric acid was the almost universal choice in the industry, but the decreasing price of hydrochloric in recent years, together with its ability to pickle faster and to retain pickling effectiveness to lower concentrations than sulfuric, have made hydrochloric acid the usual choice for new plants and converted lines. Hydrochloric acid is also said to give a "cleaner" finish.

Whether hydrochloric or sulfuric acid is used on a pickling line makes a substantial difference to the spent liquor disposal problem. For one thing, spent sulfuric acid may contain as much as 10% free acid, while spent liquor from hydrochloric lines may contain as little as 0.3% free acid. For another, many of the commonly used neutralization processes (see below) produce precipitates that may or may not be susceptible to settling and further processing, depending on which acid is used for pickling.

## Disposal

There are essentially six types of approaches that can be taken to the disposal of spent pickle liquor:

• **Discharge to a waterway.** This method is obviously very cheap, but it is losing favor for the very good reason that water pollution control authorities are clamping down hard. Nevertheless, several companies still practice it, albeit under official blessing. Direct discharge into Lake Erie, for instance, has been justified on the basis of the lake's alkalinity, which presumably can take care of the neutralization job (but, of course, cannot touch the more serious iron problem).

• **Deep well disposal.** Increasingly, steel companies have been favoring this method of disposal. As long as state authorities are willing to agree with a company that geological considerations minimize any chance of ground water contamination, the use of deep wells is likely to remain first choice for steel executives. Operating cost is not high—

perhaps $1/4$–1 cent per gallon—and wells may cost from $50,000–75,000 to drill and case. However, there is a critical need for good filtration equipment so that the solid particles in the liquor do not clog the pores in the well.

• **Neutralization.** This approach is intended to raise the pH to about 7, so that a neutral liquid can be discharged. Problems arise, however, in the disposal of the solid portion of the liquor—it may not be so solid, for instance. Ferrous hydroxide in the neutralized mixture is extremely difficult to settle and the sludge must be pumped into lagoons where it has to be kept indefinitely. This is no small problem: the pickling of 1 million tons of steel can result—if neutralization is used on the spent liquor—in the production of 200,000 tons of sludge that will not dewater. The sludge requires 150 acre ft. of permanent fill volume. Despite these disadvantages, many companies that have ample site area are willing to use up the space. Furthermore, simple neutralization may cost only 1–4 cents per gallon.

More sophisticated neutralization processes are available, however, and are generally used as a first step in some type of recovery process (see below).

• **Hauling by contractor.** Although using a contractor may be an easy way to get rid of a problem, it tends to be more expensive than simple on-site treatment—maybe 2–5 cents per gallon. Where the contractor is reliable and technically sophisticated, hauling may be very satisfactory, but, if he is not, repercussions may occur which the steel company might regret. (ES&T, March 1970, fully describes waste disposal under contract on page 195.)

• **Recovery processes.** These are designed to recover just the free acid in the spent liquor and do not attempt to regenerate acid from the iron salts (sulfates or chlorides). For this reason, recovery is used mainly on sulfuric acid liquors, in which the free acid content is relatively high—usually more than 5%. An effective recovery process, marketed by Wean Industries Inc. (Youngstown, O.), involves cooling the liquor from around 180° F. to ambient temperature

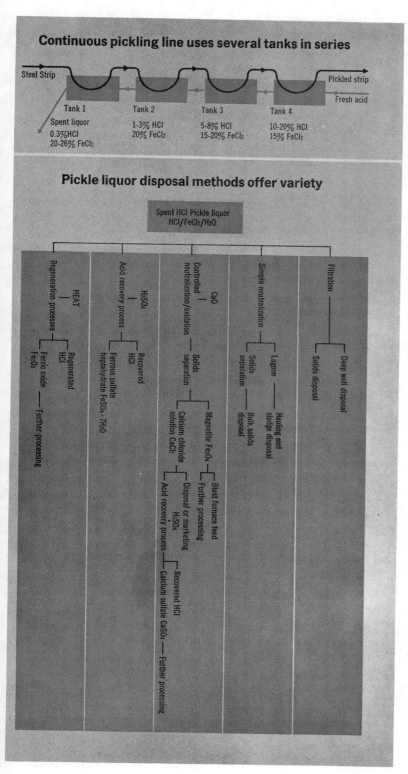

at a controlled rate. Green crystals of iron sulfate heptahydrate ($FeSO_4 \cdot 7H_2O$) separate out and can easily be removed from the free acid which is then recycled to the pickling process. The heptahydrate has found some use as a flocculent in municipal sewage treatment plants and is also used in paint and fertilizer manufacture. If no direct outlet can be found for it, Wean suggests further processing in a controlled neutralization/oxidation step in which magnetite ($Fe_3O_4$) and gypsum (hydrated $CaSO_4$) are produced, or roasting of the crystals to yield ferric oxide ($Fe_2O_3$) and sulfuric acid.

• **Regeneration processes.** Complete resue of the acid and iron values in spent pickle liquor is, in many ways, a most satisfactory method of treatment, because all of the acid, not just the free acid, is returned to the pickling line. However, because regeneration processes are rather sophisticated compared with simpler neutralization and recovery processes, they are obviously more expensive to install and operate. Richard J. Lackner, sales manager for Wean Industries' Environmental Control Division, which is licensed to build many of today's regeneration processes, says that fear of the large investment necessary (anywhere from $100,000 to $2 million) and a basic lack of desire on the part of steel mills have kept the more sophisticated processes from being widely applied.

Nevertheless, numerous such processes are available, and the aim of all, according to Lackner, is to pay for themselves through savings in reuse of the regenerated acid and recovery of useful byproducts.

A Du Pont process uses controlled neutralization and oxidation with air and yields a mixture of magnetite and calcium sulfate, together with some other solids that can easily be dewatered and landfilled. The calcium sulfate may find a market as wallboard. A variant of the Du Pont process, used with sulfuric acid liquor, is the Interlake-Du Pont process which produces magnetite and calcium chloride solution from spent hydrochloric acid liquor. The magnetite can be magnetically separated from the calcium sulfate, and, when sintered, makes good blast furnace feed. Although it is possible, in a Wean Industries' process, to convert the calcium chloride to reusable hydrochloric acid and calcium sulfate, Wean is developing a process for direct addition of sulfuric acid to the spent hydrochloric

**Pickle line.** *Modern high speed strip pickling lines are adaptable for use with either hydrochloric or sulfuric acids*

liquor. This latter process yields hydrochloric acid for reuse and makes possible the recovery of ferrous sulfate heptahydrate crystals by the method described previously.

Another regeneration method, which can be used with success to treat spent hydrochloric liquor, involves roasting the liquor to drive off hydrogen chloride gas, later absorbed in water to give the acid. Formed in this process is ferric oxide which is useful in the pigment industry and which also is finding increasing use in magnetic tape manufacture.

### Steel industry practice

Methods of disposal of spent pickle liquors in the steel industry vary considerably among the different steel firms and also vary from mill to mill. The differences stem largely from three main factors:

• Whether the mill is old or new and whether hydrochloric or sulfuric acid is used for pickling (the trend in newer mills is to hydrochloric).

• Regulations in the states and municipalities in which the various steel plants are located.

• The opinions of company management about the long-term viability of deep well disposal as a permissible way to dispose of liquid wastes.

The nation's largest steelmaker, U.S. Steel, uses deep well disposal at its largest mill (Gary, Ind.). Spent acids, both sulfuric and hydrochloric, are discharged to a 4400 ft. deep well after the solid matter has been filtered off. At other plants in Allegheny and Bucks Counties, Pa., U.S. Steel uses a simple lime neutralization process; the sludge is hauled to disposal sites. At its Pittsburgh (Calif.) plant, the company has a permit to dispose of spent pickle liquors at a specific location in the Pacific Ocean.

Republic Steel hauls the waste liquor from its Cleveland (O.) plant for disposal by contractor but is presently reevaluating costs and may go over to deep well disposal. At Republic's Gadsden (Ala.) mill, a process which regenerates iron from a hydrochloric acid pickle liquor is being used (see discussion of regeneration process above).

Bethlehem Steel's large plant at Sparrow's Point, Md., discharges spent liquor into the Patapsco River with permission of the State of Maryland; it is now constructing a lagoon to treat waste waters from the rolling mills and pickling rinse waters. At its plant in Burns Harbor, Ind., Bethlehem originally used a Du Pont neutralization process to treat sulfuric acid pickle liquor, but since switching to hydrochloric acid it has used a deep well. A Bethlehem spokesman says that the contractor is expected to remove the waste pickle liquor this year, and to recover magnetite for use in magnetic tape manufacture. Bethlehem's Lackawanna (N.Y.) mill has been discharging spent liquor into Lake Erie, but the company is now arranging, with the approval of the State of New York, to use deep well disposal.

Interlake Steel has plants in Chicago and in Newport, Ky., where various neutralization processes are used. In Chicago, Interlake is operating the Interlake-Du Pont process (see above), which produces magnetite and calcium chloride solution.

Armco Steel has two deep wells for disposal of pickle liquor at its much heralded mill in Middletown, O. Also at Middletown, Armco is working on a neutralization/oxidation process to treat rinse waters, supported by a Federal Water Pollution Control Administration demonstration grant. At its other mills in Pennsylvania, Texas, and Kentucky, Armco is either using conventional neutralization and lagooning methods or has a permit to discharge to a waterway.

All steel company spokesmen contacted by ES&T were quick to emphasize that in those cases where their disposal methods might seem to the public to be evidence of gross pollution, operations were in fact sanctioned by the relevant state or local pollution control authority. The steel industry is also strongly emphasizing that all currently employed methods are under constant surveillance and that alternatives are being sought.

### Future prospects

The fact that few of the available recovery and regeneration methods are in actual use in the steel industry seems to be a reflection not so much of the technical uncertainties of these methods— most, except perhaps the roasting techniques, are not difficult to run—but of the fact that much simpler, if less effective ways of disposal are still open. The idea of being able to recover almost all the pickling acid used in a mill is appealing—in principle. But in the absence of any overriding compulsion such as stern legislation, steel executives are still likely to opt for the easiest, cheapest method at hand, consistent with obedience to whatever local regulations are in effect.

There are, however, a few storm clouds on the horizon which threaten to pour water on the quick and easy approach. For instance, Secretary of the Interior Walter J. Hickel has gone on record to describe deep well disposal as "a potential environmental Frankenstein's monster." And deep wells are obviously being depended upon to do the disposal job by many big steelmakers. Then again, the large amounts of fill and lagoon area required by techniques which produce wet sludges are cramping the style of companies whose plant perimeters are being rapidly approached by land unusable for plant expansion.

It is entirely possible that within two or three years recovery and regeneration methods will be going on stream at a rate not seriously contemplated today.

# A monitor for 22 water pollutants

*The Enviro Monitor makes a continuous record of*
*pollutants and controls unattended operations at*
*remote locations under adverse weather conditions*

Reprinted from ENVIRON. SCI. TECHNOL., **6,** 510 (June 1972)

Industrial pollution control managers who are watching for monitors to record and control processing streams from their plants should sit back and take a look at what the new Enviro Monitor of Enviro Control (Washington, D.C.) is doing for some users in different locations in the U.S. Enviro Monitor is an all-weather monitor for as many as 22 parameters. It can be rigged to give a sound or light alarm when a particular parameter exceeds a preset value. It can also provide a permanent record of all parameters monitored.

Enviro Control, Inc., a subsidiary of Diversitron, Inc., was formed in September 1970 and although the first public announcement of its monitor came only at last year's meeting of the Water Pollution Control Federation (San Francisco, Calif.), a lot has happened since then. At press time, several users had told ES&T of their experiences with the Enviro Monitor and several others told of plans to get their hands on a monitor to check it out for their monitoring and control needs.

## Who's using it?

Desert Research Institute (Las Vegas, Nev.) is using the monitor to record effluents from certain chemical and metal refining processing operations in the region. These industrial wastes are disposed of in a ditch, and the monitor has been placed at the end of the ditch to record several characteristics of the wastes. The continuous monitoring record is being established to see if there is any contamination of groundwaters which may eventually find their way to nearby Lake Meade. The monitoring program is part of a project supported by a four-year grant, totaling about $600,000, from the Environmental Protection Agency.

Alan Peckham, assistant director of the Center for Water Resources Research at the Institute, is also project director for the EPA grant. He tells ES&T that the grant has been in effect for two years and that, other than the Enviro Monitor, "we haven't found a monitor that we could operate under battery power in a remote location and leave out of doors under a wide range of climatic conditions." The parameters that are being checked in this desert application are conductivity, pH, temperature, and chloride.

Engelhard Industries (Newark, N.J.) uses the monitor to record effluents from its metallurgical separation operations. The parameters that are being continuously monitored are conductivity, copper, pH, and turbidity. Walter Drobot, engineering consultant, says Engelhard is using the monitor today for exploratory work but plans to monitor treatment processes and resultant effluent streams.

Municipal waste treatment operations are also finding the monitor useful. Levitt & Sons, one of the nation's largest home builders, uses the monitor at Winslow township (N.J.). In this application, three parameters—dissolved oxygen (DO), pH, and temperature—are checked. The DO content of the aeration basin wastes must be carefully controlled before tertiary treatment of the wastes can proceed. Stanley Dea of the Levitt office at Lake Success, N.Y. says the unit works well on a daily basis to control and monitor the extended aeration of the process. The monitor is used to control the rate of oxygen introduction into the effluent in an aeration basin. Further details of this operation were reported at last month's meeting of the Instrument Society of America in San Francisco.

Another user is Mannington Mills (Salem, N.J.), a manufacturer of floor covering using latex. This company uses the monitor to record pH and to control addition of alum to its process waste waters. At a certain pH, the latex is flocculated by the addition of alum.

The pulp and paper industry is another potential user of the monitor. Its National Council for Atmospheric and Stream Improvement Council is checking the application of the monitor to this industry. Initially, the Council is interested in total oxygen demand (TOD), pH, sodium, chloride, and temperature.

## What's ahead?

At press time, some active negotiations were under way with other users, including a food processing firm in Oregon, a leather tanner in Maryland, a canner in Delaware (interested in DO, temperature, pH, and turbidity), a refinery in Chicago (for DO, pH, conductivity, and chloride), and a plating plant in Oklahoma (for copper, cyanide, cadmium, chromium, and pH).

Roy Ricci, marketing manager for Enviro Control, says that the 22 parameters cover many of those included in the permit applications required by the Corps of Engineers.　　SSM

---

**Advantages cited for monitor**

Some of the advantages Ricci cites for the monitor are:

- range of operating temperature ($-20°$ to $130°F$)

- unattended operation

- battery-powered options

- operation in a submersible mode or a flow-through assembly

- sensor location independent of the monitor

- modular design for matching any industrial application.

# Instruments for

*Reprinted from* ENVIRON. SCI. TECHNOL., **6**, 130 (February 1972)

**Effluent.** *Waste stream monitoring may replace receiving water measurements*

Monitoring the environment is a fundamental part of the effort to define the nature and extent of pollution problems in order to develop an intelligent plan for abatement. Because of the movement of specific pollutants through the environment and the interaction between transport media, each medium must be monitored for a variety of substances and the data integrated to yield a comprehensive picture of the overall situation. Monitoring technology is defined by nature of transport media, pollutant sources, sample requirements, and necessary analytical methodology to produce an accurate result.

Methods used by the Environmental Protection Agency for measuring air pollutants have been described previously (see ES&T, August 1971, p 678). Research on analytical instruments and methods is carried out at the National Environmental Research Centers of EPA at Research Triangle Park, N.C. (air), the Perrine Primate Laboratory, Fla. (pesticides), the Western Environmental Research Center, Las Vegas, Nev. (radiation), and the National Environmental Research Center, Cincinnati, Ohio (water).

Monitoring the aquatic environment involves both measuring water quality characteristics in receiving streams and determining specific pollutants in waste discharges. A wide variety of tests are available, and the methods and equipment depend on the purpose of the monitoring program. Stream monitoring may be likened to ambient air monitoring, in that the measurements are made at a distance from the source of the discharge and are characteristic of a mixture of constituents from several unknown sources. These measurements indicate the exposure of the general population, either human or aquatic, and thus are often a good determinant of the conditions to be alleviated. Water quality standards established by the states and approved by the federal government are designed to protect legitimate water users and do not directly indict a specific waste discharger. In contrast, direct measurement of waste effluent (source sampling) identifies the pollutant and the polluter and thus permits the application of abatement procedures at the source.

## Refuse Act

Current attention in EPA is being directed toward source monitoring as a means of control. The Refuse Act of 1899 requires a permit from the U.S. Army Corps of Engineers for all industrial waste discharges to navigable waters in the United States. As a condition of the permit, the applicant must furnish data on 11 standard waste parameters [alkalinity, biochemical oxygen demand, chemical oxygen demand, total solids, total dissolved (filterable) solids, total suspended (nonfilterable) solids, total volatile solids, ammonia, Kjeldahl nitrogen, nitrate, and total phosphorus] and a selected group of additional measurements depending on the type of industry. Table I shows the additional requirements for Standard Industrial Classification (SIC) codes 26, 284, and 2871 as examples of the expanded list of waste parameters.

The required tests are designed to characterize the volume and concentration of the major waste constituents. The measurements listed above, when applied to all industrial discharges, permit comparison between treated and untreated effluents and indicate the degree of "damage" to the receiving water in terms of oxygen depletion, sedimentation, pH change, and nutrient impact. The additional measurements required (Table I) identify the waste contributions from specific industrial operations. Reporting these waste char-

## Table I. Examples of waste parameters sampled

**SIC 26. Paper and allied products**

| | |
|---|---|
| Color | Chromium |
| Specific conductance | Copper |
| Bromide | Lead |
| Chloride | Mercury |
| Sulfate | Nickel |
| Sulfite | Sodium |
| Phenols | Zinc |
| Total organic carbon | Total coliform |

**SIC 284. Soap, detergents, and cleaning preparations, perfumes, cosmetics, and other toilet preparations**

| | |
|---|---|
| Color | Cyanide |
| Specific conductance | Sulfate |
| Turbidity | Sulfide |
| Oil and grease | Phenols |
| Chloride | Surfactants |
| | Total coliform |

**SIC 2871. Fertilizers**

| | |
|---|---|
| Color | Cadmium |
| Turbidity | Chromium |
| Hardness | Copper |
| Oil and grease | Iron |
| Chloride | Lead |
| Fluoride | Manganese |
| Nitrite | Mercury |
| Pesticides | Nickel |
| Phosphorus-ortho | Zinc |
| Sulfate | Radioactivity |
| Aluminum | Total coliform |
| Arsenic | Fecal coliform |

**Dwight G. Ballinger**

*Environmental Protection Agency*
*Cincinnati, Ohio 45202*

A rundown of available instruments for industries and agencies who must monitor many aquatic parameters

# water quality monitoring

acteristics and volumes will show what is presently being discharged to U.S. waters.

Recommended analytical procedures for most of the measurements to be made in compliance with permit requirements are given in "Methods for Chemical Analysis of Water and Wastes, 1971"; microbiological procedures are given in "Standard Methods for the Examination of Water and Wastewater," 13th ed. (APHA); chlorinated hydrocarbon pesticides methods are described in "Methods for Organic Pesticides in Water and Wastewater"; while radiation measurement techniques appear in "Standard Methods" and in "ASTM Annual Book of Standards," Part 23, "Water, Atmospheric Analysis."

Aluminum, cadmium, chromium, copper, iron, lead, manganese, sodium, and zinc are analyzed by atomic absorption, although cadmium can also be measured by dithizone and zinc by colorimetric or polarographic methods. Arsenic is detected by colorimetric and nickel by polarographic methods. Cold vapor-uv is used for mercury. Some additional waste measurements and recommended methods of analysis are shown in Tables II and III.

In general, the procedures used provide reliable information on waste loadings. These methods have been in use for many years and have been thoroughly tested in many laboratories. The Environmental Protection Agency carries out an extensive program of methods research and evaluation to establish the reliability, precision, and accuracy of the procedures used in its water and waste laboratories. Examples of the range and accuracy of some methods are shown in Table IV.

The equipment and instruments required for these analyses range from simple to complex. Table V shows the

## Table II. Physical, demand, and nutrient parameters and testing methods

| Parameter | Method |
|---|---|
| Solids | Gravimetric |
| Color | Visual comparison |
| Turbidity | 90° scatter photometer |
| Radioactivity | Gross $\alpha$, $\beta$, $\nu$, tritium |
| Specific conductance | Wheatstone bridge |
| Biochemical oxygen demand | 5-day—20°C |
| Chemical oxygen demand | Dichromate reflux |
| Ammonia | Distillation-nesslerization or AutoAnalyzer |
| Kjeldahl nitrogen | Digestion-distillation or AutoAnalyzer |
| Nitrate | Brucine sulfate or AutoAnalyzer |
| Nitrite | Diazotization or AutoAnalyzer |
| Total phosphorus | Persulfate digestion or AutoAnalyzer |
| Phosphorus-ortho | Single reagent, stannous chloride, or AutoAnalyzer |

## Table III. Testing methods for organic and general parameters

| Parameter | Method |
|---|---|
| Phenols | Colorimetric |
| Oil and grease | Hexane Soxhlet extraction |
| Total organic carbon | Combustion-infrared |
| Surfactants | Methylene blue active substance |
| Pesticides | Gas chromatography |
| Alkalinity | Electrometric titration or AutoAnalyzer |
| Bromide | Colorimetric |
| Chloride | Mercuric nitrate titration or AutoAnalyzer |
| Cyanide | Silver nitrate titration or pyridine pyrazalone |
| Fluoride | SPADNS with distillation or probe |
| Hardness | EDTA titration, AutoAnalyzer, or atomic absorption |
| Sulfate | Turbidimetric or AutoAnalyzer |
| Sulfide | Titrimetric or methylene blue colorimetric |
| Sulfite | Iodide-iodate titration |

## Table IV. Range and accuracy of test methods

| | Optimum range, mg/l. | Accuracy, % bias |
|---|---|---|
| BOD | 2–500 | a |
| COD | 40–500 | −4 |
| Ammonia, nitrogen | 0.05–2.0 | −5 |
| Nitrate, nitrogen | 0.10–2.0 | +4 |
| Total phosphorus | 0.01–0.5 | +3 |
| Chloride | 10–400 | −0.5 |
| Hardness | 10–400 | −1 |
| Sulfate | 10–500 | −3 |
| Specific conductance | 0.5–2000 μmhos | −5 |
| Arsenic | 0.01–0.5 | 0 |
| Copper | 0.1–10 | −10 |
| Lead | 0.1–10 | −12 |
| Mercury | 0.2–10 μg/l. | −13 |
| Phenols | 50–5000 μg/l. | a |
| Pesticides | 0.01–0.2 μg/l. | −20 |

a True accuracy cannot be determined for these tests.

## Table VI. Analyses can indeed be costly

| Test | Analyst level | Typical charge/ sample |
|---|---|---|
| pH | | $ 3–5 |
| Alkalinity | | $ 3–12 |
| Chloride | Technician | $ 5–7 |
| Solids | | $10–30 |
| Specific conductance | | $ 3–5 |
| BOD | | $30 |
| COD | | $15–30 |
| TOC | | $30 |
| Cyanide | | $15–30 |
| Phenols | Skilled analyst | $15–30 |
| Ammonia, nitrogen | | $ 5–15 |
| Mercury | | $20–30 |
| Metals | | $ 9–25 |
| Pesticides | | >$50 |
| Radioactivity | Expert | >$50 |
| Required parameters | Skilled analyst | $129 |

## Table V. Reliability of some on-line instruments is not established

| Parameter | Laboratory instrument | On-line instrument |
|---|---|---|
| Total organic carbon | Combustion-infrared | Yes a |
| Ammonia, nitrogen | AutoAnalyzer | Yes a |
| Total phosphorus | AutoAnalyzer | No |
| Chloride | AutoAnalyzer or electrode | Yes |
| Specific conductance | Wheatstone bridge | Yes |
| Copper | AA spectrophotometer | Yes a |
| Lead | AA spectrophotometer | Yes a |
| Mercury | AA spectrophotometer or mercury meter | Yes a |
| Phenols | Vis. spectrophotometer | Yes a |
| Pesticides | Gas chromatograph | No |
| Turbidity | Turbidimeter | Yes |
| Cyanide | Vis. spectrophotometer | No |
| Radioactivity | Counters and spectrometers | No |

a Reliability has not been established.

type of tests and the instrumentation available for both laboratory and field measurement. It is apparent from this table that field equipment is lacking for the on-line measurement of many of the required waste constituents. As noted, the reliability of some of the available monitoring instruments has not been established to the satisfaction of the Environmental Protection Agency.

In contrast to air monitoring instruments, instruments for monitoring water often are designed to perform a number of test measurements simultaneously and record the results. Most of this equipment has been designed to provide stream monitoring rather than waste discharge evaluation. While the analysis method is the same in both situations, the concentration of the measured constituent is generally greater in the waste, and extreme sensitivities are not required. Interference, corrosion, and suspended solids are more significant in waste monitoring, requiring more rugged construction and modification of flow systems. It is highly desirable that manufacturers begin constructing equipment specifically for waste conditions, rather than assume that the same monitor will operate as either a stream or waste discharge instrument. Further, adapting process control devices commonly used in the chemical industry offers an excellent source of on-line waste monitoring equipment.

The primary restraint at the present time is the lack of reliable sensor systems for many of the chemical constituents to be measured in industrial wastes. A fertile field exists for developing sensors for the more complex tests necessary for proper waste characterization.

Since much of the analytical work for monitoring wastes must be done in the laboratory, consideration should be given to the cost of the tests. Table VI lists the type of determination, the relative skill of the analyst, and examples of commercial laboratory charges for performing the tests.

Producing reliable waste data requires careful application of the proper analytical methods, use of modern equipment, and an adequate program of quality control. The quality control program should provide checks on the reagents and gases used, the performance of the instrumentation, the care and skill of the

## Available on-line water-monitoring instruments

| Parameter | Manufacturer | Cost range |
|---|---|---|
| Total organic carbon | Ionics-Union Carbide<br>Raytheon/AES | $8,000–$10,000 |
| Ammonia, nitrogen | Technicon | $5,000–$10,000 |
| Copper | Fischer-Porter<br>Hach Chemical Co. | $1,900–$4,000 |
| Mercury | Olin Corp. | $20,000 |
| Phenols | Ionics-Gulf (available 1972)<br>Du Pont | $4,000–$7,000 |
| Phosphate | Hach Chemical Co.<br>Technicon | $1,600–$5,000 |
| Multiparameter monitors, including chloride, sp. conductance, turbidity, and others | Enviro Control, Inc.<br>Schneider Instrument Co.<br>Raytheon/AES<br>Ecologic Instruments<br>Ionics-Union Carbide<br>Robertshaw<br>Hach Chemical Co.<br>Delta Scientific Corp.<br>Beckman Instruments, Inc. | $5,000–$15,000 |

Mention of companies or products does not constitute endorsement by the Environmental Protection Agency.

analyst, his ability to provide accurate results, and the proper reporting of the analytical answer. A formalized program to evaluate these variables and to establish the reliability of the data is highly desirable. Quality control charts, recording daily performance, and monitoring the output of each analyst have been found to be effective means of measuring and controlling the laboratory performance. A publication titled "Control of Chemical Analyses in Water Pollution Laboratories," soon to be distributed by EPA, outlines a quality control program for laboratories performing water and waste analyses.

### Additional Reading

Hochheiser, Seymour, Burmann, F. J., and Morgan, G. B., "Atmospheric Surveillance—The Current State of Air Monitoring Technology," **ES&T,** 5, 678 (1971).

Application for Permit to Discharge or Work in Navigable Waters and Their Tributaries, Eng. Form 4345-1, Department of the Army, Corps of Engineers, Washington, D.C. 20314.

"Methods for Chemical Analysis of Water and Wastes, 1971," Analytical Quality Control Laboratory, National Environmental Research Center, Environmental Protection Agency, Cincinnati, Ohio 45268.

"Standard Methods for the Examination of Water and Wastewater," 13th ed., American Public Health Association, 1015 18th St., N.W., Washington, D.C. 20036.

"1970 Annual Book of ASTM Standards," Part 23, "Water; Atmospheric Analysis," American Society for Testing and Materials, 1916 Race St., Philadelphia, Pa. 19103.

**Analyzing.** *Tests, such as this one for extraction of organic waste samples, may become commonplace as more parameters are measured for water pollution control*

**Dwight G. Ballinger** *is director of the Analytical Quality Control Laboratory, part of* EPA's *National Environmental Research Center in Cincinnati. The laboratory is responsible for developing methods for water and waste analysis. Mr. Ballinger received his B.S. degree in chemistry from the University of Cincinnati and is the author of many publications on analytical chemistry of water and wastes.*

Monitoring is at the heart of the nation's
water quality management effort; without it, enforcement and
cleanup programs can be of only limited effectiveness

# Water quality surveillance:

*Reprinted from* ENVIRON. SCI. TECHNOL., **5**, 115 (February 1971)

**William T. Sayers**
*Water Quality Office,*
*Environmental Protection Agency,*
*Washington, D.C.*

Within the past several years, we have witnessed an immense increase in public awareness and concern over the quality of the environment. It has become obvious that we, as a nation, must control those practices that despoil our air, water, and land. The time is long past due for us to implement the necessary steps to harmonize our technological society with our natural environment.

With respect to abatement of water pollution, efforts of various size and scope have been in practice for nearly a century. Early efforts of state and federal authorities focused on the control of waterborne diseases, such as typhoid fever. Not until 1948, however, was specific water pollution control legislation enacted. Even then, it was of a temporary nature and only a weak forerunner of the Act of 1956, the basis of our present federal water pollution control program. Amendments in 1961, 1965, 1966, and 1970 further strengthened that act to provide today's far-reaching programs in water quality management.

In passing the Federal Water Pollution Control Act, Congress recognized the primary responsibilities and rights of the states to prevent and con-

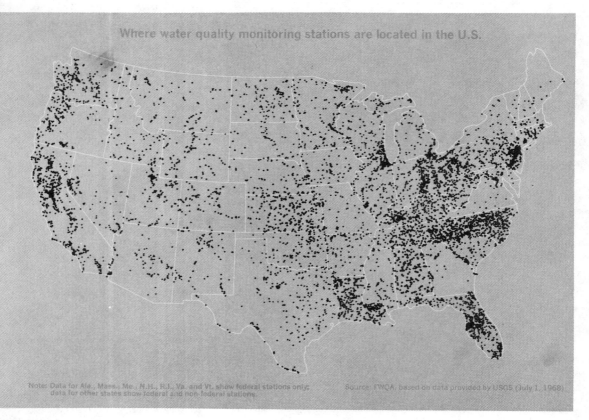

Where water quality monitoring stations are located in the U.S.

Note: Data for Ala., Mass., Me., N.H., R.I., Va. and Vt. show federal stations only; data for other states show federal and non-federal stations.

Source: FWQA, based on data provided by USGS (July 1, 1968)

# the federal-state network

## Parameter coverage and sampling frequency—U.S. water quality monitoring stations

No. of Stations Measuring Given Parameter at Indicated Frequency

| Parameter | Continuous | Daily | Weekly | Monthly | Quarterly | Annually | Other | Total |
|---|---|---|---|---|---|---|---|---|
| Temperature | 871 | 822 | 529 | 2270 | 832 | 1064 | 1276 | 7664 |
| Specific Conductance | 241 | 432 | 97 | 1636 | 491 | 1287 | 1045 | 5229 |
| Turbidity | 31 | 285 | 332 | 1053 | 615 | 57 | 623 | 2996 |
| Color | 14 | 87 | 205 | 939 | 611 | 1122 | 747 | 3725 |
| Odor | 9 | 32 | 14 | 389 | 406 | 13 | 248 | 1111 |
| pH (Field) | 77 | 55 | 156 | 745 | 372 | 122 | 721 | 2251 |
| pH (Lab) | 16 | 451 | 268 | 1914 | 749 | 1398 | 1111 | 5907 |
| EH | 18 | 0 | 0 | 7 | 0 | 0 | 1 | 26 |
| Suspended Solids | 0 | 2 | 48 | 328 | 213 | 2 | 45 | 594 |
| Other Physical Analyses | 12 | 19 | 73 | 146 | 179 | 41 | 193 | 663 |
| Dissolved Solids | 42 | 188 | 232 | 1495 | 624 | 1187 | 1063 | 4831 |
| Chloride | 10 | 121 | 118 | 1196 | 354 | 552 | 931 | 3282 |
| Nutrients (Nitrogen) | 14 | 32 | 243 | 1266 | 507 | 896 | 775 | 3733 |
| Nutrients (Phosphorus) | 23 | 58 | 135 | 853 | 609 | 889 | 918 | 3485 |
| Common Ions | 20 | 412 | 119 | 1775 | 555 | 1357 | 1288 | 5526 |
| Hardness | 6 | 362 | 225 | 1490 | 674 | 1310 | 907 | 4974 |
| Radiochemical | 21 | 3 | 37 | 339 | 448 | 31 | 265 | 1144 |
| Dissolved Oxygen | 85 | 45 | 345 | 1379 | 500 | 913 | 681 | 3948 |
| Other Gases | 10 | 86 | 13 | 34 | 0 | 5 | 18 | 166 |
| Minor Elements | 0 | 12 | 3 | 164 | 45 | 149 | 175 | 548 |
| Pesticides | 1 | 1 | 7 | 69 | 11 | 24 | 80 | 193 |
| Detergents | 0 | 1 | 85 | 363 | 463 | 33 | 114 | 974 |
| Bio. Oxygen Demand | 0 | 12 | 239 | 866 | 460 | 30 | 176 | 1783 |
| Carbon (Tot., Diss., etc.) | 0 | 4 | 4 | 22 | 0 | 31 | 4 | 65 |
| Coliforms | 13 | 240 | 423 | 1261 | 637 | 160 | 345 | 3079 |
| Other Micro-organisms | 2 | 13 | 185 | 252 | 128 | 4 | 113 | 697 |
| Biologic | 6 | 3 | 102 | 127 | 60 | 13 | 76 | 387 |
| Sediment Conc. (Susp.) | 25 | 291 | 96 | 269 | 41 | 9 | 328 | 1059 |
| Particle Size (Susp.) | 5 | 56 | 7 | 134 | 12 | 51 | 315 | 580 |
| Part. Size (Bed Load Mat.) | 0 | 0 | 0 | 10 | 2 | 6 | 33 | 42 |
| Other Sediment | 1 | 8 | 5 | 97 | 2 | 6 | 71 | 190 |

Source: FWQA, based on data provided by USGS (July 1, 1968)

trol water pollution. The federal role is designed primarily to supplement state activities. Major activities of the federal Water Quality Office (wqo) include:

• Providing grants for the construction of municipal waste treatment facilities.

• Assisting states in the development and administration of water quality standards.

• Administering a federal enforcement program against interstate pollution.

• Supporting research and development activities which seek better methods for controlling all forms of water pollution.

• Providing technical assistance on complex pollution problems.

• Encouraging effective river basin planning.

• Extending financial and other assistance to state pollution control agencies.

The success of water quality management efforts at all levels of government will, in the final analysis, be judged by our ability to provide adequate quantites of water of appropriate quality for all water uses.

Water quality and changes in it can be determined most effectively through water quality measurement. Thus, the key to effective water quality management is the ability and capability to measure substances and to understand their behavior in water. Additionally, water quality monitoring provides the only effective yardstick for evaluating the efficiency of a water quality management program.

The surveillance network described in this article aims to assess existing water quality conditions, to determine long-term trends in water quality, and to evaluate compliance with state–federal water quality standards. This information serves to identify existing and emerging problem areas and changes in water quality with time, nationally as well as locally.

## Monitoring needs and responsibilities

The effective management of water quality requires two categories of water quality monitoring. First, monitoring of individual treatment plant influent and effluent is necessary to maintain optimum treatment efficiencies. Effluent monitoring is also essential to assess the individual effects of each waste source on the waters into which they discharge. Second, the rivers, lakes, and coastal waters receiving wastes must be examined to assure attainment and maintenance of desired water quality levels consistent with criteria contained in state–federal water quality standards. Together, these two categories of monitoring will provide the information necessary for efficient management of pollution control facilities and the effective administration of water quality standards.

Monitoring of waters receiving wastes can be accomplished in two ways. One is through the operation of a vast network of strategically located long-term stations. A second approach makes use of repeated short-term surveys, each providing limited spatial coverage. Data needs can be met most efficiently by use of both approaches. A nationwide network that uses a minimal number of long-term stations can be used to identify general areas of adverse trends in water quality or violations of water quality standards. Where such conditions are found to be occurring, short-term surveys can then be conducted to define fully the extent of the problem, plus all factors contributing to the problem.

Monitoring of the individual treatment plant influent and effluent is basically the responsibility of pollution control facility managers and of local

| Uniform (100%) | Frequent (99–50%) | Less Frequent (49–20%) | Least Frequent (19–0%) |
|---|---|---|---|
| Dissolved Oxygen | Radioactivity | Arsenic | Bottom Deposits |
| pH | Parameters listed in Public Health Service Drinking Water Stds. | Barium | Chromium (trivalent) |
| Coliform | | Cadmium | Electrical Conductance |
| Temperature | Total Dissolved Solids | Chromium (hexavalent) | Ammonia |
| Floating Solids (Oil-Grease) | | Fluoride | Acidity |
| | | Lead | Alkalinity |
| Settleable Solids | | Selenium | |
| | | Silver | Carbon Chloroform Extract |
| Turbidity and/or Color | | Suspended Solids | Hydrogen Sulfide |
| Taste-Odor | | Chloride | Pesticides |
| | | Copper | Sodium |
| Toxic Substances | | Nitrate | Iron |
| | | Phenols | Plankton |
| | | Phosphate | Foaming Substances |
| | | Sulfate | Boron |
| | | Cyanide | Manganese |
| | | Median Tolerance Limit | Hardness |
| | | | Biochemical Oxygen Demand |
| | | | Methylene Blue Active Substances |
| | | | Zinc |

How frequently parameters are used in water quality criteria of state standards

and state pollution control agencies. The enormous investment in pollution control facilities required to restore, preserve, and enhance national water quality dictates that these facilities be managed at least as conscientiously as industrial production processes representing similar investments.

Monitoring of the receiving waters to evaluate and enforce water quality standards is the primary responsibility of state water pollution control agencies. The federal government also has certain obligations here, since it may be required to enforce interstate water quality standards when a state fails to exercise its responsibility.

The interdependency of these monitoring activities and their common goals require that they be closely coordinated for maximum effectiveness. We cannot afford duplication of effort. The magnitude of the monitoring effort required demands the greatest return on each dollar expended.

## Present monitoring activities

In designing a nationwide monitoring program to accomplish a specific set of objectives, all pertinent existing monitoring activities at the local, state, and federal levels should be reviewed to determine their usefulness to the proposed system. The map and table on pages 114 and 115 indicate the ex-

tent of present monitoring activities on a nationwide basis.

At the local level, most municipal water treatment facilities monitor raw water quality daily. There are about 6000 such facilities served by surface water sources. Thus, considerable information is being gathered on surface water quality in the U.S. by operators of municipal water treatment plants alone. In addition, many municipal waste water treatment programs and county agencies routinely monitor receiving waters upstream and downstream from treatment plant discharges. Many universities also regularly collect water quality data.

Most state pollution control agencies have monitoring programs for assessing surface water quality. These programs vary in scope among states and range from near-minimal to complete systems. Other water-oriented state agencies, such as conservation and geology departments, are also engaged in water data acquisition to various degrees.

More than a dozen federal agencies are engaged in the direct acquisition of water data. These activities are coordinated through the Office of Water Data Coordination of the U.S. Geological Survey, consistent with a Bureau of Budget requirement for interagency coordination to avoid duplication of effort. The budget agency (now the Office of Management and Budget) also advocates the operation of a National Network to meet the common data needs of two or more federal agencies. The U.S. Geological Survey (USGS) has been given responsibility for the management of this network. Data needs specific to any one given agency that cannot be met efficiently through the National Network will be obtained by that agency through other means.

One very significant feature of USGS's National Network will be what is referred to as its Accounting Element. This element will provide an accounting of the quantity and quality of water that flows out of 306 hydrologic basins which cover the conterminous U.S.

### Surveillance network

To achieve their goal of clean water, state and federal pollution control agencies are jointly developing a surveillance network that will serve to identify:
• Compliance and noncompliance with water quality standards.

• Water quality baselines and trends.
• Improvements in water quality produced by abatement measures being undertaken—e.g., construction of waste treatment facilities.
• Emerging water quality problems, in sufficient time to effect adequate preventive measures.

To attain these objectives as efficiently as possible, existing water data acquisition programs of all agencies will be utilized to the fullest.

The first major step in the development of this state–federal surveillance network is to complete its design. Basic steps required in the design phase are:
• Secure agreements with state water pollution control agencies on current and future monitoring locations, parameter coverage, and sampling frequencies, irrespective of political boundaries.
• Review current and proposed sampling activities of state, interstate, and federal agencies (including those of the National Network managed by USGS) to establish the need for additional stations and (or) supplementary data.
• Reach agreement among the state

and interstate agencies and the federal Water Quality Office (WQO) as to which agency will take lead responsibility in funding the new stations and supplementary activities needed.

The second major step is implementation of the jointly developed surveillance network plans. This is undertaken as the design phase is completed. Implementation involves several steps:
• Completing arrangements for acquiring data in a timely fashion from various agencies collecting data of interest to state and federal pollution control agencies.
• Arranging with USGS to operate those stations that can best be handled by it and to expand parameter coverage and (or) sampling frequency at certain existing stations in the USGS network.
• Initiating new stations and expanding existing stations operated by state pollution control agencies and the WQO. In many situations, it is expected that stations will be jointly operated by state and (or) local agencies and the WQO.
• Initiating analytical quality control procedures to assure that requisite

**Compilation.** *A* STORET *printout lists water quality data gathered over a 12-year period at a monitoring station on the Snake River in Idaho*

accuracy and precision are achieved. wqo is currently working with usgs in the development of a mutually consistent program in this area.

• Expanding data processing capability as necessary to permit timely evaluation of large volumes of data. To achieve the stated objectives of the coordinated surveillance network, it is essential that the data collected be evaluated in an expeditious manner and made readily available to all users. Only in this way can appropriate follow-up actions be taken. Timely data processing and diagnosis will be achieved through the use of wqo's existing computerized data storage and retrieval system (storet) and additional computer programs which must be developed.

A preliminary evaluation of the local, state, and federal long-term water quality monitoring effort required to provide adequate coverage of the nation's water resources indicates a need for at least 10,000 monitoring points on both interstate and intrastate waters. This may seem like a large number of stations until the length of streams, lakes, impoundments, and shoreline to be monitored is fully considered. These proposed stations, if equally spaced, would be located at 350-mile intervals. In fact, stations will be located closer together in highly developed areas and more distant in less developed areas. Obviously, there will be no overabundance of stations at this level of activity.

The majority of these stations would be well within state boundaries, and hence would most directly serve the needs of local and state water pollution control agencies. Stations providing information of key federal concern would include those along state and international boundaries, principal estuaries, mouths of major tributaries, large metropolitan complexes, and major water resource projects. There are an estimated 900 stream and 1500 open-water locations in this category.

Operational responsibility for a given new monitoring station would fall to the agency receiving greatest benefit from its operation. That is, state agencies would be expected to assume operating responsibility for most new stations within state boundaries and many new stations along state boundaries. wqo, utilizing the usgs network to achieve efficiency, would assume responsibility for those stations at the estimated 2400 loca-

**Robot monitor.** *Information from this station on the Potomac is automatically recorded by a remote* wqo *computer*

tions of key federal concern. Thus, the federal role in this area will be primarily one of providing a skeleton network of stations that serve to tie the individual state monitoring systems together into one vast nationwide pollution surveillance network. Plans call for full implementation of the network by fiscal year 1975.

### Water quality measurements

The number of stations in the network does not, alone, determine the adequacy of the network. The number and types of water quality indices measured at each station and frequency of their measurement are very important considerations. Since the choices of parameter coverage and monitoring frequencies have significant bearing on network operating cost, they should be selected with care and periodically adjusted as warranted.

The water quality indices selected for evaluation at a given station will be determined largely by the water quality criteria listed in the standards applicable to that station (see table on page 116). Certain indices, in addition to those listed in the standards, may be measured in situations where they are

considered to be of importance in establishing baseline water quality and determining trends that are required to evaluate compliance with the nondegradation clause in the water quality standards. There are still other indices that must be measured on occasion which are implicit in the standards, although not specifically listed. To thoroughly evaluate compliance with the requirement in all standards that waters be free of toxic substances, for example, would require the routine conduct of numerous tests on all samples for specific compounds known to be toxic to man, animals, and aquatic life. Obviously, the cost of doing this on even a small number of samples would be prohibitive. The plan is to evaluate all actual and potential sources of pollution in the vicinity of each monitoring station to determine which, if any, toxic substances are likely to be present, and to test for those only.

The optimum sampling frequency required to characterize water quality at a given station depends on the variability in the indices of interest. In a statistical sense, there is no answer to the question of how many samples are needed in a given time period, without foreknowledge of the variability of the constituents to be measured. Thus, a rather arbitrary frequency must be set initially for each station and adjusted at a later date, after sufficient data have been collected to make an evaluation.

To evaluate compliance with water quality standards, extreme, not average, values are of primary concern. Thus, to be *absolutely* certain that a violation has not occurred would require *continuous* sampling at each station and evaluation of *all* indices on each sample. Again, the cost of this at even a few stations would be prohibitive.

A practical approach—and one that will be followed—is to assign arbitrary sampling frequencies at each site, based on:

• The particular index of quality requiring the most frequent evaluation.

• The importance of the water uses being protected at the location and the impact of a violation on those uses.

• The potential for brief violations to occur.

Thus, samples will be collected most frequently at a station located in waters serving, for example, as a source of potable water supply for a large metropolitan area and downstream from a waste source with a potential for producing temporary violations. On the other hand, a station located in a zone subject only to minor pollution and designated for maintenance of rough fish and boating will be sampled less frequently.

Tests for those indices most subject to variation at a given station will be performed on every sample from that station. Tests for those indices less subject to variation and (or) more expensive to evaluate will be performed on alternate samples or at some other reduced frequency. Thus, the frequency of monitoring for a given water quality index may vary from continuous to four times per year, depending on the location. The frequency of monitoring at a given station will vary among the parameters being evaluated.

The proposed monitoring frequencies as well as station coverage are recognized to be near minimal. The approach being taken is to concentrate on placement and initiation of monitoring stations in the early stages of development of the surveillance network. Once the nucleus of stations is operational, greater effort will be directed toward improvement of water quality index coverage and optimization of monitoring frequencies. To assist in this endeavor, WQO has already initiated, by contract, the development of a systems analysis approach to the design of monitoring systems. This approach will include procedures for determining the optimum sampling frequency at a given location and will be applicable to streams, impoundments, and estuaries receiving any type of waste, whether from cattle feed lots, municipalities, industries, or cropland surface runoff.

Once the sampling frequency is established for a given station, the economics of manual vs. automated sample collection and (or) sensing techniques can be evaluated. In making such an evaluation, we must consider that:

• Automated devices do not completely replace field personnel since they require routine maintenance.

• Higher salaried personnel are generally required for maintenance of electronic sensors than for manual sample collection.

• Electronic sensors produce a wealth of data, which may require more elaborate and expensive data handling procedures.

Initially, perhaps as many as 15% of the stations will warrant the use of automated monitoring devices. As the population and economy continue to expand and the fixed supply of water is increasingly reused, greater and greater reliance will be placed on automated water quality sensing systems to manage water quality adequately.

To ensure the reliability of data collected by the coordinated surveillance network, an analytical quality control program will become an integral part of the overall system. All cooperating agencies will be expected to participate in such a program. The success of the pollution control effort will rest, to a great extent, upon the comparability and reliability of the information provided by the various participating agencies.

Collection of adequate quantities of accurate data is only part of the task at hand. To achieve the stated objectives, the mass of raw data received must be evaluated in a timely fashion and results made immediately available in a usable form to participating agencies for appropriate follow-up action. WQO's computerized data storage and retrieval system, STORET, is being used to meet these data handling requirements.

The STORET system, operational since 1964, utilizes many computer terminals throughout the country, all linked to a central computer. Data from four federal and about 25 state agencies are stored routinely through the remote terminals. To date, data from 30,000 stations have been stored in the central computer. The STORET system has the capability to provide a variety of statistical summaries on the stored data. When fully implemented, the STORET system will also allow comparisons, by computer, of water quality standards and the data so that apparent violations can be quickly identified for appropriate action. The system will have the capability to provide an even wider range of statistical analyses on the stored data. A variety of output formats for the raw data and statistical summaries, utilizing printers, magnetic tape, X-Y plotters, and cathode ray tubes, will be offered.

### R&D needs

In the future, water resources must be managed more adequately. The ability to acquire timely information to make the necessary operating decisions on a day-to-day basis will require more

sophisticated water quality sensing techniques and instruments than those now in use. Areas in which further research and development are needed to provide the necessary water quality management tools include:

• Adaptation of scientific management techniques for use in optimizing surveillance system designs.

• Development of automated instrumentation (portable and fixed) that is capable, with minimal maintenance, of accurately measuring a wide variety of water quality indices over long periods, and telemetering the data to a central location.

• Development of aerial noncontact sensing techniques for broad-scale evaluation of water quality conditions over vast geographical areas.

• Development of new, more inclusive water quality indices that can better lend themselves to automated sensing techniques.

Without these new techniques and systems for acquiring water data, true water quality management programs will not become a reality. Effort is presently underway in both WQO and the private sector to develop the much needed tools for water quality management. But this effort must be intensified if we, as a nation, wish to meet the challenges of tomorrow.

**William T. Sayers** *is Plans Evaluation Officer, Planning Branch, Water Quality Office (formerly* FWQA), *Environmental Protection Agency, Washington, D.C. Mr. Sayers, a C.E. (1958) and M.S. (1959) graduate of the University of Cincinnati, entered government service with the U.S. Public Health Service in 1959. He worked in Cincinnati and Dallas with the* USPHS *and was working in Minneapolis when his division became the* FWPCA. *When this article was written, Mr. Sayers was Chief, Water Quality Surveillance Systems Section,* FWQA. *The article is based on a presentation given to the* ACS *Division of Water, Air and Waste Chemistry at the Chicago National Meeting, September 1970.*

# Conductivity measurements monitor waste streams

*Reprinted from* ENVIRON. SCI. TECHNOL., 4, 116 (February 1970)

**P. A. Corrigan, V. E. Lyons, G. D. Barnes, and F. G. Hall**

*Tennessee Valley Authority, Muscle Shoals, Ala. 35660*

The seriousness of water pollution problems generally is recognized, and it is well established that the most economical and positive control method is to prevent losses from occurring and entering plant effluents. Such losses usually can be stopped if their source is known; however, some sources are difficult to locate in a complex plant. A successful pollution control program may utilize all of the several methods available for monitoring waste streams. We shall discuss only one such procedure—electrolytic conductivity measurements. Conductivity measurements have been used for decades for process control, especially in steam power plants, but only recently have they been applied to pollution abatement. Waste water from many types of plants contains large amounts of such materials as phosphates, nitrates, ammonia, and potash. Solutions of these materials conduct electricity, thus making their presence, as well as their relative quantities, easily detectable. With continuous measurements, this information is immediately available and can be used as the basis for prompt corrective action.

Conventional methods of sampling and analyzing waste streams have inherent time lags which prevent the information from being useful for control purposes. But conductivity measurements can show exactly when a loss occurs, for how long, and can give an indication of its magnitude. Conductivity recorders strategically placed help operators locate leaks and spills rapidly, and the recorder chart is a valuable aid to the supervisor responsible for pollution control.

## Theory

The ability of a solution to transmit a current depends primarily upon the number of ions per cubic centimeter. Electrolytic conductivity—also called specific conductance—is defined as the reciprocal of the resistivity, in ohm-centimeters, of a 1 cm. cube of solution at a specified temperature. The units of measurement are reciprocal ohm-cm. or mho/cm. In waste control work, the conductivities encountered are generally one millionth of this unit, or in the micromho/cm. range. For example, pure water has a conductivity of 0.055 micromho/cm., and a good quality raw water about 100-200 micromhos/cm. In contrast, seawater has a conductivity of about 50,000 micromhos/cm., and 30% sulfuric acid about 1,000,000 micromhos/cm.

An alternating current Wheatstone bridge is used widely in measuring electrolytic conductivity. The use of alternating current and electrodes coated with platinum black virtually eliminates polarization effects. In the bridge circuit, the input current is divided proportionally by the resistances of two halves of the bridge. Since all resistances are fixed except for the resistance of the conductivity cell, the cell's total resistance determines the current division. When the conductivity cell or probe is immersed in a conducting solution, there is a temporary change in its resistance which creates a potential difference across the bridge until the circuit is balanced. In the simplest conductivity meter, a variable resistance is manually repositioned to restore the balance, and this gives a measure of the change in conductivity of the solution being tested. In transmitting type meters, the change in voltage drop across the variable resistor will cause a change in voltage to a high impedance amplifier. The amplifier then sends a signal to the balancing motor proportional to the voltage change. The balancing motor mechanically repositions the sliding contact on the variable resistor, thereby causing the voltage to the amplifier to return to its original balanced condition. As the balancing motor moves, it also positions the sliding contact of a resistor in the transmitter circuit. This provides a voltage

signal output from the transmitter directly proportional to the change in conductivity of the sample cell.

### Commercial instruments

Most standard industrial conductivity instruments use a bridge voltage that ranges from a few thousandths of a volt to 50 volts. Theoretically, the Wheatstone bridge can be used with any voltage. High voltages increase the sensitivity and accuracy of the Wheatstone bridge; however, instrument manufacturers prefer to design their units for commonly used voltages to keep the cost low and to ensure the safety of operating and maintenance personnel. Most industrial users are satisfied with an accuracy of 1-2% of the full-scale range, and the voltage is chosen to provide this accuracy. Our work at Tennessee Valley Authority (TVA) used a 110 volt supply at 60 cycles per second for the instrument and 3-10 volts for the bridge.

The electrolytic conductivity of aqueous solutions usually increases with temperature at a rate of about 0.5-3% per degree Centigrade. If measurements are made at temperatures other than 25° C., the instrument circuits usually are compensated so the readings will indicate the conductivity at the reference temperature (25° C.). This is accomplished by incorporating into the Wheatstone bridge circuit a resistor which can be changed with temperature. This resistor can be either a calibrated rheostat for manual adjustment or a thermistor with the proper temperature coefficient combined with fixed resistors.

In a typical conductivity cell, the electrodes are annular bands surrounding the liquid flow opening and recessed in an insulating plastic housing. Electrode spacing is determined by the instrument design. When the cell is immersed, current flows through liquid between the electrodes and completes the Wheatstone bridge circuit.

Each cell has a cell constant that depends on the exposed electrode area, the spacing, and the material and shape of its electrodes. If the total resistance is kept in the range of 50-100,000 ohms by the proper choice of cell constant, standard low-cost components can be used to make the Wheatstone bridge. If a cell with a constant of 0.01/cm. is used to measure the conductivity of distilled water, the total resistance is $1,000,000 \times 0.01$, or 10,000 ohms, a value within the range of 50-100,000 ohms of less expensive common components. Cells with constants ranging from 0.001/cm. to 100/cm. are commercially available. Therefore, when a conductivity cell is being bought, it is important to know the micromho/cm. conductivity range of the liquid to be measured so that the cell obtained will give a total resistance within the range of 50-100,000

ohms. Conductivity indicators, transmitters, and recorders are available from most large instrument manufacturers, but only one or two U.S. companies manufacture conductivity cells in quantity.

Except for the cell, the components of conductivity instruments are common, simple parts used in many types of instruments. The components of a Wheatstone bridge are resistors, which seldom fail; other components are standard circuits which have been thoroughly engineered for long service life. Simple, portable conductivity instruments with manual temperature compensation cost about $200, and semiportable recorders with automatic temperature compensation run to $1000. Permanent installation using multipoint recorders are usually less expensive than those which use an equivalent number of semiportable recorders.

### Solution conductivities

The conductivity of a stream will vary with the rate of material loss, the kind of material, and the stream flow. A five g.p.m. flow of contaminated water might have a conductivity in excess of 10,000 micromhos/cm. and actually be carrying less contamination than a large flow (2000 g.p.m.) which might show a barely perceptible increase in conductivity over that of uncontaminated water. Therefore, before any conductivity instruments are used, a laboratory study should be made to determine the normal conductivity of the raw water used and how its conductivity changes as different amounts of the various possible contaminants are added. To be of any use, the instrument must be sensitive enough to detect that conductivity level which represents the threshold of pollution, and, often, this value is relatively low.

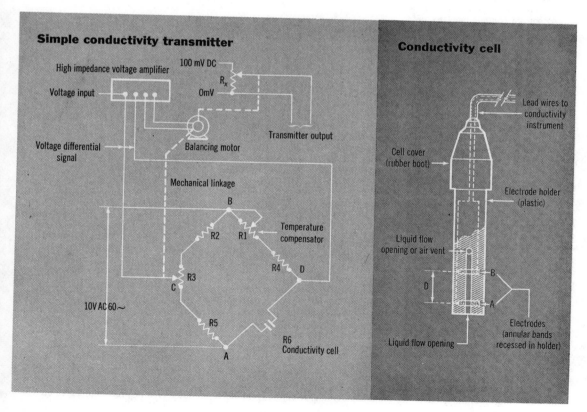

**Simple conductivity transmitter**

High impedance voltage amplifier
100 mV DC
$R_x$
Voltage input
0mV
Transmitter output
Voltage differential signal
Balancing motor
Mechanical linkage
B
R2  R1
Temperature compensator
R4  D
R3
C
10V AC 60~
R5
A
R6
Conductivity cell

**Conductivity cell**

Lead wires to conductivity instrument
Cell cover (rubber boot)
Electrode holder (plastic)
Liquid flow opening or air vent
B
D
A
Liquid flow opening
Electrodes (annular bands recessed in holder)

The electrolytic conductivities of most solutions are given in the literature, but these values are usually for pure compounds dissolved in distilled water. In manufacturing plants, uncontaminated raw water has an appreciable background conductivity, and the materials which may contaminate it are not pure. For these reasons, each plant must prepare its own conductivity vs. concentration curves, using its raw or treated water to prepare the solutions to be measured.

Conductivity curves may be prepared by successively diluting a portion of a stock solution of known compositions with raw water and measuring the conductivity of the solution after each dilution.

Fertilizer materials made at TVA's National Fertilizer Development Center include phosphoric and nitric acids, anhydrous ammonia, diammonium phosphate, and several solid and fluid fertilizers containing ammonia, nitrate, and phosphate in varying proportions. In addition, there are several process or waste materials such as sulfuric acid and fluorine compounds. These materials can contaminate waste streams either singly or in combination. In several manufacturing operations, waste streams can be contaminated by only one material. In that case, the concentration of the contaminant can be estimated reliably from the proper conductivity vs. concentration curve.

Most of the conductivity curves for possible contaminants at TVA plants are straight lines or uninflected curves. An inflection in the ammonia curve occurs at a concentration of about 4% $NH_4OH$, but this is unimportant in detecting contamination of raw water because such contamination undoubtedly would have been detected before the concentration reached this value.

An inflection also occurs in the phosphoric acid curve between 0-75 p.p.m. of $P_2O_5$. This can complicate leak detection. The literature reports the conductivity inflection for ammonia found in the TVA work and an inflection for phosphoric acid at a concentration between 50-70% $H_3PO_4$, but no inflection has been reported for concentrations between 0-75 p.p.m. $P_2O_5$. This inflection is possibly caused by a reaction of the phosphoric acid with some constituent in water used at TVA.

### Types of meters

Portable conductivity instruments weigh about four pounds, are dry cell

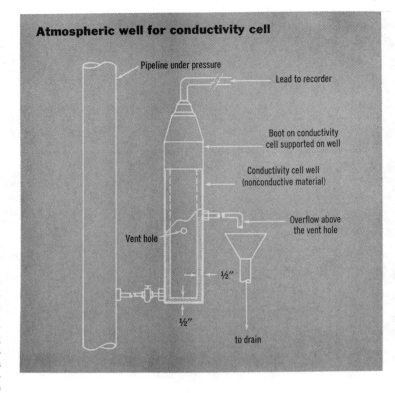

**Atmospheric well for conductivity cell**

Pipeline under pressure

Lead to recorder

Boot on conductivity cell supported on well

Conductivity cell well (nonconductive material)

Overflow above the vent hole

Vent hole

½″

½″

to drain

powered, and utilize a center reading null meter or an electron ray eye tube as balance indicators. A 25 foot cell cable is a convenient length, although longer cables are available. If the cable length is changed significantly on any instrument, the bridge must be recalibrated. The instruments normally use manual temperature compensation, but automatic compensation can be obtained.

Semiportable instruments weigh about 25 pounds, are battery operated, splashproof, and have a circular chart recorder. They are designed to operate on two 6 volt lantern batteries and, thus, can be portable. Our experience has shown that the lantern batteries have a short service life and usually are replaced with a 12 volt storage battery in a protective case. This makes the unit semiportable, requiring a car or truck for movement. It is worthwhile to purchase an instrument model which has a watertight windowed door. In damp, corrosive atmospheres, a pullover shroud of pliable transparent plastic will protect the aluminum case and door hinges and still permit the chart to be inspected without exposing the case or the electronic components to the atmosphere. If exposed, the instru-

ment should be supported above the ground against electrical grounding. The cell cable should be the maximum length obtainable, and automatic temperature compensation should be specified if the streams to be monitored routinely vary in temperature more than about 10° F.

Semiportable conductivity recorders are useful for preliminary surveys of waste systems and for troubleshooting. In addition, they are quite valuable in determining whether waste discharges are cyclic, sporadic, or continuous.

The semiportable instrument also is valuable when used as part of an operating plant. It can be quickly and inexpensively converted to operate on a 110 volt circuit as a permanent installation at a strategic point in the plant. It can be reconverted easily to battery operation for troubleshooting.

Temporary installations are most useful in making surveys of chemical sewers and drains to determine in which general areas of the plant pollutants are being discharged, and to detect sources of unsuspected contamination. Permanent installations usually are restricted to those plant areas where large amounts of possible pollutants are stored or processed. These installations are designed for immediate detection of spills and consist of conductivity cells located in surface drainage ditches from storage tank areas (with each cell monitoring only a few tanks), and cells located in the discharge lines for cooling water from heat exchangers, reactors, and other units. Each cell is connected to a recorder, usually the multipoint type, equipped with an audible alarm.

### Cell choice

The type of cell chosen and its installation for each application are critical if contamination is to be detected reliably. For use in surface drainage ditches where the effluent may contain a large proportion of suspended solids, especially during rains, the cell should be installed so as to minimize deposition of solids in the cell chamber, and should be readily removable for cleaning. This can be done with a settling chamber immediately upstream from the cell and by installing the cell facing the flow with open end pointing downward. To assure a flow through the cell when ditch flows are low, the ditch should be dammed and the cell installed through the dam an inch below the top with the vent below the dam. The cell should be of a type that can be cleaned easily and thoroughly without affecting the electrodes. Cells of one piece epoxy construction with platinum ring electrodes flush with the axial bore are rugged and dependable and

can be cleaned easily with a soft brush.

To monitor flows under pressure, it is best to bring a sample of the flow to a conductivity well containing the measuring cell at atmospheric pressure rather than to put the cell in the line under pressure. Inspection, calibration, and cleaning of a cell in an atmospheric well are much easier than for one installed in a pipeline.

Conductivity cells installed in drain lines have to be carefully located or suction and venturi effects may prevent a sample from entering the cell.

### Location

Efficiency of the cell system depends heavily on proper location. The cells should be located in branch streams rather than in the main stream below the entering side streams; this makes the system more sensitive and pinpoints the source of contamination better. The cells should be located close to the points of potential contamination, but far enough downstream so that the contaminant will be thoroughly mixed with the waste water. Otherwise, the contaminant may slip by the cell undetected, especially in waste water streams where the flow tends to be laminar. Under laminar conditions, water soluble liquids having a specific gravity much greater than one (such as superphosphoric acid and concentrated fluid fertilizers) tend to travel along the bottom of sewers before becoming mixed well enough for detection.

If the water overflows from a large basin or a jacketed vessel over a weir in one side, the flow at each end of the weir will represent only the water coming from that side of the basin or jacket. The conductivity cell must be located in the jacket drainpipe at a point where these essentially separate streams are mixed, or a leak may go undetected.

Frequently, it is difficult to properly locate a conductivity cell in such drains. In this case, it may be expedient to install two conductivity cells wired in parallel in the jacket, one on each side of the discharge weir. This will result in the recorder indicating greater background conductivity for that point than one cell would indicate, but the response to contamination also will be greater. This arrangement will permit the detection of leaks into either half of the jacket. If the cells are wired in series, the indicated background and response will be less than that obtained from one cell. All cells should be located where they can be inspected and cleaned safely and conveniently.

A permanent installation usually is designed to serve a plant area where several points of possible contamination are relatively close. At TVA, the installation usually is limited to monitoring about 12 points, one of which is uncontaminated raw water to provide background measurements or baseline conductivity. Deviation from the baseline by any of the other points indicates contamination, except for small deviations that occur because of differences in nominal cell constants. Temperature differences in streams being monitored also will produce variations unless compensation is provided.

### TVA's installation

A permanent conductivity installation is installed in TVA's fluid fertilizer manufacturing area where superphosphoric acid and ammonia gas are combined in a reactor to produce fluid fertilizer. The heat generated is removed by recycling the hot liquid through a water-cooled shell-and-tube heat exchanger which cools it further. The area contains storage tanks for phosphoric acid and the fluid fertilizer product. There is no storage for ammonia, since it is piped into the area.

The procedure we used to design the permanent conductivity installation in the area was as follows:

• The **optimum number and location of the detection cells** were determined. It was desirable to monitor small flows

**Conductivity of H₃PO₄ in raw water**

(Graph: y-axis labeled "p.p.m. P₂O₅" ranging from 0 to 350; x-axis labeled "Net conductivity (micromhos/cm.)" ranging from -50 to 450.)

(less than 100 g.p.m.) so that small losses would cause high concentrations and, therefore, easily detected conductivities. A survey of the area showed that all points of possible loss could be monitored by installing detection cells at 10 locations; two were in the outlet water lines of the two heat exchangers, and eight were in surface drains. An eleventh cell was installed in a raw water stream to provide baseline information.

• **Normal and maximum expected flow rates** were determined for each stream to be monitored. Use of rainfall data enabled prediction of the flows in outside drainage ditches.

• The **level of contamination** to be detected was determined. In this area, phosphoric acid and mixtures of ammonia and phosphoric acid were the only possible contaminants. We decided that each of the 10 installations in the fluid fertilizer area should detect $P_2O_5$ loss at the rate of 200 pounds per day. The nitrogen loss would be less, because the $N:P_2O_5$ ratio of the fertilizer being manufactured was about 1:3.

• The **net conductivity increase** was calculated for each stream by use of these data, and the conductivity vs. concentration relationships previously determined for the contaminants involved.

• The **type of instrumentation required** was selected. Each of the eight streams in the surface drainage ditches had high conductivity ratios and remained within 10° F. of raw water temperature. Cells in these streams were connected to a multipoint recorder equipped with one manual temperature compensator that provided equal compensation for all cells. Cooling water leaving the two shell-and-tube heat exchangers required more precise instrumentation.

Usual variations of 5-10 micromhos/cm. in the conductivity of the raw water would represent a significant change from the small net conductivities of 20 and 105 to be detected. The 35° F. temperature difference between the inlet and outlet water at each exchanger also required compensation. The temperature problem was eliminated by heating a sample stream of the inlet water to the temperature of the outlet water by conducting the sample stream of inlet water through several feet of tubing wound around the outlet pipe and covered with heat transfer cement; the pipe and tubing then were overlaid with insulation.

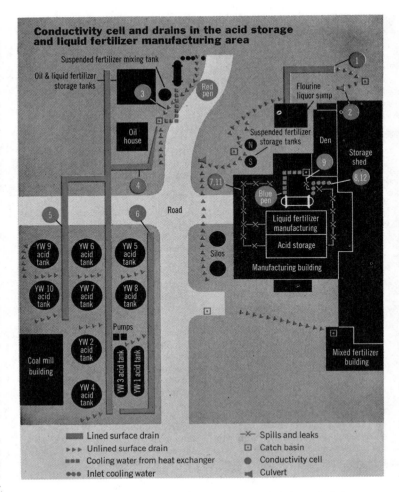

**Conductivity cell and drains in the acid storage and liquid fertilizer manufacturing area**

Legend:
- Lined surface drain
- Unlined surface drain
- Cooling water from heat exchanger
- Inlet cooling water
- Spills and leaks
- Catch basin
- Conductivity cell
- Culvert

We did not use thermistor-type temperature compensation because it is impractical to insert both the second cell and thermistor in the Wheatstone bridge circuit. Variations in the conductivity of the raw water were compensated for by selecting an instrument which measures only the ratio of the conductivities of the inlet and outlet waters; this ratio will be one if there is no contamination of the outlet water and will be greater than one if contamination is present. Matched cells must be used because differences between cells can significantly decrease the sensitivity of the installation.

The conductivity ratio technique also can be used to monitor streams where phosphoric acid is the only possible contaminant and the net conductivity to be detected is in the inflected portion of the concentration vs. conductivity curve. In this case, contamination causes the conductivity ratio to be less than one.

The physical layout of the system was designed for efficient inspection and maintenance. All conductivity cells were made readily accessible and easily removable from their mountings for inspection and cleaning.

## Maintenance

To properly maintain a conductivity cell installation, each cell and recorder should be checked daily. The cells should be removed, checked, visually, and if necessary, cleaned. The cells may be carefully cleaned by using a soft brush. When the cell is removed from the monitored stream, the recorder should be checked to make sure that the indicated conductivity is zero. Once a week, or as often as necessary, each cell should be checked against a solution of known conductivity, such as raw water. Those cells which are located in surface drainage ditches should be inspected for stoppage at least once a shift by the chemical plant operator.

## Data for design of electrolytic conductivity type contamination detection system for fluid fertilizer manufacturing area

| Stream No. | Source of flow | Possible contaminant | Estimated flow, g.p.m. | Parts per million $P_2O_5$ to give 0.1 ton per day | Net conductivity to be detected micromhos/cm.[a] | Ratio, net conductivity of stream to conductivity of raw water[b] |
|---|---|---|---|---|---|---|
| 1 | Raw water (baseline) | — | — | — | — | 1 |
| 2 | Surface drainage north of building | Fluid fertilizer | 100 | 167 | 200 | 1.11 |
| 3 | Surface drainage north of building | Fluid fertilizer | 100 | 167 | 200 | 1.11 |
| 4 | Drainage from mixing tank area | Fluid fertilizer | 20 | 835 | 1200 | 6.67 |
| 5 | Drainage from west acid tanks | Phosphoric acid | 5 | 3335 | 5300 | 29.44 |
| 6 | Drainage from middle acid tanks | Phosphoric acid | 5 | 3335 | 5300 | 29.44 |
| 7 | Drainage from east acid tanks | Phosphoric acid | 10 | 1670 | 3100 | 17.22 |
| 8 | Drainage in west side of building | Fluid fertilizer | 10 | 1670 | 2200 | 12.22 |
| 9 | Drainage in east side of building | Fluid fertilizer | 10 | 1670 | 2200 | 12.22 |
| 10 | Water from primary cooler | Fluid fertilizer | 550 | 30 | 20 | 0.11 |
| 11 | Water from secondary cooler | Fluid fertilizer | 150 | 111 | 105 | 0.58 |

[a] At 25° C.
[b] Conductivity of raw water was 180 micromhos/cm. at 25° C.

They also should be inspected after each rain and cleaned if necessary.

One point on the multipoint recorder should be wired so that it gives a calibration check reading. Any deviation from this constant check reading indicates recorder malfunction.

At the TVA National Fertilizer Development Center, we found that the electroconductivity technique is valuable for control of material losses. Five permanent installations are in use, as well as five semiportable and two portable units, for monitoring intermittent loss source. By showing when and where losses are occurring, this equipment has greatly simplified execution of our pollution abatement program, and savings that result from rapid detection of losses of valuable materials quickly repay the cost of installation and surveillance.

**P. A. Corrigan** *is senior project leader, process engineering branch, Tennessee Valley Authority's division of chemical development. He received his B.S. from the University of Maine, and has been with TVA since 1940.*

**V. E. Lyons** *is a chemical engineer on the technical staff of TVA's division of chemical operations. He joined TVA in 1936 after receiving his B.S. from Georgia Institute of Technology.*

**G. D. Barnes** *was project leader, process engineering branch, TVA division of chemical development, prior to his death during preparation of this article. He received his B.S. from Auburn University (1941).*

**F. G. Hall** *is instrument engineer, design branch, TVA's division of chemical development, a position he has held for seven years. Previously, he was with Taylor Instrument Co. He received his B.S. from Georgia Institute of Technology.*

EPA's Office of Research and Monitoring
is conducting a nationwide survey of more
than 1100 lakes in order to answer . . .

# How fast are U.S. lakes aging and why?

*Reprinted from* ENVIRON. SCI. TECHNOL., **7**, 198 (March 1973)

The National Eutrophication Survey is the largest water field-sampling program in the history of the Environmental Protection Agency. Literally, as many as 200,000 water samples will be collected from about 1100 lakes in the U.S. According to the best information available today, there are some 3800 municipal sewage treatment plants that have an impact on these lakes, and the survey is the first nation-wide attempt to learn what effect they have on the aging of our lakes. All told, the survey will cost about $5 million over four years, with all samples to be collected by the end of calendar year 1975.

Can you recall all the controversy over phosphate removal from detergents that gained national headlines a few years ago? Do you remember the joint CEQ, EPA, FDA, HEW position on the matter? Members of the Washington press corps certainly do. It was at that time, September 15, 1971, that EPA administrator William Ruckelshaus committed his agency to the fact-finding task of identifying those lakes in the U.S. that were being threatened by the addition of phosphorus-containing materials from municipal sewage treatment plants. Although the lakes survey was not officially announced until three months later, even then, on December 15, 1971, the agency had neither the manpower nor the funds to begin the program.

Within the early months of 1972 plans for the survey were generated within the EPA Office of Research and Monitoring and then submitted to the Office of Management and Budget (OMB). EPA received the go-ahead sign on March 28 for funds and 50 personnel. Of the 50 people approved for EPA, two are assigned to the Office of Water Programs, three to the Office of Research and Monitoring at EPA headquarters in Washington, D.C., 22 at the Corvallis National Environmental Research Center (NERC), leaving 23 persons at the Las Vegas NERC, including 14 members of the helicopter team that surveys and collects samples.

Basically, the major objective of this ambitious survey is to determine, once and for all, if increased phos-

phate removal at sewage treatment plants would significantly improve the water quality of a lake or impoundment that receives sewage effluent. But before this fundamental and legitimate question can be answered, three tasks must be performed:
- to determine the trophic conditions of the lakes and impoundments
- to determine the growth-controlling nutrient in each lake
- to identify the influence of nutrients on each lake

Obviously the logistics of the survey and sampling strategy are staggering. Robert Payne, EPA's program element manager for the survey says, "They could not all be done immediately." Therefore, the strategy calls for dividing the lakes in the U.S. into three distinct geographic areas —one for each of the three calendar years 1972, 1973, and 1974. The initial survey operation for 1972 was limited to 10 states in the northeast —New York, Michigan, Wisconsin, Minnesota, and the six New England states. Within this area, efforts on some 220 lakes were begun in 1972, with results to be in final form by this June.

In the 1973 survey area, now getting under way, 300–350 lakes in the remaining 17 states east of the Mississippi River will be surveyed. At press time, eight of the 17—Ala-

bama, Florida, Georgia, Kentucky, Mississippi, North Carolina, South Carolina, and Tennessee—had their plans ready to go.

Then beginning in 1974, the survey moves west of the Mississippi River and although the schedule is not final, the remaining 21 of the contiguous 48 states will have their lakes and impoundments surveyed in the next year or two.

### How it's done

There are really several parts of the survey. Samples are collected by a helicopter team, National Guardsmen, and operators at municipal sewage treatment plants. Although most samples are analyzed at the Corvallis NERC, and some work is done at the Las Vegas NERC, the number of analyses is still staggering. In fact, part of the information collected on the first 220 lakes fills a computer printout book that is more than half a foot thick!

To undertake the first task—to determine the trophic condition of the lake—the survey has three Huey helicopters on loan from the Department of the Army. Each is equipped with sophisticated water-sampling and analytical equipment.

The helicopter team drops an instrument package into the lake, and the instruments record depth, temperature, pH, dissolved oxygen, and conductivity. Data on these parameters are relayed to the helicopter where an on-board analog recording device automatically plots any one parameter against any other. In this way, the plots of temperature vs. depth, dissolved oxygen vs. depth, and the like are generated. On the average, about a half dozen sites are sampled on each lake, and three complete circuits are made on each site for any one lake. In addition, water samples can be taken at different depths in order to get a composite water sample to send back to the Las Vegas NERC for additional analyses—usually total phosphorus, dissolved phosphorus, nitrogen as ammonia, nitrate-nitrite, Kjeldahl, and alkalinity.

A remarkable fact about this helicopter team of 14 dedicated personnel, including helicopter pilots, lim-

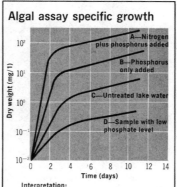

### Algal assay specific growth

A—Nitrogen plus phosphorus added

B—Phosphorus only added

C—Untreated lake water

D—Sample with low phosphate level

Dry weight (mg/1)

$10^2$

$10^1$

$10^0$

$10^{-1}$

$10^{-2}$

0   2   4   6   8   10   12   14

Time (days)

**Interpretation:**
If line B is close to line C then phosphate does not limit growth. If line B is well above line C, then phosphate does limit growth. Line D suggests amount of growth expected if phosphate is removed from lake.

*Algae. In 1972, EPA's helicopter team surveyed 220 lakes; growth is readily visible on Lake Tahoe, Calif.*

nologists (lake scientists), and others is that in a mere 35 days after OMB approved the lake survey program (March 28), the team sampled its first lake in New York state on May 2, 1972! Then from early May through early November (the team even took off in a snow storm in Michigan) of last year, the team collected samples and data from dawn to dusk six to seven days a week and sometimes holidays. In recognition of their contribution to the EPA effort, the team members each received a gold group medal award at EPA's second annual award ceremony. Leslie Dunn at the Las Vegas NERC was responsible for coordinating much of the initial effort.

Basically, the team surveyed the first 220 lakes, obtaining the data necessary to determine the trophic state of the lakes. This year, the helicopter survey team will cover 300–350 lakes in the Southeast, and will wing its way westward sometime early next year.

### The second element

The second charge—to identify the controlling nutrients in each lake—is fulfilled by the 22-member team at the Corvallis NERC. One of the samples collected by the helicopter is a large composite sample of water from each lake. The composite is made up of an integrated sample from different depths, at the various half dozen sites, taken during the spring- or fall-sampling circuit.

At Corvallis, the composite is analyzed by the algal assay procedure that was developed jointly by EPA and The Soap and Detergent Association. The composite is split into a number of samples, each of which is inoculated with a test species of algae. Phosphate is added to one sample, phosphate plus nitrogenous materials are added to another, a third acts as a control (no additions), and the fourth one is stripped of any

phosphorus materials by passage through an alum column. Then each of the four cultures is allowed to grow under controlled conditions in the laboratory for a period of 14 days. From the curves (see art) one can readily see the effect that phosphate has on the ability of that particular lake to grow algae.

### Other information needs

As well as checking on nutrients which enter streams via sewage treatment plant effluents, the survey team also wants to determine how other nutrients are entering U.S. lakes. The team does this in two ways: one involves the National Guard and another centers around plant operators at the sewage treatment plants affecting the lakes.

The first program involves sampling the mouths of the major streams entering the lake at a monthly frequency for one year. Additional sampling (twice monthly) is performed during the two months of highest runoff. This, too, is an ambitious sampling plan since there are, on the average, five to seven streams coming into each lake.

The National Guard play a key role here. Endorsement to use the Guard on a state by state basis was obtained from the Office of the Secretary of Defense and the National Guard Bureau. Their first duty, a pilot test program, occurred in Vermont last July 15, and later, approval for their use became possible in the first 10 states.

The Guard volunteers are collecting samples from streams entering each of the 220 lakes in the first 10-state area of 1972. All told, more than 1100 water samples are being collected each month by the National Guardsmen. Again, the samples are being analyzed by continuous techniques for total phosphorus and ortho-phosphate and nitrogen—ammonia, nitrate-nitrite, and Kjeldahl.

Last month, when the 1973 survey sampling got under way in 17 states east of the Mississippi River, it was anticipated that the National Guard would help with the survey, although approval for their use must be given on a state by state basis. Nevertheless, sampling started in Florida last month; and the 17-state program is scheduled to end in the spring of 1974. Six states—Alabama, Florida, Georgia, Kentucky, North Carolina, and South Carolina—have already endorsed the use of their National Guard to participate in the sampling.

Another series of samples is being collected by plant operators of the effluent from waste treatment plants on the particular lakes. This program has the endorsement and approval of the state water pollution control agency involved.

This year, all series of samples are being coordinated and scheduled so that sampling by the helicopter team, the National Guardsmen, and the plant operators is performed on a consistent time basis. In this way, the information is consistent for one growing season in the lake.

Despite the fact that sampling and analysis began in 1972, the results on the first 220 lakes will not appear before mid-June, shortly after completion of all sampling activities.

### How will these results be used?

As the final data are analyzed and interpreted, survey personnel will prepare comprehensive reports of findings lake by lake. These results will be received by state environmental authorities and will serve as the basic input for information of joint state-federal control strategies for lakes found to have serious water quality problems caused by municipal effluents. EPA's Office of Water Programs will aid each state in the planning and financing of upgraded waste treatment facilities where warranted.                                        **SSM**

**277**

**Richard D. Grundy**

*Professional Staff*
*U.S. Senate Committee on Public Works*
*Washington, D.C. 20510*

*Reprinted from* ENVIRON. SCI. TECHNOL., **5**,
1184 (December 1971)

# Strategies for control o

Eutrophication is a natural process which involves an increase in the biologic productivity of a body of water as a result of nutrient enrichment from natural sources. Environmental concern is for those instances where the natural aquatic growth processes are accelerated or increased. Under man's influence, excessive amounts of nutrients can enter an aquatic ecosystem, accelerating the eutrophication processes—this is known as "cultural eutrophication."

In the early stages, cultural euthrophication may produce beneficial effects—e.g., increased fish productivity. However, in the advanced stages, cultural eutrophication can eventually lower dissolved oxygen levels, interfere with recreational uses of water, affect drinking water taste, and result in summer blooms of undesirable blue-green algae.

Although all nutrient discharges into the aquatic environment contribute to the reservoir of material available to support aquatic productivity, current control efforts tend to emphasize nutrients limiting aquatic productivity, in particular, phosphorus. Although carbon and nitrogen also have been suggested as "limiting" nutrients, both have an atmospheric reservoir available to the aquatic environment. Phosphorus, however, has a sedimentary cycle, with no atmospheric reservoir.

This distinction between nutrients with, and without, a gaseous phase is basic to formulating a strategy for the control of cultural eutrophication. Nutrients with a gaseous phase (for example, carbon, nitrogen, oxygen, and hydrogen) have more or less perfect or complete biogeochemical cycles in which there is a continuous transfer of nutrients between the living (organic) and nonliving (inorganic) components of ecosystems. Nutrients without a prom-inent gaseous phase (for example, phosphorus, calcium, silica, and potassium) are circulated to a greater extent by processes such as erosion, sedimentation, and biological activity.

Controlling phosphorus alone and detergent phosphates, in particular, may retard cultural eutrophication where phosphorus limits aquatic productivity. However, long-term control strategies must reflect regional variations in limiting nutrients and other factors contributing to aquatic productivity. Regional variations are experienced in the contributions of nutrients entering the aquatic environment from municipal waste water, urban runoff, agricultural runoff from fertilized fields and livestock feedlots, and erosion. Agricultural runoff exhibits large seasonal variations.

### Phosphorus sources

Detergents are unique among consumer products as a source of nutrients. Nationally they represent about 5.5 billion lb annually, approximately 20 lb for every man, woman, and child. The phosphate fraction amounts to 2.2 billion lb or 30–40% of all the phosphorus entering the aquatic environment. The environmental implications of such quantities cannot be discounted; however, the potential nutrient properties and public health and environmental implications of alternative ingredients also must be considered. For lack of regional data, national estimates on the respective contributions of natural and man-generated phosphorus entering the aquatic environment are presented in the table at left.

## Phosphorus and nitrogen entering the environment

| Source | PHOSPHORUS | | NITROGEN | |
|---|---|---|---|---|
| | Millions of pounds/yr | % | Millions of pounds/yr | % |
| Natural | 245–711 | 64.41 | 1,035–4,210 | 21.51 |
| Man-generated | 686–1,015 | 74–57 | 3,990 | 79–46 |
| Domestic sewage | (387–446) | | (1,330) | |
| Runoff from | | | | |
| Urban land | (19) | | (200) | |
| Cultivated land | (110–380) | | (2,040) | |
| Livestock areas | (170) | | (420) | |
| Total | 931–1,726 | 100 | 5,025–8,200 | 100 |

Source: Ferguson 1968

Long-term solutions to cultural eutrophication will need to go beyond control of detergent phosphates and even beyond the control of phosphates in general

# nan-made eutrophication

Principal sources of phosphorus in domestic sewage are human excrement and detergents. The fraction from human excreta varies with diet, averaging 1.0 to 1.2 lb/person annually. The contribution from detergents has increased with expanded use to approximately 3.0 to 3.3 lb/person.

Significantly, some 40% of the increase in phosphorus in municipal waste waters between 1957 and 1968 resulted from an increase in the number of people discharging into municipal treatment systems rather than increased phosphorus usage. Obviously, water pollution control strategies could be more comprehensive.

Industrial utilization of detergents, processing phosphate and phosphorus products, and natural sources (such as decomposition products, bottom sediments, and erosion) also contribute to aquatic nutrient availability; however, there is insufficient data to estimate the quantities involved.

Although nitrogen has a gaseous phase, it also is worthy of consideration in control strategies, particularly where estuaries are involved. Data in the table on the previous page indicate that as much as 80% of the nitrogen entering the aquatic environment nationally comes from man-generated sources. Because of man's tremendous impact, an integrated control strategy for cultural eutrophication should consider both phosphorus and nitrogen.

## Scientific arguments

In an undisturbed lake, eutrophication is a slow, natural aging process, eventu-ally terminating in the disappearance of the lake itself. Because a lake acts as a settling basin, the input of organic and inorganic matter directly affects its productivity and life-span. Under man's influence, this process can be vastly accelerated by the discharge of excessive amounts of nutrients.

Some 15 to 20 nutrients play an essential part in eutrophication. These nutrients include carbon, nitrogen, phosphorus, silica, iron, potassium, and trace metals introduced from natural sources or as a result of man's activities. Vitamins also may act as growth stimulants after being introduced by man's activities or biological synthesis within the receiving waters themselves. Physical characteristics of a water body, such as increased light penetration and temperature rises, also affect aquatic productivity where appropriate nutrients are available.

Continuous transfers of nutrients between the living and nonliving components of an ecosystem are known as biogeochemical cycling. Nutrients with a gaseous phase such as nitrogen, carbon, oxygen, and hydrogen are important in this continuous flow between inorganic and organic states. These nutrients are considered to have "complete" cycles.

In contrast, nutrients without a prominent gaseous phase, such as phosphorus, calcium, silicon, and potassium, have cycles which are less perfect or complete. These nutrients are naturally circulated by a "sedimentary cycle" involving erosion, sedimentation, and biological activity.

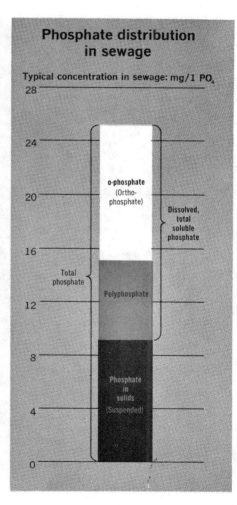

## Phosphate distribution in sewage

Typical concentration in sewage: mg/1 PO$_4$

Various compensating mechanisms tend to prevent major changes in the distribution, abundance, or availability of nutrients with a perfect cycle. Nutrients with a sedimentary cycle, however, are released into the aquatic environment more slowly by erosion, runoff, and physicochemical and biological processes such as the weathering of plant litter, soil minerals, and bedrock. They accumulate in deep lake and ocean sediments and are, for the most part, unavailable to support aquatic productivity.

Control strategies have focused on three nutrients—carbon, nitrogen, and phosphorus—as nutrients limiting eutrophication. Phosphorus, however, is believed to be the nutrient most frequently limiting aquatic productivity. It also has a sedimentary cycle and lacks the atmospheric source available to carbon and nitrogen. Therefore, it is considered the preferable nutrient for control.

Recent data for estuaries and oceans suggest that, although phosphate concentrations are very high, normally phosphorus does not limit productivity. Instead, nitrogen is apt to limit eutrophication. In such circumstances, replacement of detergent phosphates with a nitrogen-containing material might very well enhance eutrophication.

This does not mean that phosphorus could not be rendered limiting. High levels of phosphorus and carbon in estuaries result largely from river discharges. As upstream sources are brought under control, phosphorus may well become limiting.

Feasibility for controlling phosphorus is enhanced by, first, the high concentrations in domestic sewage and certain industrial waste waters, and, second, by phosphate's high reactivity and ease of precipitation by a variety of waste water treatment processes.

Phosphorus occurs in municipal waste waters in several chemical forms, including phosphates in suspended solids, polyphosphates, and orthophosphates (see figure, bottom right, p 1185).

The insoluble suspended solid fraction, human excreta and food solids, represents 30–40% of the phosphorus in municipal waste waters. Soluble polyphosphates are virtually all contributed by synthetic detergents. Soluble orthophosphates are mainly the degradation products of detergent polyphosphates, human wastes, and food solids containing phosphorus.

This distinction is significant because the suspended, insoluble forms of phosphorus are usually removed by primary or secondary treatment processes. The soluble fractions, however, are only partially removed in secondary treatment by bacterial action. The major unaffected fraction is discharged into receiving waters and available to support eutrophication, unless advanced waste

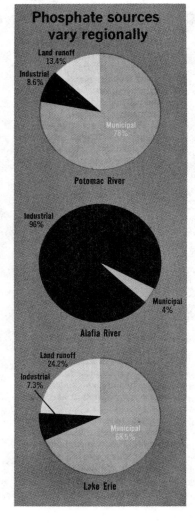

**Phosphate sources vary regionally**

Land runoff 13.4%
Industrial 8.6%
Municipal 78%

**Potomac River**

Industrial 96%
Municipal 4%

**Alafia River**

Land runoff 24.2%
Industrial 7.3%
Municipal 68.5%

**Lake Erie**

water treatment methods or land disposal are employed.

Typically, conventional secondary treatment effluent contains about 4000 parts per billion (ppb) of phosphates. This is at least 100 times the concentration for maximum rate of growth for the algal species *Chlorella pyrenoidosa*.

A dilution of greater than 100 would be required before such an effluent would have no perceptible affect on algal growth. Such dilution capabilities are rarely attainable in receiving waters in the United States.

Two complementary elements for inclusion in control strategies are advanced waste treatment and consumer product controls.

**Phosphate substitutes**

Since detergents amount to 30–40% of the phosphorus (as phosphates) in municipal sewage, they are often singled out for special attention. (Some six states and 47 communities have enacted legislation restricting phosphate levels in detergents. The majority restrict detergent phosphorus levels to 8.7% immediately and zero in 1972. Whether these restrictions can be met without substituting new public health or environmental problems remains to be seen.)

New and reformulated detergent products are being introduced so rapidly that the federal EPA recently announced suspension of its publication of data on detergent phosphorus content. There are potential human health hazards for some of the low-phosphate and nonphosphate detergent formulations. Surgeon General Jesse L. Steinfeld has advised "great care must be exercised to assure that the alternative does not create equal or greater hazards to the environment or to human health."

Some "low-phosphate" or "nonphosphate" detergents employ sodium silicate or sodium metasilicate in addition to carbonates or phosphates. The resultant products are often highly alkaline; some have required labeling under the Federal Hazardous Substances Act. Introduction of these materials under the guise of pollution control, without preassessment of their potential public health and environmental effects, may, in effect, perpetuate a fraud on the consumer.

## Estimates of national treatment expenditures for phosphates removal
### (millions of dollars)

**TOTAL TREATMENT COSTS FOR SELECTED UNIT TREATMENT (CENTS/1000 GAL)**

| %[a] | 5 Cents | 4 Cents | 3 Cents | 2.5 Cents | 2.0 Cents | 1.5 Cents |
|------|---------|---------|---------|-----------|-----------|-----------|
| 100  | 280     | 224     | 168     | 140       | 112       | 84        |
| 75   | 210     | 168     | 126     | 105       | 84        | 63        |
| 50   | 140     | 112     | 84      | 70        | 56        | 42        |
| 25   | 70      | 56      | 42      | 35        | 28        | 21        |

[a] Fraction of sewage requiring treatment

Potential public health effects identified for some highly alkaline detergents include severe damage to the throat, larynx, esophagus, or stomach and severe damage to the eyes or skin. Consumer product effects include loss of fabric strength and dye retention, loss of nonflammable characteristics for current fabrics, and corrosion of washing machines. The risks vary with alkalinity and the particular ingredient employed.

Prudent public health policy dictates that any regulatory action to ban phosphates in detergents consider the public health and environmental implications of alternative detergent formulations. Other control strategies are available to control or reverse cultural eutrophication—advanced waste water treatments, diversion, dilution, and land disposal.

### Regional differences

Phosphate input to U.S. waters from detergent sources varies regionally. Lake Washington, near Seattle, is an excellent example of a lake ecosystem where conventional secondary treatment of municipal waste waters was insufficient to control cultural eutrophication. In the mid 1960's, Seattle initiated total diversion of its treated municipal waste water around Lake Washington (although phosphate removal might, in fact, have been sufficient). Since then, cultural eutrophication in the lake apparently has reversed.

The Potomac River Basin exemplifies a river ecosystem where municipal waste waters are the primary source of phosphorus and other nutrients (see left). Approximately 65% of the waste water phosphorus is from detergent phosphates, or about 50% of the total phosphorus in the Potomac River Basin.

The Alafia River in Florida is an estuarine ecosystem where the mining industry is the primary source of phosphorus. The 14 phosphate processing plants located on the Alafia River discharge some 43,470 lb of phosphates (as phosphorus) daily into the Alafia River. Municipal and industrial waste water discharges from Tampa amount to 2700 lb of phosphorus. The overall detergent phosphorus contribution is about 4%.

Then there is Lake Erie, often cited as the worst example of cultural eutrophication. Detergent phosphates represent roughly 50 to 70% of the combined municipal and industrial sources of phosphorus. This amounts to 35 to 50% of all the phosphorus sources to Lake Erie, excluding that amount which comes from Lake Huron.

Several experts believe that control of detergent phosphates alone will probably produce no noticeable improvement in the existing conditions. Restoration of Lake Erie may very well require control of all sources of nutrients, including municipal and industrial wastes, agricultural runoff, and erosion, as well as of nutrients already in the lake. This may require harvesting of algae or treatment of lake waters in situ.

A recent five-year agreement on the Great Lakes, between Canada and the U.S., calls for accelerated municipal waste water treatment for phosphorus. With the existing level of detergent phosphorus in the U.S., 95% removal of phosphorus from municipal waste waters is necessary to achieve target loadings. Were detergent phosphorus levels to be lowered to 2–3%, around 80% removal of phosphorus would be required.

The Canadian government has adopted a dual approach combining a reduction of phosphates in detergents with phosphorus removal from municipal waste waters with a residual phosphate objective of "as near zero as possible." Detergent formulations in Canada are currently restricted to a maximum phosphate content of 20% (8.7% phosphorus).

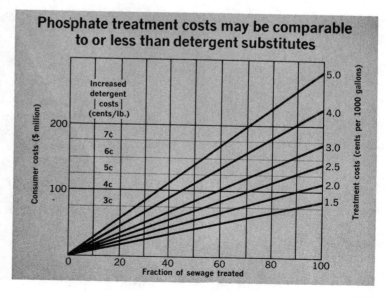

## Phosphate treatment costs may be comparable to or less than detergent substitutes

It was recognized that this action on detergents alone would have little impact in the critical Great Lakes areas. In other areas, the Canadian government believes it will make a quick and significant contribution to retarding eutrophication.

## Waste water management

An attractive alternative to banning detergent phosphorus is to reduce phosphate content to reflect regional water quality and wash water characteristics, and to combine this with waste water treatment for phosphorus. An EPA study will list those waters which have an actual eutrophication problem. Long-term solutions to cultural eutrophication may also require waste management practices which dispose of treated liquid effluents and sludges on the land.

Control of erosion and agricultural runoff must rely on improved soil conservation practices and the proper use of fertilizers. Waste management practices for feedlot wastes appear to be land disposal rather than waste treatment.

When waste water treatment is considered as a control alternative, the following technical points emerge:

• The economics of nutrient removal in general, and phosphorus removal in particular have been successfully demonstrated

• Chemical precipitation processes not only remove phosphorus but improve removal of BOD, toxicants, and other nutrients

• The amount of chemical required is frequently independent of the concentration of phosphorus in the waste water.

The cost for chemical methods applicable to existing primary and secondary treatment plants is comparable to anticipated increased detergent costs for formulating a nonphosphate detergent.

## Economics

An important consideration in any solution to cultural eutrophication is

## State and local detergent legislation on product restriction
(passed through 9/29/71)

| Area | % P | Date | No phosphate date |
|---|---|---|---|
| **New York State** | 8.7 | 1/1/72 | 6/1/73 |
| **Connecticut** | 8.7 | 2/1/72 | 6/30/73 |
| **Maine** | 8.7 | 6/1/72 | ... |
| **Florida** | | Enabling Act authorizing regulation of P content as of 12/31/72 | |
| **Indiana** (Suit filed 6/8/71) | | 12% "phosphate"—1/1/72 3% "phosphate"—1/1/73 | |
| **Minnesota** | | Enabling Act authorizing regulation of nutrient content, etc., effective on passage | |
| **Oregon** | | Packaging legislation— 9/30/71 | |
| **Chicago, III.** (Suit filed 5/4/71) | 8.7 | 2/1/71 | 7/1/72 |
| Hillside, III. | 8.7 | 2/1/71 | 7/1/72 |
| Elgin, III. | 8.7 | 7/1/71 | 7/1/72 |
| Aurora, III. | 8.7 | 7/1/71 | 7/1/72 |
| Park Forest, III. | 8.7 | 5/1/71 | 7/1/72 |
| Chicago, Heights, III. | 8.7 | 7/1/71 | 6/30/72 |
| Joliet, III. | 8.7 | 5/1/71 | 6/30/72 |
| Kankakee, III. | 8.7 | 10/1/71 | 1/1/73 |
| Niles, III. | 8.7 | 7/1/71 | 6/30/72 |
| Skokie, III. | 8.7 | 6/30/71 | 6/30/72 |
| Franklin Park, III. | 8.7 | 7/1/71 | 6/30/72 |
| Lombard, III. | 8.7 | 6/1/71 | 1/1/72 |
| Elmwood Park, III. | 8.7 | 6/1/71 | 12/31/71 |
| Harwood Heights, III. | 8.7 | 6/1/71 | 6/30/72 |
| **Lake County, III.** | 8.7 | 6/1/71 | 6/30/72 |
| **Laconia, N.H.** | 8.7 | 6/1/71 | 5/31/72 |
| **New Hampshire Lake area** (Nine cities in Carroll, Belknap Counties) | | | 6/1/71 |
| **Akron, Ohio** (Consent judgment accepted) | 8.7 | 8/6/71 | 7/1/72 |
| **Cleveland, Ohio area** | | | |
| Painesville | 8.7 | 7/1/71 | ... |
| Berea | 8.7 | 6/1/71 | 6/30/72 |
| Brook Park | 8.7 | 7/1/71 | 6/30/72 |
| Independence | 8.7 | 6/1/71 | 7/1/72 |
| Euclid | 8.7 | 8/1/71 | 6/30/72 |
| **Block Island, R.I.** | ... | ... | 4/1/71 |
| **Kissimmee, Fla.** | 8.7 | 1/1/72 | 1/1/72 |
| **Lake County, Fla.** (Consent judgment accepted) | 8.7 | 4/1/71 | 12/31/72 |

| Area | % P | Date | No phosphate date |
|---|---|---|---|
| **Dade County, Fla.** (Injunction against labeling till 8/1) (Suit filed to enjoin total P ban and labeling provisions) | 8.7 | 4/30/71 | 1/1/72 |
| **Cocoa Beach, Fla.** | 8.7 | 9/30/71 | |
| **Orange County, Fla.** | 8.7 | 5/31/71 | 1/1/72 |
| **Pinellas County (St. Pete), Fla.** (St. Petersburg itself excepted) | 8.7 | 9/1/71 | 1/1/72 |
| **Pembroke Pines, Fla.** (Enforcement temporarily suspended) | 8.7 | 6/19/71 | 1/1/72 |
| **Kennebunkport, Me.** | 14% phosphate | 6/2/71 | 1/1/72 |
| **Bridgton, Me.** | | | 6/1/71 |
| **Naples, Me.** | | | 7/1/71 |
| **Madison, Wis.** | 8.7 | 7/1/71 | (Law applies to city department purchases only) |
| **Erie County, N.Y.** | 8.7 | 4/30/71 | 1/1/72 |
| **Syracuse, N.Y.** | 8.7 | 7/1/71 | 7/1/72 |
| **Suffolk County, N.Y.** | | Total ban on ABS, alcohol sulfates, methylene blue active substances (nonionic amendment approved) | |
| **Bayville, N.Y.** | | Same as Suffolk County | |
| **Kalamazoo, Mich.** | 8.7 | 11/1/71 | |
| **Grand Rapids, Mich.** | 8.7 | 1/1/72 | |
| **Wyoming, Mich.** | 8.7 | 1/1/72 | |
| **Detroit, Mich. area** | | | |
| Detroit (Suit filed 5/3/71) | 8.7 | 7/1/71 | 7/1/72 |
| Flint, Michigan | | | 7/1/72 |
| Ypsilanti Township | 8.7 | 6/1/71 | 6/1/72 |
| Grosse Point Woods | 8.7 | 7/1/71 | 6/30/72 |
| Inkster | 8.7 | 7/1/71 | 7/1/72 |
| **Ypsilanti, Mich.** | | Ban on nonbiodegradable detergents containing ABS. Posting of grams P/use in stores | |
| **Prince George's County, Md.** | 8.7 | 1/1/72 | 1/1/73 |

the comparative costs for alternative control strategies. Ultimately, any costs will be borne by the consumer either as increased product costs or increased waste water treatment costs.

Estimates of increased product cost for detergent phosphate substitutes are available for nitrilotriacetic acid (NTA), although it is currently not being used. Manufacturers report a differential cost of approximately 5 cents/lb for raw materials. Complete replacement of the 2.5 billion lb of detergent phosphates manufactured each year would have increased product costs $145 million to $150 million and increased consumer costs by about $200 million to $400 million annually.

Several basic assumptions are necessary to estimate the cost for phosphate removal during waste water treatment: first, that the total U.S. pollution of 205 million people is discharging into municipal waste water treatment systems; second, that the average waste water volume entering a treatment plant from all sources is 150 gal per capita per day; third, that the average. phosphorus concentration is 10 mg/l. from all sources; and fourth, that detergent phosphates account for approximately 50% of this phosphorus.

On the basis of these assumptions, the maximum total national cost to treat municipal waste waters to remove phosphates can be calculated (table, p 1187). (Product cost increments are calculated per pound of phosphates, not per pound of detergents; treatment costs do not reflect percentage removal of phosphates.)

Assuming a phosphate replacement cost of 5 cents/lb and treatment costs of 5 cents/100 gal, 45% of municipal waste waters could be treated for costs comparable to use of a detergent phosphate substitute (figure, p 1187).

Should treatment costs be 2.5 cents/1000 gal, then almost 90% of the municipal waste water could be treated for less cost to the consumer than the increased product costs.

The significance is that waste water treatment reduces total phosphate levels while product controls affect the detergent phosphate fraction alone. Treatment also has the added benefit of increased removal of BOD and suspended solids.

## National policy

In 1969, the Council on Environmental Quality advocated a three-phase approach to the control of cultural eutrophication. Current control strate-gies tend to emphasize detergent phosphates. Complementary control efforts for municipal waste waters and non-point sources such as agricultural runoff are not receiving comparable attention.

There is cause for concern should the consumer equate control of detergent phosphates alone with control of cultural eutrophication. Even if the so-called "nonpolluting" detergents do the cleaning job as well as phosphate detergents, their efficacy and human and environmental safety often are poorly tested. Effective long-term control strategies for cultural eutrophication must reflect both consumer products and municipal waste waters as sources of nutrients. Ultimately, regional solutions will be required which reflect limiting nutrients and reduction of all sources of nutrients.

To implement such a comprehensive control strategy for cultural eutrophication, the following definitive information is needed on a regional or watershed basis:

• sources of nutrients contributing to the available environmental reservoirs of nutrients

• critical nutrient or nutrients limiting aquatic productivity

• cost to the consumer and the degree of control that can be accomplished through reduction of the critical nutrient in consumer products

• degree of control and costs to the consumer for removing critical nutrients from municipal waste waters

• increased removal of BOD, toxicants, and other nutrients that is accomplished by advanced waste water treatment

• potential uses for resultant sludge and precipitated nutrients from advanced waste water treatment

• potential public health and environmental hazards associated with some phosphate-free or low-phosphate detergents.

Additional elements affecting consumer products that may be required in an effective long-term policy are:

• federal guidelines which encourage detergent manufacturers to develop and market products on a regional or watershed basis which reflect local contributing factors to cultural eutrophication

• federal guidelines for preassessing public health and environmental implications of any alternatives.

The key concerns for both long- and short-term control strategies for cultural eutrophication are: first, controls which reflect the regional character of the problem; and second, concern for the potential public health and environ-mental implications of the detergent nonphosphate and low-phosphate detergent formulations.

Immediate control strategies must reflect the following elements:

• cost effectiveness of nutrient control for all sources, such as detergents, land runoff, and municipal waste waters

• cost effectiveness of nutrient removal from municipal waste waters with particular concern for methods applicable to existing primary and secondary treatment plants

• cost effectiveness of returning nutrient-containing treatment sludges and effluents to the land for ultimate disposal

• cost effectiveness and public health implications of reduction of detergent phosphate levels.

In the interim, man may have to accept detergent cleaning abilities less than those achievable with phosphate detergents.

### Additional reading

Environmental Protection Agency, Water Quality Office, "Cost of Clean Water, Cost Effectiveness and Clean Water, Volume II," Washington, D.C., March 1971.

Ferguson, F. A., "A Nonmyopic Approach to the Problem of Excess Algal Growths," **ES&T**, **2**, 188–93, (March 1968).

King, D. L., "The Role of Carbon in Eutrophication," JWPCF, **45**, 2035–51, (December 1970).

Likens, G. E., Bartsch, A. F., Lauff, G. H., and Hobbie, J. E., "Nutrients and Eutrophication," Science, **171**, 873–4 (1971).

Ryther, J. H. and Dunstan, W. M. "Nitrogen, phosphorus, and eutrophication in the marine environment," Science, **171**, 1008–13 (1971).

**Richard D. Grundy** *has been a member of the professional staff, U.S. Senate Committee on Public Works, since 1967. He is a 1963 MPh graduate of the University of California, in radiological health, and completed one year of predoctoral studies in sanitary-radiological engineering. Grundy was with the USPHS Bureau of Radiological Health (now part of EPA) from 1959–67. This article summarizes a staff report prepared for the Senate Committee on Public Works.*

# The great phosphorus controversy

*Arguments over the controlling
mechanisms of eutrophication have
scientists—and politicians—
all hot under the collar*

Reprinted from ENVIRON. SCI. TECHNOL., 4, 725 (September 1970)

A furious controversy over the role played by phosphorus in the excessive growth of algae in lakes and streams is currently raging within a section of the technical community. Although the arguments bandied back and forth are scientific in nature, their implications go far beyond the laboratory. The issue involved is whether phosphorus is indeed the key element controlling algal growth; the assumption that it is underlies all current efforts to remove phosphorus from sewage and to replace the condensed phosphates in household detergents with a nonphosphate substitute (see ES&T, February 1970, page 101, and July 1970, page 544).

### An accepted fact

For many years, the key importance of phosphorus (and of nitrogen) to the growth of aquatic algae was taken as absolute fact—and indeed the majority of water chemists and limnologists (scientists who study freshwaters) never did doubt that fact and do not do so now. Studies of the eutrophication (advanced biological aging) of bodies of water for many years focused on the increased amounts of phosphorus and nitrogen entering the water, which, in practically all cases, accompanied excessive algal growth. The connection has been accepted as so obvious and proven that no argument was really expected.

First hints of the furor yet to come appeared in 1967, when Willy Lange, a chemist turned botanist at the University of Cincinnati, published in *Nature* a paper entitled "Effect of Carbohydrates on the Symbiotic Growth of Planktonic Blue-Green Algae with Bacteria." Lange's thesis was that algae always exist in association with bacteria and that the association is mutually supportive: the algae utilize carbon dioxide and sunlight to produce organic matter and oxygen by

**Dieoff.** *Decaying algae disfigure Montrose Beach on Lake Michigan shorefront*

photosynthesis; the bacteria use oxygen in the decomposition of organic matter to produce carbon dioxide. Lange's experiments proved to his satisfaction that it was the presence of large amounts of organic material in water that made the production of huge amounts of carbon dioxide available for algal growth.

Lange's contentions were picked up and given added currency in 1969, when L. E. Kuentzel, a Wyandotte Chemical Corp. physical chemist, reviewed the literature on eutrophication and concluded (without ever having performed an experiment himself, as his critics are quick to point out) that carbon, not phosphorus, is the element that controls algal growth. Kuentzel followed Lange's reasoning that only bacterial action on dissolved organic matter could possibly produce the amounts of carbon dioxide needed for the algae to grow rapidly. According to Kuentzel, all the literature citations he studied pointed to the fact that there was sufficient organic matter present, together with phosphorus and nitrogen, to support his thesis concerning carbon dioxide production and utilization by algae. Furthermore, continued Kuentzel, in

many reported cases of excessive growth, dissolved phosphorus levels were exceedingly small. So small, in fact, that they were in some cases less than the 0.01 p.p.m. previously suggested as the minimum phosphorus concentration needed for abundant growth, a criterion provided by University of Wisconsin sanitary chemist Clair Sawyer in the 1940's. Kuentzel's interpretations were roundly opposed by members of what has been called, with considerable risk of oversimplification, the phosphorus-is-the-key school of thought.

Then, at the 1970 Purdue Industrial Waste Conference, Pat Kerr, a plant physiologist at the Federal Water Quality Administration (FWQA) Southeast Water Laboratory (Athens, Ga.), presented the results of work done by her and two colleagues from which she concluded that carbon was the controlling element. Miss Kerr's results were an extreme embarrassment to FWQA and to the federal government, who were gearing up (albeit somewhat tentatively) for a switch away from phosphates in detergents and were spending heavily on the development of processes for the removal of phosphorus from liquid

# Two schools of thought clash on many points

| Carbon-is-key school believes: | Phosphorus-is-key school believes: |
|---|---|
| Carbon controls algal growth. | Phosphorus controls algal growth. |
| Phosphorus is recycled again and again during and after each bloom. | Recycling is inefficient: some of the phosphorus is lost in bottom sediment. |
| Phosphorus in sediment is a vast reservoir always available to stimulate growth. | Sediments are sinks for phosphorus, not sources. |
| Massive blooms can occur even when dissolved phosphorus concentration is low. | Phosphorus concentrations are low during massive blooms because phosphorus is in algal cells, not water. |
| When large supplies of $CO_2$ and bicarbonate are present, very small amounts of phosphorus cause growth. | No matter how much $CO_2$ is present, a certain minimum amount of phosphorus is needed for growth. |
| $CO_2$ supplied by the bacterial decomposition of organic matter is the key source of carbon for algal growth. | $CO_2$ produced by bacteria may be used in algal growth, but main supply is from dissociation of bicarbonates. |
| By and large, severe reduction in phosphorus discharges will not result in reduced algal growth. | Reduction in phosphorus discharges will materially curtail algal growth. |

wastes. Swept along by wide interest in Miss Kerr's work and by a long, gutsy, and polemical attack on the whole phosphorus school in *Canadian Research and Development* magazine, battle lines were drawn. Lange, Kuentzel, and Miss Kerr were, once again for the convenience of argument, lumped together as the carbon-is-key school, and their arguments were heatedly discussed by high level groups in FWQA and the Council on Environmental Quality (CEQ).

## Counterattack

However, the phosphorus school counterattacked strongly and its arguments seem, at the moment, to have carried the day. Both the phosphorus and carbon schools agree that algae need, for growth, sources of inorganic carbon, phosphorus, nitrogen, and numerous other elements such as metals. Both schools further agree that algae and bacteria generally coexist, and the phosphorus school is willing to concede that the relationship may be symbiotic. But on almost all other points, they disagree (see table). At the very nub of the disagreement are two basic areas of contention:

• Precisely how much phosphorus do algae need for excessive growth.
• What sources of carbon are available to algae.

The carbon school maintains that only very small amounts of phosphorus are needed. It points to the low dissolved phosphorus concentrations found in the water of eutrophic lakes during algal blooms and believes that nutrients, including phosphorus, are recycled by organisms during growth and released for reuse during the periodic dieoff periods. Says the phosphorus school: On the contrary, algae require relatively substantial amounts of phosphorus and the incidence of low dissolved phosphorus concentrations during a bloom means that the phosphorus has been taken up by the algal cells.

The carbon school believes that the availability of utilizable carbon is the key and that diffusional processes are too slow to permit atmospheric $CO_2$ to support massive growth, hence its interpretation of the importance of bacteria-produced $CO_2$. The phosphorus school points to the fact that algae can use, in addition to free $CO_2$, carbon dioxide produced by the dissociation of dissolved bicarbonates. Phosphorus supporters say that the dissociation occurs so rapidly that supply of carbon dioxide cannot possibly be limiting, and they pooh-pooh the carbon school emphasis on the need for respiratory supply.

It is very easy to convey the wrong impression that all scientists fall simply into one or other of the two camps. In fact, most probably see some merit in both sets of arguments.

Phosphorus supporters, including University of Minnesota limnologist Joseph Shapiro, have told ES&T that they believe with Pat Kerr that carbon was indeed the controlling element in her studies. The reason for this, they say, is that Miss Kerr worked with the waters of several southern lakes in which dissolved bicarbonates are very low, and in a situation where nitrogen and phosphorus were very high. Miss Kerr herself is willing to concede that her results may not hold true "for all waters in all places at all times." She does feel, however, that removal of phosphorus but not of organic carbon from liquid wastes probably spells trouble. Phosphorus supporters continue to point out that most lakes, streams, and estuaries contain abundant supplies of inorganic carbon, and they stick with their belief that, in general, phosphorus is controlling. They do not believe that removal of phosphorus from wastes will halt all algal growth; they do believe, however, that growths will be much diminished.

Governmental bodies obviously are going along with the phosphorus school. In Canada, the federal government gave detergent manufacturers until August 1 to reduce the phosphate content of detergents to 20% (expressed as $P_2O_5$—roughly equivalent to 35% expressed as sodium tripolyphosphate), and is aiming toward a total ban by the beginning of 1972. The U.S. government has not gone as far, however. Rep. Henry Reuss's (D.-Wis.) bill to limit phosphorus in detergents is languishing in a Congressional committee, but FWQA scientists are working on a crash program to evaluate the ecological effects of sodium nitrilotriacetate (NTA), the most likely present substitute for phosphate in detergents. And spokesmen for both FWQA and CEQ say that they are entirely convinced of the merits of the case against phosphorus.

Whether the current furor will testify to the supremacy of science over politics, only time will tell. But one thing is sure—man has been responsible for the advanced eutrophication of lakes through something he has added to them in the course of his technological and social progress. It is not unreasonable to hope that all the work that has been lavished on the role of phosphorus in eutrophication will eventually result in ways to remove that something, whatever it eventually turns out to be. DHMB

**285**

# Ocean dumping poses growing threat

*Reprinted from* ENVIRON. SCI. TECHNOL., **4**, 805 (October 1970)

*Until U.S. moves to curb disposal of*
*wastes at sea, it's likely to become*
*an even more popular practice*

The fact that 48 million tons of wastes were disposed of in the oceans around the U.S. in 1968 is as alarming as the fact that $29 million was spent for such disposals. What is more shocking to many people is that such disposal practices have increased fourfold over the past 20 years and may increase substantially in the future. At present, the U.S. does not have an effective mechanism to curb, yet alone prevent, ocean disposal practices, and unless uniform procedures regulating ocean disposal are established, U.S.'s troubles will mount.

Oceans, like other vital resources, are limited in their capacity to survive abuse. Like commonly used solutions to other pollution problems, ocean disposal merely transfers the waste problem elsewhere. Recycling of materials is now a national goal. But until recycling becomes a reality, large urban coastal cities will continue to look to the sea for disposal of their wastes.

The extent of ocean disposal in 1968 is documented for the first time in the recent report, "An Appraisal of Oceanic Disposal of Barge-Delivered Liquid and Solid Wastes from U.S. Coastal Cities." Prepared by the Dillingham Environmental Co. (La Jolla, Calif.), under contract with the Bureau of Solid Waste Management (BSWM), the report is based on questionnaire and interview data obtained from records of District Offices of the U.S. Army Corps of Engineers. Dillingham personnel, David D. Smith and Robert P. Brown, conducted the inventory with BSWM's deputy director of research and development, Charles G. Gunnerson, as project officer. At press time, the report was being published by the Government Printing Office.

### Inventory

The Dillingham report identifies 126 ocean disposal sites—42 in the Pacific, 51 in the Atlantic, and 33 in the Gulf of Mexico—that 20 coastal cities used in 1968. Further, seven major categories of waste are delineated. Not

"Yeah, It's A Big Ocean—But Those Were Big Lakes Too"

surprisingly, harbor dredgings ranked first in both tonnage and cost (38 million tons and $15.5 million, respectively). But industrial wastes ranked second (4.7 million tons discharged at a cost of $8 million), and municipal sewage sludge third (4.5 million tons at a cost of $4.4 million).

Then, in the order of decreasing tonnage, Dillingham ranks refuse and garbage, construction and demolition debris, military explosives and chemical wastes, and a miscellaneous cate-

gory. However, the report does not include the amount of waste discharged that reaches all oceans through sewer and sewer sludge outfalls.

For three categories of wastes—dredged spoils, industrial wastes, and explosives—the number of disposal sites is about equal on the three coasts. Although prior to 1967, the Atomic Energy Commission disposed limited quantities of radioactive wastes in the sea, these wastes are now disposed on land. The majority of radioactive disposal sites were in the Atlantic. However, sewage sludge sites are limited to the Atlantic, and refuse (garbage) disposal in small amounts is limited to the Pacific. In general, the industrial and municipal waste disposal areas are located at distances from 15 to 100 miles offshore, the exact distance depending on the type of waste and the established regulatory procedure.

The average unit cost for disposal of the different types of wastes ranged from a low of $0.40 per ton for dredged spoils to a high of $24 per ton for containerized industrial wastes. But a figure to $600 per ton was recorded for the disposal of some miscellaneous wastes off the Pacific Coast.

Industrial wastes, the second largest category with respect to both tonnage and expenditure of disposal funds, are made up as follows (in thousands of tons):

| | |
|---|---|
| • Waste acids | 2700 |
| • Refinery wastes | 560 |
| • Pesticide wastes | 330 |
| • Paper mill wastes | 140 |
| • Others | <1 |

Although most industrial wastes originate from coastal cities, increasing amounts are being barged from interior areas of the nation. The Dillingham report warns that such practices are likely to increase.

The severity of the disposal practice is particularly true on the East Coast. For example, exclusive of dredged spoils and explosives, the discharge of industrial wastes and sewage sludges in the Atlantic is five times greater than similar operations in the Pacific and the Gulf combined. Further, industrial wastes and sewage sludges are cited in the report as the largest factor contributing to the 27.5% increase in tonnage over two five-year periods (1959–63 compared with 1964–68). In 1968, industrial wastes totaled 4.7 million tons, nearly doubling the 1959 disposal figure. Similarly, sludge disposal in 1968 was approximately 4.5 million tons, up from 2.8 million in 1959.

## Effects

Little is known about the immediate effects of ocean disposal, let alone the cumulative effect of years of disposal practices. Some studies are underway to determine the short-term effects of industrial waste disposal as well as long-term effects of municipal sewage sludge disposal practices. But if we

wait until adverse effects can be demonstrated, it may be too late to curb the existing disposal practices, particularly with respect to the growing segment involving industrial and sewage sludge disposal.

Two recent studies point out the need for stopping sewage sludge disposal in the New York Bight area. The studies were called to the attention of the U.S. Senate Subcommittee on Air and Water Pollution last March. Instituted by the Corps of Engineers, the first study is being conducted by the Bureau of Sport Fisheries and Wildlife Laboratory at Sandy Hook (N.J.), and the second by the Marine Sciences Research Center of the State University of New York at Stony Brook. Final reports are due next year.

Sen. Harrison A. Williams (D.-N.J.) told the Subcommittee that the disposal of sewage sludge was harmful to the health and welfare of New York and New Jersey residents. "It is causing a significant pollution problem. It is adversely affecting the aquatic life of the New York Bight area," Williams added.

At the hearings, Sen. Gaylord Nelson (D.-Wis.) summed up the sad state of affairs with: "The Corps of Engineers confirmed that no one really knows how many people are dumping what kind of waste into the oceans. The Corps has no tally now for permits that it has issued for the first three miles of offshore ocean waters, the state's territorial sea. Further, the Corps has never rescinded a discharge permit for an offshore disposal permit, even when the polluter had clearly violated it."

## Remedies

At present, there is neither legal provision for controlling disposal nor properly enforced methods beyond the

### Extent of U.S. ocean dumping in 1968

| | Cost | Tonnage |
|---|---|---|
| Dredging spoils | 53% | 80% |
| Bulk industrial wastes | 27% | 10% |
| Sewage sludge | 15% | 9% |
| Construction debris | 1% | 1% |
| Containerized industrials | 1% | 1% |
| Refuse and garbage | <1% | <1% |
| Explosives | <1% | <1% |
| Miscellaneous | <1% | <1% |

three-mile limit. Beyond the outer limits of a coastal nation's territorial sea (12-mile limit), international law dictates that the high seas shall remain free for the use of all nations.

What is needed is specific legislation to prevent pollution of the seas from the discharge of waste beyond the three-mile limit (up to which the U.S. now exerts its authority).

Proposals that have been introduced into the legislative hopper include S. 3484, the Marine Environment and Pollution Control Act, and S. 3488, the National Marine Waters Pollution Control and Quality Enhancement Act of 1970. The purpose of these bills, amending the Federal Water Pollution Control Act, is to establish a broad new environmental management program for the long-range protection of a portion of the ocean environment under the jurisdiction of the federal government.

Introduced by Sen. Nelson, S. 3484 would ban all dumping of solid wastes into our coastal waters, including the Great Lakes, by June 30, 1975. The only type of disposal permanently exempted from the proposed 1975 ban would be liquid wastes which contain no suspended or other solid material, which are nontoxic, or which are treated to a level equivalent to that of the natural quality of the receiving waters.

S. 3488, introduced by Sen. Harrison A. Williams (D.-N.J.), would permit discharges of wastes only beyond a 100-nautical mile limit. "There is, of course, nothing magic about 100 miles, except that it affords sufficient protection to shore areas for an interim period not now provided," Williams says.

At the Senate hearings, N.J.'s Governor William T. Cahill suggested incorporating sludge disposal facilities in the design of all new sewage treatment plants.

What is needed now is establishment by the federal government of uniform application and review procedures for marine waste disposal permits. Minimum standards for baseline surveys, monitoring procedures, and related data management systems are also needed and long overdue. Hopefully, these provisions and considerations will be included in further amendments of the Federal Water Pollution Control Act. But ocean dumping is not merely a national problem, it's an international one. SSM

# Are you drinking biorefractories too?

These materials may be in your drinking water;
they are not removed by conventional treatment,
and industries are casting about for controls

*Reprinted from* ENVIRON. SCI. TECHNOL., **7**, 14 (January 1973)

Bio what? Biorefractories, that's what. They're any number of different chemicals which can cause a taste and odor problem in your drinking water. What's more, these chemicals can and have caused seafood tainting problems in some places. Where do they come from? Industrial waste streams, the prime suspects being chemical, petrochemical and related industries. Will they hurt you? No one knows for sure, but certainly the materials are not essential. Ten to one, you would live longer without them. Can they be removed or prevented from entering rivers in the first place? The problem of biorefractories raises a long list of questions.

The public's attention to the matter of biorefractories is being attracted none too early, at least considering that:

• Two independent laboratories—one the U.S. EPA, the other a private concern—have demonstrated that there are as many as 34 different chemicals in the water at the lower end of the Mississippi River. The Mississippi also turns out to be the source of drinking water supply for more than half of all the people in the state of Louisiana!

• The 92nd Congress failed to pass a Safe Drinking Water Act. (The measure, which would have required EPA to set and enforce standards for drinking water, passed in the Senate but died in the House.)

• Studies showed that a major organic contaminant of the Evansville, Ind., municipal drinking water could be traced to an industrial outfall 150 miles upstream on the Ohio River (ES&T, November 1972, p 1036).

The trouble with biorefractories stems from the fact that biological treatment of these materials never completely removes them from a waste stream. Chemically speaking, these materials are low-molecular-weight chemicals—chlorinated aliphatic and aromatic hydrocarbons—and one property that they share in common is low volatility. Since biological treatment doesn't work, obviously other industrial waste treatment techniques must be found.

**What's being done**

In Louisiana, the problem is being addressed by the state's Department of Commerce and Industry; DC&I's executive director Charles Smith realizes that future growth and successful business depend in part on environmental protection. Recipient of a federal EPA grant, one of the largest grants in the industrial pollution control area, DC&I has attracted five chemical industries into a project to develop pollution control technologies to handle these refractory materials. The department has been in the throes of getting another three industries involved. DC&I's Vernon Strickland serves as coordinator on this particular project.

The principal technical contact for the DC&I study is the Gulf South Research Institute (GSRI), a private, nonprofit corporation. GSRI began operations in 1965 and performs contract research for others. Elias Klein, scientific director of the Lake Pontchartrain lab (one of three GSRI labs) has been heading up treatability studies on processing wastes from the industries involved.

Basically, GSRI is looking at five techniques—ozonation, carbon adsorption, solvent extraction, anaerobic/aerobic treatment, and air stripping, according to GSRI's chemical engineer James Mayes. Mayes notes that some of the work—the biotreatment studies and the carbon adsorption work—is nearing completion.

**Who's in**

Five companies—Dow, Hooker, Monochem, PPG, and Vulcan Chemical—are participating in the project. Four of these industries chlorinate ethylene and subsequently make a raft of low-molecular-weight, chemical products like chloroform, vinyl chloride, and perchlorethylene.

Monochem, the fifth, produces

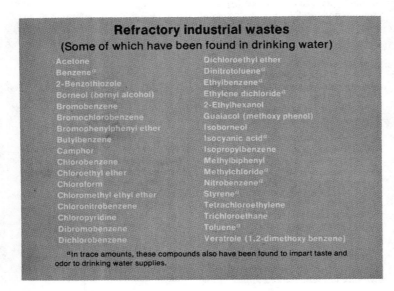

## Refractory industrial wastes
### (Some of which have been found in drinking water)

| | |
|---|---|
| Acetone | Dichloroethyl ether |
| Benzene[a] | Dinitrotoluene[a] |
| 2-Benzothiozole | Ethylbenzene[a] |
| Borneol (bornyl alcohol) | Ethylene dichloride[a] |
| Bromobenzene | 2-Ethylhexanol |
| Bromochlorobenzene | Guaiacol (methoxy phenol) |
| Bromophenylphenyl ether | Isoborneol |
| Butylbenzene | Isocyanic acid[a] |
| Camphor | Isopropylbenzene |
| Chlorobenzene | Methylbiphenyl |
| Chloroethyl ether | Methylchloride[a] |
| Chloroform | Nitrobenzene[a] |
| Chloromethyl ethyl ether | Styrene[a] |
| Chloronitrobenzene | Tetrachloroethylene |
| Chloropyridine | Trichloroethane |
| Dibromobenzene | Toluene[a] |
| Dichlorobenzene | Veratrole (1,2-dimethoxy benzene) |

[a]In trace amounts, these compounds also have been found to impart taste and odor to drinking water supplies.

acetylene; in its production of this basic chemical by the OXO reaction it comes up with a large amount of a by-product carbon waste, estimated at 20 million pounds per year. Monochem is looking at ways to heat treat its wastes to produce a source of activated carbon—if and when carbon adsorption processes take hold in the treatment scheme. Monochem is evaluating a new furnace for activation of the waste carbon; the furnace uses infrared heat treatment and is being developed by Shirco, Inc. (Dallas, Tex.). Located about 20 miles south of Baton Rouge in the Geismar area, Monochem might become a central supplier of activated carbon.

Klein indicates that three other industries, Allied, BASF/Wyandotte, and Foster Grant Co., are interested in joining the project. Allied in Baton Rouge, a manufacturer of ethylene dichloride and vinyl chloride, opts for the stripping technique; BASF/Wyandotte in the Geismar area produces polyethers and TDI (toluene diisocyanate), a basic material for the manufacture of polymers such as polyurethane-foamed materials; Foster Grant Co., also in Baton Rouge, makes ethyl benzene and styrene.

The first five industries that joined the project are located near each other. For example, the Dow plant is just south of Baton Rouge; Hooker, Monochem, and Vulcan are neighbors in the central Geismar area. But PPG is not on the river at all, it's near Lake Charles in southwest Louisiana.

The processing equipment to handle the industrial wastes is now coming on line. The ozonation equipment, with a gallon per minute capacity, was fabricated by Professor Davisson of the Texas A&M chemical engineering department. Professor Davisson also came up with the specification for the solvent extraction unit; the spec has been turned over to GSRI who will let a contract for its fabrication. The anaerobic/aerobic treatment unit, which is truck-mounted, was fabricated under an EPA contract with Rex Chainbelt, Inc. (Milwaukee, Wis.). Two units with one gallon per minute capacities were fabricated under a $112,150 contract. The units were scheduled to be delivered last month.

### Where do we go from here?

During the past 15 years, the industrial-urban complex from Baton Rouge to New Orleans has grown considerably. Today, more than 60 major industries are located between St. Francisville (upriver from North Baton Rouge) and the Belle Chasse-Scarsdale free ferry crossing at New Orleans. This complex includes industries of all types—chemical, pe-

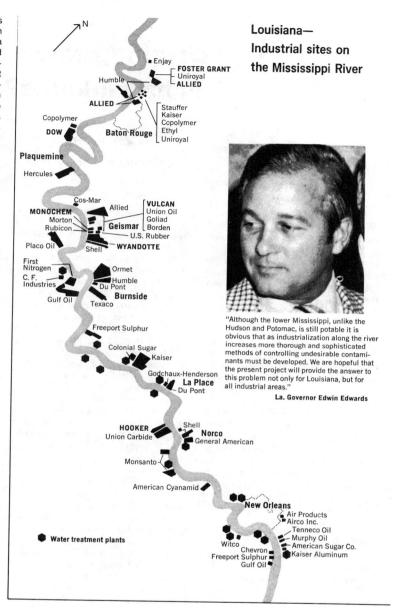

## Louisiana— Industrial sites on the Mississippi River

"Although the lower Mississippi, unlike the Hudson and Potomac, is still potable it is obvious that as industrialization along the river increases more thorough and sophisticated methods of controlling undesirable contaminants must be developed. We are hopeful that the present project will provide the answer to this problem not only for Louisiana, but for all industrial areas."

**La. Governor Edwin Edwards**

troleum refining, sugar, rubber, paper, metals, to mention a few. It has been conservatively estimated that about one third of these plants are a source of biorefractories; at the same time, it is important to point out that refineries and sugar plants, for example, categorically are *not* the source of biorefractories.

Nevertheless, in aggregate, these plants use and discharge some 13 billion gallons of river water daily. Despite the fact that there are some 60-odd industrial plant operations along this stretch of the Mississippi, there are also a number of idle in-

dustrial tracts—formerly plantations—which are waiting for the day when operations will begin. Obviously, before they start up, ways must be found to remove the odor, taste, and taint-producing materials from their industrial waste streams.

Other high-density industrial-urban areas are beginning to experience the odor, taste, and tainting problem. In the U.S., areas such as the Houston Ship Channel and the Cuyahoga River come immediately to mind. Obviously, the problem is not unique to Louisiana, nor the contiguous U.S. for that matter. SSM

# Drinking water:
# is it drinkable?

*Sure it is, say water suppliers,*

*but government officials disagree*

Reprinted from ENVIRON. SCI. TECHNOL., **4**, 811 (October 1970)

"There is a need for national concern. We aren't delivering good quality water today, particularly to the smaller communities. The nation is not maintaining high standards; we aren't training the operators; we aren't maintaining the physical facilities; states aren't maintaining good, strong state programs which are so vital for providing technical assistance and guidance to local utilities," warns James H. McDermott, director of the Bureau of Water Hygiene, the government agency responsible for the quality of the nation's water supply.

Speaking for the nation's water suppliers, Eric F. Johnson, executive director of the American Water Works Association (AWWA), states, "Public water suppliers are doing their job. At the present time we feel that our technology is adequate to meet at least the known contaminants of water supply. Technology is constantly improving—both as to the quality of water it will deliver and to the reduction of the expense of the process."

### Drinking water standards

To clarify the situation and somewhat conflicting views, deeper probing reveals substance in the arguments for both sides of the controversy. What is the origin of standards for drinking water quality? Congress passed a law more than 50 years ago which stated that interstate quarantine regulations would be established by the Surgeon General, who then established the U.S. Public Health Service (PHS) Drinking Water Standards. These standards are, in effect, performance specifications—instead of providing the means by which standards can be met, they merely state the goals. This federal law is applicable to communicable diseases between states and territories. As a result, only about 650 places in the U.S., where interstate carriers take

water on board, are subject to federal inspection, laws, and regulations. The federal government "has no direct authority relating to drinking water in towns or urban dwellings," says McDermott, and "No one has challenged the standards and the use of them. However, the states have the authority and responsibility to provide for the well-being of all of the people in each of the states."

The first standards, promulgated in 1914, covered bacteriological quality, in standing with federal authority over communicable diseases. Since then, revisions (1925, 1942, 1946) have included inorganic chemicals—potential toxicants such as lead and copper, and esthetic undesirables including iron and manganese—and, most recently (1962), organic chemicals. "Now over 10,000 potentially dangerous chemicals are in use in the environment," McDermott explains. In 1962, the standards were revised to include the specification that the carbon chloroform extract (a measurement of synthetic organic chemicals in the environment) shall not exceed 200 parts per billion (p.p.b.). As late as 1967, as a followup to the 1962 standards, an effort was made to amend and include certain pesticides. This cannot be done in a legal sense since pesticides are not entailed in the "communicable disease" category. If these specifications are met, "We have all but eliminated some of the problems that are the scourge of a good part of the world," McDermott summarizes.

In reference to the PHS Drinking Water Standards, Johnson remarked, "Those are the basic minimum standards that we have accepted as an association (AWWA) for the water supply industry. In these standards, we have both mandatory and recommended limits. The mandatory ones have to do with the safety of water supplies; the

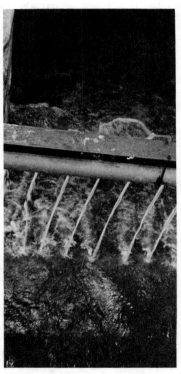

**Treatment.** *Addition of alum to raw water forms floc for sedimentation*

recommended values deal with esthetic and chemical values such as taste and odor."

"Beyond the PHS standards, the AWWA has water quality goals, much higher quality standards . . . dealing mostly with esthetic considerations," Johnson continues. These quality goals are "targets of water utilities to provide much better water. We think that water that just meets the standards is not adequate. The cost of increasing the quality from just the minimum to these goals is approximately 1 cent per person per day."

## Unsafe

The Bureau of Water Hygiene has just completed a survey of 969 representative public water supply systems located in nine areas of the country, seeking to determine whether established "standards of good practice are being applied to assure the quality and dependability of water delivered to consumers' faucets . . . and what needs to be done to assure adequate quantities of safe drinking water in the future on a national scale."

Results show that *on an average* basis, 59% of 969 systems investigated are receiving good water, leaving 41% of the systems delivering water of inferior quality. According to the PHS Drinking Water Standards, of these inferior 41%:

• 36% of 2600 individual tap water samples contained one or more bacteriological or chemical constituents exceeding PHS limits.

• 56% of the systems had physical deficiencies [poorly protected groundwater sources, inadequate disinfection capacity, inadequate clarification capacity, and (or) inadequate system pressure].

• 77% of the plant operators were inadequately trained in fundamental water microbiology.

• 46% were deficient in chemistry relating to plant operation.

• 79% of the systems were not inspected by state or county authorities in 1968; and in 50% of these cases, the plant officials did not remember when the state or county health department last surveyed the supply.

"It cannot be maintained that all of our drinking water is safe," Charles C. Johnson, Jr., administrator of the Environmental Health Service (EHS), told the AWWA convention in June. Continues McDermott, whose Bureau of Water Hygiene is a part of EHS, "The vast majority of the systems in the U.S. don't treat water other than perhaps to disinfect. We measure the quality in these many systems around the country, and we come up with a variety of statistics on the quality of the constituents in the water. Based on the existing standards, we find that we're not doing a very good job because people were getting water that exceeded our recommendations." McDermott further explains that exceeding recommendations is comparable to driving through a caution light—it's not good, but not serious. However, exceeding the mandatory standards is equivalent to running through a red

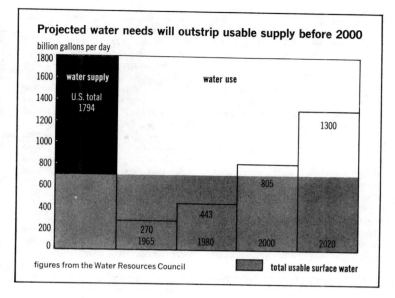

**Projected water needs will outstrip usable supply before 2000**

figures from the Water Resources Council — total usable surface water

light—taking your life in your hands. "We have got to encourage a conservative approach here—don't exceed the mandatory. Even your best run sewage treatment plant doesn't remove and was not built to remove all chemicals, and they function on the average. Sewage treatment plants can fail; we have not developed fail-safe procedures for our sewage treatment plants. This shows the significance and importance of that water treatment plant—it is your last line of defense. It cannot fail. When it does, people get sick," McDermott adds.

### Safe

"No immediate crisis exists. There are no widespread incidents of disease being transmitted by our water supplies. The situation is simply this—we have received adequate warning. The public water suppliers of the nation had better take stock of themselves," says Henry J. Graeser, superintendent of the Dallas City Water Works. "AWWA supports enthusiastically a greatly strengthened water-supply program aimed at the federal and state levels."

"Our (AWWA) basic purpose is to provide better water service to the public," contends Eric Johnson. "New problems or contaminants keep appearing; we never know when they're going to be significant as far as water supply is concerned." Within the last

decade, thousands of new chemical compounds have been developed and eventually get into streams. Checking their sanitary significance is a major undertaking. "It hasn't been done by the PHS, and they should be doing it," says E. F. Johnson. Both at federal and state levels, "top water-supply personnel have been lost to the pollution effort."

The PHS report points out certain deficiencies in some utilities, particularly in small water utilities in satellite areas bordering cities. Some of the deficiencies are in the procedure, according to E. F. Johnson. For example, some small water utilities don't take the adequate number of samples to satisfy the PHS standards. Potentially, it is not a good practice not to take the samples. If the 23,000 water utilities in the U.S. were reduced to 2300 or 230, then "we would be able to have the kind of systems for proper engineering and management," Johnson adds.

To correct deficiencies, the federal government needs more research and training. Also, the state and county health departments are responsible for, and have not maintained, adequate surveillance of the public water supply. "Noncompliance with details doesn't prove water supplies are unsafe," Graeser states. "The deficiencies seem to be more significant than they really are," Johnson adds. "Research in water resources should provide knowledge that we can work into the community where we can realize some benefit to the public," continues E. F. Johnson.

## Water supply limits

"The economically available amount of fresh water is approximately 600 billion gallons per day out of a total run-off of 1200 billion gallons per day," says E. F. Johnson. "The economically available amount increases as the price paid for water goes up. As you raise your economic sights, some other water is treatable. Furthermore," he adds, " we are using at the present time approximately 25% of fresh water economically available. If this were to double in the next 20 to 30 years, we would be using only 50%." "It's a small percentage that is now being utilized," agrees Robert A. Canham, executive secretary of the Water Pollution Control Federation (WPCF). "There are certain sites, such as the west, where there are water shortages. I don't agree with the statements made that we are close to running out of water; the matter of managing water is most important."

On the other hand, McDermott states, "We're never going to run out of water; what we may run out of is good quality water in the sense of the best available source of supply, exclu-sive of sewage. Availability is one thing, but economical feasibility in terms of delivery is quite another. Within given regions (the northeast and northwest), people may continue to have ample quantities of available and usable water, but how about the people in other sections of the country (the southwest)? If you look at it from the standpoint of averages, you can get the wrong answer. Of the 200 million people in the U.S., the average person may be all right, but that can still mean that there are 90 to 95 million others who aren't."

## Pollution control and water supply

Where does the responsibility for good water lie—in pollution control organizations, the water supply industry, or in both? The primary mission of WPCF is in water pollution control, which means collection of waste water in the sewer system, treatment of the waste water, and the final disposal. With higher degrees of waste water treatment, the situation is approaching a closed cycle, which could mean the addition of a drinking water plant to a waste water treatment facility. "Technologically, we can produce drinking water from waste water," says Canham, mentioning also the closed cycle systems successfully in use in South Africa. The feasibility and cost of reusing municipal waste waters is presently under study in five Texas cities, Secretary of the Interior Walter J. Hickel announced in July. Since the water that waste water treatment plants put into the stream is withdrawn further downstream for use as drinking water, "You can't separate the responsibilities by any means," continues Canham. McDermott agrees by saying that the water pollution control people and the water hygiene and supply people must work together.

The water suppliers control the water from point of intake to the tap—collection, treatment, and distribution. Water pollution control and water treatment are separate functions with separate approaches. The AWWA has asked for a separate and identifiable water supply function within the new Environmental Protection Agency (EPA), rather than having this function completely submerged in pollution. "The emphasis on pollution we don't object to, but what we do object to is this emphasis at the expense of the public water supply," says Johnson. Regardless of pollution control, water still must be treated before use, even if pollution is due to natural causes alone.

## Future water supplies

Drinking water supplies in the future will depend upon necessity or economics. At present, there are three distinct possibilities:

• **Desalting** includes processes from the refinement of ocean water to the treatment of brackish water. This method involves expense and transportation problems.

• **Complete recycling** is coming closer everyday, and is presently used in certain areas. However, the chances for error increase substantially with even fail-safe systems being marginal.

• **The dual-supply concept** is a feasible alternative and is not new to military, merchant, or private ships. The cost of installation is one main concern. Health officials have discouraged the dual system, not because of economics, but because of the high level of risk associated with cross-connections. According to this concept, one source of water is used for bodily contact—drinking, cooking, and bathing, while the other source, comprising 80% of water used by each person, is used for other purposes. The selection of alternatives, if necessary, will be dependent upon man's technology and his use of the environment.    CEK

**Atypical.** *Not all water treatment centers are this sophisticated. The Bureau of Water Hygiene reports physical deficiencies in 56% of U.S. plants examined*

# Injection wells pose a potential threat

*The unknowns of subsurface storage of wastes far outnumber the knowns; until the U.S. moves to curb such storage, it will likely become an even more popular practice*

Reprinted from ENVIRON. SCI. TECHNOL., **6**, 120 (February 1972)

Not only is subsurface storage of wastes one of the newest techniques of waste disposal, it is, ironically, a form of potential environmental pollution. Industries, knowing full well that they no longer can put wastes into waterways, are turning to subsurface storage as one alternative.

It would seem that the public's attention to the practice is none too early, at least considering the facts:

• All told, there are tens of thousands of injection wells, more than 200 of which are used for injection of industrial wastes

• Each day about 10 million barrels of oil field brines are being injected into underground reservoirs from which no fluids are being withdrawn

• Each day the chemical industry is disposing of some 1 million barrels of aqueous waste solutions into 175 deep wells.

Disposal of man-made wastes underground has had various labels attached to it. They range from a "Frankenstein monster" to an "unappraised resource." Without question, the number of such industrial subsurface disposal sites has doubled in the past decade and will more than likely redouble in the next. No federal regulation at present covers the practice; a handful of states either have or are considering regulations.

But information on which regulations might be based is lacking in the whole area of subsurface storage. At least, this was the consensus of attendees at the first national symposium on the subject. Billed as "Underground Waste Management and Environmental Implications," and held last December in Houston, Texas, the symposium was sponsored by the American Association of Petroleum Geologists (AAPG) and the U.S. Geological Survey (USGS). Additional support came from 20 technical associations, five federal agencies—including the Department of the Interior and the Environmental Protection Agency (EPA)—and one state agency—the Texas Water Quality Board (TWQB).

The symposium was attended by approximately 600 people, and 35 papers were presented on various aspects of the injection practices, the majority of them dealing with petroleum geology and oil brine injection experiences. Some were limited to radioactive disposal, while others dealt with the engineering aspects of construction of deep wells.

## Scope

There are literally tens of thousands of wells for injection of oil field brines and for stimulation of oil flow from depleted wells. There are more than 200 wells for industrial waste disposal, although a precise estimate is not yet available.

"Underground space is an extremely important resource," said Vincent McKelvey, the newly appointed USGS director, in his keynote speech to the symposium.

"What must be realized is that underground space is not vacant space," said McKelvey. When wastes are injected they must, of necessity, displace other liquids and gases.

Two main questions that the new USGS director posed are:

• Can subsurface storage be used to solve our fluid waste problems?

• Does the use of such subsurface space require the public oversight and federal, state, and local regulations? The answers are still in the making.

Basically, the uses of subsurface storage are divided into two categories. The first group includes materials for which later withdrawal is contemplated, such as natural gas and freshwater recharge. Of more obvious concern as a potential environmental threat are the other materials for which withdrawal is not contemplated. The materials in this second group are principally industrial wastes.

Regardless of possible hazards, underground space is being used today. In the eastern, midwestern, and Rocky Mountain states it is used for natural gas storage. In the Southwest, in Long Island, N.Y., and in California, fresh water harvested from runoff surpluses is used for recharge. In coastal areas, underground space is used for injection of treated or partially treated waters to stem the tide of saltwater intrusion.

Nevertheless, the unknowns of the practice far outnumber the knowns. The state of knowledge is primitive, at best. The simple fact remains that we cannot, with any sense of accuracy, calculate where the injected material is going, at least without undertaking extensive and costly tests. No one knows with any certainty where wastes have been injected.

As a start, USGS in fiscal 1972 is performing a study of present practices. The national survey on the status of subsurface storage and other basic research in cause-effect relationships

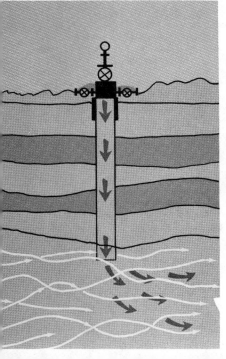

will cost $1.2 million, but first results will not be available until the end of 1972.

Robert Stallman of USGS said in his technical overview at the symposium: "The available theories are either so simplified that they do not represent the real system adequately, or they are so complex that they have not been tested." Stallman elaborated, "I believe that the fundamental model of dispersion, as related to permeability and the potential field, is now the greatest deficiency in theory needed for prediction of effects due to waste disposal. A third area of technological weakness is the mathematical construction of predictive equations. Literally, all of our equations are based on the microscopic scale. Models based on the macroscale are needed."

Perhaps the most recent compilation of information on deep-well disposal is the USGS publication, "Water Supply Paper No. 2020." It contains 700 references on the subject and is available from the U.S. Government Printing Office, Washington, D. C. for $1.50.

### Federal policy

What is important is recognition of the fact that underground space is an extremely valuable resource, McKelvey reiterates. "It can be renewable or nonrenewable depending on how it is used. For example, the consignment of subsurface space to storage of waste may make it a nonrenewable or a nonreusable resource. In contrast, the underground space would be both renewable and reusable when used for the temporary storage of an inert material."

Already there are federal regulations regarding subsurface storage on federal lands. For example, subsurface space for natural gas storage is sold or rented. McKelvey notes that under the Mineral Licensing Act, agreements have been reached for storage of natural gas on federal land in 15 locations in the U.S. "Others are being negotiated," he says.

On private lands, it is a different story. But it is becoming abundantly clear that the use of underground space must be regulated and monitored to protect the interest of others and to guard against damage to other resources.

Stanley Greenfield, the EPA assistant administrator for Research and Monitoring, noted in his symposium banquet address that the first mention of subsurface disposal in federal legislation occurs in the Senate-passed water amendments of last year (S. 2770). "In all current laws concerning water quality, the federal interest and author-

ity is only implicit in provisions concerning surface waters and public health service standards for drinking water supplies," he said.

Greenfield told the banquet attendees that "at the present time, we in EPA neither oppose nor promote deep-well waste injection. We regard it as an important technique with great possibility for both benefit and harm. And we feel that each injection project must be judged on its individual merits."

The present EPA policy on industrial injection wells is contained in a 2-page document. It was promulgated in October 1970, before EPA was formed, but Greenfield says that it is the policy that is strictly followed today and that he endorses it. Basically, the policy says not to use injection wells for wastes that can be treated. What can be injected are wastes that cannot be disposed of by an alternative treatment method.

It is important to note that the federal government is on record as opposing such waste disposal—"Without strict controls and a clear demonstration" that such disposal will not harm "present or potential subsurface water supplies . . . or otherwise damage the environment." The policy statement, Greenfield says, "concluded with a ringing declaration that we recognize subsurface injection as a limited technique—limited in space and in time—to be used with great caution and only until better methods of disposal are developed."

The assistant administrator explains that the EPA water quality research groups at Ada, Okla., and elsewhere are working out elaborations of the policy statement. In Greenfield's words, "They are trying to spell out the nuts-and-bolts specifications, to define what is adequate, practical, well-designed, and so on—all the policy's ambiguous, weasel words which everyone agrees to because they mean different things to different people."

### State regulations

In the absence of federal regulations, a handful of states have enacted their own. But many state regulations are simply inadequate to control the underground space for wastes.

Only four states—Michigan, Ohio, Texas, and West Virginia—of those states that are members of the Interstate Oil Compact Commission have specific laws governing subsurface disposal of industrial wastes. Other member states, including Arkansas, Florida, Illinois,

*"The uses of underground space are already diverse and are likely to become more so with the passage of time."*

**USGS Director McKelvey**

Kansas, Missouri, Oregon, and Pennsylvania, have improved statutes. Sixteen other states are in the throes of developing policy statements and regulations. But the symposium attendee consensus was that no state law is exemplary. None would serve as the model for other states to emulate.

States that are nonmembers of the Interstate Oil Compact Commission are also making headway. These include California, New York, and Pennsylvania. In fact, ORSANCO—the Ohio River Valley Water Sanitation Commission (ES&T, January 1971, p 22)—is busily developing a law which may serve as a model for future regulation of injection wells.

Texas, a state with not only the largest number of petroleum geologists but more subsurface wells than any other state in the union, has experience in subsurface implacement of materials dating back to 1936 with the East Texas Salt Water Disposal Co. In 1961, the Texas legislature passed the Injection Well Act, which was subsequently amended in 1965 and 1969.

At present in Texas, permits for injection wells must be obtained from the Texas Water Quality Board. Along with a $25 application fee, the TWQB requires:

• a treatability study
• a list of adjacent landowners
• technical report including data on the geology, waste data, and aspects of well construction. Although the act does not require public hearings, the

**"EPA of necessity is watching you."**

**EPA Asst. Administrator Greenfield**

TWQB routinely holds public hearings on its permits.

Since passage of the act, 100 permits have been issued in the past decade; 50 are active today, 10 were canceled, and the remaining 40 either have not been drilled or are not presently in operation.

Robert B. Hill of the TWQB estimates that "more than 14 million gallons of industrial wastes are injected daily." He adds, "This represents a miniscule amount of the total industrial and municipal waste produced daily and discharged to surface drainageways."

Hill says that TWQB receives 10–12 new applications each year. It renews its permits every five years. Recently, TWQB has been asking permit holders to upgrade the use of wells—questioning whether the wastes could be treated, but this practice is just beginning.

Monsanto has spent $3 million in subsurface injection in Texas alone. It started its first well in 1962 and a second one in 1965. Dow has five wells in Texas. Its first was put into operation in 1969 and there have been one or two new ones every year since.

Monsanto Pollution Abatement Manager Ronald Sadow says that his company's industrial wastes are pretreated prior to injection. Usually, a pretreatment operation involves waste storage, separation of oil and/or suspended solids through flotation or gravity means, filtration through coarse sand or fine cartridge and diatomaceous earth, chemical fixation of pH, treatment to correct for corrosiveness or biological growths, followed by additional storage, and final pumping to the well.

Another Monsanto experience involves its injection well near Pensacola,

Fla. D. A. Goolsby of the USGS says, "Since 1963, more than 4 billion gallons of acidic industrial waste has been injected into a limestone aquifer." Although no pressure change has been noted in the overlying aquifer, pressure changes from 90 to 108 psi have been observed in two outlying monitor wells in the injection formation.

### Effects

Although the final consequences of deep-well injection practices are far from clear, there are known experiences of earthquakes, land subsidence, and mishaps with existing wells. However, no generalization is possible; every case is unique (ES&T, August 1970, p 642). There have undoubtedly been mishaps, but it would seem that the notoriety of a few wells far surpasses the relatively innocuous experiences of others. For instance, tens of thousands of wells have been injecting oil brines without mishaps. And more than 200 industrial wells are operating without obvious problems today.

One of the first deep wells that received public attention was the 2-mile deep Rocky Mountain Arsenal well, near Denver, Colo. The well was drilled by the Corps of Engineers to dispose of wastes for which no other disposal procedure was available. One of the materials, for example, was a by-product from the production of a nerve gas. But the stimulation of earthquakes in the Denver area, for which the well was blamed (1962, 1963, and again in 1964–66) was certainly unpredicted.

The Earthquake Research Center (Menlo Park, Calif.) now has confirmed earlier suspicions that the Denver earthquakes were a function of time and injection. Still, the center has no predictive capability for forecasting future earthquakes, although studies are being made which may allow prediction.

Perhaps the most notorious mishap occurred at the Hammermill Paper Co. well at Erie, Pa. Between 1964 and 1968, approximately 55,000 barrels of spent sulfite wastes containing fibrous materials were pumped into the well under pressures between 1100–1300 psi. The well blew one Easter Sunday morning and spewed 150,000 gallons of wastes for several days into Lake Erie. What happened? The injection tubing corroded for one thing, and the so-called chemical heat problem was a second factor implicated in the failure of this well.

Another effect of wells is land subsidence—the settling of land masses.

In fact, land subsidence due to fluid withdrawal has been reported from many parts of the world. USGS spokesman J. F. Poland told the symposium attendees that subsidence develops most commonly in overdrawn groundwater basins, but subsidence of serious proportions also has been reported in several oil and gas fields.

The San Joaquin Valley of California is the area in the U.S. of most intensive land subsidence. More than 4000 sq miles are affected; in 1969, the maximum subsidence was 29 ft. Nevertheless, the rate of subsidence has been considerably reduced by injecting surface water in underground strata, according to this USGS spokesman.

Between 1917 and 1968, Poland continued, overpumping of groundwater caused as much as 13 ft of subsidence in the Santa Clara Valley and a decline in artesian head of some 180 ft. Water injected underground here has raised the water table 70 ft in four years and reduced the rate of subsidence to a few hundredths of a foot per year.

### Looking ahead

As was made clear in the symposium, every subsurface injection case is unique. But the questions remain: How many wells should there be? What materials should be injected? It seems reasonable, although it may not be economical, that no industrial waste should be injected if it can be treated. But other questions, too, raise matters of policy. For example, human wastes logically might be put down wells in arctic environments. At least the current alternatives—filling drums with the wastes and scattering the filled drums across the arctic tundra—seem even less acceptable. Oil field brines have been injected for years with no mishap; certainly there would seem no need to halt this practice. And where else but underground could you store radioactive wastes? The AEC stores high-level wastes in containers that can be continuously monitored and in 1967 banned ocean dumping of such materials.

Under certain conditions, subsurface implacement of materials may be practical and feasible, at least when all aspects of the interdisciplinary practices are known—the hydrology, the geology, and the engineering aspects of well construction. In the meantime, we must proceed with great care. In the long term, subsurface storage may yet prove to be the best, perhaps the only, way to dispose of the undisposables.　SSM

**Charles A. Caswell**
*Gurnham and Associates, Inc.*
*Chicago, Ill.*

# Underground waste disposal:

*Reprinted from* ENVIRON. SCI. TECHNOL., **4**, 642 (August 1970)

Waste injection wells are becoming popular but the field
is still plagued by some fundamental misunderstandings

Deep-well disposal has become a controversial topic, at least to the environmental control engineer, his associates, and much of the general public. However, in view of the wide potential for useful application of subsurface fluid, gas, and solid handling techniques to water and air resource conservation problems, the subject should be of interest to every practicing engineer, as well as to those engaged in other phases of environmental control work. Deep-well disposal of liquid wastes is really only one small part of a wider field—the application of subsurface geologic technology to natural resource conservation.

While there will be some simple arithmetic presented here, I will not attempt the usual translation of logic into extensive engineering mathematics for computerized application. Unquestionably, computer technology is a tremendous asset to the engineer and enables us to construct, almost instantaneously, a massive, inverted, pyramid of rapidly expanding confusion balanced neatly upon any initial point of irrelevancy that can be expressed by a numerical symbol. But it will be my purpose here only to point out some of the broader, overall aspects of the subject of subsurface considerations. A few points of technical relevancy will be illustrated, and thereafter those who so desire will be free to construct their own pyramids.

Practically speaking, deep-well disposal is, at the moment, the one facet of subsurface technology that most engineers are probably familiar with, and my discussion will start with some comments on this specific method, particularly regarding areas in which a review of current literature indicates some fundamental misconceptions.

## Origins

The process actually originated in the petroleum industry over 40 years ago as a means of disposing of salt water that commonly accompanies oil taken from producing wells. The original reason for this was probably the well-known fact, at least in the oil field, that salt water has an amazing affinity for any expensive cows! Later, the concept of using this water to increase oil recoveries developed and then the process of "secondary recovery," or "water-flooding," was born. Geologists then developed a broad spectrum of engineering data regarding the injection of fluids into subsurface geologic formations of various lithologic character. Concurrently the concepts of repressuring exhausted oil fields with gas came along, and once again a backlog of engineering data on injecting gases into geologic formations began to develop. This has been further accelerated by the development of underground gas storage facilities. Both these techniques may well have application in alleviating air pollution problems in the future, and research should be initiated on this possibility.

Within the last 15 years, with the growing problems of industrial waste, the use of these subsurface techniques for industrial and municipal waste disposal began to develop. The first industry outside the petroleum companies to use this technique was the chemical process industry, probably because operation of specialized oil field service divisions allowed chemical process engineers to become acquainted with these concepts. By about 1965, the boom was on. Because of a rapid proliferation of waste disposal problems, particularly with toxic or difficult-to-treat wastes, underground disposal was attempted by industries of all types. Some of these attempts were successful, others were not.

No accurate figures are available, due to variations between different states and companies in reporting requirements, but it appears there are about 40,000 saltwater injection wells operated by the petroleum industry in the U.S. either for brine disposal or secondary recovery purposes. There may be another 1100 fluid injection wells involving waste disposal, groundwater recharge, and protection against saltwater invasion operated by various industries and municipalities. In addition, the petroleum industry uses about 20,000 gas injection wells for reservoir maintenance purposes, secondary recovery, or underground gas storage.

If we consider the additional tens of thousands of wells that have been involved in extracting oil, water, gas, sulfur, and salt from beneath the earth's surface, it's apparent that the massive backlog of data regarding subsurface geologic techniques can be of primary interest and a useful tool to all involved in the problems of natural resource protection under many combinations of circumstances.

## Pros and cons

As is true of any engineering system, there are advantages and disadvantages involved in the use of these methods. It is the engineer's job to weigh the factors involved in coming up with a workable solution. Advantages of subsurface techniques to water resource problems under good geologic circumstances are:

- A potential method of ultimate disposal, in the sense that wastes untreatable by other means may

# concepts and misconceptions

be permanently removed from our immediate environment, in most cases, and for very long times in others. With proper system design, the length of anticipated storage time can be relied upon to provide neutralization of the waste by the continuous normal geochemical and geohydrological processes long before the waste would ever migrate to the surface.

- Protection of fresh groundwater supplies from saltwater invasion.
- Underground storage of fresh water in arid regions to reduce evaporation losses, or to store intermittent freshwater supplies.
- Groundwater recharge in areas heavily dependent upon groundwater sources as a water supply.
- Solids storage and disposal under some conditions.
- Effective use in conjunction with other water supply or waste disposal processes. This may be particularly true when a waste of small volume is highly dangerous or toxic, and the cost of treating the waste by usual means may make effective treatment of the total flow impossible. In such cases, it is frequently feasible to separate the high pollutant level waste stream from the main waste flow and dispose of it underground at less cost than trying to pay for treating the entire flow on the basis of the combined pollutant levels. Where the water supply problem is more critical than waste disposal, it is frequently feasible to apply additional treatment to effluent from a secondary waste treatment plant and inject it back for later use.
- Potential solution to some air pol-

**Drilling.** *Oilfield technology often can be used for natural resource protection*

lution problems, as injection of toxic or otherwise obnoxious fumes into underground storage reservoirs.

The foregoing advantages are sometimes offset by several disadvantages, the most important of which might be:

- Existence of unfavorable geologic circumstances in the vicinity of the problem. This factor alone places some rather severe limitations in many areas.
- Fluid incompatability, in case of liquid injections, between the natural formation fluid and the fluid being injected.
- Possible loss of control of the waste liquids after injection.
- Legal problems that may arise from such activity. For example: when does underground trespass start? What are the degree and extent of the injector's liability in the event of surface or subsurface damage to nearby properties? Who owns water injected in groundwater recharge? There are, of course, many other problems of this type that will arise as usage of these methods increases. Obviously, the use of subsurface techniques for industrial wastes has not been widespread enough to have built up much in the way of legal precedent; many of the petroleum and water rights laws on the books are rarely precisely applicable to the industrial waste disposal problem.

Finally, there is a tremendous problem involved in educating regulatory bodies, the body politic, and the industrial users to understanding the design and operational principles and limitations inherent in the application of these subsurface techniques to the field of water and air resources. Sometimes it seems that the necessity for the application of a certain amount of thought and intelligence to utilization of underground techniques is probably their greatest disadvantage.

### Well design

The field of underground fluid or gas handling has two major divisions—the surface factors and the subsurface factors. To put it simply, the first basic requirement for underground disposal is a usable hole in the ground. This involves a geologic study for determination of a specific location and the operation of drilling equipment. In most cases, these factors

can be handled much more effectively by independent contractors intimately familiar with them, than by the usual design and construction engineer who normally specializes in surface structure construction. The drilling, testing, and casing program of a well, if properly done, will establish the volumes of fluid it can effectively accept and the pressures required to maintain these acceptable volumes at the desired rates of injection. Supervision of these phases of the project should be handled by a consulting geologist familiar with subsurface conditions. Once these data are acquired, the design of the treating and pumping equipment to prepare and handle the fluid at the surface, prior to injection, and the additional surface pumping equipment necessary to inject the fluid into the formation, can be handled by any capable engineer familiar with the basic principles of handling waste fluids or gases.

However, in the consideration of the subsurface factors, the average industrial project engineer is, usually, at somewhat of a disadvantage by virtue of background training and normal experience. For this reason, illustration of a few basic principles involved in the subsurface functions of the fluid handling system might be of both interest and value to the engineer involved in the overall supervision of any environmental control problem involving underground disposal, even though this phase of the project should normally be handled by a consultant.

The areas in which current literature indicates some misunderstandings are injection pressure requirements, hydraulic fracturing techniques, formation fluid capacities, and fluid migration conditions. An attempt will be made to use, as illustrations, examples involved in these areas. So we must go "down the hole," so to speak, and take a worm's eye view of this system in order to discuss some of the mechanics involved at the bottom end of our "usable hole in the ground."

### Migration rates

First, let's consider briefly the question of fluid well capacities and fluid migration in the geological formations. By assuming some typical parameters for a disposal well (*see inserts*), it can be calculated that a square mile of formation around the well could contain as much as 5,217,200,000 gallons

of fluid. If a waste fluid were injected at a rate of 10,000 gallons per day, it would take 521,720 days, or about 1429 years to fill up this space—and still move only one-half mile from the well site. By assuming a 50% saturation of the well by the original formation fluid, these figures would, of course, be cut in half.

Since a finely engineered disposal well would not use a receptor formation that outcrops nearby, it's apparent that under normal conditions of fluid migration it would take a long, long time for this material to reach the surface. Let us take, for example, an improbable situation in which our hypothetical formation might outcrop, or come to the surface, ten miles away from the disposal well. Using the 50% formation saturation figure, so that we only need 714.5 years to move the first half mile, we then find we still need 14,290 years for the injection fluid to arrive at the surface of the outcrop. Going one step further, just to be conservative, we might say that these figures are in error by 90% due to unknown factors affecting the migration rate, and that the actual migration occurs in 10% of our computed time. We still have a little over 71 years to move the first half mile and 1429 years to reach the outcrop.

If we add a dilution factor to account for mixing waste with the formation fluid, we are forced to conclude that it will take an extremely long-lived waste to reach an outcrop only ten miles away in its original form. In view of these factors above, it would seem wise to give more consideration to the use of underground formation space for storage, treatment, and ultimate reduction of some types of pollutants.

Obviously, this total available space may not be available for storage at any particular location having these assumed porosity conditions. The determining factors here will be the percentage of saturation of the formation by naturally occurring fluid, and whether the acceptance of the injected fluid is achieved by displacement, diffusion, or a combination of both. The original formation fluid hydraulic conditions, hydrostatic or hydrodynamic, will also affect this factor of ultimate usable storage volume. It might be added here that observation of many core analyses and electric log data taken from drilling wells,

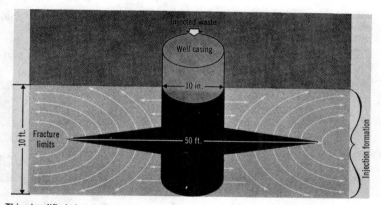

This simplified sketch illustrates several salient points about injection wells.

**Well capacity:** A formation of the dimensions indicated and with typical porosity (25%) and permeability (10 millidarcies) has a tremendous storage capacity, up to $8.2 \times 10^6$ gallons per acre of formation. As mentioned in the text, such a formation, in the absence of large amounts of naturally occurring fluids, can receive significant amounts of waste for literally hundreds of years without the fluid advancing more than a few miles from the well site.

**Injection pressure:** Injection pressure is a function of the effective pore space exposed to the advancing front of the injection fluid. As the fluid moves radially outward from the well, the surface area exposed to the fluid front (effective porosity) per unit volume of injected fluid increases. As a result, there is a rapid decrease in the injection pressure per given injected fluid volume and rate.

**Fracturing techniques:** If the formation surrounding the original bore hole is fractured, the net effect is to increase the initial surface area available for injection. For the 50-foot fracture indicated, each face of the fracture becomes a disk having 282,000 sq. in. of surface area. When this is compared with the initial surface area of 3768 sq. in. for the original 10-in. bore hole, it is easy to see why injection wells frequently "go on a vacuum" when fractured. The arrows show the fluid dispersion patterns around the fracture, from which it can be seen that channeling induced by fracturing will normally occur far beyond the limits of the original fracture.

a water well, which looks something like this:

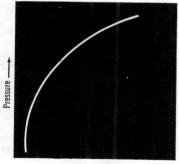

In other words, pressure required for a given fluid volume to flow at a given rate increases with decreasing distance from the well. The pressure curve for an injection well appears reversed but actually expresses the operation of the same physical factors:

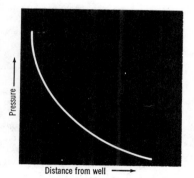

Here the greatest pressure required for flow is nearest the bore hole with a rapid decrease in required pressure away from the bore. To demonstrate the skin effect let us consider the fluid as moving through a series of ever-enlarging concentric cylinders as distance from the bore hole increases. If we assume an initial bore hole cylinder of 10 in. in diameter and a formation depth of 10 ft., the surface area of the bore hole cylinder is 3768 sq. in.

By the time the fluid has moved to a diameter of 20 in. from the axis of the bore hole, the total surface has become 7536 sq. in.; at 30 in. from the axis we have acquired an area of 11,304 sq. in. It can easily be seen that the steady increase in the cross-sectional area available to carry the fluid results in a very rapid decrease

ranging in depth from 1000 to 24,000 ft., has rarely shown formation fluid saturations in the 100% range.

The original formation bottom hole pressure will also have an effect on both injection pressure and the ultimate usable storage volume by virtue of the effect on the injection pressures required. Observation of many hundred drill stem tests at varying depths and across most of the Mid-Continent region has rarely indicated bottom hole pressures that even approached theoretical calculated hydrostatic heads for the depths involved. However, frequently in the unconsolidated sediments along the Gulf Coast, we do find theoretical hydrostatic and theoretical overburden pressures close to recorded pressures.

### Fluid flow

Permeability, in practical terms, refers to the ability of the fluid to move

through existing porosity, and thus into, or out of, the bore hole and on through the formation. This, in turn, tends to become a direct function of pore geometry by virtue of the normal friction resistance to fluid flow. Damage to the formation face in the bore hole by action of the drilling bit may also cause considerable modification of pore geometry as it affects fluid flow. As a matter of fact, in most cases most of the pressure required to inject fluid into a porous and permeable formation is really being used to overcome this combination of factors, causing high friction resistance only near the bore hole. This total friction resistance to flow is referred to as the "skin effect."

One of the great advantages of fracturing techniques lies in their use to alleviate this skin effect. Most of you are, no doubt, familiar with the appearance of the drawdown curve on

of pressure required to move a given volume of fluid, at any given rate, into the formation. This condition will continue until the limits of the formation's geographic dimensions are reached or as long as the fluid is moving. Also, note that the pressure, volume, and velocity effects on the fluid flow are going to be constantly changing with the changing radius of flow around the bore hole. This point is very significant in terms of the potential of a disposal well to trigger earthquakes. Due to this rapid pressure drop it would take a very special set of circumstances for a disposal well to trigger a quake.

There are other factors involved in the flow mechanics that are pertinent to the skin effect, since they affect permeability, or "transmissibility:"

- Pore geometry as it relates to turbulence near the bore hole, and surface tension and viscosity of the fluid.
- Bottom hole pressure of the original formation, which in some cases, exerts a back pressure on the injected fluid.
- Hydrologic conditions, whether static or dynamic.
- Percentage saturation of the original porosity by the in-place fluid, if any.
- Damage to the formation face by the original drill bit or infiltration of drilling fluid.

An interesting example of the combined effects on these factors occurred at the recent completion of a 3500-ft. well. In this case, an attempt was being made to hydraulically fracture a potential producing formation after a pipe had been set on top of the formation. The hole had been cleaned out, washed with mud acid, and the fracturing procedure instituted. Pressure against the formation increased to 8000 p.s.i. before the geologist gave the word to shut down because of the possibility of equipment damage. Not one drop of fluid had moved into the formation. The electric log and microlog indicated suitable porosity and permeability, and a drill stem test recovered only a few feet of oil-cut drilling mud, with no salt water. Thus, because a good shut-in-bottom hole pressure built up from 0 to 200 p.s.i., it was felt that the possibility of a blocked formation face existed.

For these reasons, it was decided to attempt an open hole perforation with large caliber expendable jet charges. The fracturing equipment was removed from the well, and when the perforating job was completed, another attempt made to fracture the formation was successful. The formation began to take fluid slowly at 900 p.s.i., fractured at 1650 p.s.i., and then took 2000 g.p.m. of sand slurry at an injection pressure of 300 p.s.i. The well was completed as a good pumping oil well for about 10 b.b.l. per hour with an operating bottom hole pressure of 420 p.s.i.g. Obviously the initial failure to pump into, or "fracture," the formation effectively was one due entirely to skin effect factors.

## Misleading tests

The foregoing case is merely cited as an example of how great the effect of these factors can be when they are present in just the right combination. Pump tests on an injection well can also be misleading if initial pressure is simply accepted at face value, as evidence of the ability of a formation to accept fluids or the operating injection pressures required. In general, fracturing can cause a tremendous increase in fluid acceptability of a formation.

Unfortunately, hydraulic fracturing and its effects seem to be misunderstood factors in the underground disposal field. When interest in underground disposal began to spread, the first thing engineers apparently did was to research the petroleum field with respect to water injection wells. But there is a great deal of difference between a water injection well and a disposal well. The requirements are entirely different. The water injection well is used only in water-flooding for the purpose of secondary recovery. Under these conditions, the advance of the flood face must be kept as uniform as possible to provide a clean sweep of the in-place oil. This usually requires relatively low initial injection pressures, slow injection rates, and the absence of fracturing to avoid channeling and consequent bypassing of the oil in place in the vicinity of the well bore.

A saltwater disposal well is very different. Here it makes no difference whether or not the flood front is kept uniform; the problem here is simply to put as much fluid as possible into the formation at as low a cost as possible. This condition is much more similar to waste disposal than is water flooding. Under these conditions hy-

**Derrick.** *Drilling tests are critical steps in planning for waste disposal wells*

draulic fracturing is commonly used to increase formation fluid acceptability and reduce required injection pressures. Saltwater disposal wells are almost routinely fractured, and in the first year or two operate on a vacuum or gravity flow, and injection pressure buildup thereafter is at a very slow rate. In a waste disposal well a uniformly advancing floodfront is not needed, and the greatly increased expense required to maintain it can hardly be justified. (It should be added that hydraulically induced fractures do not close if the pressure drops. They are kept open by propping agents, such as sand or small beads, injected in a slurry as a part of the fracturing process.) Another very important misconception shows up in the often repeated comment that water produced with oil is usually injected back into the producing formation, thus more space is available. Quite the contrary is true, which points up how the difference in terminology between petroleum technologists and those not so familiar with the field can create misunderstanding. Where significant amounts of water are produced with oil, the oil is coming from a water-driven reservoir, already under natural flood, and injecting additional water into these sands would only cause possible channeling and consequent loss of oil. Thus, rather than waste money to gild the lily, water produced with oil from these reservoirs is normally disposed of by injection into wells other than the producing horizon.

## Water injection

The only place where water is normally injected into the same formation from which oil is produced is in the case of "water-floods" and these are only used to recover oil from reservoirs that contain no natural water drive. As a matter of fact, they normally contain very little or no waste at all. The water produced is that which has been previously injected. In this case water for flooding has to be obtained by drilling water supply wells to horizons other than the producing formation.

A misunderstanding of this key point seems to have led to many erroneous computations in the literature regarding available storage space and required injection pressures. It might well be said, as mentioned earlier, that failure to differentiate among some of these basic conditions appears to have led to the creation of some interesting pyramids of confusion, based on initial points of irrelevancy, with respect to the use of fracturing, available storage space, and calculated injection pressure requirements. There are various techniques of perforating, "shooting," acidizing, and hydraulic fracturing that can be used both to overcome the skin effect and to improve and maintain injective capability. It is, however, important to realize that the pressure required initially to inject fluid into an untreated formation is not necessarily a good index, of itself, of the pressure ultimately required for fluid disposal into that formation.

Of course, getting the fluid into the formation is only part of the problem—then comes the question of who is responsible for it. Although a considerable body of law has been built up regarding underground fluids and their migration, most of it has developed from our frontier day concepts of riparian rights and private property that have since been held to be applicable to petroleum and subsurface waters. Since underground waste disposal is a very new concept in legal terms, there has been little as yet in the way of precedents established for degrees of responsibility concerning migrated industrial waste.

There are, however, several mechanical factors known to be involved in any consideration of fluid migration in subsurface formations:

- Degree of formation cementation and formation porosity.
- Lateral extent of effective permeability and porosity.
- Characteristics of beds overlying and underlying the injection formation.
- The presence of faults and (or) fracture patterns in the area.
- Earthquake occurrence, frequency, and intensity in the area.
- Freshwater—saltwater contact levels are important in groundwater recharge, underground water storage, prevention of saltwater invasion, and finally, in regard to decisions specifying depth for projected waste disposal wells.
- Structural attitude of the injection formation.
- Hydrological characteristics of the disposal formation.
- Effectiveness of casing and casing cementing program.

It might be well to point out that this last factor is a point in injection well planning where great care must be exercised. While also true for producing wells, it is far more critical in injection wells due to pressure gradients resulting from the reversal of the direction of flow which the normal well casing program is designed to handle. The case of a disposal well "blow-out" at Lake Erie is well-known.

A review of published literature indicates that many disposal wells allow only for an inch of diameter differential between hole size and casing size. For example, 7.5- or 8-inch casing will frequently be set in a 9-inch hole. This allows a thin half inch film of cement between the casing and the formation. Since the cement-to-casing bond and the cement-to-formation bond usually represent the weakest link in the pressure retention chain, it would be well to specify, as a minimum, at least 2 inches of diameter differential between the casing collar o.d. and the hole for waste injection wells.

## More knowledge

With regard to the present state-of-the-art of application of subsurface techniques to water resource problems and planning, there is an urgency for additional research in the foregoing areas to assist in establishing logical legal and operational principles by people familiar with both petroleum technology and industrial waste disposal techniques. For example, the field of suspended solids injection, an

entirely new utilization of subsurface techniques, is an area in which there is still much to be learned. Some techniques have been evolved for the disposal of radioactive wastes, but even here much improvement is needed.

In addition, some successful experimentation is being done on the injection of sewage sludge solids into certain favorably constituted formations. The types of porosity and permeability required for high solids content fluid injection need to be further defined and determined. For example, the "lost circulation" zones, which oil well drillers unhappily experience, might well be feasible for waste sludge or slurry injection programs.

We might sum up by saying that subsurface injection has been used, either experimentally or on a large scale with varying degrees of success for a wide range of applications, shows that subsurface geologic techniques are important tools for the environmental engineer in his continuing struggle to meet the needs of society. It will pay him well to recognize the circumstances under which he may increase the effectiveness of his natural resource control and protection programs. After all, man is a very unique animal, the only one that is capable of destroying his own environment—and with it, himself. Whether he does this, or not, will ultimately depend upon engineers, scientists, and political leaders who must create, design, and build society's structures to protect the environment.

**Charles A. Caswell** *is a senior associate with Gurnham and Associates, Ltd., where he is involved in problem, data, and cost analysis for water and waste treatment. Mr. Caswell was awarded a B.S. in petroleum engineering and paleontology from the University of Oklahoma in 1940. He has over 20 years experience in every area of petroleum exploration and production and has published numerous articles on these subjects. Mr. Caswell is a member of the Water Pollution Control Federation.*

# The case for higher rate waste water treatment

No "universal" design standard for biological waste water treatment plants exists in the United States. Each state has its own body of regulations, design guidelines, or statutes which set forth in greater or lesser detail acceptable design criteria. Often, even where detailed design guides exist, these are subject to judgmental modification and interpretation by regulatory agency officials.

Perhaps the nearest approach to a universal design standard is the "Recommended Standards for Sewage

*Even so-called "overloaded" plants may in fact be capable of treating larger flows; modified operation is the key to better utilization of design capacity*

*Reprinted from* ENVIRON. SCI. TECHNOL., **6,** 794 (September 1972)

Works," published and revised periodically by the Great Lakes–Upper Mississippi River Board (GLUMRB) of State Sanitary Engineers. This document is frequently referred to as the "Ten States Standards" since the Board consists of Illinois, Indiana, Iowa, Michigan, Minnesota, Missouri, New York, Ohio, Pennsylvania, and Wisconsin. Other states follow these recommended standards at least to some extent and, in any event, these standards are fairly representative of regulatory agency thinking and practices generally.

**Brian L. Goodman**

*Black & Veatch*
*Kansas City, Mo. 64114*

With reference to the "permissible loadings" for several variants of the activated sludge process of waste water treatment, GLUMRB permits organic loadings to vary from a low of 12.5 lb BOD (signifies BOD₅) per 1000 ft³ of aeration tank capacity (extended aeration process) to a high of 50 lb BOD/1000 ft³ (contact stabilization or step aeration processes). For plants designed to treat 500,000 gal./day of waste water or less (5000-person contributing population or less), the loading range is 12.5–30 lb BOD/1000 ft³, depending on the process variant selected.

The contention exists (with supporting data) that substantially higher loadings should be permitted. It has also been contended (and demonstrated) that utilizing higher design loadings in properly designed waste water treatment plants would not result in deterioration of plant performance or effluent quality. Furthermore, "volumetric loading" criteria such as "lb BOD/1000 ft³ of aeration tank volume/day" are meaningless both from a practical and a fundamental standpoint. Such criteria ignore completely the fact that the food-to-microorganisms ratio governs characteristics and performance of biological waste water treatment systems. In short, volumetric criteria leave unanswered the question, "How many, if any, microorganisms are in the aeration tank with the BOD?"

## Case for higher loadings

Data supporting the soundness of higher loading levels abound in the published literature and have been freely available to regulatory agencies and consulting engineers alike for many years. In the early 1950's for example, the University of Florida studied a waste water treatment system designed on a 3.8-hr aeration period. A "typical activated sludge" was produced when the unit was operated as a combined chemical–biological treatment system, the Florida researchers reported in *Sewage Ind. Wastes* (February 1952). During this phase of the study, water-softening sludge (principally calcium carbonate) was added at various dose rates ranging from 25–200 mg/l. (as

CaCO₃). At air inputs of 0.28–0.30 ft³/day (waste-treated basis) BOD removals averaged 91%, and suspended solids removals averaged 94%.

Experiments (begun in 1937) conducted by the Federal Institute of Water Supply, Sewage Purification and Water Pollution Control, Swiss Federal Institute of Technology, Zurich, Switzerland, covered a range of aeration periods of from 1.07–9 hr and a range of aeration tank loadings of from 23.1–202 lb BOD/1000 ft³ (*Sewage Ind. Wastes*, January 1954). Mixed liquor suspended solids ranged between 2830 and 3930 mg/l. Treatment system effluent BOD values ranged from 8–32 mg/l. while effluent suspended solids values ranged from 8–47 mg/l. Effluent suspended solids and BOD concentrations of 20 mg/l. or less were achieved at aeration tank loadings of 160 lb/1000 ft³ or less.

Experiments from 1948–1951 by the Research Institute for Public Health Engineering, National Health Research Council, The Hague, Netherlands, and the Institute for Sewage Purification indicated that if sufficient aeration capacity was available, the aeration period could be greatly reduced (*Sewage Ind. Wastes*, February 1954). Loadings of 6–8 times "normal" (normal being 20–60 lb BOD/1000 ft³) could be treated by activated sludge. At a loading of 300 lb BOD/1000 ft³ of aeration tank capacity, an average effluent BOD level of 39 mg/l. was achieved with individual values as low as 16 mg/l. At a loading of 150 lb BOD/1000 ft³, an average effluent BOD of 15 mg/l. was obtained in one run and 11 mg/l. during another. Effluent BOD averaged 23 mg/l. at a loading of 197 lb BOD/1000 ft³. At 275 lb. BOD/1000 ft³, effluent BOD concentration averaged 15 mg/l. As explained in "Biological Treatment of Sewage and Industrial Waste" (Reinhold, 1956), treatment of a beet sugar process waste at an aeration tank loading of 625 lb/1000 ft³ produced 90% BOD removal.

In the same publication, operating data from 65 waste water treatment plants in the U.S. support the contention, that, as long as a minimum aeration period of 1.0–1.5 hr is provided, activated sludge plant performance is in-

dependent of aeration time and is, beyond that point, dependent on loading in terms of lb BOD/lb MLSS (mixed liquor suspended solids). Data indicate about 93% removal of applied BOD up to at least a loading of 0.5 lb BOD/lb MLSS/day. Activated sludge plants with loadings ranging from 150–600 lb BOD/1000 ft³/day removed 89–92% of the applied BOD at plant loadings from 150–400 lb BOD/1000 ft³.

In 1959, an activated sludge waste water treatment plant utilized series aeration and clarification, described in *Water Sewage Works* (February 1959). A total aeration period of 1.6 hr at design flow was provided. The BOD loading averaged 124 lb/1000 ft³/day. The observed average effluent BOD concentration was 13 mg/l. or a removal efficiency of 92.9%. Experiments in the Netherlands showing that an activated sludge system loaded at 3746 lb BOD/1000 ft³/day produced an effluent with a 28 mg/l. BOD level are discussed in "Advances in Biological Waste Treatment" (Macmillan Co., 1963).

Experiments conducted by the Department of Sanitary Engineering, College of Technology, Hanover, Germany, described in *J. Water Pollut. Contr. Fed.* (October 1960), revealed that BOD removals averaging 87% were achieved at a loading of 200 lb BOD/1000 ft³/day. At another location, aeration tank loadings of 225–644 lb BOD/1000 ft³ resulted in effluent BOD values averaging 27 mg/l. (an average removal of 87%).

From observations and experiments conducted in Hungary, BOD removals of 90% or greater are reported in "Advances in Water Pollution Control" (Pergamon Press, 1965) at loadings ranging from 199–268 lb BOD/1000 ft³/day for 1.0–1.5-hr aeration periods. In 1965, high-rate activated sludge was recognized as a valid process variant. The successful development of this process variant depended on first developing high-intensity aeration devices (*J. Water Pollut. Contr. Fed.*, February 1965). High mixed liquor solids and short aeration periods were process characteristics. In studies by the Hydraulic Research Institute, Prague, Czechoslovakia, BOD removals of 90%

or greater were reported at the Third International Conference on Water Pollution Research (1966) for volumetric loadings of 96–324 lb BOD/1000 ft³/day.

In 10 high-rate plants in Canada, 200 samples were collected and subjected to analysis. Design volumetric loading was 125 lb BOD/1000 ft³/day with a minimum aeration tank detention time of 1.75 hr. High-rate data were compared to data from 57 conventional activated sludge plants (1200 samples) and 28 extended aeration plants (240 samples) (Ontario Water Resources Commission). The mean effluent BOD concentration for the high-rate systems was 14.4 mg/l. and the suspended solids concentration was 13.5. Effluent mean BOD levels for conventional and extended aeration plants were 17.2 and 14.8 mg/l., respectively. Both capital and operating costs for high-rate plants were less than for conventional plants.

The University of Iowa studied volumetric loadings up to 200 lb BOD/1000 ft³/day (*Water Wastes Eng.*, July, August, and October 1967). At a loading of 151 lb BOD/1000 ft³, an effluent BOD of 13 mg/l. was reported.

BOD removal of 90% or greater was reported in "Advances in Water Quality Improvement" (Univ. of Texas Press, 1968) for loadings of 53–358 lb BOD/1000 ft³/day at 13°C and 52–247 BOD/1000 ft³/day at 11°C. Data presented indicate that 90% BOD removal was associated with mixed liquor suspended solids loadings of 0.6 lb BOD/lb MLSS/day or less within the range of temperatures studied.

During 2½ years of operations at the Grand Island, Neb., waste water treatment plant, volumetric BOD loadings increased over the period from 66 lb BOD/1000 ft³/day to a high of 136. BOD removal efficiencies increased from 94% in 1967 to 98% in 1969 with a 99% removal achieved for 2½ months of 1969 (effluent BOD = 5 mg/l.) (*J. Water Pollut. Contr. Fed.*, May 1970).

An operating plant at Pentyrch, Wales, had an average volumetric loading for a six-month period of 203 lb BOD/1000 ft³/day. According to *Water Waste Treat.* (January-February 1969), aeration detention time varied between 1.17 and 2.32 hr. Effluent BOD concentration averaged 23.7 mg/l. including one high-value sample. Excluding this one sample, the average effluent BOD level was 15.7 mg/l.

**Aeration.** *The Grand Island, Nebraska complete mixing activated sludge plant above, like the two 0.5-mgd contact stabilization tanks at the right, can achieve about 98% BOD removal*

A study carried out at Rutgers, the State University (New Jersey), and reported in *J. Water Pollut. Contr. Fed.* (January 1971), determined the removal rate of soluble organics:

• Soluble TOC (total organic carbon) is reduced to a relatively stable level after 30 min of aeration. Initial concentration reduction averaged 70%.

• Soluble COD is reduced to a relatively stable level after 1 hr of aeration. Initial concentration reduction averaged 77%.

• COD:TOC ratios decrease steadily over the first hour of aeration and fluctuate irregularly thereafter.

• Over a 30-min to 1-hr aeration interval, the COD:TOC ratio decreased while the TOC concentration remained constant. This combination of events suggested the transfer of soluble organics between the aqueous phase and bacteria.

• Estimated net increase in the mean oxidation state of soluble carbon averaged 1.61 units after 1 hr and 1.67 units after 6 hr of aeration. Thus, over 96% of the apparent oxidation occurred within the first hour of treatment.

In a field study, an activated sludge plant operated at volumetric loadings ranging from 75.2–145 lb BOD/1000 ft³/day. Removal of BOD from primary clarified effluent, explained in *J. Water Pollut. Contr. Fed.* (March 1971), averaged 83%; thus, overall plant removals are presumed to have averaged over 90% (unfortunately plant influent BOD data was not collected).

### Design and loading criteria

The preceding literature review sets forth ample justification for an upward revision of the permissible design loading presently extant generally in the U.S. However, a simple increase alone will not take maximum advantage of the activated sludge process. The mixing mode must also be considered. Many

U.S. activated sludge plants have been designed in the past as plug flow or semiplug flow aeration systems. (Wastes added to the aeration tank at any given time move through the tank to the effluent end without substantial mixing with the remainder of the tank contents.) Such systems are subject to upset resulting from relatively large organic loading surges. They are even at some considerable disadvantage owing to "normal" daily peak loadings since the effect of peak or surge loads is concentrated on those microbes at and near the influent end of a plug flow tank at the time of such peak or surge. Perhaps one fourth or less of the total microbial population of the system must, under these circumstances, attempt to cope with loadings ranging from two to three times design to many times that level. The microorganisms involved suffer metabolic upset as well as inability to produce the desired removal of applied organics. Frequently, hydraulic overloading occurs during surge organic loading, thus reducing the time available to the microbes to react.

Contrasted with the "concentration of effect" inherent in plug flow systems is the "dilution of effect" inherent in completely mixed systems. In completely mixed aeration systems, organic load increases are essentially immediately dispersed throughout the aeration tank contents—the effect on any single microbe is one fourth or less that in a plug flow system. Any increase in aeration tank effluent-soluble BOD will be minimized owing both to dilution and to the fact that a far greater number of microbes are involved in the process of adsorbing, absorbing, and metabolizing the applied BOD even though the time to do so is reduced by an accompanying hydraulic surge.

The importance of complete mixing, to more fully utilize the inherent treatment capacity of the activated sludge process, is nowhere better illustrated than at Grand Island, Neb. This complete mixing activated sludge plant was

designed to treat the waste waters from a 130,000-head-of-cattle/day slaughterhouse and the City of Grand Island. The flow sheet consists of, in order, grit removal, primary sedimentation, completely mixed aeration, and final sedimentation. Over the course of the day, the BOD loading to the aeration portion of the treatment system varied from a 800-lb/hr high level during midafternoon to a 100-lb/hr low during the early morning hours. Despite this wide range of process loading levels, BOD removal efficiencies were 94% in 1967, 95% in 1968, and 98% in 1969. During the first 2½ months of 1970, 99% removal was achieved. In addition, observed plant performance followed very closely that predicted by fundamental complete mixing activated sludge relationships.

With these considerations in mind, completely mixed aeration tanks could be employed (with or without preceding primary clarification) and permissible BOD volumetric loadings could be increased to at least 60 lb BOD/1000 ft³/day (contrasted with 12.5–30.0 presently) for small plants (less than 100,000 gal./day or 1000 population equivalent daily treatment capacity). For large

plants (100,000 gal./day or 1000 population equivalent and greater) at least 125 lb BOD/1000 ft³/day (contrasted with 30–50 presently) loadings could be employed.

### Beyond activated sludge

If treatment levels are desired beyond those possible with activated sludge treatment alone, chemical and/or physical treatment units can be added to the basic biological treatment system. This, of course, raises the question of the degree of treatment possible with biological treatment systems alone. I have observed (as have American Society of Civil Engineers members and University of Kansas researchers) BOD removal rates of 173–360/day at 20°C with completely mixed activated sludge systems. Utilizing, for example, a value of 203/day, the unmetabolized BOD remaining (F) after 3 hr of aeration in a completely mixed system (assuming an influent BOD of 240 mg/l.) would be:

$$F = \frac{240}{203\,(0.125) + 1} = 9 \text{ mg/l.}$$

If the aeration period is increased to 5 hr:

$$F = \frac{240}{203\,(0.208) + 1} = 5.5 \text{ mg/l.}$$

Thus, 96–98% or greater removal of applied BOD is within the demonstrated capabilities of activated sludge systems. The failure of many activated sludge systems to achieve these levels of efficiency very often stems from a failure

to design properly the final clarification portion of the system. System effluent suspended solids concentrations of 10–15 mg/l. can result in effluent BOD concentrations 6–9 mg/l. higher than those indicated by the foregoing calculations. Observed BOD removals could then be reduced to 92% or less, depending on the actual losses of solids in the system effluent.

To ensure low effluent suspended solids levels and, consequently, low effluent BOD values, tertiary filters are more frequently employed today. By using dual media (crushed anthracite and sand) effluent filters operating at filtration rates of 3–5 gal./min/ft², filtered effluent BOD and suspended solids concentrations of 2–4 mg/l. are achieved. It is, therefore, possible to achieve 99% BOD removal through combined biological and physical treatment units.

For higher removals of applied organics, carbon adsorption columns can be added to the system following filtration. My associates and I also included phosphorus removal. Adding phosphorus-precipitating chemicals directly to the aeration tank, prior to aeration, or following the aeration step, can remove 90% or more of the applied phosphorus.

Recently, chemical–physical treatment processes have been advocated as alternatives to biological, biological–chemical, and biological–chemical–physical treatment systems. A number of laboratory and pilot plant chemical–physical treatment systems have been studied. However, to date, no full-scale chemical–physical treatment plants are in operation in the U.S. The first such plant will be operated by the City of Rosemount, Minn., when it is completed. Valid comparisons between biological and chemical–physical waste water treatment systems must await collection and analysis of routine operational data from Rosemount and other full-scale chemical–physical plants.

### Outlook

Generally accepted U.S. design loading criteria for activated sludge waste water treatment plants are shown to be "over-cautious" as indicated by the demonstrated efficiencies of higher rate treatment systems in many parts of the world. Increases in permissible design loading rates, which are modest by comparison with established practice elsewhere, are urged. In terms of process efficiency, completely mixed activated sludge waste water treatment systems are capable of BOD removals of up to 99% even when loaded at rates considerably in excess of those generally permitted.

Adopting only the proposals presented here would result in substantially decreased waste water treatment capital and operating costs. Not considered here are further reductions in capital costs which would result from compatible and rational design of other portions of the total waste water treatment system. These economies would stem from the fact that a more rational aeration system design would yield an activated sludge possessing more desirable characteristics (increased solids settling rate, sludge density index, etc.) which would permit advantageous changes in the design criteria for other portions of the treatment system.

Also, exceptionally high-quality effluents can be obtained through addition of chemical and/or physical treatment processes to the basic biological process. Such combined treatment systems are in routine operation in many parts of the U.S. today.

In an era when required land area, capital costs, and operating costs are of such importance, the opportunity significantly to reduce these factors should not be overlooked. This is especially deserving of the most serious and immediate considerations since both land area requirements and costs can be very significantly reduced without sacrificing process efficiency.

**Brian L. Goodman** *is presently head of the Process Control Section, Civil-Sanitary Division, Black & Veatch, Consulting Engineers. Mr. Goodman is experienced in administrating all phases of comprehensive research, development, and design in areas of waste water treatment and transport.*

# Index

# Z